Transformer Design Principles

Third Edition

Transformer Design Principles

Third Edition

Robert M. Del Vecchio, Bertrand Poulin,
Pierre T. Feghali, Dilipkumar M. Shah,
and Rajendra Ahuja

CRC Press
Taylor & Francis Group
Boca Raton London New York

CRC Press is an imprint of the
Taylor & Francis Group, an **informa** business

CRC Press
Taylor & Francis Group
6000 Broken Sound Parkway NW, Suite 300
Boca Raton, FL 33487-2742

First issued in paperback 2022

© 2018 by Taylor & Francis Group, LLC
CRC Press is an imprint of Taylor & Francis Group, an Informa business

No claim to original U.S. Government works

ISBN-13: 978-1-498-78753-6 (hbk)
ISBN-13: 978-1-03-233952-8 (pbk)
DOI: 10.1201/9781315155920

Library of Congress Cataloging-in-Publication Data

Names: Del Vecchio, Robert M., author.
Title: Transformer design principles / Robert M. Del Vecchio, Bertrand Poulin, Pierre T. Feghali, Dilipkumar M. Shah, and Rajendra Ahuja.
Description: Third edition. | Boca Raton : Taylor & Francis, CRC Press, 2018.
| Revised edition of: Transformer design principles / [authors], Robert M. Del Vecchio ... [et al.]. 2010. | Includes bibliographical references and index.
Identifiers: LCCN 2017011211| ISBN 9781498787536 (hardback : alk. paper) | ISBN 9781315155920 (ebook)
Subjects: LCSH: Electric transformers--Design and construction.
Classification: LCC TK2551 .T765 2018 | DDC 621.31/4--dc23
LC record available at https://lccn.loc.gov/2017011211

Visit the Taylor & Francis Web site at
http://www.taylorandfrancis.com

and the CRC Press Web site at
http://www.crcpress.com

Contents

Preface

The third edition of this book extends and further develops some of the topics in the second edition. For instance, the multiterminal transformer model is extended to include a second transformer that could be a booster or the second transformer of a 2-core phase shifter. This second transformer can also be included in an impulse simulation program.

Although the second edition discussed the linear impedance boundary method, it pointed out its deficiencies in terms of calculating eddy losses in nonlinear magnetic materials, such as tank steel. This new edition includes a section on how to correct the method for nonlinear materials.

The more complicated calculation for the directed oil flow disk thermal model in the previous edition is now replaced by a more efficient calculation based on graph theory.

Transformer design normally begins with an optimization calculation to produce a minimum cost design based on the client's requirements. Therefore Chapter 19 on optimization methods, which was included in the first edition, has been added. This calculation should produce a starter design, which can be further modified when subjected to more detailed screening by other design programs. Although the starting point for most designs, this chapter is near the end of the book. Most of the book is concerned with detailed design methods. These are based on realistic transformer models that cover specific characteristics and associated limits that the transformer must satisfy.

Since large power transformers especially have unique client specifications, a generic transformer design is usually not possible. Moreover, new materials with different material constants are being developed and used, such as natural ester oil instead of mineral oil. In addition, different physical configurations may be necessary for different designs, such as the use of wound-in-shields or interleaving for high voltage designs, or different placements of oil flow washers for cooling different designs. The models must be flexible enough to handle these. Model development in this book therefore starts from general physical principles appropriate to the model in question so that the formulas and procedures arrived at can be applied to a variety of transformers and the materials they contain.

Because the readers may come from a variety of backgrounds, as little technical jargon as possible is used. SI (MKS) units are used throughout, as well as standard terminology and symbols.

Authors

Robert M. Del Vecchio, PhD, earned his BS in physics from the Carnegie Institute of Technology, Pittsburgh, Pennsylvania; MS in electrical engineering; and PhD in physics from the University of Pittsburgh, Pennsylvania, in 1972. He served in several academic positions from 1972 to 1978. He then joined the Westinghouse R&D Center, Pittsburgh, Pennsylvania, where he worked on modeling magnetic materials and electrical devices. He joined North American Transformer (now SPX Transformer Solutions), Milpitas, California, in 1989, where he developed computer models and transformer design tools. Currently, he is a consultant.

Bertrand Poulin earned his BE in electrical engineering from École Polytechnique Université de Montréal, Quebec, Canada in 1978 and MS in high voltage engineering in 1988 from the same university. Bertrand started his career in a small repair facility for motors, generators, and transformers in Montréal in 1978 as a technical advisor. In 1980, he joined the transformer division of ASEA in Varennes, Canada, as a test engineer and later as a design and R&D engineer. In 1992, he joined North American Transformer where he was involved in testing and R&D and finally manager of R&D and testing. In 1999, he went back to ABB in Varennes where he held the position of technical manager for the Varennes facility and senior principal engineer for the Power Transformer Division of ABB worldwide. He is a member of the IEEE Power and Energy Society, an active member of the Transformers Committee, and a registered professional engineer in Québec, Canada.

Pierre T. Feghali, PE, MS, earned his bachelor's degree in electrical engineering from Cleveland State University, Ohio in 1985 and his master's degree in engineering management in 1996 from San Jose State University. He has worked in the transformer industry for more than 23 years. He started his career in distribution transformer design at Cooper Power Systems in Zanesville, Ohio. In 1989, he joined North American Transformer in Milpitas, California, where he was a senior design engineer. Between 1997 and 2002, he held multiple positions at the plant, including production control manager, quality and test manager, and plant manager. He became vice president of Business Development and Engineering at North American Substation Services, Inc. He is a Professional Engineer in the state of California and an active member of the IEEE and PES.

Dilipkumar M. Shah earned his BSEE from the M.S. University of Baroda (India) in 1964 and his MSEE in power systems from the Illinois Institute of Technology (Chicago, Illinois) in 1967. From 1967 until 1977, he worked as a transformer design engineer at Westinghouse Electric, Delta Star, and Aydin Energy Systems. He joined North American Transformer in 1977 as a senior design engineer and then became the engineering manager. He left in 2002 and has been working as a transformer consultant for utilities world wide, covering areas such as design reviews, diagnosing transformer failures, and advising transformer manufacturers on improving their designs and manufacturing practices.

Rajendra Ahuja graduated from the University of Indore in India where he earned a BEng Hons (electrical) degree in 1975. He worked at BHEL and GEC Alsthom India and was involved in the design and development of EHV transformers and in the development of wound-in-shield-type windings. He also has experience in the design of special transformers for traction, furnace, phase shifting, and rectifier applications. He joined North American Transformer (now SPX Transformer Solutions) in 1994 as a principal design engineer and became the manager of the testing and development departments. He became the vice president of engineering at SPX Transformer Solutions. He is an active member of the Power and Energy Society, the IEEE Transformers Committee, and the IEC. He is currently a consultant.

1

Introduction

1.1 Historical Background

Transformers are electrical devices that change or transform voltage levels between two circuits. In the process, current values are also transformed. However, the power transferred between the circuits is unchanged, except for a typically small loss that occurs in the process. This transfer only occurs when alternating current (a.c.) or transient electrical conditions are present. Transformer operation is based on the principle of induction, formulated by Faraday in 1831. He found that when a changing magnetic flux links a circuit, a voltage or electro-motive force (emf) is induced in the circuit. The induced voltage is proportional to the number of turns linked by the changing flux. Thus, when two circuits are linked by a common flux and there are different linked turns in the two circuits, there will be different voltages induced. This situation is shown in Figure 1.1 where an iron core is shown carrying the common flux. The induced voltages V_1 and V_2 will differ since the linked turns N_1 and N_2 differ.

Devices based on Faraday's discovery, such as inductors, were little more than laboratory curiosities until the advent of a.c. electrical systems for power distribution, which began toward the end of the nineteenth century. Actually, the development of a.c. power systems and transformers occurred almost simultaneously since they are closely linked. The invention of the first practical transformer is attributed to the Hungarian engineers Karoly Zipernowsky, Otto Blathy, and Miksa Deri in 1885 [Jes97]. They worked for the Hungarian Ganz factory. Their device had a closed toroidal core made of iron wire. The primary voltage was a few kilovolts and the secondary about 100 V. It was first used to supply electric lighting.

Modern transformers differ considerably from these early models but the operating principle is still the same. In addition to transformers used in power systems, which range in size from small units that are attached to the tops of telephone poles to units as large as a small house and weighing hundreds of tons, there are a myriad of transformers used in the electronics industry. The latter range in size from units weighing a few pounds, which are used to convert electrical outlet voltage to lower values required by transistorized circuitry, to micro-transformers, which are deposited directly onto silicon substrates via lithographic techniques.

Needless to say, we will not be covering all of these transformer types here in any detail, but will instead focus on the larger power transformers. Nevertheless, many of the issues and principles discussed are applicable to all transformers.

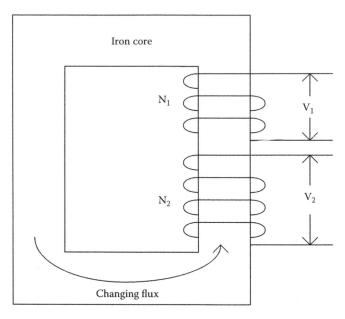

FIGURE 1.1
Transformer principle illustrated for two circuits linked by a common changing flux.

1.2 Uses in Power Systems

The transfer of electrical power over long distances becomes more efficient as the voltage level rises. This can be shown by considering a simplified example. Suppose we wish to transfer power P over a long distance. In terms of the voltage V and line current I, this power can be expressed as

$$P = VI \tag{1.1}$$

Let's assume that the line and load at the other end are purely resistive so that V and I are in phase, that is, V and I are real quantities for the purposes of this discussion. For a line of length L and cross-sectional area A, its resistance is given by

$$R = \rho \frac{L}{A} \tag{1.2}$$

where ρ is the electrical resistivity of the line conductor. The line or transmission losses are

$$\text{Loss} = I^2 R \tag{1.3}$$

and the voltage drop in the line is

$$\text{Voltage drop} = IR \tag{1.4}$$

Substituting for I from (1.1), the loss divided by the input power and voltage drop divided by the input voltage are

$$\frac{\text{Loss}}{P} = \frac{PR}{V^2}, \quad \frac{\text{Voltage drop}}{V} = \frac{PR}{V^2} \tag{1.5}$$

Since P is assumed given, the fractional loss and voltage drop for a given line resistance are greatly reduced as the voltage is increased. However, there are limits to increasing the voltage, such as the availability of adequate and safe insulation structures and the increase of corona losses.

Looking at (1.5) from another point of view, we can say that for a given input power and fractional loss or voltage drop in the line, the line resistance increases as the voltage squared. From (1.2), since L and ρ are fixed, an increase in R with V implies a wire area decrease so that the wire weight per unit length decreases. This implies that power at higher voltages can be transmitted with less weight of line conductor at the same line efficiency as measured by line loss divided by power transmitted.

In practice, long distance power transmission is accomplished with voltages in the range of 100–500 kV and more recently with voltages as high as 765 kV. These high voltages are, however, incompatible with safe usage in households or factories. Thus, the need for transformers is apparent to convert these to lower levels at the receiving end. In addition, generators are, for practical reasons such as cost and efficiency, designed to produce electrical power at voltage levels of ~10 to 40 kV. Thus, there is also a need for transformers at the sending end of the line to boost the generator voltage up to the required transmission levels. Figure 1.2 shows a simplified version of a power system with actual voltages indicated. GSU stands for generator step-up transformer.

In modern power systems, there is usually more than one voltage step-down from transmission to final distribution, each step-down requiring a transformer. Figure 1.3 shows a transformer situated in a switch yard. The transformer takes input power from a high voltage line and converts it to lower voltage power for local use. The secondary power could be further stepped down in voltage before reaching the final consumer. This transformer could supply power to a large number of smaller step-down transformers. A transformer of the size shown could support a large factory or a small town.

There is often a need to make fine voltage adjustments to compensate for voltage drops in the lines and other equipment. These voltage drops depend on the load current, so they vary throughout the day. This is accomplished by equipping transformers with tap changers.

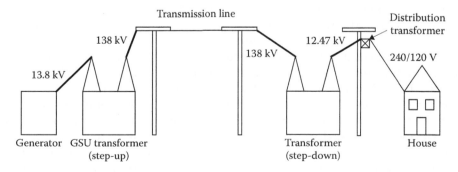

FIGURE 1.2
Schematic drawing of a power system.

FIGURE 1.3
Transformer located in a switching station, surrounded by auxiliary equipment. (Courtesy of Waukesha Electric Systems, Waukesha, WI.)

These are devices that add or subtract turns from a winding, thus altering its voltage. This process can occur under load conditions or with the power disconnected from the transformer. The corresponding devices are called, respectively, load or de-energized tap changers.

Load tap changers are typically sophisticated mechanical devices that can be remotely controlled. Tap changes can be made to occur automatically when the voltage levels drop below or rise above certain predetermined values. Maintaining nominal or expected voltage levels is highly desirable since much electrical equipment is designed to operate efficiently and sometimes only within a certain voltage range. This is particularly true for solid-state equipment. De-energized tap changing is usually performed manually. This type of tap changing can be useful if a utility changes its operating voltage level at one location or if a transformer is moved to a different location where the operating voltage is slightly different. Thus, it is done infrequently. Figure 1.4 shows three load tap changers and their connections to three windings of a power transformer. The same transformer can be equipped with both types of tap changers.

Most power systems today are 3-phase systems, that is, they produce sinusoidal voltages and currents in three separate lines or circuits with the sinusoids displaced in time relative to each other by 1/3 of a cycle or 120 electrical degrees as shown in Figure 1.5. At any instant of time, the three voltages sum to zero. Such a system made possible the use of generators and motors without commutators, which were cheaper and safer to operate. Thus, transformers that transformed all three phase voltages were required. This could be accomplished by using three separate transformers, one for each phase, or more commonly by combining all three phases within a single unit, permitting some economies particularly in the core structure. A sketch of such a unit is shown in Figure 1.6. Note that the three fluxes produced by the different phases are, like the voltages and currents, displaced in

FIGURE 1.4
Three load tap changers attached to three windings of a power transformer. These tap changers were made by the
Maschinenfabrik Reinhausen Co., Germany.

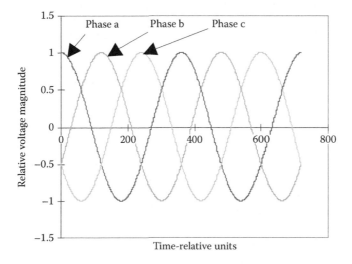

FIGURE 1.5
Three-phase voltages versus time.

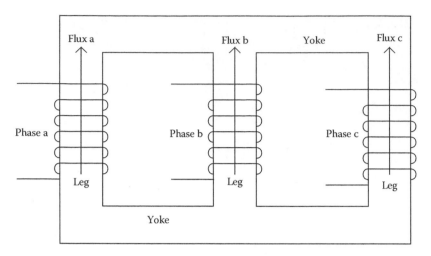

FIGURE 1.6
3-phase transformer utilizing a 3-phase core.

time by 1/3 of a cycle relative to each other. This means that, when they overlap in the top or bottom yokes of the core, they cancel each other out. Thus the yoke steel does not have to be designed to carry more flux than is produced by a single phase.

At some stages in the power distribution system, it is desirable to furnish single-phase power. For example, this is the common form of household power. To accomplish this, only one of the output circuits of a 3-phase unit is used to feed power to a household or group of households. The other circuits feed similar groups of households. Because of the large numbers of households involved, on average each phase will be equally loaded.

Because modern power systems are interconnected so that power can be shared between systems, sometimes voltages do not match at interconnection points. Although tap changing transformers can adjust the voltage magnitudes, they do not alter the phase angle. A phase angle mismatch can be corrected with a-phase-shifting transformer. This inserts an adjustable phase shift between the input and output voltages and currents. Large power phase shifters generally require two 3-phase cores housed in separate tanks. A fixed phase shift, usually of 30°, can be introduced by suitably interconnecting the phases of standard 3-phase transformers, but this is not adjustable.

Transformers are fairly passive devices containing very few moving parts. These include tap changers and cooling fans, which are needed on most units. Sometimes pumps are used on oil-filled transformers to improve cooling. Because of their passive nature, transformers are expected to last a long time with very little maintenance. Transformer lifetimes of 25–50 years are common. Often, units will be replaced before their useful life is up because of improvements in losses, efficiency, and other aspects over the years. Naturally, a certain amount of routine maintenance is required. In oil-filled transformers, the oil quality must be checked periodically and filtered or replaced if necessary. Good oil quality insures sufficient dielectric strength to protect against electrical breakdown. Key transformer parameters such as oil and winding temperatures, voltages, currents, and oil quality as reflected in gas evolution are monitored continuously in many power systems. These parameters can then be used to trigger logic devices to take corrective action should they fall outside of acceptable operating limits. This strategy can help prolong the useful operating life of a transformer. Figure 1.7 shows the end of a transformer tank where a control

FIGURE 1.7
End view of a transformer tank showing the control cabinet at the bottom left, which houses the electronics. The radiators are shown on the far left. (Courtesy of Waukesha Electric Systems, Waukesha, WI.)

cabinet is located, which houses the monitoring circuitry. Also shown projecting from the sides are radiator banks equipped with fans. This transformer is fully assembled and is being moved to the testing location in the plant.

1.3 Core-Form and Shell-Form Transformers

Although transformers are primarily classified according to their function in a power system, they also have subsidiary classifications according to how they are constructed. As an example of the former type of classification, we have generator step-up transformers, which are connected directly to the generator and raise the voltage up to the line transmission level or distribution transformers, which is the final step in a power system, transferring single-phase power directly to the household or customer. As an example of the latter type of classification, perhaps the most important is the distinction between core-form and shell-form transformers.

The basic difference between a core-form and a shell-form transformer is illustrated in Figure 1.8. In a core-form design, the coils are wrapped or stacked around the core. This lends itself to cylindrical coils. Generally, high-voltage and low-voltage coils are wound concentrically with the low-voltage coil inside the high-voltage one. In the shell-form design, the core is wrapped or stacked around the coils. This lends itself to flat oval-shaped coils called pancake coils, with the high- and low-voltage windings stacked side by side, generally in more than one layer each in an alternating fashion.

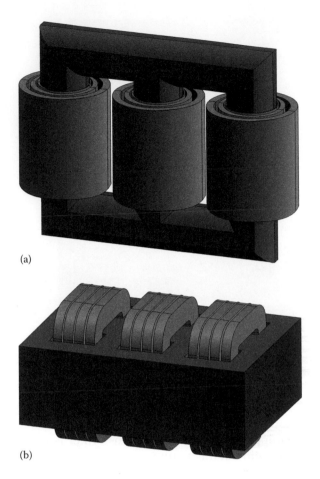

(a)

(b)

FIGURE 1.8
3-phase core-form (a) and shell-form (b) transformers contrasted.

Each of these types of constructions has its advantages and disadvantages. Perhaps the ultimate determination between the two comes down to a question of cost. In distribution transformers, the shell-form design is very popular because the core can be economically wrapped around the coils. For moderate to large power transformers, the core-form design is more common, possibly because short-circuit forces can be better managed with cylindrically shaped windings.

1.4 Stacked and Wound Core Construction

In both core-form and shell-form types of constructions, the core is made of thin layers or laminations of electrical steel, especially developed for its good magnetic properties. The magnetic properties, however, are best along a particular direction called the rolling direction because this is the direction in which the hot steel slabs move through the rolling mill, which squeeze them down to thin sheets. Thus, this is the direction the flux should naturally want to take in a good core design. The laminations can be wrapped around the coils or stacked. Wrapped or wound cores have few, if any, joints so they carry flux nearly uninterrupted by gaps. Stacked cores have gaps at the corners where the core steel changes direction. This results in poorer magnetic characteristics than for wound cores. In larger power transformers, stacked cores are much more common, while in small distribution transformers, wound cores predominate. The laminations for both types of cores are coated with an insulating coating to prevent large eddy current paths from developing, which would lead to high losses.

In one type of wound core construction, the core is wound into a continuous "coil." The core is then cut so that it can be inserted around the coils. The cut laminations are then shifted relative to each other and reassembled to form a staggered stepped type of joint. This type of joint allows the flux to make a smoother transition over the cut region than would be possible with a butt type of joint where the laminations are not staggered. Very often, in addition to cutting, the core is reshaped into a rectangular shape to provide a tighter fit around the coils. Because the reshaping and cutting operations introduce stress into the steel, which is generally bad for the magnetic properties, these cores need to be re-annealed before use to help restore these properties. A wound core without a joint would need to be wound around the coils or the coils would need to be wound around the core. Techniques for doing this are available but somewhat costly.

In stacked cores for core-form transformers, the coils are circular cylinders, which surround the core legs. In smaller transformers, up to approximately 10 MVA 3-phase rating or 5 MVA single-phase, the cross section of the core legs may be circular or rectangular, based either on economical reasons or manufacturer's preference and technology. In larger ranges, the preferred cross section of the core is circular since this will maximize the flux carrying area. In practice, the core is stacked in steps, which approximates a circular cross section as shown in Figure 1.9. Note that the laminations are coming out of the paper and carry flux in this direction, which is the sheet rolling direction. The space between the core

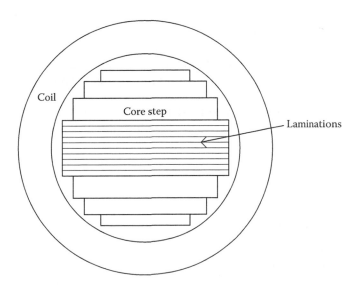

FIGURE 1.9
Stepped core used in core-form transformers to approximate a circular cross section.

FIGURE 1.10
3-phase stepped core for a core-form transformer without the top yoke.

and innermost coil is needed to provide insulation clearance for the voltage difference between the winding and the core, which is at ground potential. It is also used for the cooling medium such as oil to cool the core and inner coil. Structural elements to prevent the coil from collapsing under short-circuit forces may also be placed in this region.

In practice, because only a limited number of standard sheet widths are kept in inventory and because stack heights are also descretized, at least by the thickness of an individual sheet, it is not possible to achieve ideal circular coverage. Figure 1.10 shows a 3-phase stepped core for a core-form transformer without the top yoke. This is added after the coils are inserted over the legs. The bands around the legs are used to facilitate the handling of the core. They could also be used to hold the laminations together more tightly and prevent them from vibrating in service. Such vibrations are a source of noise.

1.5 Transformer Cooling

Because power transformers are usually more than 99% efficient, the input and output power are nearly the same. However, because of the small inefficiency, there are losses inside the transformer. The sources of these losses are I^2R losses in the conductors, losses in the electrical steel due to the changing magnetic flux, and losses in metallic tank walls and other metallic structures caused by the stray time varying flux, which induces eddy currents. These losses lead to temperature rises, which must be controlled by cooling.

The primary cooling media for transformers are oil and air. In oil-cooled transformers, the coils and core are immersed in an oil-filled tank. The oil is then circulated through radiators or other types of heat exchanger so that the ultimate cooling medium is the surrounding air or possibly water for some types of heat exchangers. In small distribution transformers, the tank surface in contact with the air provides enough cooling surface so that radiators are not needed. Sometimes, in these units the tank surface area is augmented by means of fins or corrugations.

The cooling medium in contact with the coils and core must provide adequate dielectric strength to prevent electrical breakdown or discharge between components at different voltage levels. For this reason, oil immersion is common in higher voltage transformers since oil has a higher breakdown strength than air. Often, one can rely on the natural convection of oil through the windings driven by buoyancy effects to provide adequate cooling so that pumping isn't necessary. Air is a more efficient cooling medium when it is blown by means of fans through the windings for air-cooled units.

In some applications, the choice of oil or air is dictated by safety considerations such as the possibility of fire. For units inside buildings, air cooling is common because of the reduced fire hazard. While transformer oil is combustible, there is usually little danger of fire since the transformer tank is often sealed from the outside air or the oil surface is blanketed with an inert gas such as nitrogen. Although the flash point of oil is quite high, if excessive heating or sparking occurs inside an oil-filled tank, combustible gasses could be released.

Another consideration in the choice of cooling is the environment. Mineral oil used in transformers is known to be detrimental to the environment in case of an accident. For transformers such as those used on planes or trains or units designed to be transportable for emergency use, air cooling is preferred. For units not so restricted, oil is the preferred cooling medium, so one finds oil-cooled transformers in general use from large generator or substation units to distribution units on telephone poles.

There are other cooling media, which find limited use in certain applications. Among these is sulfur hexafluoride gas, usually pressurized. This is a relatively inert gas, which has a higher breakdown strength than air and finds use in high voltage units where oil is ruled out for reasons such as those mentioned earlier and where air doesn't provide enough dielectric strength. Usually when referring to oil-cooled transformers, one means that the oil is standard transformer oil. However, there are other types of oil, which find specialized usage. One of these is silicone oil. This can be used at a higher temperature than standard transformer oil and at a reduced fire hazard. Other types finding increasing use for environmental reasons are the natural ester-based oils. These are made from edible seeds and are biodegradable. They also have low flammability and are nontoxic.

1.6 Winding Types

For core-form power transformers, there are two main methods of winding the coils. These are sketched in Figure 1.11. Both types are cylindrical coils, having an overall rectangular cross-section. In a disk coil, the turns are arranged in horizontal layers called disks, which are wound alternately out-in, in-out, etc. The winding is usually continuous so that the last inner or outer turn gradually transitions between the adjacent layers. When the disks have only one turn, the winding is called a helical winding. The total

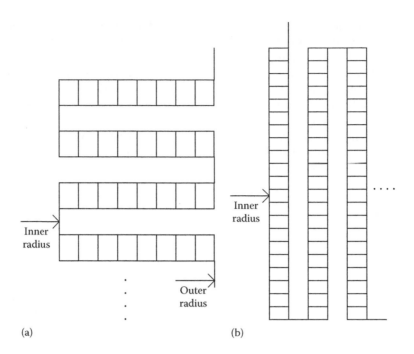

FIGURE 1.11
Two major types of coil construction for core-form power transformers: (a) disk coil and (b) layer coil.

number of turns will usually dictate whether the winding will be a disk or helical winding. The turns within a disk are usually touching so that a double layer of insulation separates the metallic conductors. The space between the disks is left open, except for structural separators called radial spacers or key spacers. This allows room for cooling fluid to flow between the disks, in addition to providing clearance for withstanding the voltage difference between them.

In a layer coil, the coils are wound in vertical layers, top-bottom, bottom-top, etc. The turns are typically wound in contact with each other in the layers, but the layers are separated by means of spacers so that cooling fluid can flow between them. These coils are also usually continuous with the last bottom or top turn transitioning between the layers.

Both types of winding are used in practice. Each type has its proponents. In certain applications, one or the other type may be more efficient. However, in general they can both be designed to function well in terms of ease of cooling, ability to withstand high voltage surges, and mechanical strength under short-circuit conditions.

If these coils are wound with more than one wire or cable in parallel, then it is necessary to insert cross-overs or transpositions, which interchange the positions of the parallel cables at various points along the winding. This is done to cancel loop voltages induced by the stray flux. Otherwise, such voltages would drive currents around the loops formed when the parallel turns are joined at either end of the winding, creating extra losses.

The stray flux also causes localized eddy currents in the conducting wire whose magnitude depends on the wire cross-sectional dimensions. These eddy currents and their associated losses can be reduced by subdividing the wire into strands of smaller cross-sectional dimensions. However, these strands are then in parallel and must therefore be transposed to reduce the loop voltages and currents. This can be done during the winding process when the parallel strands are wound individually. Wire of this type, consisting of

FIGURE 1.12
Disk winding shown in position over the inner windings and core. Clamping structures and leads are also shown. (Courtesy of Waukesha Electric Systems, Waukesha, WI.)

individual strands covered with an insulating paper wrap, is called magnet wire. The transpositions can also be built into the cable. This is called continuously transposed cable and generally consists of a bundle of 5–83 strands or more, each covered with a thin enamel coating. One strand at a time is transposed along the cable so that all the strands are eventually transposed at every turn around the core. The overall bundle is then sheathed in a paper wrap.

Figure 1.12 shows a disk winding situated over the inner windings and core and clamped at either end via the insulating blocks and steel structure shown. Short horizontal gaps are visible between the disks. Vertical columns of radial or key spacer projections are also visible. This outer high voltage winding is center-fed so that the top and bottom halves are connected in parallel. The leads feeding this winding are on the left.

1.7 Insulation Structures

Transformer windings and leads must operate at high voltages relative to the core, tank, and structural elements. In addition, different windings and even parts of the same winding operate at different voltages. This requires that some form of insulation between these various parts be provided to prevent voltage breakdown or corona discharges. The surrounding oil or air, which provides cooling also has some insulating value. The oil is of a special composition and must be purified to remove small particles and moisture. The type of oil most commonly used, as mentioned previously, is called transformer oil.

Further insulation is provided by paper covering the wire or cables. When saturated with oil, this paper has a high insulation value. Other types of wire covering besides paper are sometimes used, mainly for specialty applications such as higher temperature operation. Other insulating structures, which are generally present in sheet form, often wrapped into a cylindrical shape, are made of pressboard. This is a material made of cellulose fibers, which are compacted together into a fairly dense and rigid matrix. Key spacers, blocking material, and lead support structures are also commonly made of pressboard or wood.

Although normal operating voltages are quite high, 10–500 kV, the transformer must be designed to withstand even higher voltages, which can occur when lightning strikes the electrical system or when power is suddenly switched on or off in some part of the system. However infrequently these occur, they could permanently damage the insulation, disabling the unit, unless the insulation is designed to withstand them. Usually such events are of short duration. There is a time-dependence to how the insulation breaks down. A short duration high voltage pulse is no more likely to cause breakdown than a long duration low voltage pulse. This means that the same insulation that can withstand normal operating voltages, which are continuously present, can also withstand the high voltages arising from lightning strikes or switching operations, which are present only briefly. In order to insure that the abnormal voltages do not exceed the breakdown limits determined by their expected durations, lightning or surge arrestors are often used to limit them. These arrestors thus guarantee that the voltages will not rise above a certain value so that breakdown will not occur, assuming their durations remain within the expected range.

Because of the different dielectric constants of oil or air and paper, the electric stresses are unequally divided between them. Since the oil dielectric constant is about half that of paper and air is an even smaller fraction of paper's, the electric stresses are generally higher in oil or air than in the paper insulation. Unfortunately, oil or air has a lower breakdown stress than paper. In the case of oil, it has been found that subdividing the oil gaps by means of thin insulating barriers, usually made of pressboard, can raise the breakdown stress in the oil. Thus large oil gaps between the windings are usually subdivided by multiple pressboard barriers as shown schematically in Figure 1.13. This is referred to as the major insulation structure. The oil gap thicknesses are maintained by means of long vertical narrow sticks glued around the circumference of the cylindrical pressboard barriers. Often the barriers are extended by means of end collars, which curves around the ends of the windings to provide subdivided oil gaps at either end of the windings to strengthen these end oil gaps against voltage breakdown.

The minor insulation structure consists of the smaller oil gaps separating the disks and maintained by the key spacers, which are narrow insulators, usually made of pressboard and spaced radially around the disk's circumference as shown in Figure 1.13b. Generally, these oil gaps are small enough that subdivision is not required. In addition, the turn to turn insulation, usually made of paper, can be considered as part of the minor insulation structure. Figure 1.14 shows a winding during construction with key spacers and vertical sticks visible.

The leads, which connect the windings to the bushings or tap changers or to other windings, must also be properly insulated since they are at high voltage and pass close to tank walls or structural supports, which are grounded. They also can pass close to other leads at different voltages. High stresses can be developed at bends in the leads, particularly if they are sharp, so that additional insulation may be required in these areas. Figure 1.15 shows a rather extensive set of leads along with structural supports made of pressboard or wood. The leads pass close to the metallic clamps at the top and bottom and will also be near the tank wall when the core and coil assembly is inserted into the tank.

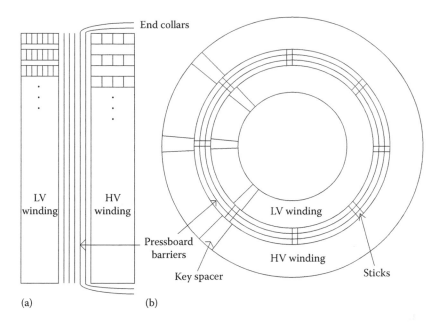

FIGURE 1.13
Major insulation structure consisting of multiple barriers between windings. Not all the key spacers or sticks are shown: (a) side view and (b) top view.

FIGURE 1.14
View of a partially wound winding showing the key spacers and vertical sticks.

FIGURE 1.15
Leads and their supporting structure emerging from the coils on one side of a 3-phase transformer. (Courtesy of Waukesha Electric Systems, Waukesha, WI.)

Although voltage breakdown levels in oil can be increased by means of barrier subdivisions, there is another breakdown process, which must be guarded against. This is breakdown due to creep. It occurs along the surfaces of the insulation. It requires sufficiently high electric stresses directed along the surface as well as sufficiently long uninterrupted paths over which the high stresses are present. Thus, the barriers themselves, sticks, key spacers, and lead supports can be a source of breakdown due to creep. Ideally, one should position these insulation structures so that their surfaces conform to voltage equipotential surfaces to which the electric field is perpendicular. Thus, there would be no electric fields directed along the surface. In practice, this is not always possible, so a compromise must be reached.

The major and minor insulation designs, including overall winding to winding separation and the number of barriers as well as disk-to-disk separation and paper covering thickness, are often determined by design rules based on extensive experience. However, in cases of newer or unusual designs, it is often desirable to do a field calculation using a finite-element program or other numerical procedure. This can be especially helpful when the potential for creep breakdown exists. Although these methods can provide accurate calculations of electric stresses, the breakdown process is not as well understood, so there is usually some judgment involved in deciding what level of electrical stress is acceptable.

1.8 Structural Elements

Under normal operating conditions, the electromagnetic forces acting on the transformer windings are quite modest. However, if a short-circuit fault occurs, the winding currents can increase 10–30-fold, resulting in forces of 100–900 times normal since the forces increase as the square of the current. The windings and supporting structure must be designed to withstand these fault current forces without permanent distortion of the windings or supports. Because current protection devices are usually installed, the fault currents are interrupted after a few cycles, but that is still long enough to do some damage if the supporting structure is inadequate.

Faults can be caused by falling trees that hit power lines, providing a direct current path to ground or by animals or birds bridging across two lines belonging to different phases, causing a line to line short. These should be rare occurrences, but over the 20–50 year lifetime of a transformer, their probability increases, so sufficient mechanical strength to withstand these is required.

The coils are generally supported at the ends with pressure rings. These are thick rings of pressboard or other material that cover the winding ends. The center opening allows the core to pass through. The rings are in the range of 30–180 mm for large power transformers. Some blocking made of pressboard or wood is required between the tops of the windings and the rings since all of the windings are not necessarily of the same height. Additional blocking is usually placed between the ring and the top yoke and clamping structure to provide some clearance between the high winding voltages and the grounded core and clamp. These structures can be seen in Figure 1.12. The metallic clamping structure can also be seen.

The top and bottom clamps are joined by vertical tie plates, sometimes called flitch plates, which pass along the sides of the core. The tie plates are solidly attached at both ends, so they pull the top and bottom clamps together by means of tightening bolts, compressing the windings. These compressive forces are transmitted along the windings via the key spacers, which must be strong enough in compression to accommodate these forces. The clamps and tie plates are generally made of structural steel. Axial forces that tend to elongate the windings when a fault occurs will put the tie plates in tension. The tie plates must also be strong enough to carry the gravitational load when the core and coils are lifted as a unit since the lifting hooks are attached to the clamps. The tie plates are typically about 10 mm thick and of varying width depending on the expected short-circuit forces and transformer weight. The width is often subdivided to reduce eddy current losses. Figure 1.16 shows a top view of the clamping structure.

The radial fault forces are countered inwardly by means of the sticks, which separate the oil barriers, and by means of additional support next to the core. The windings themselves, particularly the innermost one, are often made of hardened copper or bonded cable to provide additional resistance to the inward radial forces. The outermost winding is usually subjected to an outer radial force, which puts the wires or cables in tension. The material itself must be strong enough to resist these tensile forces since there is no supporting structure on the outside to counter these forces. A measure of the material's strength is its proof stress. This is the stress required to produce a permanent elongation of 0.2% (sometimes 0.1% is used). Copper of specified proof stress can be ordered from the wire or cable supplier.

The leads are also acted on by extra forces during a fault. The forces are produced by the stray flux from the coils or from nearby leads, interacting with the lead's current. The leads are therefore braced by means of wood or pressboard supports, which extend from the clamps. This lead support structure can be quite complicated, especially if there are many leads and interconnections. It is usually custom made for each unit. Figure 1.15 is an example of such a structure.

The assembled coil, core, clamps, and lead structure are placed in a transformer tank. The tank serves many functions, one of which is to contain the oil for an oil-filled unit. It also provides protection not only for the coils and other transformer components but for

FIGURE 1.16
Top view of clamping structure for a 3-phase transformer.

FIGURE 1.17
Large power transformer showing tank and attachments.

personnel from the high voltages present inside since the tank is grounded. If made of soft (magnetic) steel, it keeps stray flux from getting outside the tank. The tank is usually made airtight so that air doesn't enter and oxidize the oil.

Aside from being a containment vessel, the tank also has numerous attachments such as bushings for getting the electrical power into and out of the unit, an electronic control and monitoring cabinet for recording and transferring sensor information to remote processors or receiving remote control signals, and radiators with or without fans to provide cooling. On certain units, there is a separate tank compartment for tap changing equipment. Also some units have a conservator attached to the tank cover or to the top of the radiators. This is a large, usually cylindrical, structure, which contains oil in communication with the main tank oil. It also has an air space, which is separated from the oil by a sealed diaphragm. Thus, as the tank oil expands and contracts due to temperature changes, the flexible diaphragm accommodates these volume changes while maintaining a sealed oil environment. Figure 1.17 shows a large power transformer with a cylindrical conservator visible on top. Bushings mounted on top of the tank are visible. Also shown are the radiators with fans. Technicians are working on the control box.

1.9 Modern Trends

Changes in power transformers tend to occur very slowly. Issues of reliability over long periods of time and compatibility with existing systems must be addressed by any new technology. A major change, which has been ongoing since the earliest transformers, is the improvement in core steel. The magnetic properties, including losses, have improved dramatically over the years. Better stacking methods, such as stepped lapped construction, have resulted in lower losses at the joints. The use of laser or mechanical scribing has also helped lower the losses in these steels. Further incremental improvements in all of these areas can be expected.

The development of amorphous metals as a core material is relatively new. Although these materials have very low losses, lower than the best rolled electrical steels, they also have a rather low saturation induction (~1.5 T versus 2 T for rolled steels). They are also rather brittle and difficult to stack. This material has tended to be more expensive than rolled electrical steel and, since expense is always an issue, has limited their use. However, this could change with the cost of no-load losses to the utilities. Amorphous metals have found use as wound cores in distribution transformers. However, their use as stacked cores in large power transformers is not commonplace.

The development of improved wire types, such as transposed cable with epoxy bonding, is an ongoing process. Newer types of wire insulation covering such as Nomex are finding use. Nomex is a synthetic material, which can be used at higher temperatures than paper. It also has a lower dielectric constant than paper, so it produces a more favorable stress level in the adjacent oil than paper. Although it is presently a more expensive material than paper, it has found a niche in air-cooled transformers or in the rewinding of older transformers. Its thermal characteristics would probably be underutilized in transformers filled with transformer oil because of the limitations on the oil temperatures.

Pressboard insulation has undergone improvements over time such as precompressing to produce higher density material, which results in greater dimensional stability in transformer applications. This is especially helpful in the case of key spacers, which bear the compressional forces acting on the winding. Also, preformed parts made of press-board, such as collars at the winding ends and high voltage lead insulation assemblies, are becoming more common and are facilitating the development of higher voltage transformers.

Perhaps the biggest scientific breakthrough, which could revolutionize future transformers, is the discovery of high temperature superconductors. These materials are still in the early stage of development. They could operate at liquid nitrogen temperatures, which is a big improvement over the older superconductors that operate at liquid helium temperatures. It has been exceedingly difficult to make these new superconductors into wires of the lengths required in transformers. Nevertheless, prototype units are being built and technological improvements can be expected [Meh97].

A big change, which is occurring in newer transformers, is the increasing use of on-line monitoring devices. Fiber optic temperature sensors are being inserted directly into the windings to monitor the hottest winding temperature. This can be used to keep the transformer's loading or overloading within appropriate bounds so that acceptable insulation and adjacent oil temperatures are not exceeded and the thermal life is not too negatively impacted. Gas analysis devices are being developed to continuously record the amounts and composition of gasses in the cover gas or dissolved in the oil. This can provide an early indication of overheating or of arcing so that corrective action can be taken before the situation deteriorates too far. Newer fiber optic current sensors based on the Faraday effect are being developed. These weigh considerably less than present current sensors and are much less bulky. Newer miniaturized voltage sensors are also being developed. Sensor data in digitized form can be sent from the transformer to a remote computer for further processing. Newer software analysis tools should help to more accurately analyze fault conditions or operational irregularities.

Although tap changers are mechanical marvels, which operate very reliably over hundreds of thousands of tap changing operations, as with any mechanical device, they are subject to wear and must be replaced or refurbished from time to time. Electronic tap changers, using solid-state components, have been developed. Aside from essentially eliminating the wear problem, they also have a much faster response time than mechanical tap changers, which could be very useful in some applications. Their expense relative to mechanical tap changers has been one factor limiting their use. Further developments perhaps resulting in lower cost can be expected in this area.

As mentioned previously, there are incentives to transmit power at higher voltages. Some of the newer high voltage transmission lines operate in a d.c. mode. In this case, the conversion equipment at the ends of the line which change a.c. to d.c. and vice versa requires a transformer. However, this transformer does not need to operate at the line voltage. For high voltage a.c. lines, however, the transformer must operate at these higher voltages. At present, transformers, which operate in the range of 750–800 kV, have been built. Even higher voltage units have been developed. A better understanding of high voltage breakdown mechanisms, especially in oil, is needed to spur growth in this area.

2

Magnetism and Related Core Issues

2.1 Introduction

Transformer cores are constructed predominantly of ferromagnetic material. The most common material used is iron, with the addition of small amounts of silicon and other elements that help improve the magnetic properties and/or lower losses. Other materials, which find use in electronic transformers, are the nickel–iron alloys (permalloys) and the iron oxides (ferrites). The amorphous metals, generally consisting of iron, boron, and other additions, are also finding use as cores for distribution transformers. These materials are all broadly classified as ferromagnetic and, as such, have many properties in common. Among these are saturation magnetization or induction, hysteresis, and a Curie temperature above which they cease to be ferromagnetic.

Cores made of silicon steel (~3% Si) are constructed of multiple layers of the material in sheet form. The material is fabricated in rolling mills from hot slabs or ingots. Through a complex process of multiple rolling, annealing, and coating stages, it is formed into thin sheets of 0.18–0.3 mm thickness and up to a meter wide. The material has its best magnetic properties along the rolling direction, and a well-constructed core will take advantage of this. The good rolling direction magnetic properties are due to the underlying crystalline orientation, which is called a Goss or cube-on-edge texture as shown in Figure 2.1. The cubic crystals have the highest permeability along the cube edges. The visible edges pointing along the rolling direction are highlighted in the figure. Modern practice can achieve crystal alignments of >95%. The permeability is much lower along the cube diagonals or cube face diagonals. The latter are pointing in the sheet width direction.

In addition to its role in aiding crystal alignment, the silicon helps increase the resistivity of the steel from about 25 $\mu\Omega$ cm for low-carbon magnetic steel to about 50 $\mu\Omega$ cm for 3% Si–Fe. This higher resistivity leads to lower eddy current losses. Silicon also lowers the saturation induction from about 2.1 T for low-carbon steel to about 2.0 T for 3% Si–Fe. Silicon confers some brittleness on the material, which is an obstacle to rolling to even thinner sheet thickness. At higher silicon levels, the brittleness increases to the point where it becomes difficult to roll. This is unfortunate because at 6% silicon content, the magnetostriction of the steel disappears. Magnetostriction is a length change or strain, which is produced by the induction in the material. At a.c. frequencies, this contributes to the noise level in a transformer.

The nickel–iron alloys or permalloys are also produced in sheet form. Because of their malleability, they can be rolled extremely thin. The sheet thinness results in very low eddy current losses so that these materials find use in high-frequency applications. Their saturation induction is lower than that for silicon steel.

Ferrite cores are made of sintered powder. They generally have isotropic magnetic properties. They can be cast directly into the desired shape or machined after casting.

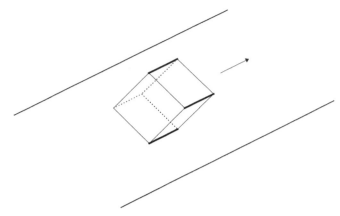

FIGURE 2.1
Goss or cube-on-edge crystalline texture for silicon steel.

They have extremely high resistivities, which permits their use in high-frequency applications. However, they have rather low saturation inductions.

Amorphous metals are produced by directly casting the liquid melt onto a rotating, internally cooled drum. The liquid solidifies extremely rapidly, resulting in the amorphous (noncrystalline) texture of the final product. The material comes off the drum in the form of a thin ribbon with controlled widths, which can be as high as ~250 mm. The material has a magnetic anisotropy determined by the casting direction and subsequent magnetic anneals so that the best magnetic properties are along the casting direction. Their saturation induction is about 1.5 T. Because of their thinness and composition, they have extremely low losses. These materials are very brittle, which has limited their use to wound cores. Their low losses make them attractive for use in distribution transformers, especially when no-load loss evaluations are high.

Ideally, a transformer core would carry the flux along a direction of highest permeability and in a closed path. Path interruptions caused by joints, which are occupied by low-permeability air or oil, lead to poorer overall magnetic properties. In addition, the cutting or slitting operations can introduce localized stresses that degrade the magnetic properties. In stacked cores, the joints are often formed by overlapping the laminations in steps to facilitate flux transfer. Nevertheless, the corners result in regions of higher loss. This can be accounted for in design by multiplying the ideal magnetic circuit losses, usually provided by the manufacturer on a per-unit-weight basis, by a building factor of >1. Another, possibly better, way to account for the extra loss is to apply a loss multiplying factor to the steel occupying the corner or joint region only. More fundamental methods to account for these extra losses have been proposed, but these tend to be too elaborate for routine use. Joints also give rise to higher exciting current, that is, the current in the coils necessary to drive the required flux around the core.

2.2 Basic Magnetism

The discovery by Oersted that currents give rise to magnetic fields led Ampere to propose that material magnetism results from localized currents. He proposed that large numbers of small current loops, appropriately oriented, could create the magnetic fields associated with magnetic materials and permanent magnets. At the time, the atomic nature of matter was not understood. With the Bohr model of the atom, where electrons are in orbit around a small massive nucleus, the localized currents could be associated with the moving electron. This gives rise to an orbital magnetic moment, which persists even though a quantum description has replaced the Bohr model. In addition to the orbital magnetism, the electron itself was found to possess a magnetic moment that cannot be understood simply from the circulating current point of view. Atomic magnetism results from a combination of both orbital and electron moments.

In some materials, the atomic magnetic moments either cancel or are very small so that little material magnetism results. These are known as paramagnetic or diamagnetic materials, depending on whether an applied field increases or decreases the magnetization. Their permeabilities relative to vacuum are nearly equal to 1. In other materials, the atomic moments are large and there is an innate tendency for them to align due to quantum mechanical forces. These are the ferromagnetic materials. The alignment forces are of very short range, operating only over atomic distances. Nevertheless, they create regions of aligned magnetic moments, called domains, within a magnetic material. Although each domain has a common orientation, this orientation differs from domain to domain. The narrow separations between domains are regions where the magnetic moments are transitioning from one orientation to another. These transition zones are referred to as domain walls.

In nonoriented magnetic materials, the domains are typically very small and randomly oriented. With the application of a magnetic field, the domain orientation tends to align with the field direction. In addition, favorably orientated domains tend to grow at the expense of unfavorably oriented ones. As the magnetic field increases, the domains eventually all point in the direction of the magnetic field, resulting in a state of magnetic saturation. Further increases in the field cannot orient more domains so the magnetization does not increase but is said to saturate. From this point on, further increases in induction are due to increases in the field only.

The relation between induction, **B**, magnetization, **M**, and field, **H**, in SI units is (boldface symbols are used to denote vectors)

$$\mathbf{B} = \mu_0 \left(\mathbf{H} + \mathbf{M} \right) \tag{2.1}$$

where $\mu_0 = 4\pi \times 10^{-7}$ H/m in SI units. For many materials, **M** is proportional to **H**:

$$\mathbf{M} = \chi \mathbf{H} \tag{2.2}$$

where χ is the susceptibility, which need not be a constant. Substituting into (2.1),

$$\mathbf{B} = \mu_0 \left(1 + \chi \right) \mathbf{H} = \mu_0 \mu_r \mathbf{H} \tag{2.3}$$

where $\mu_r = 1 + \chi$ is the relative permeability. We see directly in (2.1) that as **M** saturates because all the domains are similarly oriented, **B** can only increase due to increases in **H**. This occurs at fairly high **H** or exciting current values, since **H** is proportional to the exciting current. At saturation, since all the domains have the same orientation, there are no domain walls. Since **H** is generally small compared to **M** for high-permeability ferromagnetic materials up to saturation, the saturation magnetization and saturation induction are nearly the same and will be used interchangeably.

As the temperature increases, the thermal energy begins to compete with the alignment energy and the saturation magnetization begins to fall until the Curie point is reached

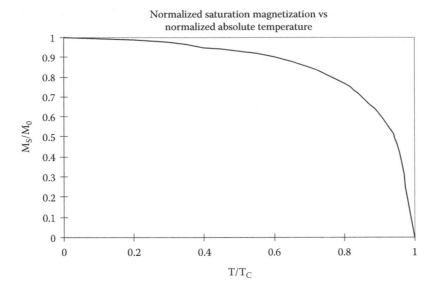

FIGURE 2.2
Relationship between saturation magnetization and absolute temperature, expressed in relative terms, for pure iron. This also applies reasonably well to other ferromagnetic materials containing predominately iron, nickel, or cobalt. M_S is the saturation magnetization at absolute temperature T and M_0 the saturation magnetization at T = 0 K. T_C is the Curie temperature in K.

where ferromagnetism completely disappears. For 3% Si–Fe, the saturation magnetization or induction at 20°C is 2.0 T (Tesla) and the Curie temperature is 746°C. This should be compared with pure iron where the saturation induction at 20°C is 2.1 T and the Curie temperature is 770°C.

The fall off with temperature follows fairly closely a theoretical relationship between the ratio of saturation induction at absolute temperature T to saturation induction at T = 0 K to the ratio of absolute temperature T to the Curie temperature expressed in K. For pure iron, this relationship is graphed in Figure 2.2 [Ame57]. This same normalized graph also applies rather closely to other iron-containing magnetic materials such as Si–Fe as well as to nickel- and cobalt-based magnetic materials.

Thus, to find the saturation magnetization of 3% Si–Fe at a temperature of 200°C = 473 K, since the Curie temperature is 746°C, the ratio $T/T_C = 473/1019 = 0.464$. From the graph, this corresponds to $M_S/M_0 = 0.94$. On the other hand, we know that at 20°C, where $T/T_C = 0.287$, $M_S/M_0 = 0.98$. Thus, $M_0 = 2.0/0.98 = 2.04$ and M_S (T = 200°C) = 0.94 × 2.04 = 1.92 T. This is only a 4% drop in the 20°C saturation magnetization. Considering that core temperatures are unlikely to reach 200°C, temperature effects on magnetization should not be a problem in transformers under normal operating conditions.

Ferromagnetic materials typically exhibit the phenomenon of magnetostriction, that is, a length change or strain resulting from the induction or flux density that they carry. Since this length change is independent of the sign of the induction, for an a.c. induction at frequency f, the length oscillations occur at frequency 2f. These length vibrations contribute to the noise level in transformers. Magnetostriction is actually a fairly complex phenomenon and can exhibit hysteresis as well as anisotropy.

Another source of noise in transformers, often overlooked, results from the transverse vibrations of the laminations at unsupported free ends. This can occur at the outer

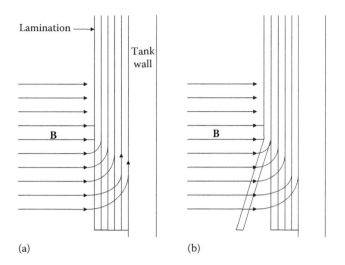

FIGURE 2.3
Geometry for obtaining a qualitative understanding of the force on a loose end lamination: (a) no loose laminations and (b) loose end lamination.

surfaces of the core and shunts if these are not constrained. This can be shown qualitatively by considering the situation shown in Figure 2.3. In Figure 2.3a, we show a leakage flux density vector **B** impinging on a packet of tank shunt laminations that are flat against the tank wall and rigidly constrained. After striking the lamination packet, the flux is diverted into the packet and transported upward in the figure since we are looking at the bottom end. In Figure 2.3b, the outer laminations are constrained from a certain distance upward, below which they are free to move. Part of the flux density B is diverted along this outer packet, and a reduced flux density impinges on the portion of the packets that are free to move. The magnetic energy associated with Figure 2.3b is lower than that associated with Figure 2.3a. A force is associated with this change in magnetic energy. This force acts to pull the outer lamination outward. It is independent of the sign of B so that if B is sinusoidal of frequency f, the force will have a frequency of 2f. Thus, it contributes to the transformer noise at the same frequency as magnetostriction. If this frequency is near the mechanical vibration frequency of the cantilevered (free) portion of the packets, a resonance situation may develop.

2.3 Hysteresis

Hysteresis, as the name implies, means that the present state of a ferromagnetic material depends on its past magnetic history. This is usually illustrated by means of a B–H diagram. Magnetic field changes are assumed to occur slowly enough that eddy current effects can be ignored. We assume the B and H fields are collinear although, in general, they need not be. Thus, we can drop the vector notation. If we start out with a completely demagnetized specimen (this state requires careful preparation) and increase the magnetic field from 0, the material will follow the initial curve as shown in Figure 2.4a. This curve can be continued to saturation, B_s. If, at some point along the initial curve, the field is reversed and

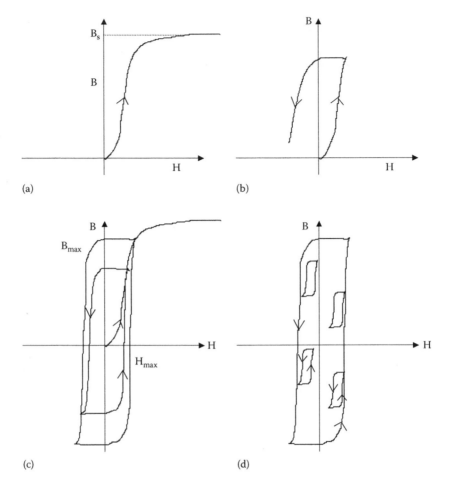

FIGURE 2.4
Hysteresis processes: (a) initial curve, (b) following a normal hysteresis loop after reversal from the initial curve, (c) family of normal hysteresis loops, and (d) minor or incremental hysteresis loops.

decreased, the material will follow a normal hysteresis loop as shown in Figure 2.4b. If the magnetic field is cycled repeatedly between $\pm H_{max}$, the material will stay on a normal hysteresis loop determined by H_{max}. There is a whole family of such loops as shown in Figure 2.4c. The largest loop occurs when B_{max} reaches B_s. This is called the major loop. These loops are symmetrical about the origin. If at some point along a normal loop, other than the extreme points, the field is reversed and cycled through a smaller cycle back to its original value before the reversal occurred, a minor or incremental loop is traced as shown in Figure 2.4d. Considering the many other possibilities for field reversals, the resulting hysteresis paths can become quite complicated.

Perhaps the most important magnetic path is a normal hysteresis loop since this is traced in a sinusoidal cyclic magnetization process. Several key points along such a path are shown in Figure 2.5. As the field is lowered from H_{max} to zero, the remaining induction is called the remanence, B_r. As the field is further lowered into negative territory, the absolute value of the field at which the induction drops to zero is called the coercivity, H_c. Because the loop is symmetrical about the origin, there are corresponding points on the

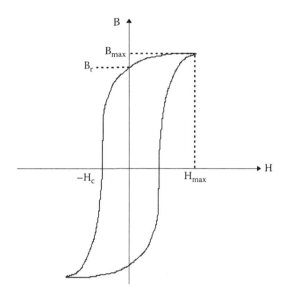

FIGURE 2.5
Key points along a normal hysteresis loop.

negative branch. At any point on a magnetization path, the ratio of B to H is called the permeability, while the slope of the B–H curve at that point is called the differential permeability. Other types of permeability can be defined. The area of the hysteresis loop is the magnetic energy per unit volume and per cycle. This energy is dissipated over each hysteresis cycle.

In oriented Si–Fe, the relative permeability for inductions reasonably below saturation is so high that the initial curve is close to the B axis if B and H are measured in the same units as they are in the Gaussian system or if B versus $\mu_o H$ is plotted in the SI system. In addition, the hysteresis loops are very narrow in these systems of units and closely hug the initial curve. Thus, for all practical purposes, we can assume a single-valued B–H characteristic for these high-permeability materials coinciding with the initial curve. Since inductions in transformer cores are kept well below saturation in normal operation to avoid high exciting currents, the effects of saturation are hardly noticeable and the core, for many purposes, can be assumed to have a constant permeability.

2.4 Magnetic Circuits

In stacked cores and especially in cores containing butt joints as well as in gapped reactor cores, the magnetic path for the flux is not through a homogeneous magnetic material. Rather, there are gaps occupied by air or oil or other nonmagnetic materials of relative permeability equal to 1. In such cases, it is possible to derive effective permeabilities by using a magnetic circuit approximation. This approximation derives from the mathematical similarity of magnetic and electrical laws. In resistive electric circuits, the conductivities of the wires and circuit elements are usually so much higher than the surrounding medium (usually air) that little current leaks away from the circuit. However, in magnetic circuits where the flux corresponds to the electric current, the circuit permeability, which is the analog of conductivity, is not many orders of magnitude higher than that of the surrounding medium so that flux leakage does occur. Thus, whereas the circuit approach is nearly exact for electric circuits, it is only approximate for magnetic circuits [Del94].

Corresponding to Kirchhoff's current law at a node,

$$\sum_i I_i = 0 \tag{2.4}$$

where I_i is the current into a node along a circuit branch i (positive if entering the node, negative if leaving), we have the approximate magnetic counterpart

$$\sum_i \Phi_i = 0 \tag{2.5}$$

where Φ_i is the flux into a node. Whereas (2.4) is based on conservation of current and ultimately charge, Equation 2.5 is based on conservation of flux. Kirchhoff's voltage law can be expressed as

$$\sum_i V_i = \oint \mathbf{E} \cdot \mathbf{d}\ell = \text{emf} \tag{2.6}$$

where $\mathbf{E}$ is the electric field and the voltage drops V_i are taken around the loop in which an electromotive force (emf) is induced. The integrals in this equation are path integrals and ℓ the path length. With $\mathbf{E}$ corresponding to $\mathbf{H}$, the magnetic analogy is

$$\sum_i (\text{NI})_i = \oint \mathbf{H} \cdot \mathbf{d}\ell = \text{mmf} \tag{2.7}$$

where $(\text{NI})_i$ are the amp-turns and mmf is the magnetomotive force.

In terms of the current density $\mathbf{J}$ and the flux density $\mathbf{B}$, we have

$$I = \int \mathbf{J} \cdot \mathbf{dA}, \quad \Phi = \int \mathbf{B} \cdot \mathbf{dA} \tag{2.8}$$

so that $\mathbf{J}$ and $\mathbf{B}$ are corresponding quantities in the two systems. In (2.8), the integrals are surface integrals and $\mathbf{A}$ the vectorial cross-sectional area perpendicular to $\mathbf{J}$ or $\mathbf{B}$. Ohm's law in its basic form can be written as

$$\mathbf{J} = \sigma \mathbf{E} \tag{2.9}$$

where σ is the conductivity. This corresponds to

$$\mathbf{B} = \mu \mathbf{H} \tag{2.10}$$

in the magnetic system, where μ is the permeability.

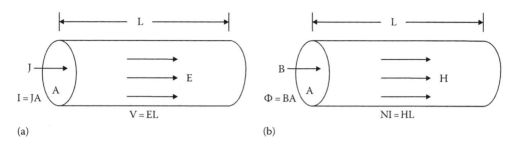

FIGURE 2.6
Geometries for simple resistance and reluctance calculation: (a) simple resistive electrical element and (b) simple reluctive magnetic element.

To obtain resistance and corresponding reluctance expressions for simple geometries, consider a resistive or reluctive element of length L and uniform cross-sectional area A as shown in Figure 2.6. Let a uniform current density J flow through the resistive element and a uniform flux density B flow through the magnetic element normal to the surface area A. Then

$$I = JA, \quad \Phi = BA \tag{2.11}$$

Let a uniform electric field E drive the current and a uniform magnetic field H drive the flux. Then

$$V = EL, \quad NI = HL \tag{2.12}$$

Using (2.9), we have

$$\frac{I}{A} = \sigma \frac{V}{L} \implies V = \left(\frac{L}{\sigma A}\right)I \tag{2.13}$$

where the quantity in parenthesis is recognized as the resistance. Similarly from (2.10),

$$\frac{\Phi}{A} = \mu \frac{NI}{L} \implies NI = \left(\frac{L}{\mu A}\right)\Phi \tag{2.14}$$

where the quantity in parenthesis is called the reluctance. A similar analysis could be carried out for other geometries using the basic field correspondences.

As an application of the circuit approach, consider a simple magnetic core with an air gap as shown in Figure 2.7. A flux Φ is driven around the circuit by a coil generating an mmf of NI. The path through the magnetic core of permeability $\mu = \mu_o \mu_r$ has mean length L and the path in the air gap of permeability μ_o has length L_o. The reluctances in the magnetic material and air gap are

$$R_{mag} = \frac{L}{\mu_o \mu_r A}, \quad R_{air} = \frac{L_o}{\mu_o A} \tag{2.15}$$

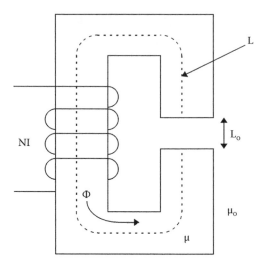

FIGURE 2.7
Air gap magnet driven by mmf = NI.

Since these reluctances are in series, we have

$$NI = \Phi\left(\frac{L}{\mu_o\mu_r A} + \frac{L_o}{\mu_o A}\right) = \Phi\frac{L}{\mu_o\mu_r A}\left(1 + \mu_r\frac{L_o}{L}\right) \tag{2.16}$$

Thus, since μ_r can be quite large, the reluctance with air gap can be much larger than without so that more mmf is required to drive a given flux. Note that we ignored fringing in the air gap. This could be approximately accounted for by letting the area, A, be larger in the air gap than in the core material.

In general, for two reluctances in series having the same cross-sectional area A but lengths L_1, L_2 and permeabilities μ_1, μ_2, the total reluctance is

$$R_{tot} = \frac{L_1}{\mu_1 A} + \frac{L_2}{\mu_2 A} = \frac{L}{A}\left(\frac{f_1}{\mu_1} + \frac{f_2}{\mu_2}\right) \tag{2.17}$$

where $L = L_1 + L_2$, $f_1 = L_1/L$, and $f_2 = L_2/L$. Thus, the effective permeability of the combination is given by

$$\frac{1}{\mu_{eff}} = \frac{f_1}{\mu_1} + \frac{f_2}{\mu_2} \tag{2.18}$$

In (2.18), the μ's can be taken as relative permeabilities since μ_o factors out. This could be extended to more elements in series. An identical relationship holds for the effective conductivity of a series of equal cross-sectional area resistive elements. As a numerical example, suppose material 1 is magnetic steel with a relative permeability of 10,000 and material 2 is air with a relative permeability of 1. For a small air or oil joint, let $f_2 = 0.001$ and therefore $f_1 = 0.999$. From (2.18), we see that $1/\mu_{eff} = 0.0000999 + 0.001 = 0.0010999$, so that $\mu_{eff} = 909$. Thus, this small air gap results in a fairly large decrease in the circuit relative permeability from that of a circuit consisting of only magnetic steel. According to (2.18), for μ_1 very large and $\mu_2 = 1$, μ_{eff} approaches $1/f_2$ so the overall relative permeability is determined by the relative size of the gap in the magnetic circuit.

Thus, in a transformer core with joints, the effective permeability is reduced relative to that of an ideal core without joints. This requires a higher exciting current to drive a given flux through the core. The core losses will also increase mainly due to flux distortion near the joint region. For linear materials, the slope of the B–H curve will decrease for a jointed core relative to an unjointed core. For materials that follow a hysteresis loop, the joints will have the effect of skewing or tilting the effective hysteresis loop of the jointed core away from the vertical compared with the unjointed ideal core. This can be shown by writing (2.7) as

$$NI = H_1 \ell_1 + H_2 \ell_2 = H_{eff} \ell \tag{2.19}$$

where
  H_1 is the field in the core steel
  H_2 is the field in the gap
  ℓ_1 and ℓ_2 are core and gap lengths
  ℓ is the total length
  H_{eff} is an effective applied field

Assuming the gap is linear with permeability μ_o (relative permeability 1), we have

$$B_2 = \mu_2 H_2 \tag{2.20}$$

Also assume that there is no leakage, the core and gap have the same cross-sectional area, and the flux is uniformly distributed across it. Then $B_1 = B_2 = B$ from flux continuity. Thus, we get for the H field in the core material

$$H_1 = H_{eff} \frac{\ell}{\ell_1} - \frac{B\ell_2}{\mu_o \ell_1} \tag{2.21}$$

Since $\ell \approx \ell_1$, H_1 is more negative than H_{eff} when B is positive. To see the effect this has on the hysteresis loop of the jointed core, refer to Figure 2.8. H_{eff} is proportional to the mmf by (2.19), and H_1 is the field in the core that traces the intrinsic hysteresis loop. On the top part

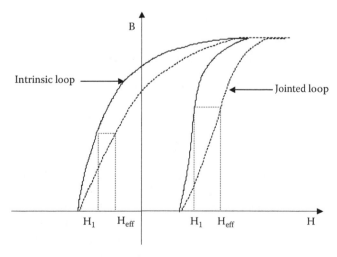

FIGURE 2.8
Effect of air or oil gaps on the hysteresis loop.

of the loop, B > 0, so $H_1 < H_{eff}$. This means that the intrinsic hysteresis loop is shifted to the left of the loop associated with the jointed core. Since the shift is proportional to B, it is greater at the top of the loop and vanishes when B = 0. This has the effect of tilting the jointed core loop to the right. On the lower half of the loop where B is negative, $H_1 > H_{eff}$. This means that the intrinsic loop is shifted to the right of the jointed core loop, the shift again proportional to B. Thus, the jointed core is tilted to the left at the bottom, as expected by symmetry. Notice that the remanence falls for the jointed loop because according to (2.21), when $H_{eff} = 0$, $H_1 < 0$ and the associated B value is lower down on the jointed loop. A very narrow hysteresis loop, approximating the initial magnetization curve, would also be tilted for a jointed core.

2.5 Inrush Current

When a transformer is disconnected from a power source, the current is interrupted and the magnetic field or mmf driving flux through the core is reduced to zero. As we have seen in the preceding section, the core retains a residual induction, which is called the remanence when the hysteresis path is on the positive descending (negative ascending) branch of a normal loop. In order to drive the core to the zero magnetization state, it would be necessary to gradually lower the peak induction while cycling the field.

Since the intrinsic normal hysteresis loops for oriented Si–Fe have fairly flat tops (or bottoms), the remanence is close to the peak induction. However, the presence of gaps in the core reduces this somewhat. When the unit is reenergized by a voltage source, the flux change must match the voltage change according to Faraday's law:

$$V = -N \frac{d\Phi}{dt} \tag{2.22}$$

For a sinusoidal voltage source, the flux is also sinusoidal:

$$\Phi = \Phi_p \sin(\omega t + \varphi) \tag{2.23}$$

where Φ_p is the peak flux. Substituting into (2.22), we get

$$V = -N\omega\Phi_p \cos(\omega t + \varphi) \tag{2.24}$$

Hence, the peak voltage, using $\omega = 2\pi f$, where f is the frequency in Hz, is

$$V_p = 2\pi f N \Phi_p \tag{2.25}$$

Thus, to follow the voltage change, the flux must change by $\pm\Phi_p$ over a cycle. If, in a worst-case scenario, the voltage source is turned on at the bottom of a cycle (negative Φ_p) and the remanent induction has a positive value of B_r with an associated core flux of $B_r A_c$, where A_c is the core area, then the flux will rise to $2\Phi_p + B_r A_c$ when the voltage reaches a value corresponding to $+\Phi_p$. Since, under normal conditions, $\Phi_p = B_p A_c$, then we see that the core induction would rise to $2B_p + B_r$. Since B_p is usually ~10% to 20% below saturation in typical power transformers, this means that the core will be driven strongly into saturation, which requires a very high exciting current. This exciting current is called the inrush current and can be many times the normal load current in a transformer.

Actually as saturation is approached, the flux will no longer remain confined to the core but will spill into the air or oil space inside the coil that supplies the exciting current. Thus, beyond saturation, the entire area inside the exciting coil, including the core, must be considered the flux-carrying area with an incremental relative permeability of 1. Let B_r be the residual induction in the core, which, without loss of generality, we can take to be positive. In the following, we will assume that this is the remanence. Thus, the residual flux is $\Phi_r = B_r A_c$. Let $\Delta\Phi$ be the flux change required to bring the voltage from its turn-on point up to its maximum value in the same sense as the residual flux. $\Delta\Phi$ could be positive, negative, or zero, depending on the turn-on point. We assume it is positive here.

Part of the increase in $\Delta\Phi$ will simply bring the induction up to the saturation level, entailing the expenditure of little exciting power or current. This part is given approximately by $(B_s - B_r)A_c$, where B_s is the saturation induction. Beyond this point, further increases in $\Delta\Phi$ occur with the expenditure of high exciting current since the incremental relative permeability is 1. Since beyond saturation, the core and air or oil have the same permeability, the incremental flux density will be the same throughout the interior of the coil, ignoring end effects. Letting the interior coil area up to the mean radius R_m be A ($A = \pi R_m^2$), the incremental flux density is

$$B_{inc} = \frac{\Delta\Phi - (B_s - B_r)A_c}{A} \tag{2.26}$$

The incremental magnetic field inside the coil, ignoring end effects, is $H_{inc} = NI/h$, where NI are the exciting amp-turns and h the coil height. We are ignoring the exciting amp-turns required to reach saturation since these are comparatively small. Since the permeability for this flux change is μ_o, we have $B_{inc} = \mu_o H_{inc} = \mu_o NI/h$, which implies from (2.26)

$$NI = \frac{h}{\mu_o A}\left[\Delta\Phi - (B_s - B_r)A_c\right] \tag{2.27}$$

Assuming, in the worst case, that the voltage is turned on at a corresponding flux density that is at the most negative point in the cycle, we have, using (2.25),

$$\Delta\Phi_{max} = 2\Delta\Phi_p = \frac{2V_p}{2\pi fN} = 2B_p A_c \tag{2.28}$$

In the third equality, we are computing this flux change mathematically as if it all occurred in the core as it would under normal conditions. Thus, Equation 2.27 becomes, substituting the maximum flux change $\Delta\Phi_{max}$ for $\Delta\Phi$ for a worst-case situation,

$$NI_{max} = \frac{hA_c}{\mu_o A}\left(2B_p + B_r - B_s\right) \tag{2.29}$$

As a numerical example, consider transformers with stacked cores and step-lapped joints where $B_r \approx 0.9B_p$. Taking h = 2 m, A_c/A = 0.5, B_p = 1.7 T, B_r = 0.9, B_p = 1.53 T, and B_s = 2 T, we obtain $NI_{max} = 2.33 \times 10^6$ amp-turns. For N = 500, I_{max} = 4660 amps. This is very high for an exciting current.

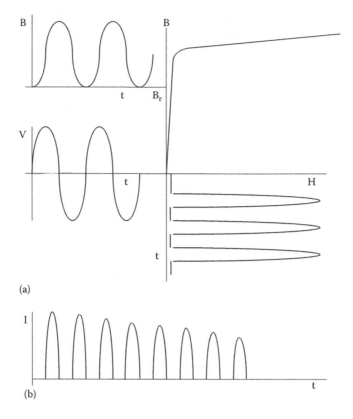

FIGURE 2.9
Distortion of exciting current due to saturation: (a) B–H curve indirectly relating voltage to exciting current and (b) exciting current proportional to H vs time.

While the voltage is constrained to be sinusoidal, the exciting current will be distorted due mainly to saturation effects. Even below saturation, there is some distortion due to nonlinearity in the B–H curve. Figure 2.9a illustrates the situation on inrush. The sinusoidal voltage is proportional to the incremental induction, which is displaced by the remanent induction. It requires high H values near its peak and comparatively small to zero H values near its trough. This is reflected in the exciting current, which is proportional to H. This current appears as a series of positive pulses separated by broad regions of near zero value as shown in Figure 2.9a. The positive current pulses appear undistorted because they are associated with the saturated portion of the B–H curve where the relative permeability is close to 1. The high exciting inrush current will damp out with time as suggested in Figure 2.9b. This is due to resistive effects.

2.6 Fault Current Waveform and Peak Amplitude

For comparison purposes, we now examine the time dependence of the fault current. We will ignore the load current at the time of the fault and assume the transformer is suddenly grounded at t = 0. The equivalent circuit is shown in Figure 2.10, which will be derived

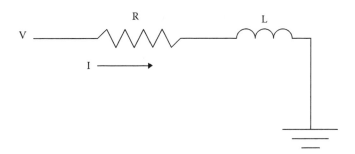

FIGURE 2.10
Circuit for fault current analysis.

later. Here R and L are the resistance and leakage reactance of the transformer, respectively, including any contributions from the system. The voltage is given by

$$V = V_p \sin(\omega t + \varphi) \tag{2.30}$$

where φ is the phase angle, which can have any value, in general, since the fault can occur at any time during the voltage cycle. The circuit equation is

$$V_p \sin(\omega t + \varphi) = \begin{cases} RI + L\dfrac{dI}{dt}, & t \geq 0 \\ 0, & t < 0 \end{cases} \tag{2.31}$$

Using Laplace transforms, the current transform is given by [Hue72]

$$I(s) = \frac{V_p(s \sin\varphi + \omega\cos\varphi)}{(R+Ls)(s^2 + \omega^2)} \tag{2.32}$$

Taking the inverse transform, we obtain

$$I(t) = \frac{V_p}{R\sqrt{1+(\omega L / R)^2}} \left[\sin(\beta - \varphi)e^{-(R/L)t} + \sin(\omega t + \varphi - \beta) \right] \tag{2.33}$$

where $\beta = \tan^{-1}(\omega L / R)$.

The steady-state peak current amplitude is given by

$$I_{p,ss} = \frac{V_p}{R\sqrt{1+(\omega L / R)^2}} \tag{2.34}$$

Using this and letting $\tau = \omega t$, $\nu = \omega L / R$, we can rewrite (2.33), using trigonometric identities, as

$$
\begin{aligned}
I(\tau) &= I_{p,ss} \left[\sin(\beta - \varphi)e^{-(\tau/\nu)} + \sin(\tau + \varphi - \beta) \right] \\
&= I_{p,ss} \left[\sin(\beta - \varphi)\left(e^{-(\tau/\nu)} - \cos\tau\right) + \cos(\beta - \varphi)\sin\tau \right]
\end{aligned} \tag{2.35}
$$

where $\beta = \tan^{-1}\nu$.

To find the maximum amplitude for a given φ, we need to solve

$$\frac{\partial I}{\partial \tau} = \sin(\beta - \varphi)\left(\frac{e^{-(\tau/\nu)}}{\nu} - \sin\tau\right) - \cos(\beta - \varphi)\cos\tau = 0 \qquad (2.36)$$

In addition, if we wish to determine the value of φ that produces the largest fault current, we need to solve

$$\frac{\partial I}{\partial \varphi} = \cos(\beta - \varphi)\left(e^{-(\tau/\nu)} - \cos\tau\right) - \sin(\beta - \varphi)\sin\tau = 0 \qquad (2.37)$$

From (2.36) and (2.37), we see that

$$\frac{\sin(\beta - \varphi)}{\cos(\beta - \varphi)} = \frac{\cos\tau}{\left(\left(e^{-(\tau/\nu)}/\nu\right) - \sin\tau\right)} = \frac{\nu\cos\tau}{\left(e^{-(\tau/\nu)} - \nu\sin\tau\right)}$$

$$\frac{\sin(\beta - \varphi)}{\cos(\beta - \varphi)} = \frac{\left(e^{-(\tau/\nu)} - \cos\tau\right)}{\sin\tau}$$

Equating the right-hand sides of both equations, we get

$$\frac{\left(e^{-(\tau/\nu)} - \cos\tau\right)}{\sin\tau} = \frac{\nu\cos\tau}{\left(e^{-(\tau/\nu)} - \nu\sin\tau\right)} \quad \Rightarrow \quad \left(e^{-(\tau/\nu)} - \cos\tau\right)\left(e^{-(\tau/\nu)} - \nu\sin\tau\right) = \nu\sin\tau\cos\tau$$

Multiplying out the product in the last equation, we get

$$e^{-(2\tau/\nu)} - e^{-(\tau/\nu)}\nu\sin\tau - e^{-(\tau/\nu)}\cos\tau + \nu\sin\tau\cos\tau = \nu\sin\tau\cos\tau \quad \Rightarrow$$

$$e^{-(2\tau/\nu)} - e^{-(\tau/\nu)}\nu\sin\tau - e^{-(\tau/\nu)}\cos\tau = 0 \quad \Rightarrow \quad e^{-(\tau/\nu)} - \nu\sin\tau - \cos\tau = 0$$

From the last equation, we see that

$$e^{-(\tau/\nu)} - \cos\tau = \nu\sin\tau \quad \Rightarrow \quad \frac{\sin(\beta - \varphi)}{\cos(\beta - \varphi)} = \frac{\left(e^{-(\tau/\nu)} - \cos\tau\right)}{\sin\tau} = \frac{\nu\sin\tau}{\sin\tau} = \nu$$

Thus, we find that solving (2.36) and (2.37) simultaneously leads to

$$\tan(\beta - \varphi) = \nu = \tan\beta \qquad (2.38)$$

$$e^{-(\tau/\nu)} - \nu\sin\tau - \cos\tau = 0 \qquad (2.39)$$

Equation 2.38 shows that $\varphi = 0$ produces the maximum amplitude and the time at which this maximum occurs is given by the solution of (2.39) for $\tau > 0$. Substituting into (2.35), we obtain

$$I_{max} = I_{p,ss} \left[\sin(\beta) v \sin \tau + \cos(\beta) \sin \tau \right] = \cos(\beta) \left(\tan(\beta) v \sin \tau + \sin \tau \right)$$
$$= \cos(\beta) \left(1 + v^2 \right) \sin \tau$$

But since $\cos \beta = 1/\sqrt{[1 + (\tan \beta)^2]} = 1/\sqrt{(1 + v^2)}$, we get

$$\frac{I_{max}}{I_{p,ss}} = \sqrt{1 + v^2} \sin \tau \tag{2.40}$$

This is the asymmetry factor over the steady-state peak amplitude with τ obtained by solving (2.39).

The asymmetry factor is generally considered with respect to the steady-state rms current value. This new ratio is called K in the literature and is thus given by substituting $I_{p,ss} = \sqrt{2} I_{rms,ss}$ in (2.40):

$$K = \frac{I_{max}}{I_{rms,ss}} = \sqrt{2 \left(1 + v^2 \right)} \sin \tau, \quad v = \frac{\omega L}{R} = \frac{x}{r} \tag{2.41}$$

We have expressed the ratio of leakage impedance to resistance as x/r. x and r are normalized quantities, that is, the leakage reactance and resistance divided by a base impedance value, which cancels out in the ratio. Equation 2.41 has been parametrized as [IEE57]

$$K = \sqrt{2} \left\{ 1 + e^{-(r/x)(\phi+(\pi/2))} \sin \phi \right\}, \quad \phi = \tan^{-1} \left(\frac{x}{r} \right) = \beta \quad \text{and} \quad \frac{x}{r} = v \tag{2.42}$$

This parametrization agrees with (2.41) to within 0.7%. Table 2.1 shows some of the K values obtained by the two methods.

At $\varphi = 0$ where the asymmetry is greatest, Equation 2.35 becomes

$$I(\tau) = \frac{\sqrt{2} I_{rms,ss}}{\sqrt{1 + v^2}} \left[v \left(e^{-(\tau/v)} - \cos \tau \right) + \sin \tau \right] \tag{2.43}$$

This is graphed in Figure 2.11 for $v = 10$.

TABLE 2.1

Comparison of Exact with Parametrized K Values

$v = x/r$	K (Exact)	K (Parametrized)
1	1.512	1.509
2	1.756	1.746
5	2.192	2.184
10	2.456	2.452
50	2.743	2.743
1000	2.828	2.824

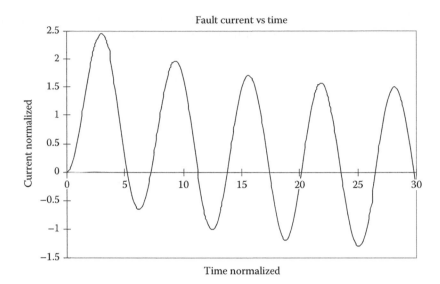

FIGURE 2.11
Fault current versus time for the case of maximum offset and $\nu = x/r = 10$.

The lowest Fourier coefficients of (2.35) are

$$\frac{a_o}{I_{p,ss}} = \sin(\beta - \varphi)\frac{\nu}{2\pi}\left(1 - e^{-(2\pi/\nu)}\right)$$

$$\frac{a_1}{I_{p,ss}} = \sin(\beta - \varphi)\left[\frac{\nu}{\pi(1+\nu^2)}\left(1 - e^{-(2\pi/\nu)}\right) - 1\right]$$

$$\frac{a_2}{I_{p,ss}} = \sin(\beta - \varphi)\frac{\nu}{\pi(1+4\nu^2)}\left(1 - e^{-(2\pi/\nu)}\right) \quad (2.44)$$

$$\frac{b_1}{I_{p,ss}} = \sin(\beta - \varphi)\frac{\nu^2}{\pi(1+\nu^2)}\left(1 - e^{-(2\pi/\nu)}\right) + \cos(\beta - \varphi)$$

$$\frac{b_2}{I_{p,ss}} = \sin(\beta - \varphi)\frac{2\nu^2}{\pi(1+4\nu^2)}\left(1 - e^{-(2\pi/\nu)}\right)$$

The ratio of second harmonic to fundamental amplitude, $\left(\sqrt{a_2^2 + b_2^2}\right)/\left(\sqrt{a_1^2 + b_1^2}\right)$, is tabulated in Table 2.2 for a range of φ and ν values.

For power transformers, x/r is typically >20. We see from Table 2.2 that the second harmonic content relative to the fundamental is <4.3% for $x/r > 20$. On the other hand, the second-to-first harmonic ratio of the inrush current for power transformers is generally >8%. Although inrush current waveforms are not as simple to characterize, a glance at Figure 2.9 suggests that it would have a high second harmonic content.

TABLE 2.2

Ratio of Second Harmonic to Fundamental Amplitude of Fault Current for Various Values of the Reactance to Resistance Ratio and Voltage Phase Angle

$\nu = x/r$	$\varphi = 0°$	$\varphi = 30°$	$\varphi = 45°$	$\varphi = 90°$	$\varphi = 120°$	$\varphi = 135°$
1	0.0992	0.0357	0	0.119	0.168	0.166
2	0.128	0.0752	0.0440	0.0752	0.152	0.167
3	0.127	0.0831	0.0566	0.0475	0.123	0.145
4	0.118	0.0819	0.0589	0.0323	0.101	0.124
5	0.109	0.0778	0.0576	0.0232	0.0543	0.107
6	0.0996	0.0731	0.0550	0.0175	0.0735	0.0941
7	0.0916	0.0684	0.0521	0.0137	0.0644	0.0835
8	0.0845	0.0641	0.0493	0.0110	0.0572	0.0750
9	0.0784	0.0601	0.0465	0.0090	0.0515	0.0680
10	0.0730	0.0565	0.0440	0.0075	0.0467	0.0622
15	0.0540	0.0432	0.0341	0.0036	0.0320	0.0434
20	0.0427	0.0347	0.0276	0.0022	0.0242	0.0333
25	0.0352	0.0290	0.0232	0.0014	0.0195	0.0270
50	0.0188	0.0158	0.0128	0.0004	0.0099	0.0138
100	0.0097	0.0083	0.0067	0.0001	0.0050	0.0070
500	0.0020	0.0017	0.0014	0	0.0010	0.0014
1000	0.0010	0.0008	0.0007	0	0.0005	0.0007

Thus, a comparison of this ratio is sometimes used to distinguish inrush from fault currents. In fact, x/r would need to fall below 10 before this method breaks down.

2.7 Optimal Core Stacking

As discussed in Section 1.4, cores for power transformers are normally constructed of stacks of thin laminations. These are often arranged to approximate a circular shape. Because of the limited number of lamination widths available, the circular shape consists of a finite number of stacks or steps, with all the laminations in a given step having the same width. Thus, the circular shape is not completely filled with flux-carrying steel. It is important to optimize the stacking pattern for a given number of steps in order to fill the circular area with as much core steel as possible. In addition, because of the insulating, nonferromagnetic coating applied to the laminations, the area of the lamination stacks must be corrected for this coating in order to arrive at the true flux-carrying area of the core. This correction or stacking factor is normally about 96%, that is, the actual core-carrying area is about 96% of the geometric stack area.

For a given number of steps, one can maximize the core area to obtain an optimal stacking pattern. Figure 2.12 shows the geometric parameters, which can be used in such an optimization, namely, the x and y coordinates of the stack corners, which touch the circle of radius R. Only 1/4 of the geometry is modeled due to symmetry considerations.

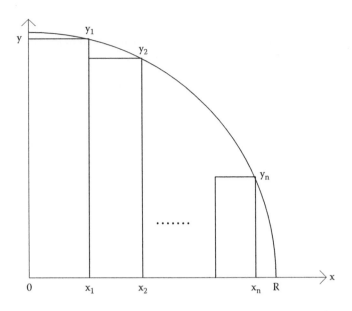

FIGURE 2.12
Geometric parameters for finding the optimum step pattern.

The corner coordinates must satisfy

$$x_i^2 + y_i^2 = R^2 \tag{2.45}$$

For a core with n steps, where n refers to the number of stacks in half the core cross section, the core area, A_n, is given by

$$A_n = 4\sum_{i=1}^{n}(x_i - x_{i-1})y_i = 4\sum_{i=1}^{n}(x_i - x_{i-1})\sqrt{R^2 - x_i^2} \tag{2.46}$$

where $x_0 = 0$. Thus, the independent variables are the x_i since the y_i can be determined from them using (2.45). To maximize A_n, we need to solve the n equations

$$\frac{\partial A_n}{\partial x_i} = 0, \quad i = 1,\ldots,n \tag{2.47}$$

We can show that

$$\frac{\partial^2 A_n}{\partial^2 x_i} < 0 \tag{2.48}$$

so that the solution to (2.47) does represent a maximum. Applying (2.47) to (2.46), note that, in taking the derivative with respect to x_i, this term occurs in two terms, one with x_i and x_{i-1} and one with x_{i+1} and x_i. Isolating these terms in the sum in (2.46), we need to differentiate

$$A_n(x_{i-1}, x_i, x_{i+1}) = 4\left((x_i - x_{i-1})\sqrt{R^2 - x_i^2} + (x_{i+1} - x_i)\sqrt{R^2 - x_{i+1}^2}\right)$$

Differentiating with respect to x_i and setting the result to 0, we get

$$\frac{\partial A_n}{\partial x_i} = 0 = \sqrt{R^2 - x_i^2} + (x_i - x_{i-1})\left(\frac{-x_i}{\sqrt{R^2 - x_i^2}}\right) - \sqrt{R^2 - x_{i+1}^2}$$

Multiplying this last expression by the radical in the denominator, we get

$$R^2 - x_i^2 - x_i(x_i - x_{i-1}) - \sqrt{(R^2 - x_i^2)(R^2 - x_{i+1}^2)} = 0$$

Transferring the radical to the right-hand side of the equation and squaring, we get, after some algebraic manipulation,

$$(R^2 - x_i^2)\left[x_{i+1}^2 - x_i(3x_i - 2x_{i-1})\right] + x_i^2(x_i - x_{i-1})^2 = 0, \quad \text{for } i = 1, \dots, n. \tag{2.49}$$

In the first and last equations ($i = 1$ and $i = n$), we need to set $x_0 = 0$ and $x_{n+1} = R$.

Since (2.49) represents a set of nonlinear equations, an approximate solution scheme such as a Newton–Raphson iteration can be used to solve them. At this point, it may be worthwhile to say a few words about this iteration method since it can be very useful in general in solving a set of nonlinear equations. Thus, given a set of n functions, f_i in n unknowns, x_i, we wish to solve

$$f_i(\mathbf{x}) = 0, \quad i = 1, \dots, n \tag{2.50}$$

where we have put $\mathbf{x}$ in boldface type to indicate that it is an n-dimensional vector. Since $\mathbf{x}$ is unknown, we wish to solve these equations iteratively by letting $\mathbf{x}$ change by $\Delta\mathbf{x}$ on each iteration and continue the iterations until $\Delta\mathbf{x}$ is less than some small value. To determine $\Delta\mathbf{x}$, we set

$$f_i(\mathbf{x} + \Delta\mathbf{x}) = 0, \quad i = 1, \dots, n \tag{2.51}$$

We now expand the function in a Taylor series to first order, obtaining

$$f_i(\mathbf{x}) + \sum_{j=1}^{n} \frac{\partial f_i}{\partial x_j} \Delta x_j = 0, \quad i = 1, \dots, n \tag{2.52}$$

This last equation can be rearranged to

$$\sum_{j=1}^{n} \frac{\partial f_i}{\partial x_j} \Delta x_j = -f_i(\mathbf{x}), \quad i = 1, \dots, n \tag{2.53}$$

This can be written in matrix form as

$$\begin{pmatrix} \dfrac{\partial f_1}{\partial x_1} & \cdots & \dfrac{\partial f_1}{\partial x_n} \\ \vdots & \ddots & \vdots \\ \dfrac{\partial f_n}{\partial x_1} & \cdots & \dfrac{\partial f_n}{\partial x_n} \end{pmatrix} \begin{pmatrix} \Delta x_1 \\ \vdots \\ \Delta x_n \end{pmatrix} = - \begin{pmatrix} f_1(\mathbf{x}) \\ \vdots \\ f_n(\mathbf{x}) \end{pmatrix} \tag{2.54}$$

TABLE 2.3

Normalized x Coordinates That Maximize the Core Area for a Given Number of Steps

Number of Steps, N	Fraction of Circle Occupied, $A_n/\pi R^2$	Normalized x Coordinates, x_i/R
1	0.6366	0.7071
2	0.7869	0.5257, 0.8506
3	0.8510	0.4240, 0.7070, 0.9056
4	0.8860	0.3591, 0.6064, 0.7951, 0.9332
5	0.9079	0.3138, 0.5336, 0.7071, 0.8457, 0.9494
6	0.9228	0.2802, 0.4785, 0.6379, 0.7700, 0.8780, 0.9599
7	0.9337	0.2543, 0.4353, 0.5826, 0.7071, 0.8127, 0.9002, 0.9671
8	0.9419	0.2335, 0.4005, 0.5375, 0.6546, 0.7560, 0.8432, 0.9163, 0.9723
9	0.9483	0.2164, 0.3718, 0.4998, 0.6103, 0.7071, 0.7921, 0.8661, 0.9283, 0.9763
10	0.9534	0.2021, 0.3476, 0.4680, 0.5724, 0.6648, 0.7469, 0.8199, 0.8836, 0.9376, 0.9793

Applying this general procedure to the functions in (2.49), we note that each row of this matrix has only three entries for the derivatives of x_{i-1}, x_i, and x_{i+1}. These are

$$\frac{\partial f_i}{\partial x_{i-1}} = 2x_i\left(R^2 - x_i^2\right) - 2x_i^2\left(x_i - x_{i-1}\right)$$

$$\frac{\partial f_i}{\partial x_i} = \left(R^2 - x_i^2\right)\left(-6x_i + 2x_{i-1}\right) - 2x_i\left[x_{i+1}^2 - x_i\left(3x_i - 2x_{i-1}\right)\right]$$

$$+ 2x_i\left(x_i - x_{i-1}\right)\left(2x_i - x_{i-1}\right)$$

$$\frac{\partial f_i}{\partial x_{i+1}} = 2x_{i+1}\left(R^2 - x_i^2\right)$$

together with $x_0 = 0$ and $x_{n+1} = R$.

To start the iterations, let the x's be equally spaced initially. Note that these equations can be normalized by dividing by R^4, so that the normalized solution coordinates x_i/R are independent of R. Table 2.3 gives the normalized solution for various numbers of steps.

In practice, because only a limited number of standard sheet widths are kept in inventory and because stack heights are also discretized, at least by the thickness of an individual sheet, it is not possible to achieve the ideal coverage given in the table. A practical method of handling this is to optimize, assuming continuous variables as was done here, and then to readjust the stack dimensions to correspond to the nearest discretized choices available. This must be done in such a way that the stacks do not fall outside the core circle. In addition, it is often necessary to allow room for the tie or flitch plates on the outside of the core stack. This is accomplished by fixing the value of x_n in the equations given earlier. Simply set x_n equal to a value necessary to accommodate the tie plates and drop the nth equation in (2.49). Core cooling ducts can also be accommodated by reducing the thickness of one of the neighboring stacks, near where the cooling ducts are to be located, by the thickness of the cooling duct. These adjustments will lower the core area from the maximized area but are necessary in practice.

3

Circuit Model of a 2-Winding Transformer with Core

3.1 Introduction

Circuit models are often used when a transformer is part of a larger circuit such as a utility network. In this case, only the terminal characteristics of the transformer are usually of interest. We will primarily be concerned with transformers operating at steady-state power frequencies, that is, generally 50 or 60 Hz. In this case, capacitive effects can usually be ignored. When the core is operating sufficiently below saturation, core nonlinearity effects can also be ignored. For transient processes such as those occurring during switching or lightning strikes, a more detailed transformer model is needed. We will start by developing a circuit model of the core before considering the effects of adding the windings.

3.2 Circuit Model of the Core

We will consider a single-phase core and transformer. Usually, 3-phase cores and transformers can be analyzed as three separate single-phase units. Interconnections among the phases can be added later. For simplicity, only a 2-winding transformer will be considered here, having a primary and secondary winding. The primary winding is normally understood as attached to the input power source, while the secondary winding feeds a load or loads.

If the secondary winding were open circuited, then the transformer would behave like an inductor with a high-permeability closed iron core. It would therefore have a high inductance so that little exciting current would be required to generate the primary voltage or back emf. Some I^2R loss will be generated by the exciting current in the primary winding; however, this will be small compared with the load current losses. There will, however, be losses in the core due to the changing flux. These losses are, to a good approximation, proportional to the square of the induction, B^2. Hence, they are also proportional to the square of the voltage across the core. Thus, these losses can be accounted for by putting an equivalent resistor across the transformer voltage and ground, where the resistor has the value

$$R_c = \frac{V_{rms}^2}{W_c} \tag{3.1}$$

where
  V_{rms} is the rms phase voltage
  W_c is the core loss

Here we are assuming that the primary voltage is sinusoidal, which is usually the case. The open-circuited inductance can be obtained from

$$V = L_c \frac{dI_{ex}}{dt} \quad \Rightarrow \quad L_c = \frac{V_{rms}}{\omega I_{ex,rms}} \tag{3.2}$$

where I_{ex} is the inductive component of the exciting current, which is assumed to be sinusoidal with angular frequency ω. In general, this current can be sinusoidal or nonsinusoidal, depending on whether the core is operating linearly or nonlinearly. Since the voltage V is generally sinusoidal, the first expression in (3.2) would require L_c to be a nonlinear function of I_{ex} when the core is operating nonlinearly. The second expression in (3.2) is valid when I_{ex} is sinusoidal with angular frequency $\omega = 2\pi f$, with f being the frequency in Hz. So long as the core is operating reasonably far from saturation, it is generally a good assumption that I_{ex} is sinusoidal and L_c is a constant inductance. Thus, the circuit so far will look like that shown in Figure 3.1. The resistance R_p is the resistance of the primary (or excited) winding. As saturation is approached, the inductance L_c and the resistance R_c will become nonlinear. If necessary, capacitive effects can be included by putting an equivalent capacitance in parallel with the core inductance and resistance. Since the core losses are supplied by the input power source, there is a component of the total excitation current, $I_{ex,tot}$, which generates the core loss. We labeled it I_c in the figure. It will be in quadrature, that is, at a 90° phase angle, with I_{ex}.

A useful circuit model for the core when the excitation is sinusoidal and nonlinearities are ignored is shown in Figure 3.2.

Here R_c is the core resistance, accounting for the core loss, and $X_c = \omega L_c$ is the magnitude of the core reactance, accounting for the magnetization of the core. j is the imaginary unit, which mathematically produces the 90° phase relation between resistance and reactance.

We will show how the parameters of this circuit model can be extracted from test data and then use them to obtain the core power factor, that is, the fraction of the core power that is dissipated as loss. As before, the resistance can be obtained from the calculated or

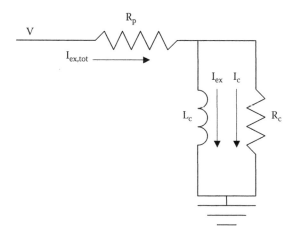

FIGURE 3.1
Transformer core circuit model with secondary open circuited.

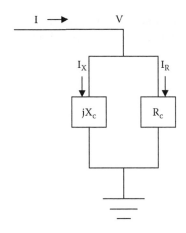

FIGURE 3.2
Transformer core circuit model with sinusoidal excitation.

tested core losses as given by (3.1). This, in turn, allows the calculation of the current, I_R, from the formula

$$I_R = \frac{V}{R_c} \tag{3.3}$$

where V and I are assumed to be rms quantities.

Letting Z_c equal the parallel combination of X_c and R_c, we have

$$\frac{1}{Z_c} = -\frac{j}{X_c} + \frac{1}{R_c} \tag{3.4}$$

Hence,

$$\frac{1}{|Z_c|^2} = \frac{1}{X_c^2} + \frac{1}{R_c^2} \tag{3.5}$$

Rewriting and taking square roots, we have

$$\frac{1}{X_c} = \sqrt{\frac{1}{|Z_c|^2} - \frac{1}{R_c^2}} \tag{3.6}$$

From a knowledge of V, taken as the reference phasor (a real quantity), and the magnitude of I, the input current, which is available from test data, we have

$$Z_c = \frac{V}{|I|} \tag{3.7}$$

Thus, the magnitude of Z_c can be obtained by inputting the required test data into the last formula. This allows us to calculate X_c from the previous formula. Knowing X_c, we can get the current I_X from the formula

$$I_X = \frac{V}{jX_c} \qquad (3.8)$$

The magnitude of the power into the core is given by $V|I|$, where $|I|$ is the magnitude of the input current. The power dissipated in losses is given by VI_R. Thus, the ratio of the power loss to the input power is the power factor:

$$\text{Power factor} = \frac{I_R}{|I|} \qquad (3.9)$$

This is usually denoted by $\cos\theta$, where θ is the angle between the total current I and the real current I_R in the complex plane.

As a numerical example, assuming a single-phase core, let $V = 40{,}000$ V, $W = 2000$ W, and $I = 0.25$ amps. Then we get

$$R_c = \frac{V^2}{W} = 800{,}000 \ \Omega$$

$$I_R = \frac{V}{R_c} = 0.05 \text{ amps}$$

$$Z_c = \frac{V}{|I|} = 160{,}000 \ \Omega$$

$$X_c = 163{,}300 \ \Omega$$

$$I_X = -\frac{jV}{X_c} = -j0.245 \text{ amps}$$

$$\text{Power factor} = \cos\theta = \frac{I_R}{|I|} = 0.200$$

3.3 2-Winding Transformer Circuit Model with Core

When the secondary circuit is connected to a load, the emf generated in the secondary winding by the changing core flux will drive a current through the secondary circuit. This additional current (amp-turns) would alter the core flux unless equal and opposite amp-turns flow in the primary winding. Since the core flux is determined by the impressed primary voltage, the net amp-turns must equal the small exciting amp-turns. Hence, the primary and secondary amp-turns due to load current, excluding the exciting current, must cancel out.

Figure 3.3 shows a schematic of the flux pattern in a 2-winding transformer under load. The currents are taken as positive when they flow into a winding, and the dots on the terminals indicate that the winding sense is such that the induced voltage is positive at that terminal relative to the terminal at the other end of the winding when positive exciting current flows into the transformer. Notice that the bulk of the flux flows through the core and links

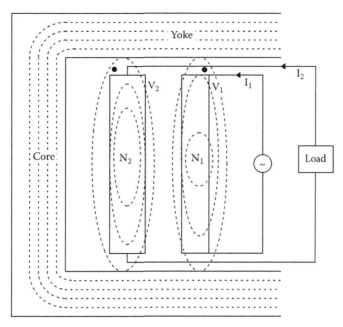

FIGURE 3.3
Schematic of a 2-winding single-phase transformer with leakage flux.

both windings. However, some of the flux links only one winding. When referring to flux linkages, we assume partial linkages are included. Some of these can be seen in the figure.

Faraday's law for the voltage induced in a coil is based on the concept of flux linkages. Up to now, we have assumed that a flux Φ passing through a coil of N turns links all N turns so that the flux linkage is $N\Phi$. As seen in the figure, except for the core flux, some of the flux linking a coil only links some of the turns or possibly only a fraction of a turn. At this point, we will assume that Φ is the flux linkage per turn so that $N\Phi$ does, in fact, equal the flux linkage, λ. Turning this around, we have $\Phi = \lambda/N$.

As Figure 3.3 shows, some of the flux linking one of the coils is due to flux created by current in the second coil. We will define Φ_{ij} as the flux per turn linking coil i due to current in coil j with a corresponding interpretation of λ_{ij}. Thus, Φ_{11} is the flux per turn in coil 1 linking coil 1 due to its own current and Φ_{12} is the flux per turn linking coil 1 due to current in coil 2. The corresponding flux linkages are λ_{11} and λ_{12}. We will use λ_i with a single subscript to refer to the total flux linkage of coil i. Thus, we have

$$\begin{aligned} \lambda_1 &= \lambda_{11} + \lambda_{12} = N_1\left(\Phi_{11} + \Phi_{12}\right) \\ \lambda_2 &= \lambda_{22} + \lambda_{21} = N_2\left(\Phi_{22} + \Phi_{21}\right) \end{aligned} \tag{3.10}$$

It is useful to define the leakage flux per turn as

$$\begin{aligned} \Phi_{\ell 1} &= \Phi_{11} - \Phi_{21} \\ \Phi_{\ell 2} &= \Phi_{22} - \Phi_{12} \end{aligned} \tag{3.11}$$

In the following, we will usually refer to Φ, with or without subscripts, as a flux, but it should be understood that it is a flux per turn. Thus, the leakage flux for coil 1 is the self-flux of coil 1 minus the flux linking coil 2 due to current in coil 1. Since the core flux links both coils, this flux is excluded from the leakage flux. The same applies to the leakage flux for coil 2. The word leakage flux gives the impression that this is flux that has leaked out of the core but in reality it is simply flux that does not link both coils. Most of this flux is in the medium surrounding the core, usually oil or air. Both oil and air are linear materials with relative permeabilities close to 1. Thus, the inductances associated with this flux, which are given by

$$L_{\ell 1} = \frac{N_1 \Phi_{\ell 1}}{I_1}, \quad L_{\ell 2} = \frac{N_2 \Phi_{\ell 2}}{I_2} \qquad (3.12)$$

are nearly unaffected by core nonlinearities. These are called single-winding leakage inductances and are essentially constant.

Combining (3.10) and (3.11), we have

$$\lambda_1 = N_1 \left(\Phi_{\ell 1} + \Phi_{21} + \Phi_{12} \right), \quad \lambda_2 = N_2 \left(\Phi_{\ell 2} + \Phi_{12} + \Phi_{21} \right) \qquad (3.13)$$

Let $\Phi_c = \Phi_{12} + \Phi_{21}$. As can be seen from (3.13), this is a common flux linking both coils and is therefore predominately core flux, Φ_c. From (3.12) and (3.13), we see that

$$\lambda_1 = L_{\ell 1} I_1 + N_1 \Phi_c, \quad \lambda_2 = L_{\ell 2} I_2 + N_2 \Phi_c \qquad (3.14)$$

The voltage equations for the two windings are

$$V_1 = R_1 I_1 + \frac{d\lambda_1}{dt}, \quad V_2 = R_2 I_2 + \frac{d\lambda_2}{dt} \qquad (3.15)$$

where λ_1 and λ_2 are the total flux linkages for windings 1 and 2 and R_1, R_2 their resistances. Substituting from (3.14), the voltage equations become

$$\begin{aligned} V_1 &= R_1 I_1 + L_{\ell 1} \frac{dI_1}{dt} + N_1 \frac{d\Phi_c}{dt} \\ V_2 &= R_2 I_2 + L_{\ell 2} \frac{dI_2}{dt} + N_2 \frac{d\Phi_c}{dt} \end{aligned} \qquad (3.16)$$

Although Φ_c was obtained by adding fluxes produced by each coil acting separately, in reality, this common flux is produced by the simultaneous action of both coils. When the resultant flux is nonlinear, it can no longer be regarded as the simple addition of separate fluxes. Thus, when nonlinear effects are important, Φ_c requires a broader interpretation.

Figure 3.4 shows a circuit diagram for (3.16). It contains an ideal transformer with a turns ratio of N_1/N_2. This is the same as the voltage ratio between the primary and secondary sides as seen in the figure and given by

$$\frac{E_1}{E_2} = \frac{N_1}{N_2} \qquad (3.17)$$

The voltages E_1 and E_2 are voltages induced in the primary or secondary windings by the changing common flux, as required by Faraday's law. Separating out this common flux provides the physical basis for the ideal transformer.

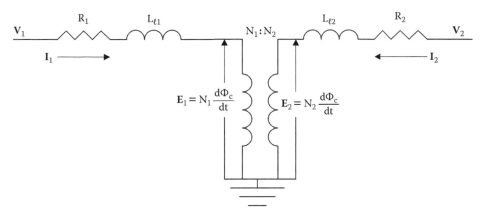

FIGURE 3.4
Circuit model of a loaded 2-winding transformer, including a possibly nonlinear ideal transformer.

The 2-winding circuit model so far has ignored details of the core excitation. We note that this excitation must come from the difference between the amp-turns of windings 1 and 2 and is typically small. To account for this, it is customary to siphon off this amp-turn difference from one of the windings and let it flow into the core circuit model of Figure 3.1, considered as a shunt branch off one of the windings. Typically, this branch is taken off the primary winding. For physical reasons, this core shunt branch should be placed after the primary winding resistance and leakage reactance since the exciting current flows in the primary winding. The core shunt branch would then directly determine the core voltage E_1.

Since this exciting current is small, the approximation is often made to place the core shunt branch at the terminal itself. This will be expedient for later developments, where it is desirable to have amp-turn balance in the main transformer circuit. This approximation can be improved by adding the winding resistance and leakage reactance to the core's resistance and reactance. This circuit model with core is shown in Figure 3.5.

In this figure, the current into terminal 1 would be $I_{ex,tot} + I_1$. If desired, capacitive effects can be included by placing a capacitor in parallel with the core circuit.

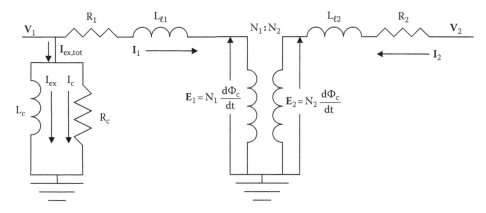

FIGURE 3.5
Circuit model of a loaded 2-winding transformer, including a possibly nonlinear core and ideal transformer.

3.4 Approximate 2-Winding Transformer Circuit Model without Core

At this point, our main concern will be with the transformer circuit beyond the core shunt branch so we will drop this branch from further discussion here. We recognize that while V_1 is still the voltage at the primary terminal, I_1 is not the primary terminal current although it is very close to it. Provided V_1 is sinusoidal, I_1 will also be sinusoidal as long as the exciting current is small.

Amp-turn balance in the ideal transformer requires that

$$N_1 I_1 + N_2 I_2 = 0 \tag{3.18}$$

Notice that our current convention that currents into the terminals are positive requires that I_1 or I_2 must be negative for (3.18) to hold. Since current is assumed to flow into the primary winding because this is taken as the power source, the current I_2 must be negative. A negative current into a terminal is the same as a positive current out of the terminal.

Rewriting (3.16), we have

$$V_1 = R_1 I_1 + L_{\ell 1} \frac{dI_1}{dt} + E_1$$
$$V_2 = R_2 I_2 + L_{\ell 2} \frac{dI_2}{dt} + E_2 \tag{3.19}$$

Using (3.17) and (3.18), we can transform (3.19) to

$$V_1 = \frac{N_1}{N_2} V_2 + \left[R_1 + \left(\frac{N_1}{N_2} \right)^2 R_2 \right] I_1 + \left[L_{\ell 1} + \left(\frac{N_1}{N_2} \right)^2 L_{\ell 2} \right] \frac{dI_1}{dt} \tag{3.20}$$

The circuit model for (3.20) is shown in Figure 3.6. Note that the ideal transformer is now a device for transforming the secondary voltage V_2 to the primary side, accounting for the first term on the right-hand side of (3.20).

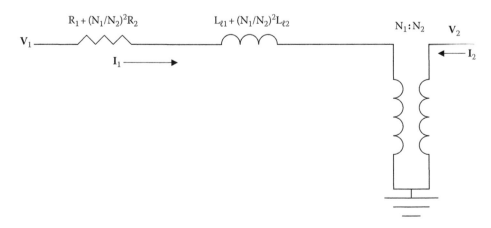

FIGURE 3.6
Circuit model of a 2-winding transformer under load referred to the primary side.

Here the resistances and inductances have been combined to give equivalent quantities:

$$R_{1,eq} = R_1 + \left(\frac{N_1}{N_2}\right)^2 R_2, \quad L_{1,eq} = L_{\ell 1} + \left(\frac{N_1}{N_2}\right)^2 L_{\ell 2} \tag{3.21}$$

These are the 2-winding resistance and leakage inductance referred to the primary side. Although the equivalent resistance accounts for the I^2R losses in the windings, it can also be modified to account for losses caused by the stray flux since these are proportional to the square of the current to a good approximation.

Transformers, and especially power transformers, generally operate with currents and voltages that are sinusoidal in time. These are often treated as complex quantities, called phasors. These will be treated in more detail in a later chapter. For now, we note that the time derivative of a phasor of angular frequency ω is given by

$$\frac{d\mathbf{I}}{dt} = j\omega\mathbf{I} \tag{3.22}$$

where boldfaced quantities are used to denote phasors. Applying this to the time derivative in (3.20) and using (3.21), we get

$$\mathbf{V}_1 = \frac{N_1}{N_2}\mathbf{V}_2 + R_{1,eq}\mathbf{I}_1 + j\omega L_{1,eq}\mathbf{I}_1 \tag{3.23}$$

We can now define an equivalent impedance (2-winding impedance) by

$$Z_{1,eq} = R_{1,eq} + j\omega L_{1,eq} \tag{3.24}$$

Using this, Equation 3.23 simplifies to

$$\mathbf{V}_1 = \mathbf{I}_1 Z_{1,eq} + \mathbf{E}_1, \quad \mathbf{V}_2 = \mathbf{E}_2 \tag{3.25}$$

Here $\mathbf{E}_1$ and $\mathbf{E}_2$ simply denote voltages across the ideal transformer, with $\mathbf{E}_2 = \mathbf{V}_2$ and $\mathbf{E}_1 = (N_1/N_2)\mathbf{E}_2$. These are not the same voltages across the ideal transformer as those in Figure 3.4, although the same symbol is used. They obey the voltage ratio equation (3.17). Of course, the amp-turn balance equation (3.18) also holds for this ideal transformer. The equivalent circuit for (3.25) is shown in Figure 3.7.

Using (3.17) and (3.18), we can transform (3.25) to

$$\mathbf{V}_2 = \frac{N_2}{N_1}\mathbf{V}_1 + \left(\frac{N_2}{N_1}\right)^2 Z_{1,eq}\mathbf{I}_2 \tag{3.26}$$

Letting

$$E_1' = V_1, \quad E_2' = \left(\frac{N_2}{N_1}\right)E_1' \tag{3.27}$$

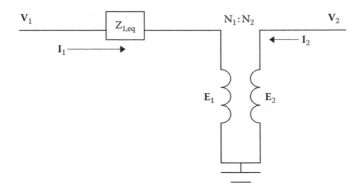

FIGURE 3.7
Simplified circuit model of a 2-winding transformer under load referred to the primary side.

and defining

$$Z_{2,eq} = \left(\frac{N_2}{N_1}\right)^2 Z_{1,eq} = R_{2,eq} + j\omega L_{2,eq} \tag{3.28}$$

where

$$R_{2,eq} = R_2 + \left(\frac{N_2}{N_1}\right)^2 R_1, \quad L_{2,eq} = L_{\ell 2} + \left(\frac{N_2}{N_1}\right)^2 L_{\ell 1} \tag{3.29}$$

we can express (3.26) as

$$V_2 = Z_{2,eq}I_2 + E_2', \quad V_1 = E_1' \tag{3.30}$$

We have placed a prime on the **E**'s to emphasize that these are different from the **E**'s in (3.26); however, they still obey the voltage ratio equation (3.17) for an ideal transformer. The circuit model for (3.30) is shown in Figure 3.8.

Note the symmetry between Figures 3.7 and 3.8. This is especially clear from a comparison of (3.21) and (3.29). We could have initially transferred impedances to the secondary

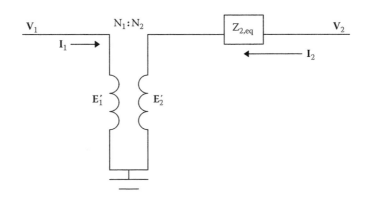

FIGURE 3.8
Simplified circuit model of a 2-winding transformer under load referred to the secondary side.

side and arrived at (3.29) directly. From both of these equations, as well as from (3.28), we see that impedances are transferred across an ideal transformer by the square of the turns ratio or its reciprocal, depending on which way the transfer occurs.

3.5 Vector Diagram of a Loaded Transformer with Core

For increased accuracy, it is useful to add the core excitation circuit to the circuit model developed in the last section, where the leakage impedances were transferred to one side of the ideal transformer. This is a good approximation since the core excitation current is small compared with the load currents. In Figure 3.5, we placed the core excitation circuit at terminal 1. However, this could also be placed on the terminal 2 side of the ideal transformer. This, along with the 2-winding leakage impedance, will be placed on the terminal 2 side, since we want to explore the relationship between the load current and voltage and the excitation current and voltage. This leads to the circuit diagram shown in Figure 3.9.

For the vector (complex plane) diagram, we will assume that the positive sense of the secondary current is out of terminal 2, as it normally is. We are only looking at the secondary side, so this will not conflict with our normal sign convention. We also assume that the secondary ideal transformer voltage E_2' is the reference voltage and that all currents and voltages are sinusoidal. We note that the voltage drop, V, across a reactance, X, and current, I, through a reactance, are related by $V = jXI$. Since multiplication by j in the complex plane corresponds to a counterclockwise rotation of 90°, the voltage drop across a reactance is in a perpendicular direction to the current direction. On the other hand, for a resistor, R, the voltage drop is $V = IR$ so this voltage drop is in the direction of the current. For an impedance with resistive and reactive components, the voltage drop direction is determined by the relative magnitudes of the resistive and reactive components.

We will assume that the load impedance in Figure 3.9 is mostly resistive and is given by $Z_L = R_L + jX_L$, whereas the leakage impedance is entirely reactive. The vector diagram corresponding to Figure 3.9 is shown in Figure 3.10.

This diagram is not to scale. Usually, the exciting current is a much smaller vector relative to the load current. We should note that the core flux is in the direction of the exciting

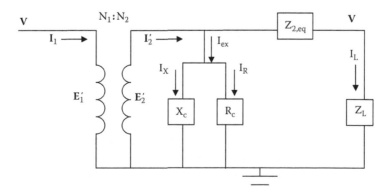

FIGURE 3.9
Circuit diagram of a loaded transformer with core excitation circuit on the secondary side.

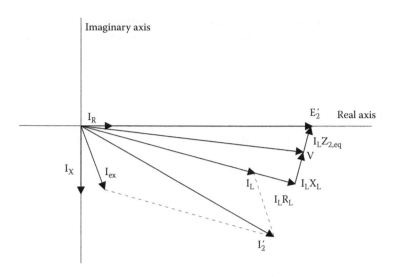

FIGURE 3.10
Vector diagram showing the relationships between the voltages and currents in a loaded 2-winding transformer with core.

current, I_X, whereas the leakage flux is in the direction of the load current, I_L. These flux densities can overlap near the center of the core legs. For predominantly resistive load currents, their vector addition does not increase the magnitude of the resultant flux density very much. But for reactive load currents, the flux densities add directly. When reactive load currents are high, some consideration must be given to the possibility of saturating the center of the core legs due to this phenomenon.

3.6 Per-Unit System

Transformer impedances, along with other quantities such as voltages and currents, are often expressed in the per-unit (p.u.) system. In this system, these quantities are expressed as a ratio with respect to the transformer's nominal or rated phase quantities. Thus, if the rated or base phase voltages are V_{b1}, V_{b2} and the base currents are I_{b1}, I_{b2}, where 1 and 2 refer to the primary and secondary sides, then the base impedances are

$$Z_{b1} = \frac{V_{b1}}{I_{b1}}, \quad Z_{b2} = \frac{V_{b2}}{I_{b2}} \tag{3.31}$$

The rated or base voltages, currents, and impedances are assumed to transfer from one side to the other by means of the ideal transformer relationships among voltages, currents, and impedances. These base quantities are all taken to be positive. Thus, the minus sign is neglected in the base current transfer across sides. From this, it can be shown that the base power, P_b, is the same on both sides of an ideal 2-winding transformer:

$$P_b = V_{b1}I_{b1} = V_{b2}I_{b2} \tag{3.32}$$

In an actual transformer, the real power into a transformer is nearly the same as that leaving it on the secondary side. This is because the transformer losses are a small fraction of the power transferred.

The p.u. base quantities of interest are current, voltage, and power. From (3.32), we see that these quantities are interdependent. Only 2 independent base quantities are needed in most applications. Of the 3, it is usual to specify the base power, P_b, which is the same on both sides of an ideal 2-winding transformer. It normally refers to the rated power per phase into a transformer and remains fixed regardless of the number of secondary windings, whether loaded or not. The other base quantity is the base voltage. This will differ from winding to winding or terminal to terminal. It is usually taken to mean the rated or no-load voltage of the winding or terminal, whichever is being considered.

The primary side voltage, V_1, current, I_1, and impedance, Z_1, are expressed in the p.u. system, assuming power and voltage as base quantities, by

$$V_{1,pu} = v_1 = \frac{V_1}{V_{b1}}, \quad I_{1,pu} = i_1 = I_1 \frac{V_{b1}}{P_b}, \quad Z_{1,pu} = z_1 = Z_1 \frac{P_b}{V_{b1}^2} \quad (3.33)$$

and similarly for the secondary quantities. As indicated in (3.33), we use lowercase letters to denote p.u. quantities.

Often, the p.u. values are multiplied by 100 and expressed as a percentage. However, care must be taken in using percentage values in calculations, particularly in multiplications, since this can lead to errors.

Since the base voltages, currents, and impedances transfer across the ideal transformer in the same manner as their corresponding circuit quantities, the p.u. values of the circuit quantities are the same on both the primary and secondary sides. Although voltages, currents, and impedances for transformers of greatly different power ratings can differ considerably, their p.u. values tend to be very similar. This can facilitate calculations since one develops a pretty good feel for the magnitudes of p.u. quantities such as resistances or reactances for any size transformer. Thus, the 2-winding leakage reactances when expressed in the p.u. system are generally in the range of 5%–15% for all power transformers. The exciting currents of modern power transformers are typically ~0.1% in the p.u. (%) system. This is also their percentage of the rated load current since this has the value of 100% in the p.u. system. The 2-winding resistances, which account for the transformer losses, can be obtained in the p.u. system by noting that modern power transformers are typically >99.5% efficient. This means that <0.5% of the rated input power goes into losses. Thus, we have

$$\frac{\text{Losses}}{\text{Rated Power}} = \frac{R_1 I_{b1}^2}{P_b} = \frac{R_1 P_b}{V_{b1}^2} = \frac{R_1}{Z_{b1}} = r < 0.5\% \quad (3.34)$$

We will use the symbol X for reactance magnitudes, where $X = \omega L$, with L the inductance. Thus, x will be a p.u. reactance. Using (3.34) and the estimate of the leakage reactance given earlier in the p.u. system for power transformers, we can estimate the x/r ratio:

$$\frac{x}{r} = \frac{0.05 - 0.15}{< 0.005} > 10 - 30 \quad (3.35)$$

Thus, 10 is probably a lower limit and, as previously shown, is high enough to allow discriminating inrush from fault current on the basis of second harmonic analysis.

3.7 Voltage Regulation

At this point, it is useful to discuss the topic of voltage regulation as an application of the transformer circuit model just developed and the p.u. system. In this context, the core characteristics do not play a significant role so we will use the simplified circuit model of Figure 3.8. Voltage regulation is defined as the change in the magnitude of the secondary voltage between its open-circuited value and its value when loaded divided by the value when loaded with the primary voltage held constant. We can represent the load by an impedance, Z_L. The relevant circuit is shown in Figure 3.11. In the figure, we have shown a load current, I_L, where $I_L = -I_2$.

It is convenient to perform this calculation in the p.u. system, using lowercase letters to represent p.u. quantities. Thus, since base quantities transfer across the ideal transformer like their corresponding physical quantities, we have

$$
\begin{aligned}
\mathbf{e}_1' &= \frac{\mathbf{E}_1'}{V_{b1}} = \frac{(N_1/N_2)\mathbf{E}_2'}{(N_1/N_2)V_{b2}} = \frac{\mathbf{E}_2'}{V_{b2}} = \mathbf{e}_2' = \mathbf{e} \\
\mathbf{i}_1 &= \frac{\mathbf{I}_1}{I_{b1}} = -\frac{(N_2/N_1)\mathbf{I}_2}{(N_2/N_1)I_{b2}} = -\frac{\mathbf{I}_2}{I_{b2}} = -\mathbf{i}_2 = \mathbf{i}_L
\end{aligned}
\tag{3.36}
$$

Therefore, the ideal transformer can be eliminated from the circuit in the p.u. system since voltages and currents in this system are the same on both sides of the ideal transformer. Also, since $z_{2,eq} = z_{1,eq}$, we can drop the numerical subscript and denote the equivalent transformer p.u. impedance by

$$
z = r + jx
\tag{3.37}
$$

where
  r is the p.u. 2-winding resistance
  x the p.u. 2-winding leakage reactance

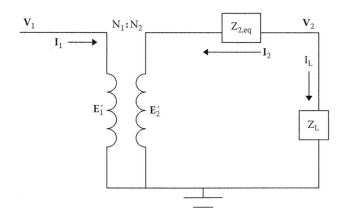

FIGURE 3.11
Circuit model of a 2-winding transformer with a load on the secondary side.

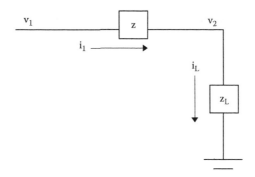

FIGURE 3.12
The circuit of Figure 3.11 shown in the p.u. system.

Similarly, we write

$$z_L = \frac{Z_L}{Z_{b2}} = r_L + jx_L \tag{3.38}$$

The p.u. circuit is shown in Figure 3.12.

Thus, when the secondary terminal is open circuited so z_L and its ground connection are missing in Figure 3.12, we obtain for the open-circuited value of v_2

$$v_{2,oc} = v_1 \tag{3.39}$$

We assume that $\mathbf{v}_1$ is a reference phasor (zero-phase angle) and v_1 is therefore its magnitude. When the load is present, we have

$$\mathbf{v}_2 = \mathbf{i}_L z_L, \quad \mathbf{v}_1 = \mathbf{i}_L (z + z_L) \tag{3.40}$$

From (3.40), we have

$$\mathbf{v}_2 = \frac{z_L \mathbf{v}_1}{(z + z_L)} = \frac{\mathbf{v}_1}{(1 + z/z_L)} \quad \Rightarrow \quad |\mathbf{v}_2| = \frac{v_1}{|1 + z/z_L|} \tag{3.41}$$

Hence, the voltage regulation is given by, using (3.39),

$$\frac{v_{2,oc} - |\mathbf{v}_2|}{|\mathbf{v}_2|} = \left|1 + \frac{z}{z_L}\right| - 1 \tag{3.42}$$

At this point, it is customary to express the complex number z_L in its polar form:

$$z_L = |z_L| e^{j\theta}, \quad |z_L| = \sqrt{r_L^2 + x_L^2}, \quad \theta = \tan^{-1}\left(\frac{x_L}{r_L}\right) \tag{3.43}$$

We note, from (3.40), that the magnitude of z_L is also given by

$$|z_L| = \left|\frac{\mathbf{v}_2}{\mathbf{i}_L}\right| = \frac{v_2}{i_L}$$

so that

$$z_L = \frac{v_2}{i_L} e^{j\theta}$$

where the magnitudes of v_2 and i_L are represented by the symbols in ordinary type. Keeping the Cartesian representation of a complex number for z, we have for the ratio

$$
\begin{aligned}
\frac{z}{z_L} &= (r+jx)\left(\frac{i_L}{v_2}\right)e^{-j\theta} = \left(\frac{i_L}{v_2}\right)(r+jx)(\cos\theta - j\sin\theta) \\
&= \left(\frac{i_L}{v_2}\right)(r\cos\theta + x\sin\theta) + j(x\cos\theta - r\sin\theta)
\end{aligned}
\tag{3.44}
$$

Substituting (3.44) into (3.42), we get

$$
\begin{aligned}
\text{Regulation} &= \sqrt{\left[1+\left(\frac{i_L}{v_2}\right)(r\cos\theta + x\sin\theta)\right]^2 + \left[\left(\frac{i_L}{v_2}\right)(x\cos\theta - r\sin\theta)\right]^2} - 1 \\
&= \sqrt{1+2\left(\frac{i_L}{v_2}\right)(r\cos\theta + x\sin\theta) + \left(\frac{i_L}{v_2}\right)^2(r^2 + x^2)} - 1
\end{aligned}
\tag{3.45}
$$

Since z is generally small compared with z_L, the terms other than unity in (3.45) are small compared to unity. Using the approximation for small ε,

$$\sqrt{1+\varepsilon} = 1 + \frac{1}{2}\varepsilon - \frac{1}{8}\varepsilon^2 \tag{3.46}$$

the regulation is given to second order in z/z_L by

$$\text{Regulation} = \left(\frac{i_L}{v_2}\right)(r\cos\theta + x\sin\theta) + \frac{1}{2}\left(\frac{i_L}{v_2}\right)^2(x\cos\theta - r\sin\theta)^2 \tag{3.47}$$

4

Reactance and Leakage Reactance Calculations

4.1 Introduction

The leakage reactance calculations performed here apply to a single phase unit or to one phase of a 3-phase transformer. The phase can have multiple windings interconnected in such a way that only 2 or 3 terminals (external or buried) result. Thus autotransformers, with or without tertiary, are included as well as transformers with separate tap windings. The calculations for windings that are interconnected are based on 2-winding leakage reactances. These reactances can be obtained from a simple analytical formula, which will be presented here, from finite element calculations or from more complicated analytical methods, such as Rabins' method, which will be presented in a later chapter. We will also be giving the per-unit expressions for these reactances.

It is usually desirable to design some reactance into a transformer in order to limit any fault current. These currents flow in the transformer windings when, for example, a terminal is shorted to ground. When this happens, the internal impedance of a transformer, which is mainly reactive, is primarily responsible for limiting this current. In addition, these reactances determine the voltage regulation of the unit. Hence, it is desirable at the design stage to be able to calculate these reactances based on the geometry of the coils and core and the nature of the (single phase) winding interconnections.

Although impedances have a resistive component, winding resistances are usually not too difficult to calculate, especially d.c. resistances. Complications arise from eddy current contributions, generated usually by time-varying flux from neighboring windings. Eddy currents produced by the current within the winding itself are usually kept small, if necessary, by subdividing the cross-sectional area of the winding into multiple parallel strands and then transposing these. For power transformers, the resistances are usually small compared with leakage reactances but should be taken into account when greater accuracy is desired.

We begin by discussing the general theory of inductance calculations. Reactances are obtained from the inductances by multiplying by the angular frequency under sinusoidal conditions, which are assumed here. The following references have been used in this chapter: [Del94], [MIT43], [Lyo37], [Blu51], [Wes64].

4.2 General Method for Determining Inductances and Mutual Inductances

Self- and mutual inductances are usually defined in terms of flux linkages between circuits. However, it is often more convenient to calculate them in terms of magnetic energy. In this section, we show the equivalence of these methods and develop useful formulas for the determination of these inductances in terms of the magnetic field values.

Consider a set of stationary circuits as shown in Figure 4.1a. We assume that we slowly increase the currents in them by means of batteries with variable control. In calculating the work done by the batteries, we ignore any I^2R or dissipative losses or, more realistically, we treat these separately. Thus, the work we are interested in is the work necessary to establish the magnetic field. Since losses are assumed to be zero, this work will equal the magnetic energy stored in the final field. Because of the changing flux linking the different circuits, electro-motive forces (emf's) will be induced in them by Faraday's law. We assume that the battery controllers are adjusted so that the battery voltages just balance the induced voltages throughout the process. Because the circuits can have finite cross-sectional areas as shown in the figure, we imagine subdividing them into infinitesimal circuits or tubes carrying an incremental part of the total current, dI, as in Figure 4.1b.

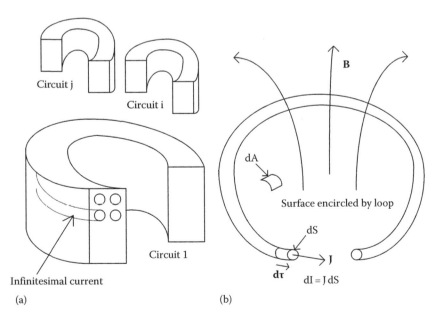

FIGURE 4.1
Method of circuit subdivision used to calculate self- and mutual inductances: (a) collection of circuits subdivided into infinitesimal current loops and (b) infinitesimal current loop which is part of a system of circuits.

4.2.1 Energy by Magnetic Field Methods

The incremental energy, dW, which the batteries supply during a time interval, dt, is

$$dW = dt \int V dI = dt \int_{\substack{\text{cross-sectional} \\ \text{area of circuits}}} V J dS \tag{4.1}$$

Here, the integrand is the infinitesimal power flowing through the infinitesimal circuit or loop carrying current dI and along which a voltage V is induced. We have also substituted the current density J times the infinitesimal cross-sectional area dS for the incremental current in the loop so that the last integral is over the cross-sectional area of the circuits. From Faraday's law for a single-turn circuit,

$$V = \frac{d\Phi}{dt} \tag{4.2}$$

where Φ is the flux linked by the circuit. Substituting into (4.1)

$$dW = \int_{\substack{\text{cross-sectional} \\ \text{area of circuits}}} d\Phi J dS \tag{4.3}$$

By definition,

$$\Phi = \int_{\substack{\text{area of loop}}} \mathbf{B} \cdot \mathbf{n} dA = \int_{\substack{\text{area of loop}}} (\nabla \times \mathbf{A}) \cdot \mathbf{n} dA = \oint_{\substack{\text{along loop}}} \mathbf{A} \cdot d\tau \tag{4.4}$$

In (4.4) we have introduced the vector potential, **A**, where

$$\mathbf{B} = \nabla \times \mathbf{A} \tag{4.5}$$

We have also made use of Stoke's theorem [Pug62]. We are denoting vectors with bold-faced quantities here. Since we will not be using phasors in this chapter, no confusion should arise.

In (4.4), the surface integral is over the surface encircled by the wire loop, and the line integral is along the loop as shown in Figure 4.1b. **n** is the unit normal to the surface encircled by the loop, and $d\tau$ is the infinitesimal distance vector along the loop. Note that $d\tau$ points in the same direction as **J** considered as a vector. Since the current loops are fixed, we find by taking the differential of (4.4)

$$d\Phi = \oint_{\substack{\text{along loop}}} d\mathbf{A} \cdot d\tau \tag{4.6}$$

Substituting this into (4.3), we get

$$dW = \int_{\substack{\text{cross-sectional} \\ \text{area of circuits}}} \left(\oint_{\substack{\text{along loop}}} d\mathbf{A} \cdot d\tau \right) J dS \tag{4.7}$$

Since $\mathbf{J}$ points along $\mathbf{d}\boldsymbol{\ell}$ and since the volume element $dV = d\tau\,dS$, we can rewrite (4.7)

$$dW = \int_{\substack{\text{cross-sectional} \\ \text{area of circuits}}} \left(\oint_{\text{along loop}} d\mathbf{A} \cdot \mathbf{J} d\tau dS \right) = \int_{\substack{\text{volume} \\ \text{of circuits}}} d\mathbf{A} \cdot \mathbf{J} dV \qquad (4.8)$$

In changing to a volume integral, we are simply recognizing the fact that the integral over the cross-sectional area of the circuits combined with an integral along the lengths of the infinitesimal circuits amounts to an integration over the volume of the circuits. The total work done to establish the final field values, recognizing that the circuit remains stationary during this process, is

$$W = \int_0^A dW = \int_{\substack{\text{volume} \\ \text{of circuits}}} dV \int_0^A \mathbf{J} \cdot d\mathbf{A} = \frac{1}{2} \int_{\substack{\text{volume} \\ \text{of circuits}}} \mathbf{A} \cdot \mathbf{J} dV \qquad (4.9)$$

where the last equality follows if $\mathbf{A}$ increases proportionally with $\mathbf{J}$ as we build up the fields.

It is sometimes more convenient for calculational purposes to express (4.9), or more generally (4.8), in terms of the field values directly. First of all the volume integration in both of these equations could be taken over all space since the current density is zero everywhere except within the circuits. (By all space we mean the solution space of the problem of interest.) For this, we use one of Maxwell's equations in SI units, ignoring displacement currents:

$$\nabla \times \mathbf{H} = \mathbf{J} \qquad (4.10)$$

where $\mathbf{H}$ is the magnetic field. Substituting into (4.8), we get

$$dW = \int_{\text{all space}} d\mathbf{A} \cdot (\nabla \times \mathbf{H}) dV \qquad (4.11)$$

Using the vector identity for general vector fields $\mathbf{P}$, $\mathbf{Q}$ [Pug62],

$$\nabla \cdot (\mathbf{P} \times \mathbf{Q}) = \mathbf{Q} \cdot (\nabla \times \mathbf{P}) - \mathbf{P} \cdot (\nabla \times \mathbf{Q}) \qquad (4.12)$$

we have, upon substitution into (4.11) with the identification $\mathbf{P} \to \mathbf{H}$, $\mathbf{Q} \to d\mathbf{A}$,

$$dW = \int_{\text{all space}} \mathbf{H} \cdot (\nabla \times d\mathbf{A}) dV + \int_{\text{all space}} \nabla \cdot (\mathbf{H} \times d\mathbf{A}) dV$$

$$= \int_{\text{all space}} \mathbf{H} \cdot d\mathbf{B} dV + \int_{\text{boundary surface}} (\mathbf{H} \times d\mathbf{A}) \cdot \mathbf{n} dS \qquad (4.13)$$

In this equation, we have used the definition of A given in (4.5) and the Divergence Theorem to convert the last volume integral into a surface integral. The surface integral generally vanishes since the boundary surface is usually at infinity where the fields drop to zero. Thus we can drop it to get

$$dW = \int\limits_{all\ space} \mathbf{H} \cdot d\mathbf{B} dV \qquad (4.14)$$

The total work done to establish the final field values is

$$W = \int_0^B dW = \int\limits_{all\ space} dV \int_0^B \mathbf{H} \cdot d\mathbf{B} = \frac{1}{2} \int\limits_{all\ space} \mathbf{H} \cdot \mathbf{B} dV \qquad (4.15)$$

where the last equality follows for the situation where **B** increases proportionally with **H**. Either (4.9) or (4.15) can be used to calculate the magnetic energy.

4.2.2 Energy from Electric Circuit Methods

We now show how the magnetic energy can be related to inductances or leakage inductances from circuit considerations. The voltages induced in the circuits of Figure 4.1a can be written in terms of inductances and mutual inductances as

$$V_i = \sum_j M_{ij} \frac{dI_j}{dt} \qquad (4.16)$$

where M_{ij} is the mutual inductance between circuits i and j. When j = i, $M_{ii} = L_i$, where L_i is the self-inductance of circuit i. The incremental energy for the circuits of Figure 4.1a can be expressed, making use of (4.16) as

$$dW = \sum_i V_i I_i dt = \sum_{i,j} M_{ij} I_i dI_j = \sum_i L_i I_i dI_i + \sum_{i \neq j} M_{ij} I_i dI_j \qquad (4.17)$$

For linear systems, it can be shown that $M_{ij} = M_{ji}$. Thus the second sum in (4.17) can be written as

$$\sum_{j \neq i} M_{ij} I_i dI_j = \frac{1}{2} \left(\sum_{i \neq j} M_{ij} I_i dI_j + \sum_{i \neq j} M_{ji} I_j dI_i \right)$$

$$= \frac{1}{2} \sum_{i \neq j} M_{ij} \left(I_i dI_j + I_j dI_i \right) = \sum_{i < j} M_{ij} d\left(I_i I_j \right) \qquad (4.18)$$

The first equality in this equation is a matter of changing the index labels in the sum, and the second equality results from the symmetric nature of M_{ij} for linear systems. The third equality results from the double counting, which happens when i and j are summed independently. Substituting (4.18) into (4.17), we get

$$dW = \frac{1}{2} \sum_i L_i d\left(I_i^2 \right) + \sum_{i < j} M_{ij} d\left(I_i I_j \right) \qquad (4.19)$$

Integrating this last equation, we obtain

$$W = \frac{1}{2}\sum_i L_i I_i^2 + \sum_{i<j} M_{ij} I_i I_j \tag{4.20}$$

Thus, (4.20) is another expression for the magnetic energy in terms of inductances and mutual inductances. As an example, consider a single circuit for which we want to know the self-inductance. From (4.20),

$$L_1 = \frac{2W}{I_1^2} \tag{4.21}$$

where W can be calculated from (4.9) or (4.15) with only the single circuit in the geometry. W is usually obtainable from finite element codes. If we have 2 circuits, then (4.20) becomes

$$L_1 I_1^2 + L_2 I_2^2 + 2M_{12} I_1 I_2 = 2W \tag{4.22}$$

If these two circuits are the high (label 1) and low (label 2) voltage coils of a 2-winding ideal transformer with N_1 and N_2 turns respectively, then we have $I_2/I_1 = -N_1/N_2$ so that (4.22) becomes

$$L_1 + \left(\frac{N_1}{N_2}\right)^2 L_2 - 2\left(\frac{N_1}{N_2}\right)M_{12} = \frac{2W}{I_1^2} \tag{4.23}$$

The expression on the left is the 2-winding leakage inductance referred to the high-voltage side [MIT43]. The magnetic energy on the right side can be calculated from (4.9) or (4.15) with the 2 windings in the geometry and with the Ampere-turns balanced. Multiplying by $\omega = 2\pi f$ produces the leakage reactance.

The mutual inductance between two circuits is needed in detailed circuit models of transformers where the circuits of interest are sections of the same or different coils. We can obtain this quantity in terms of the vector potential. By definition, the mutual inductance between circuits 1 and 2, M_{12}, is the flux produced by circuit 2, which links circuit 1, divided by the current in circuit 2. Circuits 1 and 2 could be interchanged in this definition without changing the value of the mutual inductance for linear systems. We use the infinitesimal loop approach as illustrated in Figure 4.1 to obtain this quantity. Each infinitesimal loop represents an infinitesimal fraction of a turn. For circuit 1 with a single turn, an infinitesimal loop carrying current dI_1 represents a fractional turn dI_1/I_1. Therefore, the flux linkage between this infinitesimal turn due to the flux produced by circuit 2, $d\lambda_{12}$, is

$$d\lambda_{12} = \frac{dI_1}{I_1} \int_{\substack{\text{surface enclosed by} \\ \text{infinitesimal loop in circuit 1}}} \mathbf{B}_2 \cdot \mathbf{n} dS_1 = \frac{dI_1}{I_1} \int_{\substack{\text{surface enclosed by} \\ \text{infinitesimal loop in circuit 1}}} (\nabla \times \mathbf{A}_2) \cdot \mathbf{n} dS_1$$

$$= \frac{dI_1}{I_1} \oint_{\substack{\text{along infinitesimal} \\ \text{loop in circuit 1}}} \mathbf{A}_2 \cdot \mathbf{d}\tau_1 \tag{4.24}$$

where Stokes' theorem has been used. Note also that the flux linkage λ is flux times turns and this is what is used in Faraday's law when dealing with circuits.

Integrating this expression over the cross-sectional area of circuit 1 and using the fact that $dI_1 = J_1 \, dS_1$ and that $\mathbf{J}_1$ as a vector points along $d\boldsymbol{\tau}_1$, we can write

$$\lambda_{12} = \frac{1}{I_1} \int\limits_{\substack{\text{cross-sectional} \\ \text{area of circuit 1}}} J_1 dS_1 \left(\oint\limits_{\substack{\text{along infinitesimal} \\ \text{loop in circuit 1}}} \mathbf{A}_2 \cdot d\boldsymbol{\tau}_1 \right) = \frac{1}{I_1} \int\limits_{\substack{\text{volume} \\ \text{of circuit 1}}} \mathbf{A}_2 \cdot \mathbf{J}_1 dV_1 \qquad (4.25)$$

where we have used the same integration devices as were used previously. Thus, the mutual inductance between these single-turn circuits is given by

$$M_{12} = \frac{\lambda_{12}}{I_2} = \frac{1}{I_1 I_2} \int\limits_{\substack{\text{volume} \\ \text{of circuit 1}}} \mathbf{A}_2 \cdot \mathbf{J}_1 dV_1 \qquad (4.26)$$

This formula reduces to the formula for self-inductance when 1 and 2 are the same circuit and letting $M_{11} = L_1$.

We should note that the mutual inductance calculation (4.26), together with the self-inductance calculation (4.21), could be used to determine the leakage inductance from formula (4.23). However, when an iron core is involved, L_1, L_2, and M_{12} tend to be large quantities so that the subtractions in (4.23) can lead to large errors unless the inductances are determined with high accuracy.

4.3 2-Winding Leakage Reactance Formula

Although the 2-winding leakage reactance is needed to model a 2-winding transformer, the 2-winding leakage reactances between pairs of windings in a multi-winding transformer are useful for modeling these transformers as well. Therefore, this is an important quantity to determine. This can be calculated by advanced analytical techniques such as Rabins' method [Rab56] or finite element methods, which solve Maxwell's equations directly. These methods are especially useful if the distribution of amp-turns along the winding is nonuniform, due, for example, to tapped out sections or thinning in sections of windings adjacent to taps in neighboring windings. However, simpler idealized calculations have proven adequate in practice, particularly at the early design stage. Such a calculation will be discussed here.

The simple reactance calculation assumes that the amp-turns are uniformly distributed along the windings. It also treats the windings as if they were infinitely long solenoids insofar as the magnetic field is concerned, although a correction for fringing at the ends is included in the final formula.

The magnetic field inside an infinitely long solenoid can be found from symmetry considerations using Maxwell's equation (4.10) expressed in integral form:

$$\oint \mathbf{H} \cdot d\boldsymbol{\tau} = I \qquad (4.27)$$

where the integral is around a closed loop enclosing a current I through the surface of the loop. For an ideal, infinitely tall solenoid with N turns per height h uniformly distributed,

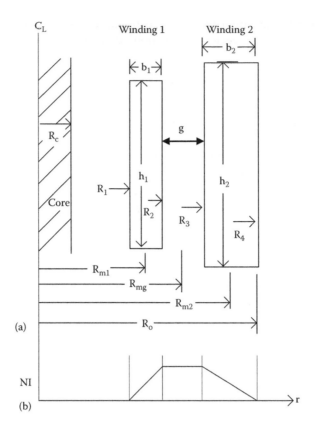

FIGURE 4.2
Parameters used in 2-winding leakage reactance calculation: (a) geometric parameters and (b) electrical parameters.

an application of this formula, together with symmetry considerations, tells us that there is a uniform axially directed field inside the solenoid with the magnitude $H = NI/h$ and a zero field outside the solenoid.

We can apply this formula to the 2-winding transformer considered here. The parameters of interest are shown in Figure 4.2. The windings have a finite thickness so the amp-turns are a function of the radius parameter r. The total amp-turns are assumed to sum to zero for the 2 windings as is necessary for a leakage reactance calculation. Then the magnetic field as a function of radius can be found by applying (4.27) to a rectangular loop in a plane that passes through the center line of the solenoid, which encloses both windings with the vertical side of the rectangle outside the windings kept fixed, while the inner vertical side is gradually shifted towards the outer side. This results in a magnetic field distribution proportional to the amp-turn distribution shown in Figure 4.2b, that is, in the SI system:

$$H(r) = \frac{NI(r)}{h} \tag{4.28}$$

where $NI(r)$ is the function of r shown in the figure, linearly increasing from 0 through winding 1, remaining constant in the gap, and decreasing to 0 through winding 2.

H is independent of the z-coordinate in this model and points vectorially in the z-direction. The flux density is therefore

$$B(r) = \mu_o H(r) = \mu_o \frac{NI(r)}{h} \tag{4.29}$$

The permeabilities of the materials in or between the winding are essentially that of vacuum, $\mu_o = 4\pi \times 10^{-7}$ in the (SI) system. B also points in the z-direction. We can take $h = (h_1 + h_2)/2$ as an approximation.

For calculation purposes, we need to express B as a function of r analytically. First of all, since the total amp-turns is zero, we have, ignoring the signs of the currents, $N_1 I_1 = N_2 I_2 = NI$. Therefore, we get for B(r), referring to Figure 4.2 for an explanation of the symbols,

$$B(r) = \mu_o \frac{NI}{h} \begin{cases} \dfrac{(r-R_1)}{(R_2-R_1)}, & R_1 < r < R_2 \\ 1, & R_2 < r < R_3 \\ \dfrac{(R_4-r)}{(R_4-R_3)}, & R_3 < r < R_4 \end{cases} \tag{4.30}$$

The 2-winding leakage inductance of winding 1 with respect to winding 2, $L_{\ell 12}$, can be obtained from the magnetic energy in the leakage field by means of the expression

$$\frac{1}{2} L_{\ell 12} I_1^2 = \frac{1}{2\mu_o} \int\limits_{space} B^2 dV \tag{4.31}$$

Here the order of the subscripts is important. Substituting (4.30) into the integral, we get

$$\begin{aligned}
\frac{1}{2\mu_o} \int\limits_{space} B^2 dV &= \frac{(\mu_o NI)^2}{2\mu_o h^2} \times 2\pi h \left[\int_{R_1}^{R_2} \left(\frac{r-R_1}{R_2-R_1}\right)^2 r dr + \int_{R_2}^{R_3} r dr + \int_{R_3}^{R_4} \left(\frac{R_4-r}{R_4-R_3}\right)^2 r dr \right] \\
&= \frac{\pi\mu_o (NI)^2}{h} \times \left[\frac{(R_2^2-R_1^2)}{6} + \frac{(R_2-R_1)^2}{12} + \frac{(R_3^2-R_2^2)}{2} + \frac{(R_4^2-R_3^2)}{6} - \frac{(R_4-R_3)^2}{12} \right]
\end{aligned} \tag{4.32}$$

In terms of mean radii R_m, thicknesses b, and gap g shown in Figure 4.2, Equation 4.32 can be written as

$$\frac{1}{2\mu_o} \int\limits_{space} B^2 dV = \frac{\pi\mu_o (NI)^2}{h} \left[\frac{R_{m1}b_1}{3} + \frac{b_1^2}{12} + R_{mg}g + \frac{R_{m2}b_2}{3} - \frac{b_2^2}{12} \right] \tag{4.33}$$

Thus using (4.31), we find for the leakage inductance

$$L_{\ell 12} = \frac{2\pi\mu_o N_1^2}{h} \left[\frac{R_{m1}b_1}{3} + \frac{R_{m2}b_2}{3} + R_{mg}g + \frac{b_1^2}{12} - \frac{b_2^2}{12} \right] \tag{4.34}$$

Often, the last two terms in (4.34) are dropped since they approximately cancel. The leakage reactance magnitude is $X_{\ell12} = 2\pi f L_{\ell12}$, so we get

$$X_{\ell12} = \frac{(2\pi)^2 \mu_o f N_1^2}{h}\left[\frac{R_{m1}b_1}{3} + \frac{R_{m2}b_2}{3} + R_{mg}g + \frac{b_1^2}{12} - \frac{b_2^2}{12}\right] \qquad (4.35)$$

The per-unit impedance of winding 1 is

$$Z_{b1} = \frac{V_{b1}^2}{(VI)_{b1}} = \frac{N_1^2(V_{b1}/N_1)^2}{(VI)_{b1}} \qquad (4.36)$$

where
 $(VI)_{b1}$ is the base volt-amps (power) per phase
 V_{b1}/N_1 is the base Volts/turn for winding 1

Letting x denote the per-unit reactance, where $x = X_{\ell12}/Z_{b1}$, we get

$$x = \frac{(2\pi)^2 \mu_o f (VI)_{b1}}{(V_{b1}/N_1)^2 h}\left[\frac{R_{m1}b_1}{3} + \frac{R_{m2}b_2}{3} + R_{mg}g + \frac{b_1^2}{12} - \frac{b_2^2}{12}\right] \qquad (4.37)$$

We left the subscripts off x because the per-unit reactance does not depend on the order of the windings. This is because the base power rating $(VI)_b$ and the base volts per turn (V_b/N) are the same for both windings. In contrast the leakage reactance, as opposed to the per-unit value, is proportional to the square of the winding turns in the first winding as seen in (4.35).

In order to correct for flux fringing at the winding ends, it has been found that a good approximation is to increase h by the amount

$$s = 0.32(R_o - R_c) \qquad (4.38)$$

where
 R_o is the outer radius of the outermost coil
 R_c is the core radius

A fringing correction factor has also been developed by W. Rogowski that is widely used [Blu51]. Thus we obtain

$$x = \frac{(2\pi)^2 \mu_o f (VI)_b}{(V_b/N)^2 (h+s)}\left[\frac{R_{m1}b_1}{3} + \frac{R_{m2}b_2}{3} + R_{mg}g + \frac{b_1^2}{12} - \frac{b_2^2}{12}\right] \qquad (4.39)$$

for the per-unit 2-winding reactance.

These analytical formulas such as (4.39) are useful for providing insight into achieving the required transformer reactances. For example, if one wants to increase the high to low-voltage windings reactance in order to reduce the fault currents and hence the forces on these windings during a fault, this could be achieved by increasing the high to low gap, g, or reducing the overall winding height, h. Similarly if a tertiary winding is present, which could experience a fault, it may be desirable to increase the high- or low-voltage winding to tertiary winding reactance to mitigate the effects of this fault. In this case the height of the tertiary winding could be reduced without affecting the high to low-voltage winding reactance. This is generally possible since tertiary windings usually have fewer turns than the high or low-voltage windings. However, caution is necessary here since the fringing field of the tertiary winding, which produces losses in the tie plate or core, would now be inside the innermost vertical oil channel where the cooling is probably less efficient than if it occurred outside this channel for a taller winding.

4.4 Ideal 2-, 3-, and Multi-Winding Transformers

Since leakage reactances are part of a circuit, which includes an ideal transformer, it is useful to discuss the ideal transformer first, particularly one which can have more than two windings. We will consider a single phase ideal transformer with 2 or more windings per phase. The phase could be one phase of a 3-phase transformer, provided the phases can be treated independently. Although much of the discussion concerns 2- or 3-winding transformers, the extension to ideal multi-winding transformers is straightforward. We also look at the ideal autotransformer, where the voltages of 2 windings are combined to produce the high-voltage at one terminal with the voltage of one of the windings providing the low voltage at the other terminal.

For a 2-winding transformer, the ideal conditions are

$$\frac{E_1}{E_2} = \frac{N_1}{N_2}, \quad N_1 I_1 + N_2 I_2 = 0 \tag{4.40}$$

where
   the E's are the voltages across the 2 windings with turns N_1 and N_2
   the I's are the rated currents through the windings

We use the convention that the currents are positive into the winding. From (4.40), we can show that

$$E_1 I_1 + E_2 I_2 = 0 \tag{4.41}$$

Since EI is the power, this says that the net power into the ideal transformer is zero, that is, the power into winding 1 equals the power out of winding 2. Thus an ideal transformer is lossless.

For a 3-winding transformer, Equation 4.40 generalizes to

$$\frac{E_1}{N_1} = \frac{E_2}{N_2} = \frac{E_3}{N_3}, \quad N_1 I_1 + N_2 I_2 + N_3 I_3 = 0 \tag{4.42}$$

From (4.42), it follows that

$$E_1 I_1 + E_2 I_2 + E_3 I_3 = 0 \tag{4.43}$$

that is, the net power into a 3-winding ideal transformer is zero. These latter equations extend readily to multi-winding transformers.

We also note that impedances are transferred across an ideal transformer by the equation

$$Z_1 = \left(\frac{N_1}{N_2}\right)^2 Z_2 \tag{4.44}$$

An ideal 2-winding transformer is depicted in Figure 4.3.

It is often convenient to work with per-unit quantities. That is, we choose base values for voltages, currents, etc. and express the actual voltages, currents, etc. as ratios with respect to these base values. Only two independent base values need to be chosen and these are usually taken as power or VI and voltage V. The base power is normally taken as the VI rating of the unit per phase and so has a fixed value for a given transformer. The base voltage is taken as the open-circuit-rated phase voltage of each terminal. Thus we have a $(VI)_b = (VI)_{b1} = (VI)_{b2}$ and V_{b1}, V_{b2}, where b denotes base value. From these, we derive base currents of

$$I_{b1} = \frac{(VI)_b}{V_{b1}}, \text{ etc.} \tag{4.45}$$

and base impedances

$$Z_{b1} = \frac{V_{b1}^2}{(VI)_{b1}}, \text{ etc.} \tag{4.46}$$

Letting small letters denote per-unit quantities, we have

$$v_1 = \frac{V_1}{V_{b1}}, \text{ etc.} \quad i_1 = \frac{I_1}{I_{b1}}, \text{ etc.} \quad z_1 = \frac{Z_1}{Z_{b1}}, \text{ etc.} \tag{4.47}$$

Note that the base voltages are the actual voltages under rated conditions for the ideal transformer. Therefore, they satisfy the voltage equations in (4.40) and (4.42) so that we have

$$\frac{V_{b1}}{V_{b2}} = \frac{N_1}{N_2}, \text{ etc.} \quad \frac{I_{b1}}{I_{b2}} = \frac{N_2}{N_1}, \text{ etc.} \quad \frac{Z_{b1}}{Z_{b2}} = \left(\frac{V_{b1}}{V_{b2}}\right)^2 = \left(\frac{N_1}{N_2}\right)^2, \text{ etc.} \tag{4.48}$$

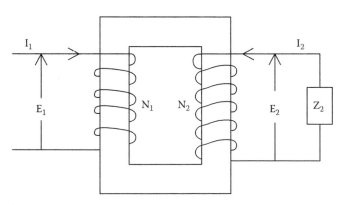

FIGURE 4.3
Ideal 2-winding transformer with a load on its secondary terminals.

since everything is on a common VI base. Note that the base quantities are always positive. The signs in the per-unit quantities are determined by the signs of the underlying quantity. Since all the windings are put on the same power base, Equations 4.41 and 4.43 are not applicable to base quantities.

In terms of per-unit quantities, we have for an ideal transformer

$$e_1 = \frac{E_1}{V_{b1}} = 1, \quad e_2 = \frac{E_2}{V_{b2}} = 1, \text{etc.} \tag{4.49}$$

and, from (4.42) and (4.48)

$$N_1 \frac{I_1}{I_{b1}} + N_2 \frac{I_2}{I_{b1}} + N_3 \frac{I_3}{I_{b1}} = N_1 \frac{I_1}{I_{b1}} + N_2 \frac{I_2}{I_{b2}} \left(\frac{I_{b2}}{I_{b1}} \right) + N_3 \frac{I_3}{I_{b3}} \left(\frac{I_{b3}}{I_{b1}} \right)$$
$$= N_1 (i_1 + i_2 + i_3) = 0 \tag{4.50}$$

Therefore,

$$i_1 + i_2 + i_3 = 0 \tag{4.51}$$

Thus, an ideal 3-circuit transformer can be represented by a one-circuit description as shown in Figure 4.4, if per-unit values are used. Equation 4.51 can also be extended to more than three windings.

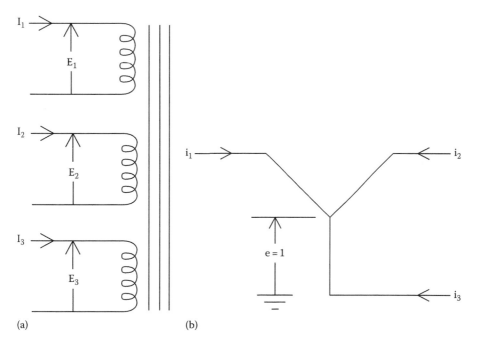

(a) (b)

FIGURE 4.4
Ideal 3-circuit transformer schematic: (a) standard quantities and (b) per-unit quantities.

4.4.1 Ideal Autotransformer

The ideal 2-terminal autotransformer is shown in Figure 4.5. The two coils, labeled s for series and c for common, are connected together so that their voltages add to produce the high-voltage terminal voltage E_1. Thus $E_1 = E_s + E_c$. The secondary or low-voltage terminal voltage is $E_2 = E_c$. Similarly, from the figure, $I_1 = I_s$ and $I_2 = I_c - I_s = I_c - I_1$. Using the expressions for a 2-winding unit, which are true regardless of the interconnections involved,

$$\frac{E_s}{N_s} = \frac{E_c}{N_c}, \quad I_s N_s + I_c N_c = 0 \tag{4.52}$$

we find

$$E_1 = E_c \left(1 + \frac{N_s}{N_c} \right) = E_2 \left(\frac{N_c + N_s}{N_c} \right)$$

$$I_s N_s + I_c N_c = I_1 N_s + \left(I_1 + I_2 \right) N_c = I_1 \left(N_c + N_s \right) + I_2 N_c = 0 \tag{4.53}$$

Thus

$$\frac{E_1}{E_2} = \left(\frac{N_c + N_s}{N_c} \right), \quad \frac{I_1}{I_2} = -\left(\frac{N_c}{N_c + N_s} \right) \tag{4.54}$$

so that the effective turns ratio as seen by the terminals is $n = (N_c + N_s)/N_c$. Any impedances on the low-voltage terminal could be transferred to the high-voltage circuit by the square of this turns ratio.

The co-ratio is defined as

$$r = \frac{E_1 - E_2}{E_1} = 1 - \frac{E_2}{E_1} = 1 - \frac{1}{n} = \frac{n-1}{n} \tag{4.55}$$

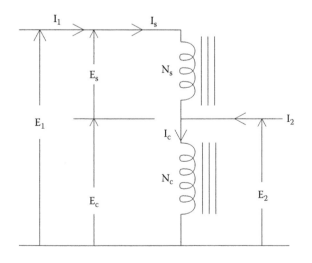

FIGURE 4.5
Ideal autotransformer.

where $r \leq 1$. We note here that, since $E_1 = E_s + E_c$ and $E_2 = E_c$, an alternative definition of r is

$$r = \frac{E_s}{E_1}$$

The terminal power rating of an autotransformer is $E_1 I_1 = E_2 I_2$, but the power rating of each coil is $E_s I_s = E_c I_c$. The ratio of these power ratings is

$$\frac{E_s I_s}{E_1 I_1} = \frac{E_1 - E_2}{E_1} = r \tag{4.56}$$

Thus, since $r \leq 1$, the terminal rating is always greater than or equal to the single-coil rating. Thus, for a conventional transformer with the same terminal rating as an autotransformer, the windings of an autotransformer have a lower rating than the conventional transformer windings, and this can result in smaller and less expensive windings.

4.5 Leakage Reactance for 2-Winding Transformers Based on Circuit Parameters

Although a 2-winding circuit model was developed previously from magnetic flux considerations, we consider here a model based on circuit parameters such as inductances and mutual inductances. This will allow us to extend the 2-winding model to 3 and eventually to a multi-winding model.

We will start with the terminal equations developed previously in Chapter 3:

$$V_1 = I_1 R_1 - \frac{d\lambda_1}{dt}, \quad V_2 = I_2 R_2 - \frac{d\lambda_2}{dt} \tag{4.57}$$

where
λ_1 is the total flux linkage of coil 1
R_1 its resistance and similarly for coil 2

If we let Φ_c = the common flux linking all turns of both coils, which is mainly core flux, we can write

$$\lambda_1 = (\lambda_1 - N_1 \Phi_c) + N_1 \Phi_c, \quad \lambda_2 = (\lambda_2 - N_2 \Phi_c) + N_2 \Phi_c \tag{4.58}$$

The quantity $(\lambda_i - N_i \Phi_c)$ is the leakage flux of coil i. It exists mainly in the oil or air and conductor material but not in the core to any great extent. Thus it exists in nonmagnetic, that is, linear, materials and therefore should depend linearly on the currents. Thus we can write very generally

$$-\frac{d(\lambda_1 - N_1 \Phi_c)}{dt} = L_1 \frac{dI_1}{dt} + M_{12} \frac{dI_2}{dt} \tag{4.59}$$

or, assuming sinusoidal quantities,

$$-\frac{d(\lambda_1 - N_1\Phi_c)}{dt} = jI_1X_{11} + jI_2X_{12} \tag{4.60}$$

where

I_1 and I_2 are phasors

$X_{11} = 2\pi f\, L_1$ and $X_{12} = 2\pi f\, M_{12}$, where f is the frequency, and j is the imaginary unit

We do not distinguish phasors by boldfaced type here, since they all have the same phase. Using similar expressions for λ_2, Equation 4.57 becomes

$$V_1 = I_1R_1 + jI_1X_{11} + jI_2X_{12} + E_1$$
$$V_2 = I_2R_2 + jI_2X_{22} + jI_1X_{12} + E_2 \tag{4.61}$$

where $E_1 = -N_1 d\Phi_c/dt$ is the no-load terminal voltage of terminal 1, etc., for E_2 so that $E_1/E_2 = N_1/N_2$. We have also used the fact that $X_{12} = X_{21}$ for linear systems.

We are going to ignore the exciting current of the core since this is normally much smaller than the load currents. Thus, using the amp-turn balance condition in (4.40), we can rewrite (4.61)

$$V_1 = I_1\left[R_1 + j\left(X_{11} - \frac{N_1}{N_2}X_{12}\right)\right] + E_1$$
$$V_2 = I_2\left[R_2 + j\left(X_{22} - \frac{N_2}{N_1}X_{12}\right)\right] + E_2 \tag{4.62}$$

or, more succinctly,

$$V_1 = I_1Z_1 + E_1$$
$$V_2 = I_2Z_2 + E_2 \tag{4.63}$$

where Z_1 and Z_2 are single-winding leakage impedances. These last equations correspond to those encountered in the last chapter. We have two impedances on either side of an ideal transformer. We can transform one of the impedances across the transformer by means of (4.40). In particular, we let $V_2 = E_2' = I_2Z_2 + E_2$. Then, since $E_1/E_2 = N_1/N_2$ and $I_1/I_2 = -N_2/N_1$, $E_2' = -(N_1/N_2)I_1Z_2 + (N_2/N_1)E_1$ so that solving for E_1,

$$E_1 = \frac{N_1}{N_2}E_2' + \left(\frac{N_1}{N_2}\right)^2 I_1Z_2 \tag{4.64}$$

Substituting into the first equation in (4.63), we get

$$V_1 = I_1\left[Z_1 + \left(\frac{N_1}{N_2}\right)^2 Z_2\right] + \frac{N_1}{N_2}E_2', \quad V_2 = E_2' \tag{4.65}$$

The impedances in (4.63) have been transformed into a single impedance called the 2-winding leakage impedance, Z_{12}, which depends on which of the single-winding impedances was transferred across the ideal transformer so that $Z_{12} \neq Z_{21}$. Simplifying the notation in (4.65), we have

$$V_1 = I_1 Z_{12} + E_1', \quad V_2 = E_2' \tag{4.66}$$

where we defined

$$E_1' = \frac{N_1}{N_2} E_2'$$

We can convert this equation to a per-unit basis by the following process:

$$\frac{V_1}{V_{b1}} = \frac{I_1 Z_{12}}{V_{b1}} + \frac{E_1'}{V_{b1}} = \frac{I_1}{V_{b1}} \left[\frac{V_{b1}^2}{(VI)_{b1}} \right] \left[\frac{(VI)_{b1}}{V_{b1}^2} \right] Z_{12} + \frac{E_1'}{V_{b1}} = \frac{I_1}{I_{b1}} \frac{Z_{12}}{Z_{b1}} + \frac{E_1'}{V_{b1}}$$

$$\frac{V_2}{V_{b2}} = \frac{E_2'}{V_{b2}}$$

or

$$v_1 = i_1 z_{12} + 1, \quad v_2 = 1 \tag{4.67}$$

The equivalent circuit for (4.66) is shown in Figure 4.6 along with the per-unit version.

We note that $E_1' / E_2' = N_1 / N_2$ but that $E_1' \neq E_1$ and $E_2' \neq E_2$ except at no-load.

Thus a 2-winding transformer is characterized by a single value of leakage impedance Z_{12}, which has both resistive and reactive components:

$$Z_{12} = R_{12} + X_{\ell 12}$$

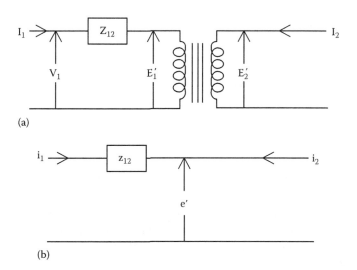

(a)

(b)

FIGURE 4.6
Circuit models of a 2-winding transformer with leakage impedance: (a) circuit with a single effective 2-winding leakage impedance and (b) single circuit using per-unit quantities.

where

$$R_{12} = R_1 + \left(\frac{N_1}{N_2}\right)^2 R_2 \tag{4.68}$$

and

$$X_{\ell 12} = X_1 + \left(\frac{N_1}{N_2}\right)^2 X_2$$

referring to the primary winding. We can obtain expressions for referring quantities to the secondary winding by interchanging 1 and 2 in the said formulas. In large power transformers, $X_{\lambda 12} \gg R_{12}$ so we are normally concerned with obtaining leakage reactances.

In terms of previously defined quantities from (4.62) and (4.63),

$$X_{\ell 12} = X_{11} + \left(\frac{N_1}{N_2}\right)^2 X_{22} - 2\left(\frac{N_1}{N_2}\right) X_{12} \tag{4.69}$$

Note that the leakage reactance of (4.69) corresponds to the leakage inductance of (4.23) by multiplying by ω. We will usually use the symbol Z_{12} when referring to leakage reactance and ignore the resistive component. The leakage reactance is often calculated by finite element energy methods, using (4.23) and multiplying by ω.

We should note that another method of obtaining the 2-winding leakage impedance, which corresponds with how it is measured, is to short-circuit terminal 2 and perform an impedance measurement at terminal 1. Thus

$$Z_{12} = \frac{V_1}{I_1}\bigg|_{V_2=0} \tag{4.70}$$

Setting $V_2 = 0$ in (4.63), we get $E_2 = -I_2 Z_2$. Substituting into the V_1 equation in (4.63) and using (4.40), we find

$$V_1 = I_1\left[Z_1 + \left(\frac{N_1}{N_2}\right)^2 Z_2\right] \tag{4.71}$$

so that, from (4.70), we get $Z_{12} = Z_1 + (N_1/N_2)^2 Z_2$ as before.

4.5.1 Leakage Reactance for a 2-Winding Autotransformer

The circuit model for a 2-winding autotransformer can be constructed from separate windings as shown in Figure 4.7. From the method of obtaining leakage impedance given in (4.70), we measure the impedance at the H terminal with the X terminal shorted. But this will yield the same leakage impedance as that of an ordinary 2-winding transformer so that $Z_{HX} = Z_{12}$, that is, the terminal leakage impedance of a 2-winding autotransformer is the same as that of a transformer with the same windings but not autoconnected.

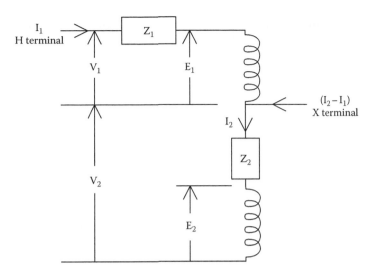

FIGURE 4.7
Circuit model of a 2-winding autotransformer with leakage impedance, based on a separate windings circuit model.

On a per-unit basis, however, the terminal leakage impedance of an autotransformer differs from that of a regular 2-winding transformer. We see from (4.46) that the base impedance is equal to the voltage squared divided by the rated power. Keeping the power base the same, since this is usually determined by the external circuit, the base impedance is proportional to the terminal voltage squared. For an autotransformer, the autoconnected terminal voltage is the voltage of the series winding divided by the coratio, r, which is the terminal voltage of a nonautoconnected transformer primary winding divided by r. Thus, the base impedance of an autotransformer is $1/r^2$ times that of a regular transformer. Since the per-unit impedance is the impedance in ohms divided by the base impedance, this equals r^2 times the per-impedance of a regular 2-winding transformer on the same power base.

4.6 Leakage Reactances for 3-Winding Transformers

We can go through the same arguments for a 3-winding transformer, isolating the flux common to all coils, Φ_c, and expressing the leakage flux for each coil, which exists in nonmagnetic materials, in terms of self- and mutual inductances, which are constants. We obtain

$$
\begin{aligned}
V_1 &= I_1 R_1 + j\left(I_1 X_{11} + I_2 X_{12} + I_3 X_{13}\right) + E_1 \\
V_2 &= I_2 R_2 + j\left(I_2 X_{22} + I_3 X_{23} + I_1 X_{12}\right) + E_2 \\
V_3 &= I_3 R_3 + j\left(I_3 X_{33} + I_1 X_{13} + I_2 X_{23}\right) + E_3
\end{aligned}
\tag{4.72}
$$

where the order of the suffixes doesn't matter for linear materials (constant permeability) since $X_{ij} = X_{ji}$.

Using the same assumptions as before concerning the neglect of core excitation, Equation 4.42, we substitute $I_3 = -(N_1/N_3)I_1 - (N_2/N_3)I_2$ in the first 2 equations and $I_2 = -(N_1/N_2)I_1 - (N_3/N_2)I_3$ in the third equation of (4.72) to obtain

$$V_1 = I_1\left[R_1 + j\left(X_{11} - \frac{N_1}{N_3}X_{13}\right)\right] + jI_2\left(X_{12} - \frac{N_2}{N_3}X_{13}\right) + E_1$$

$$V_2 = I_2\left[R_2 + j\left(X_{22} - \frac{N_2}{N_3}X_{23}\right)\right] + jI_1\left(X_{12} - \frac{N_1}{N_3}X_{23}\right) + E_2 \qquad (4.73)$$

$$V_3 = I_3\left[R_3 + j\left(X_{33} - \frac{N_3}{N_2}X_{23}\right)\right] + jI_1\left(X_{13} - \frac{N_1}{N_2}X_{23}\right) + E_3$$

Now add and subtract $jI_1(N_1/N_2)(X_{12} - (N_1/N_3)X_{23})$ from the first equation, add and subtract $jI_2(N_2/N_1)(X_{12} - (N_2/N_3)X_{13})$ from the second, and add and subtract $jI_3(N_3/N_1)(X_{13} - (N_3/N_2)X_{12})$ from the third equation of (4.73) to obtain

$$V_1 = I_1\left[R_1 + j\left(X_{11} - \frac{N_1}{N_3}X_{13} - \frac{N_1}{N_2}X_{12} + \frac{N_1^2}{N_2 N_3}X_{23}\right)\right]$$
$$+ \left\{jI_1\frac{N_1}{N_2}\left(X_{12} - \frac{N_1}{N_3}X_{23}\right) + jI_2\left(X_{12} - \frac{N_2}{N_3}X_{13}\right) + E_1\right\}$$

$$V_2 = I_2\left[R_2 + j\left(X_{22} - \frac{N_2}{N_3}X_{23} - \frac{N_2}{N_1}X_{12} + \frac{N_2^2}{N_1 N_3}X_{13}\right)\right] \qquad (4.74)$$
$$+ \left\{jI_1\left(X_{12} - \frac{N_1}{N_3}X_{23}\right) + jI_2\frac{N_2}{N_1}\left(X_{12} - \frac{N_2}{N_3}X_{13}\right) + E_2\right\}$$

$$V_3 = I_3\left[R_3 + j\left(X_{33} - \frac{N_3}{N_2}X_{23} - \frac{N_3}{N_1}X_{13} + \frac{N_3^2}{N_1 N_2}X_{12}\right)\right]$$
$$+ \left\{jI_1\left(X_{13} - \frac{N_1}{N_2}X_{23}\right) + jI_3\frac{N_3}{N_1}\left(X_{13} - \frac{N_3}{N_2}X_{12}\right) + E_3\right\}$$

Substituting for $I_3 = -(N_1/N_3)I_1 - (N_2/N_3)I_2$ in the term in braces in the last equation, we obtain

$$V_3 = I_3\left[R_3 + j\left(X_{33} - \frac{N_3}{N_2}X_{23} - \frac{N_3}{N_1}X_{13} + \frac{N_3^2}{N_1 N_2}X_{12}\right)\right]$$
$$+ \left\{jI_1\left(\frac{N_3}{N_2}X_{12} - \frac{N_1}{N_2}X_{23}\right) + jI_2\left(\frac{N_3}{N_1}X_{12} - \frac{N_2}{N_1}X_{13}\right) + E_3\right\} \qquad (4.75)$$

Comparing the terms in braces of the resulting V_1, V_2, and V_3 equations, we find, using (4.42),

$$\frac{\{\ \}_1}{N_1} = \frac{\{\ \}_2}{N_2} = \frac{\{\ \}_3}{N_3} \qquad (4.76)$$

Therefore, labeling the terms in brackets E_1', E_2', E_3', we obtain

$$V_1 = I_1 Z_1 + E_1', \quad V_2 = I_2 Z_2 + E_2', \quad V_3 = I_3 Z_3 + E_3' \quad (4.77)$$

where

$$Z_1 = R_1 + j\left(X_{11} - \frac{N_1}{N_3} X_{13} - \frac{N_1}{N_2} X_{12} + \frac{N_1^2}{N_2 N_3} X_{23} \right)$$

$$Z_2 = R_2 + j\left(X_{22} - \frac{N_2}{N_3} X_{23} - \frac{N_2}{N_1} X_{12} + \frac{N_2^2}{N_1 N_3} X_{13} \right) \quad (4.78)$$

$$Z_3 = R_3 + j\left(X_{33} - \frac{N_3}{N_2} X_{23} - \frac{N_3}{N_1} X_{13} + \frac{N_3^2}{N_1 N_2} X_{12} \right)$$

and

$$\frac{E_1'}{N_1} = \frac{E_2'}{N_2} = \frac{E_3'}{N_3} \quad (4.79)$$

Here, Z_1, Z_2, and Z_3 are the single-winding leakage impedances and the applicable multi-circuit model is shown in Figure 4.8a along with the per-unit circuit in Figure 4.8b.

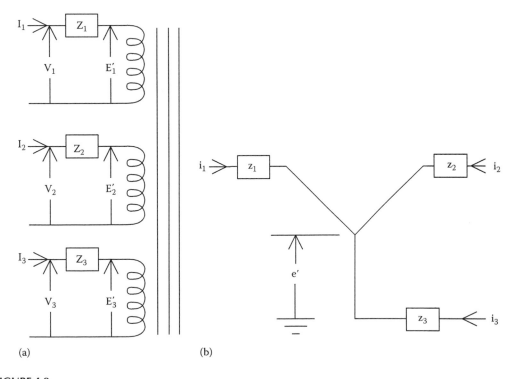

(a) (b)

FIGURE 4.8
Circuit models of a 3-winding transformer with leakage impedance: (a) separate circuit description and (b) single circuit description in terms of per-unit values.

This is possible because (4.79) implies $e_1' = e_2' = e_3' = e'$ and by Equation 4.51. This figure should be compared with Figure 4.4.

Since the single-winding leakage impedances are not directly measured or easily calculated, it is desirable to express these in terms of 2-winding leakage impedances. We will refer to these 2-winding leakage impedances as Z_{12}, Z_{13}, and Z_{23}, which correspond to the notation Z_{12} used previously for the 2-winding case. Thus, to measure the 2-winding leakage impedance between windings 1 and 2, we short-circuit 2, open circuit 3, and measure the impedance at terminal 1:

$$Z_{12} = \frac{V_1}{I_1} \bigg|_{\substack{V_2=0 \\ I_3=0}} \tag{4.80}$$

From (4.77), we see that this implies $I_2 Z_2 + E_2' = 0$. Using the voltage and current equations in (4.42), we find for Z_{12} and similarly for Z_{13} and Z_{23}

$$Z_{12} = Z_1 + \left(\frac{N_1}{N_2}\right)^2 Z_2$$

$$Z_{13} = Z_1 + \left(\frac{N_1}{N_3}\right)^2 Z_3 \tag{4.81}$$

$$Z_{23} = Z_2 + \left(\frac{N_2}{N_3}\right)^2 Z_3$$

We should note that the resistive and reactive components of (4.81) are the same as the corresponding components of the 2-winding resistance and leakage reactance formulas of (4.68) and (4.69) by substituting corresponding subscripts. The subscript ordering is chosen so that the second subscript refers to the shorted winding. The expression changes if we reverse subscripts according to

$$Z_{21} = \left(\frac{N_2}{N_1}\right)^2 Z_{12} \tag{4.82}$$

Solving (4.81) for the Z_i's, we get

$$Z_1 = \frac{1}{2}\left[Z_{12} + Z_{13} - \left(\frac{N_1}{N_2}\right)^2 Z_{23} \right]$$

$$Z_2 = \frac{1}{2}\left(\frac{N_2}{N_1}\right)^2 \left[Z_{12} + \left(\frac{N_1}{N_2}\right)^2 Z_{23} - Z_{13} \right] \tag{4.83}$$

$$Z_3 = \frac{1}{2}\left(\frac{N_3}{N_1}\right)^2 \left[Z_{13} + \left(\frac{N_1}{N_2}\right)^2 Z_{23} - Z_{12} \right]$$

These equations are very useful for determining the single-winding leakage impedances from the 2-winding leakage impedances for a 3-winding transformer, since the 2-winding impedances are fairly easily obtainable. This is not possible for a 2-winding transformer since there are two single-winding leakage impedances but only one 2-winding leakage impedance.

Using per-unit values, where Z_{b1} is the base impedance of circuit 1 so that $z_{12} = Z_{12}/Z_{b1}$, etc., and $z_1 = Z_1/Z_{b1}$, etc., we find, using (4.48), that (4.83) can be expressed as

$$z_1 = \frac{z_{12} + z_{13} - z_{23}}{2}$$

$$z_2 = \frac{z_{12} + z_{23} - z_{13}}{2} \tag{4.84}$$

$$z_3 = \frac{z_{13} + z_{23} - z_{12}}{2}$$

Similarly (4.81) becomes, in per-unit terms,

$$z_{12} = z_1 + z_2, \quad z_{13} = z_1 + z_3, \quad z_{23} = z_2 + z_3 \tag{4.85}$$

In the case of per-unit 2-winding leakage impedances, the order of the subscripts doesn't matter. Again, small letters are used for per-unit values.

4.6.1 Leakage Reactance for an Autotransformer with a Tertiary Winding

The autotransformer with tertiary winding can be obtained by interconnecting elements of the 3-winding transformer circuit model as shown in Figure 4.9. Here, the notation corresponds to that of Figure 4.8a. The problem is to re-express this in terms of terminal quantities. Thus, the appropriate terminal H voltage is $V_1 + V_2$, and the appropriate terminal X current is $I_2 - I_1$. Using (4.77),

$$V_H = V_1 + V_2 = I_1 Z_1 + I_2 Z_2 + E_1' + E_2' \tag{4.86}$$

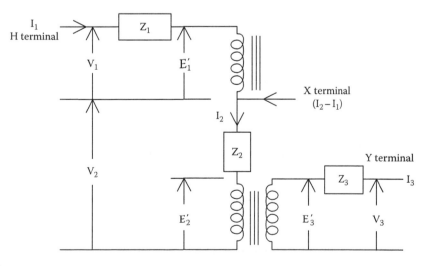

FIGURE 4.9
3-Winding autotransformer circuit model, derived from the three separate winding circuit models.

Substituting for I_2 from the amp-turn balance equation in (4.42) into this equation, we obtain

$$V_H = I_1\left(Z_1 - \frac{N_1}{N_2}Z_2\right) + \left[-\frac{N_3}{N_2}I_3Z_2 + E_1' + E_2'\right] \tag{4.87}$$

Similarly, using the amp-turn balance equation in (4.42), we can express $I_X = I_2 - I_1$. Then we have $I_X = I_2(1 + (N_2/N_1)) + (N_3/N_1)I_3$. Solving this for I_2 and substituting into the V_2 equation in (4.77), we obtain

$$V_X = V_2 = I_X\frac{N_1Z_2}{(N_1+N_2)} + \left[-\frac{N_3}{N_1+N_2}I_3Z_2 + E_2'\right] \tag{4.88}$$

By adding and subtracting $N_3^2I_3Z_2/N_2(N_1+N_2)$ from the V_3 equation in (4.77), we get

$$V_Y = V_3 = I_3\left(Z_3 + \frac{N_3^2Z_2}{N_2(N_1+N_2)}\right) + \left[-\frac{N_3^2}{N_2(N_1+N_2)}I_3Z_2 + E_3'\right] \tag{4.89}$$

We see that the terms in brackets in (4.87) through (4.89) satisfy

$$\frac{[\quad]_H}{N_1+N_2} = \frac{[\quad]_X}{N_2} = \frac{[\quad]_Y}{N_3} \tag{4.90}$$

Labeling these terms in brackets E_H, E_X, E_Y, we can rewrite (4.87) through (4.89)

$$V_H = I_HZ_H + E_H, \quad V_X = I_XZ_X + E_X, \quad V_Y = I_YZ_Y + E_Y \tag{4.91}$$

where $V_H = V_1 + V_2$, $V_X = V_2$, $V_Y = V_3$, $I_H = I_1$, $I_X = I_2 - I_1$, $I_Y = I_3$ are terminal quantities and

$$\begin{aligned} Z_H &= Z_1 - \frac{N_1}{N_2}Z_2 \\ Z_X &= \frac{N_1}{(N_1+N_2)}Z_2 \\ Z_Y &= Z_3 + \frac{N_3^2}{N_2(N_1+N_2)}Z_2 \end{aligned} \tag{4.92}$$

Thus, the circuit model shown in Figure 4.10a looks like that of Figure 4.8a with H, X, and Y substituted for 1, 2, and 3.

In terms of the measured 2-terminal impedances, we have, as before,

$$Z_{HX} = \left.\frac{V_H}{I_H}\right|_{\substack{V_X=0 \\ I_Y=0}} \tag{4.93}$$

Rewriting the amp-turn balance equation in (4.42), we obtain

$$N_1I_1 + N_2(I_2 - I_1) + N_2I_1 + N_3I_3 = (N_1 + N_2)I_H + N_2I_X + N_3I_Y = 0 \tag{4.94}$$

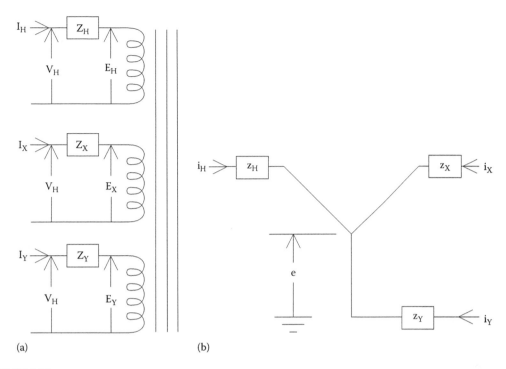

FIGURE 4.10
Circuit models of a 3-winding autotransformer based on terminal parameters: (a) separate circuit description and (b) single circuit description in terms of per-unit values.

From (4.93) with (4.90), (4.91), and (4.94), we obtain

$$Z_{HX} = Z_H + \left(\frac{N_1 + N_2}{N_2}\right)^2 Z_X$$

$$Z_{HY} = Z_H + \left(\frac{N_1 + N_2}{N_3}\right)^2 Z_Y \qquad (4.95)$$

$$Z_{XY} = Z_X + \left(\frac{N_2}{N_3}\right)^2 Z_Y$$

At this point, it is worthwhile to revert to per-unit terminal quantities. Because of the auto-connection, we have, again choosing as base quantities the VI per phase rating of the unit and the rated (no-load) terminal voltages per phase, $(VI)_b$, V_{bH}, V_{bX}, V_{bY},

$$\frac{V_{bH}}{V_{bX}} = \frac{N_1 + N_2}{N_2}, \quad \frac{V_{bH}}{V_{bY}} = \frac{N_1 + N_2}{N_3}, \quad \frac{V_{bX}}{V_{bY}} = \frac{N_2}{N_3}$$

$$\frac{I_{bH}}{I_{bX}} = \frac{N_2}{N_1 + N_2}, \quad \frac{I_{bH}}{I_{bY}} = \frac{N_3}{N_1 + N_2}, \quad \frac{I_{bX}}{I_{bY}} = \frac{N_3}{N_2} \qquad (4.96)$$

$$\frac{Z_{bH}}{Z_{bX}} = \left(\frac{N_1 + N_2}{N_2}\right)^2, \quad \frac{Z_{bH}}{Z_{bY}} = \left(\frac{N_1 + N_2}{N_3}\right)^2, \quad \frac{Z_{bX}}{Z_{bY}} = \left(\frac{N_2}{N_3}\right)^2$$

since $(VI)_b$ is the same for each terminal. From (4.94), we have on a per-unit basis,

$$i_H + i_X + i_Y = 0 \qquad (4.97)$$

Similarly from (4.90), on a per-unit basis,

$$e_H = e_X = e_Y = e \tag{4.98}$$

On a per-unit basis, Equation 4.95 becomes, using (4.96),

$$z_{HX} = z_H + z_X, \quad z_{HY} = z_H + z_Y, \quad z_{XY} = z_X + z_Y \tag{4.99}$$

where the same VI base is used for all the terminals and small letters denote per-unit values. Note that the order of the subscripts for the 2-winding per-unit terminal impedances doesn't matter. Solving these for z_H, z_X, and z_Y, we obtain a set of equations similar to (4.84):

$$z_H = \frac{z_{HX} + z_{HY} - z_{XY}}{2}$$

$$z_X = \frac{z_{HX} + z_{XY} - z_{HY}}{2} \tag{4.100}$$

$$z_Y = \frac{z_{HY} + z_{XY} - z_{HX}}{2}$$

The per-unit circuit model is depicted in Figure 4.10b.

Equation 4.92 contains terminal impedances and single-coil impedances and the bases are different for these. It is sometimes convenient to express the per-unit terminal impedances in terms of the per-unit winding impedances. We can rewrite (4.92)

$$\frac{Z_H}{Z_{bH}} = \frac{Z_1}{Z_{b1}}\left(\frac{Z_{b1}}{Z_{bH}}\right) - \frac{N_1}{N_2}\frac{Z_2}{Z_{b2}}\left(\frac{Z_{b2}}{Z_{bH}}\right)$$

$$\frac{Z_X}{Z_{bX}} = \left(\frac{N_1}{N_1 + N_2}\right)\frac{Z_2}{Z_{b2}}\left(\frac{Z_{b2}}{Z_{bX}}\right) \tag{4.101}$$

$$\frac{Z_Y}{Z_{bY}} = \frac{Z_3}{Z_{b3}}\left(\frac{Z_{b3}}{Z_{bY}}\right) + \frac{N_3^2}{N_2(N_1 + N_2)}\frac{Z_2}{Z_{b2}}\left(\frac{Z_{b2}}{Z_{bY}}\right)$$

Keeping $(VI)_b$ the same for both per-unit values, we have

$$\frac{Z_{b1}}{Z_{bH}} = \left(\frac{N_1}{N_1 + N_2}\right)^2, \quad \frac{Z_{b2}}{Z_{bH}} = \left(\frac{N_2}{N_1 + N_2}\right)^2$$

$$\frac{Z_{b2}}{Z_{bX}} = 1, \quad \frac{Z_{b3}}{Z_{bY}} = 1, \quad \frac{Z_{b2}}{Z_{bY}} = \left(\frac{N_2}{N_3}\right)^2 \tag{4.102}$$

Substituting (4.102) into (4.101), we get

$$z_H = \left(\frac{N_1}{N_1 + N_2}\right)^2 z_1 - \frac{N_1}{N_2}\left(\frac{N_2}{N_1 + N_2}\right)^2 z_2$$

$$z_X = \frac{N_1}{(N_1 + N_2)} z_2 \tag{4.103}$$

$$z_Y = z_3 + \left(\frac{N_2}{N_1 + N_2}\right)z_2$$

This equation expresses the per-unit terminal impedances in terms of the per-unit single-winding impedances for a 3-winding autotransformer. In terms of the terminal turns ratio defined previously, $n = (N_1 + N_2)/N_2$, these last equations can be written:

$$z_H = \left(\frac{n-1}{n}\right)^2 z_1 - \frac{(n-1)}{n^2} z_2$$

$$z_X = \left(\frac{n-1}{n}\right) z_2 \tag{4.104}$$

$$z_Y = z_3 + \frac{z_2}{n}$$

The per-unit circuit model is depicted in Figure 4.10b. Note that from (4.83) or (4.84), the autotransformer circuit parameters can be derived from 2-winding leakage impedance values, which can be obtained from finite element methods or analytic formulas.

Using (4.99) and the expressions for z_1, z_2, and z_3 in terms of per-unit 2-winding leakage reactances, we can obtain the terminal-to-terminal per-unit leakage reactances for the autotransformer in terms of the 2-winding per-unit leakage reactances. We find

$$z_{HX} = \left(\frac{n-1}{n}\right)^2 z_{12}$$

$$z_{HY} = \frac{(1-n)}{n^2} z_{12} + \left(\frac{n-1}{n}\right) z_{13} + \frac{z_{23}}{n} \tag{4.105}$$

$$z_{XY} = z_{23}$$

The first equation in (4.105) is the same as that found for the per-unit terminal impedance of a 2-terminal autotransformer in Section 4.5.1 since $(n - 1)/n$ is the coratio. The last equation expresses the fact that the terminal impedance between the X and Y windings is independent of the autoconnection.

4.6.2 Leakage Reactance between 2 Windings Connected in Series and a Third Winding

It is useful to calculate the impedance between a pair of windings connected in series and a third winding in terms of 2-winding leakage impedances, as shown in Figure 4.11. This can be regarded as a special case of an autotransformer with the X-terminal open. But this leakage impedance is just Z_{HY}. From (4.95) and (4.92), we obtain

$$Z_{HY} = Z_1 + Z_2 + \left(\frac{N_1 + N_2}{N_3}\right)^2 Z_3 \tag{4.106}$$

or, in per-unit terms, using (4.102) together with $Z_{b3}/Z_{bH} = [N_3/(N_1 + N_2)]^2$,

$$z_{HY} = \left(\frac{N_1}{N_1 + N_2}\right)^2 z_1 + \left(\frac{N_2}{N_1 + N_2}\right)^2 z_2 + z_3 \tag{4.107}$$

These can be expressed in terms of 2-winding leakage impedances by means of (4.83) or (4.84).

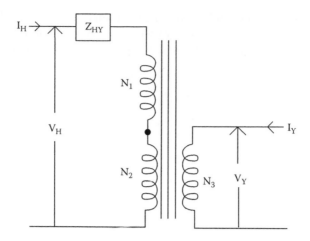

FIGURE 4.11
Leakage impedance between two series-connected windings and a third winding.

4.6.3 Leakage Reactance of a 2-Winding Autotransformer with X-Line Taps

X-line taps are taps connected to the X terminal of an autotransformer. They are used to regulate the low voltage without affecting the volts/turn of an autotransformer. Taps connected to the neutral or ground end of the LV winding would affect the number of turns in both the high- and low-voltage windings. Since the high voltage input level is assumed fixed by the external circuit, the volts/turn would change and this would change the induction level of the autotransformer. According to Faraday's law, assuming sinusoidal voltage and flux density at frequency f,

$$V_{rms} = \sqrt{2}\pi fNB_pA_c \quad \Rightarrow \quad \frac{V_{rms}}{N} = \sqrt{2}\pi fB_pA_c \qquad (4.108)$$

where
V_{rms} is the rms voltage
N is the number of turns
B_p is the peak flux density
A_c is the core area

Since f and A_c are fixed, the flux density (induction level) is determined by the volts/turn. If neutral taps are used, this characteristic must be considered in their design.

A circuit model of a 2-winding autotransformer with X-line taps constructed from the 3 separate winding circuit model is shown in Figure 4.12. The derivation of its 2-terminal leakage impedance uses (4.77) together with (4.42). In terms of terminal parameters, we have $V_H = V_1 + V_2$, $V_X = V_2 + V_3$, $I_H = I_1$, $I_X = I_3$. We also require that $I_2 = I_1 + I_3$.
Thus

$$
\begin{aligned}
V_H &= V_1 + V_2 = I_1Z_1 + \left(I_1 + I_3\right)Z_2 + E_1' + E_2' \\
&= I_1\left(Z_1 + Z_2\right) + I_3Z_2 + E_1' + E_2' \\
V_X &= V_2 + V_3 = \left(I_1 + I_3\right)Z_2 + I_3Z_3 + E_2' + E_3' \\
&= I_3\left(Z_2 + Z_3\right) + I_1Z_2 + E_2' + E_3'
\end{aligned}
\qquad (4.109)
$$

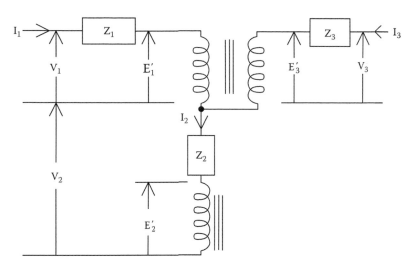

FIGURE 4.12
Circuit model of a 2-winding autotransformer with X-line taps, derived from the three separate winding circuits.

Substitute $I_3 = -(N_1/N_3)I_1 - (N_2/N_3)I_2$ into the V_H equation and $I_1 = -(N_2/N_1)I_2 - (N_3/N_1)I_3$ into the V_X equation to obtain

$$V_H = I_1 \left[Z_1 + \left(\frac{N_3 - N_1}{N_3} \right) Z_2 \right] - \frac{N_2}{N_3} I_2 Z_2 + E_1' + E_2'$$

$$V_X = I_3 \left[Z_3 + \left(\frac{N_1 - N_3}{N_1} \right) Z_2 \right] - \frac{N_2}{N_1} I_2 Z_2 + E_2' + E_3'$$

(4.110)

Add and subtract $N_2/(N_2+N_3)I_1Z_2$ from the V_H equation, and add and subtract $(N_2/(N_1+N_2))I_3Z_2$ from the V_X equation to obtain, after some algebraic manipulations,

$$V_H = I_1 \left[Z_1 + \left(\frac{N_3 - N_1}{N_2 + N_3} \right) Z_2 \right] + \left\{ \frac{N_2}{N_3} Z_2 \left[\left(\frac{N_3 - N_1}{N_2 + N_3} \right) I_1 - I_2 \right] + E_1' + E_2' \right\}$$

$$V_X = I_3 \left[Z_3 + \left(\frac{N_1 - N_3}{N_1 + N_2} \right) Z_2 \right] + \left\{ \frac{N_2}{N_3} Z_2 \left[\left(\frac{N_3 - N_1}{N_1 + N_2} \right) I_1 - \left(\frac{N_2 + N_3}{N_1 + N_2} \right) I_2 \right] + E_2' + E_3' \right\}$$

(4.111)

Notice that the terms in brackets satisfy

$$\frac{\{\ \}_H}{N_1 + N_2} = \frac{\{\ \}_X}{N_2 + N_3}$$

(4.112)

so that (4.111) can be rewritten:

$$V_H = I_H Z_H + E_H, \quad V_X = I_X Z_X + E_X$$

(4.113)

where

$$Z_H = Z_1 + \left(\frac{N_3 - N_1}{N_2 + N_3}\right)Z_2, \quad Z_X = Z_3 + \left(\frac{N_1 - N_3}{N_1 + N_2}\right)Z_2 \tag{4.114}$$

and

$$\frac{E_H}{N_1 + N_2} = \frac{E_X}{N_2 + N_3} \tag{4.115}$$

To obtain the 2-terminal leakage impedance, we use the definition

$$Z_{HX} = \left.\frac{V_H}{I_H}\right|_{V_X = 0} \tag{4.116}$$

From (4.42), we obtain

$$N_1 I_1 + N_2 (I_1 + I_3) + N_3 I_3 = (N_1 + N_2)I_1 + (N_2 + N_3)I_3$$
$$= (N_1 + N_2)I_H + (N_2 + N_3)I_X = 0 \tag{4.117}$$

The above equations are identical to those for a 2-winding transformer studied previously, except for labeling. Thus, from earlier work, we can solve (4.116) to get

$$Z_{HX} = Z_H + \left(\frac{N_1 + N_2}{N_2 + N_3}\right)^2 Z_X \tag{4.118}$$

Substituting from (4.114), this becomes

$$Z_{HX} = Z_1 + \left(\frac{N_1 - N_3}{N_2 + N_3}\right)^2 Z_2 + \left(\frac{N_1 + N_2}{N_2 + N_3}\right)^2 Z_3 \tag{4.119}$$

On a per-unit basis, using a fixed $(VI)_b$ and, noting that

$$\frac{Z_{b1}}{Z_{bH}} = \left(\frac{N_1}{N_1 + N_2}\right)^2, \quad \frac{Z_{b2}}{Z_{bH}} = \left(\frac{N_2}{N_1 + N_2}\right)^2, \quad \frac{Z_{b3}}{Z_{bH}} = \left(\frac{N_3}{N_1 + N_2}\right)^2$$

we find

$$z_{HX} = \left(\frac{N_1}{N_1 + N_2}\right)^2 z_1 + \left(\frac{N_2}{N_1 + N_2}\right)^2 \left(\frac{N_1 - N_3}{N_2 + N_3}\right)^2 z_2 + \left(\frac{N_3}{N_2 + N_3}\right)^2 z_3 \tag{4.120}$$

This is the two-terminal leakage impedance of a 2-winding autotransformer with X-line taps. It can be expressed in terms of the 2-winding leakage impedances via Equations 4.83 or 4.84.

5

Phasors, 3-Phase Connections, and Symmetrical Components

5.1 Phasors

Phasors are essentially complex numbers with a built-in sinusoidal time dependence. They are usually written as a complex number in polar form. A complex number, $C = C_{Re} + jC_{Im}$, can be expressed in polar form as

$$C = C_{Re} + jC_{Im} = |C|e^{j\theta} = |C|(\cos\theta + j\sin\theta)$$

$$\text{where } |C| = \sqrt{C_{Re}^2 + C_{Im}^2}, \quad \theta = \tan^{-1}\left(\frac{C_{Im}}{C_{Re}}\right) \tag{5.1}$$

Here $|C|$ is the magnitude of the complex number, with real and imaginary components, C_{Re} and C_{Im}.

A sinusoidal voltage has the form $V = V_p\cos(\omega t + \varphi)$, where $\omega = 2\pi f$ is the angular frequency, with f being the frequency in Hz, and φ a-phase angle. V_p is the peak value of the voltage. This can be written as

$$V = Re\left[V_p e^{j(\omega t + \varphi)}\right] \tag{5.2}$$

where Re[] means to take the real part of the expression following. In the complex plane, expression (5.2) is the projection on the real axis of the complex number given in the brackets, as illustrated in Figure 5.1.

As time increases, the complex voltage rotates counterclockwise around the origin, with the real projection varying as the cosine function. Currents and other voltages in a linear system, considered as complex numbers, will also rotate around the origin with time at the same rate since ω is the same for these. Since these numbers are all rotating together, at any snapshot in time, they will all have the same relative position with respect to each other. Therefore, it is convenient to factor out the common time dependence and consider only the phase relationships between the numbers. This is possible because the circuit equations are linear and

$$\mathbf{V} = V_p e^{j(\omega t + \varphi)} = V_p e^{j\omega t} e^{j\varphi} \tag{5.3}$$

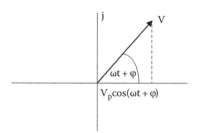

FIGURE 5.1
Voltage as the real projection of a complex number.

The complex number with time dependence is called a phasor and written in boldfaced type. Usually, however, the time dependence is understood or factored out and the quantity without the time exponential factor is also called a phasor. We will mainly discuss phasors without explicitly including the time exponential. If time dependences are required, the quantity can be multiplied by the complex time factor $e^{j\omega t}$ and, if required, the real projection taken. Usually, this is not necessary because most calculations with phasors can be done without the common time-dependent factor. However, the time-dependent factor is important when performing time differentiation or integration. We note that $d\mathbf{V}/dt = j\omega\mathbf{V}$. This implicitly makes use of the built-in time dependence of the voltage phasor $\mathbf{V}$. We also sometimes refer to phasors as vectors since complex numbers behave in many respects like vectors in the complex plane and are usually depicted as directed arrows much like vectors as in Figure 5.1. It is usually clear from the context whether a complex number or a mathematical vector is meant.

The voltage phasor in (5.3) was written in terms of its peak magnitude, V_p. Often, however, root mean square (rms) values of voltages and currents are used in power system discussions. The square of the rms value of the voltage is given by

$$V_{rms}^2 = \frac{V_p^2}{T} \int_0^T \cos^2(\omega t + \varphi)\,dt \tag{5.4}$$

where T is the period of the sinusoidal waveform, $T = 1/f = 2\pi/\omega$. Letting $\theta = \omega t$ in this integral, it becomes

$$V_{rms}^2 = \frac{V_p^2}{2\pi} \int_0^{2\pi} \cos^2(\theta + \varphi)\,d\theta = \frac{V_p^2}{2} \implies V_{rms} = \frac{V_p}{\sqrt{2}} \tag{5.5}$$

This quantity can be obtained from the phasor by the formula

$$V_{rms} = \sqrt{\frac{\mathbf{V}\mathbf{V}^*}{2}} \tag{5.6}$$

where * denotes the complex conjugate.

Another important quantity is the electrical power. This is the product of the voltage and current. At any instant in time, this gives the instantaneous power. But we are usually interested in the average power. Since the relative phase between the voltage and current doesn't change with time, we can select our time origin so that the voltage has its maximum value at t = 0, so that $\varphi_V = 0$ and the current phase leads the voltage by $\varphi_I = \varphi$. (If φ_I is negative, then it would lag the voltage.) Then the average power is

$$P_{ave} = \frac{1}{T}\int_0^T V(t)I(t)dt = \frac{V_pI_p\omega}{2\pi}\int_0^{2\pi/\omega} \cos(\omega t)\cos(\omega t+\varphi)dt \tag{5.7}$$

Letting $\theta = \omega t$, this can be written as

$$P_{ave} = \frac{V_pI_p}{2\pi}\int_0^{2\pi} \cos\theta\cos(\theta+\varphi)d\theta = \frac{V_pI_p}{2}\cos\varphi \tag{5.8}$$

In terms of rms voltage and current, $P_{ave} = V_{rms}I_{rms}\cos\varphi$. In terms of phasors, $P_{ave} = \mathrm{Re}[\mathbf{VI}^*]/2$. This leads one to the definition of complex power:

$$P = \frac{\mathbf{VI}^*}{2} = V_{rms}I_{rms}\left(\cos\varphi - j\sin\varphi\right) \tag{5.9}$$

The imaginary power is associated with reactors and capacitors in the circuit. If it is negative, it is capacitive power since the current leads the voltage in a capacitor. If positive, it is inductive power since the current lags the voltage in an inductor. Because the complex conjugate of the current is taken in (5.9), a positive φ produces a negative imaginary power and vice versa for a negative φ. Here, leading and lagging refer to the time-dependent phasors as they rotate counterclockwise with time in the complex plane. As time increases, the voltage vector passes a given point before the current vector for an inductor and vice versa for a capacitor. For a single capacitor or inductor, $\varphi = \pm\pi/2$ so that the average power vanishes according to (5.8). Capacitors and inductors store power and then return it at various times during a cycle, but they do not consume power on average. For a resistor, $\varphi = 0$ so that the imaginary power vanishes and the average power consumption assumes its maximum value.

Voltages and currents are often written as phasors having their rms magnitude. This is done since these are usually the quantities of interest in a power system and also so that the power expression in (5.9) can be written more simply as $P = \mathbf{VI}^*$ for rms voltage and current phasors. Equation 5.5 can be used to convert between these different expressions.

Power systems are 3-phase systems. This means that the voltages and currents come in triples. The voltages and currents in the 3 phases are shifted in time relative to each other by a third of a cycle. For a cycle of period T, the voltages are time shifted by 0, T/3, and 2T/3. Since T = 1/f, where f is the frequency in Hz, we have $\omega T/3 = 2\pi/3 = 120°$ and $2\omega T/3 = 4\pi/3 = 240° = -120°$. Thus, the 3 voltages can be written (in standard order) as

$$V_a(t) = V_a\cos(\omega t)$$
$$V_b(t) = V_a\cos\left[\omega\left(t+\frac{2T}{3}\right)\right] = V_a\cos\left[\omega t + \frac{2\omega T}{3}\right] = V_a\cos(\omega t - 120°) \tag{5.10}$$
$$V_c(t) = V_a\cos\left[\omega\left(t+\frac{T}{3}\right)\right] = V_a\cos\left[\omega t + \frac{\omega T}{3}\right] = V_a\cos(\omega t + 120°)$$

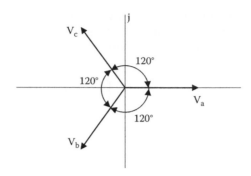

FIGURE 5.2
Phasor representation of a 3-phase voltage system.

Here, an arbitrary phase angle could be added to each. We have selected this arbitrary phase so that the a-voltage has its maximum value at time 0. They all have the same magnitude, and this is taken as the magnitude of the a-voltage. Expressed as phasors, these become

$$\mathbf{V_a} = V_a e^{j\omega t}, \quad \mathbf{V_b} = V_a e^{j(\omega t - 120°)}, \quad \mathbf{V_c} = V_a e^{j(\omega t + 120°)} \tag{5.11}$$

The phasors are written in boldfaced type but the magnitudes in regular type. At time 0, these can be represented in the complex plane as shown in Figure 5.2.

The phasors in Figure 5.2 are said to be balanced since $\mathbf{V_a} + \mathbf{V_b} + \mathbf{V_c} = 0$. This equation holds for every instant of time. These phasors are usually written with the time dependence factored out. This ordering of the phasors is called positive sequence ordering. As the phasors rotate in time, the a-phasor is the first to cross the real axis (it's already there), followed by the b-phasor and then the c-phasor. The Cartesian form for these phasors, normalized to have unit magnitude, is given by

$$1, \quad -\frac{1}{2} - j\frac{\sqrt{3}}{2}, \quad -\frac{1}{2} + j\frac{\sqrt{3}}{2} \tag{5.12}$$

In this form, it can be seen immediately that they sum to zero.

The 3-phase system is convenient for the smooth operation of generators and motors, which require rotating flux for their operation, and this rotating flux can be established with 3-phase currents. Although 2 independent phasors could also be used, apparently 3 phasors work better. Also, 2 independent phasors could not sum to zero, and this property of summing to zero has useful implications for grounding.

5.2 Y and Delta 3-Phase Connections

There are two basic types of 3-phase connections in common use, the Y (Wye) and delta (Δ) connections as illustrated schematically in Figure 5.3. In the Y connection, all 3 phases are connected to a common point, which may or may not be grounded. In the Δ connection, the phases are connected end to end with each other. In the Y connection, the line current

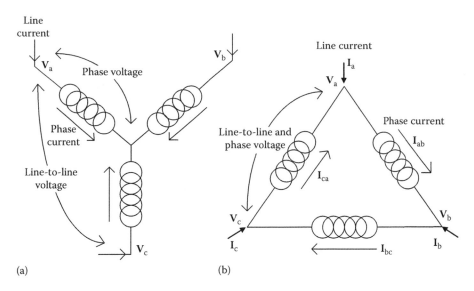

(a) (b)

FIGURE 5.3
Basic 3-phase connections: (a) Y connection and (b) Δ connection.

flows directly into the winding where it is called the winding or phase current. Note that in a balanced 3-phase system, the currents sum to zero at the common node in Figure 5.3a. Therefore, under balanced conditions, even if this point were grounded, no current would flow to ground. In the Δ connection, the line and phase currents are different. On the other hand, the line-to-line voltages in the Y connection differ from the voltages across the windings or phase voltages, whereas they are the same in the Δ connection. Note that the coils are shown at angles to each other in the figure to emphasize the phasor relationships, whereas in practice they are usually side by side and vertically oriented, as shown in Figure 5.4 for the delta connection.

The line-to-line voltages are usually specified when a power company orders a transformer. For a delta connection, these are the same as the winding voltages. For a Y connection, these can be determined from the winding voltages and vice versa using phasor algebra. Here, as in later sections of this chapter, we are considering rated voltages and ignoring impedance drops. Let V_a, V_b, V_c be the phase or winding voltages in a Y-connected set of transformer windings and let V_{ab} denote the line-to-line voltage between phases a and b, etc. for the other line-to-line voltages. Then, these line-to-line voltages are given by

$$V_{ab} = V_a - V_b, \quad V_{bc} = V_b - V_c, \quad V_{ca} = V_c - V_a \qquad (5.13)$$

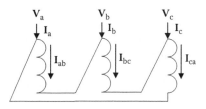

FIGURE 5.4
Delta interconnections as they would appear for windings on a 3-phase core.

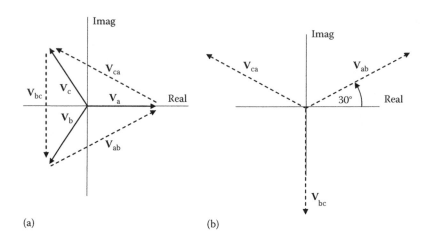

FIGURE 5.5
Phasor representation of line-to-line voltages and their relation to the phase voltages in a Y-connected set of 3-phase windings: (a) vector subtraction and (b) line-to-line voltages.

Using a phasor description, these can be readily calculated. The line-to-line phasors are shown graphically in Figure 5.5. Figure 5.5a shows the vector subtraction process explicitly, and Figure 5.5b shows the set of 3 line-to-line voltage phasors for the Y connection. The line-to-line voltage magnitude can be found geometrically as the diagonal of the parallelogram formed by equal length sides making an angle of 120° with each other. Thus, we have

$$|\mathbf{V}_{ab}| = |\mathbf{V}_a - \mathbf{V}_b| = \sqrt{|\mathbf{V}_a|^2 + |\mathbf{V}_b|^2 - 2|\mathbf{V}_a||\mathbf{V}_b|\cos 120°}$$
$$= \sqrt{|\mathbf{V}_a|^2 + |\mathbf{V}_b|^2 + |\mathbf{V}_a||\mathbf{V}_b|} = \sqrt{3}|\mathbf{V}_a| \tag{5.14}$$

We have used the fact that the different phase voltages have the same magnitude.

The graphical description of this process is useful for visualization purposes, but the math is simpler if the unit phasors of (5.12) are used. In this case, we have

$$\mathbf{V}_{ab} = V_a\left[1 - \left(-\frac{1}{2} - j\frac{\sqrt{3}}{2}\right)\right] = V_a\left(\frac{3}{2} + j\frac{\sqrt{3}}{2}\right)$$
$$= \sqrt{3}V_a e^{j\tan^{-1}\left(1/\sqrt{3}\right)} = \sqrt{3}V_a e^{j30°} \tag{5.15}$$

Thus, the magnitude of the line-to-line voltage is $\sqrt{3}$ times the phase voltage magnitude in Y-connected transformer coils, and they are rotated +30° relative to the phase voltages as shown in Figure 5.5b. Since the phase voltages are internal to the transformer and therefore impact the winding insulation structure, a more economical design is possible if these can be lowered. Hence, a Y connection is often used for the high voltage coils of a 3-phase transformer.

The relationship between the phase and line currents of a delta-connected set of windings can be found similarly. From Figure 5.3, we see that the line currents are given in terms of the phase currents by

$$\mathbf{I}_a = \mathbf{I}_{ab} - \mathbf{I}_{ca}, \quad \mathbf{I}_b = \mathbf{I}_{bc} - \mathbf{I}_{ab}, \quad \mathbf{I}_c = \mathbf{I}_{ca} - \mathbf{I}_{bc} \tag{5.16}$$

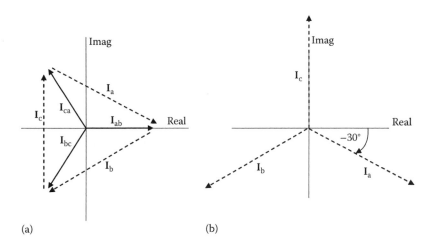

FIGURE 5.6
Phasor representation of line currents and their relation to the phase currents in a delta-connected set of 3-phase windings: (a) vector subtraction and (b) line currents.

These are illustrated graphically in Figure 5.6. Note that, as shown in Figure 5.6b, the line currents form a positive sequence set rotated $-30°$ relative to the phase currents. The magnitude relationship between the phase and line currents follows similarly as for the voltages in a Y connection. However, let us use phasor subtraction directly in (5.16) to show this. We have

$$\mathbf{I}_a = \mathbf{I}_{ab}\left[1 - \left(-\frac{1}{2} + j\frac{\sqrt{3}}{2}\right)\right] = \mathbf{I}_{ab}\left(\frac{3}{2} - j\frac{\sqrt{3}}{2}\right)$$

$$= \sqrt{3}\mathbf{I}_{ab}e^{-j\tan^{-1}\left(1/\sqrt{3}\right)} = \sqrt{3}\mathbf{I}_{ab}e^{-30°} \tag{5.17}$$

where we have used $\mathbf{I}_{ab} = \mathbf{I}_{ca}$. Non-boldfaced quantities are magnitudes.

Thus, the magnitudes of the line currents into or out of a delta-connected set of windings are $\sqrt{3}$ times the winding or phase current magnitudes. Since low voltage terminal currents are typically greater than high voltage terminal currents, it is common to connect the low voltage windings in delta because lowering the phase currents can produce a more economical design although other considerations are often the deciding factor.

The rated power of a transformer is generally taken to be the 3-phase power and is expressed in terms of terminal voltages and currents. The terminal voltage is the voltage to ground or virtual ground in the case of a delta connection, and the terminal current is the current into the terminal. The rated voltage ignores any impedance drops in the transformer, and the current assumes a rated load at unity power factor. Therefore, the voltages and currents are in phase, and we need only to deal with magnitudes. For a Y-connected set of windings, the terminal voltage to ground equals the voltage across the winding and the line current equals the phase current. Hence, the total rated power into all three phases is 3 × the phase voltage × the phase current. In terms of the line-to-line voltages and line current, we have, using (5.14), that the total rated power is $\sqrt{3}$ × the line-to-line voltage × the line current. Expressed mathematically, we have

$$\text{Power} = 3V_a I_a = 3\frac{V_{ab}}{\sqrt{3}}I_a = \sqrt{3}V_{\text{line-line}}I_{\text{line}}$$

For a delta-connected set of windings, the line-to-line voltages are the same as the phase volt-ages, whereas the line current is $\sqrt{3}$ times the phase current according to (5.17). Therefore, the power can be expressed in terms of phase quantities or terminal quintiles according to

$$\text{Power} = 3V_a I_{ab} = 3V_a \frac{I_a}{\sqrt{3}} = \sqrt{3} V_{\text{line-line}} I_{\text{line}}$$

Thus, the total rated power, whether expressed in terms of line quantities or phase quantities, is the same for Y- and delta-connected windings and is given mathematically by these equations.

We also note that the rated input power is the same as the power flowing through the windings. This may be obvious here, but there are some connections where this is not the case, in particular the autotransformer connection as discussed previously.

An interesting 3-phase connection is the open delta connection. In this connection, one of the windings of the delta is missing although the terminal connections remain the same. This is illustrated in Figure 5.7. This could be used especially if the 3 phases consist of sepa-rate units and if one of the phases is missing either because it is intended for future expan-sion or it has been disabled for some reason. Thus, instead of (5.16) for the relationship between line and phase currents, we have

$$\mathbf{I}_a = \mathbf{I}_{ab} - \mathbf{I}_{ca}, \quad \mathbf{I}_b = -\mathbf{I}_{ab}, \quad \mathbf{I}_c = \mathbf{I}_{ca} \tag{5.18}$$

But this still implies

$$\mathbf{I}_a + \mathbf{I}_b + \mathbf{I}_c = 0 \tag{5.19}$$

so that the terminal currents form a balanced 3-phase system. Likewise, the terminal voltages form a balanced 3-phase system. However, only 2 phases are present in the windings. As far as the external electrical system is concerned, the 3 phases are balanced. The total rated input or output power is, as before, $\sqrt{3}$ × line-to-line voltage × line current. But as (5.18) shows, the line currents have the same magnitude as the phase or winding currents. Since the windings are designed to carry a certain maximum current under rated conditions without overheating, this winding or phase current would be the same for a full or open delta. But for a full delta,

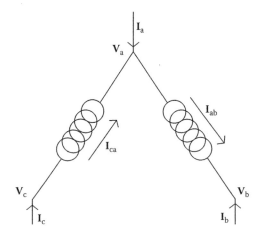

FIGURE 5.7
Open delta connection.

the line current is $\sqrt{3}$ times the winding current. This means that, for an open delta, the line currents must be reduced by $1/\sqrt{3}$ times that of a full delta in order that the same phase currents flow in the windings. Therefore, since the line-to-line voltages are the same, the rated power for an open delta is $1/\sqrt{3} = 0.577$ times that of a full delta.

5.3 Zig-Zag Connection

The zig-zag connection is often used as a grounding connection for unbalanced faults. In normal operation, it does not allow balanced 3-phase current to flow through the windings and so is an open circuit to this type of current. However, when an unbalanced fault occurs, it provides a path for currents that are all in phase with each other and have the same magnitude. The connection is shown in Figure 5.8.

The lines are oriented to indicate the direction of the voltage phasor on a phasor diagram, with the arrows indicating the direction of current flow. Only the relative orientation matters. The voltages increase from the ground point to the terminals, opposite to the direction of current flow. The lines are all of equal length so their voltage magnitudes are all the same. A zig-zag connection requires 2 windings per phase for a total of 6 interconnected windings. The unprimed voltages in the figure are for the zig windings with the label a, b, or c, corresponding to the phase to which they belong. The primed voltages are for the zag windings with the label associated with their phase. Note that the zag windings have the opposite voltage orientation to the zig windings in their phase. This winding interconnection, as it would appear for the set of 6 windings on a 3-phase core, is shown in Figure 5.9.

A phasor diagram of these voltages, choosing the V_a phasor to be the reference phasor, is given in Figure 5.10. The voltage between the a-terminal and ground is shown graphically in the figure. Using the unit phasors of (5.12) and the fact that all the phasors have the same magnitude, the phasor addition can be done algebraically:

$$\mathbf{V}_{a \text{ to ground}} = \mathbf{V}_a + \mathbf{V}_c' = V_a \left[1 - \left(-\frac{1}{2} + j\frac{\sqrt{3}}{2} \right) \right] = V_a \left(\frac{3}{2} - j\frac{\sqrt{3}}{2} \right)$$

$$= \sqrt{3} V_a e^{-j30°} \tag{5.20}$$

Thus, we see that the terminal voltage to ground is $\sqrt{3} \times$ the winding voltage and is shifted by $-30°$.

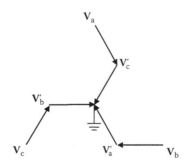

FIGURE 5.8
Zig-zag connection. The arrows indicate the direction of current flow.

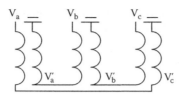

FIGURE 5.9
Zig-zag interconnected windings on a 3-phase core.

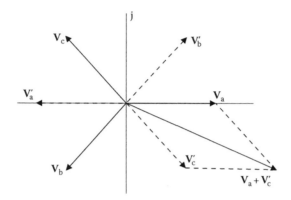

FIGURE 5.10
Phasor diagram of the zig and zag voltages.

We can see from Figure 5.8 that if a-phase current flows into the a-terminal and the a-zig winding, it would also have to flow into the zag winding of the c-phase. This cannot occur since we could not get amp-turn balance with the c-phase current flowing into the c-zig winding. Therefore, no balanced phase currents can flow into the terminals. However, currents all of the same phase could flow into the terminals. In fact, we can see from the current directions in the figure and the fact that the windings all have the same number of turns (same voltage magnitudes) that we would have amp-turn balance in all the phases in this case.

5.4 Scott Connection

The Scott connection is used to convert voltages from a 2-phase system to a 3-phase balanced system or vice versa. The 2-phase voltage phasors are assumed to be 90° apart on a phasor diagram and of equal magnitude. They are therefore independent phasors and any phasor can be constructed from these by a suitable vectorial combination. Before discussing the physical realization of this connection in terms of windings, it is useful to examine the phasor relationships directly. These are shown in Figure 5.11. The 2-phase voltages are labeled $\mathbf{V}_1$ and $\mathbf{V}_2$. The 3-phase voltages are labeled $\mathbf{V}_a$, $\mathbf{V}_b$, $\mathbf{V}_c$. These are constructed from components of the 2-phase voltages as shown in the figure.

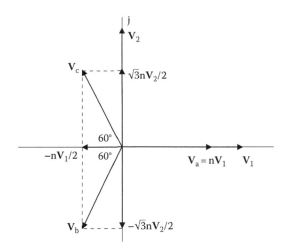

FIGURE 5.11
Scott connection from a 2-phase system of voltages, labeled 1 and 2, to a 3-phase voltage system, labeled a, b, c.

Here, the voltage (turns) ratio between V_a and V_1 is n, where $n = V_a/V_1$. This allows for the possibility that the voltage can be stepped up or down in going from 2 to 3 phases. The 3-phase voltages have components along the 2-phase voltages as shown in the figure. We are assuming that the magnitudes of $\mathbf{V}_1$ and $\mathbf{V}_2$ are the same and that $\mathbf{V}_2 = j\mathbf{V}_1$. This is expressed mathematically as

$$\mathbf{V}_a = n\mathbf{V}_1$$
$$\mathbf{V}_b = -\frac{1}{2}n\mathbf{V}_1 - \frac{\sqrt{3}}{2}n\mathbf{V}_2 \qquad (5.21)$$
$$\mathbf{V}_c = -\frac{1}{2}n\mathbf{V}_1 + \frac{\sqrt{3}}{2}n\mathbf{V}_2$$

In order to realize this with windings, we need 2 windings in the 2-phase system with voltages V_1 and V_2. These would probably have to be on separate cores since they do not form a balanced system. This means that, for conventional core designs with legs and yokes, fluxes from the 2 phases would not cancel in the core yokes as they do for a 3-phase balanced system of currents. However, there are examples where a single core could be used [Kra88]. We will use the 2-core realization given in [MIT43]. First of all, we note that the b and c-phases each have a voltage component that is ½ the magnitude of the a-phase voltage. Therefore, the a-phase could be constructed of 2 separate windings, each having the same number of turns. These windings would be on the phase 1 core. Each of the separate a-phase windings could then contribute half the a-phase voltage component to the b- and c-phase voltages. The b and c-phases also have a component along phase 2 of the same magnitude but opposite sign. These could both come from a single winding having a voltage magnitude of $\sqrt{3}/2 \times$ the a-phase voltage magnitude. This realization is shown in Figure 5.12.

As can be seen from the figure, the 3-phase voltages are composed of contributions from the secondary windings on different cores. The secondary voltage points on the different cores are labeled V_A, etc. for convenience. The subscript will also be used to label the

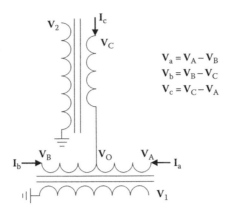

FIGURE 5.12
Scott connection for 2-phase to 3-phase transformation, using windings on 2 separate cores.

winding ends, and a pair of these will label the winding. The 3-phase system has no real ground, as is the case for a delta connection. The A–B winding on phase 1 is center tapped, with the C–O winding on phase 2 connected to the center tap position. The C–O winding must have $\sqrt{3}/2$ times the turns in the full A–B winding. Assume the winding voltages for the a, b, c system increase from B to A on the core 1 winding and from O to C on the core 2 winding. This is consistent with the V_1 and V_2 voltage directions for the 1, 2 system. The a-phase voltage starts at the B end of the winding and increases along both half windings to the A position so its voltage is the sum of the voltages from each half of the A–B winding. The b-phase voltage starts at position C and decreases along the C–O winding and then decreases from O to B along half the A–B winding. Its voltage is therefore minus the C–O winding voltage minus half the A–B winding voltage as required by vector addition on the phasor diagram. The c-phase voltage starts at the A position and decreases along half the A–B winding and then increases along the C–O winding. Its voltage is therefore minus half the A–B winding voltage plus the C–O winding voltage as required by the phasor diagram vector addition.

It is interesting to consider the winding currents in the a, b, c system. The terminal currents are shown in Figure 5.12. These should also form a balanced system and will therefore cancel at the center tap position O. Starting with the C–O winding, the c-phase current needs to balance the winding 2 current. The voltage turns ratio between these two windings is given by $V_{C-O}/V_2 = \sqrt{3}n/2$, where n was defined earlier as the turns ratio between the full A–B winding and winding 1. Therefore, for amp-turn balance, we have

$$\mathbf{I}_c = -\frac{2}{\sqrt{3}n}\mathbf{I}_2 = -j\frac{2}{\sqrt{3}n}\mathbf{I}_1$$

The currents as well as the voltages for the 2-phase system are at 90° with respect to each other on a phasor diagram. At the center node, we have

$$\mathbf{I}_a + \mathbf{I}_b = -\mathbf{I}_c = j\frac{2}{\sqrt{3}n}\mathbf{I}_1$$

For amp-turn balance on the phase 1 core, we require

$$\mathbf{I}_a - \mathbf{I}_b = -\frac{2}{n}\mathbf{I}_1$$

Solving this set of equations, we get

$$\mathbf{I}_a = \frac{\mathbf{I}_1}{n}\left(-1 + \frac{j}{\sqrt{3}}\right), \quad \mathbf{I}_b = \frac{\mathbf{I}_1}{n}\left(1 + \frac{j}{\sqrt{3}}\right), \quad \mathbf{I}_c = -j\frac{2}{\sqrt{3}n}\mathbf{I}_1$$

We see that these sum to zero and have amp-turn balance with the 1- and 2-phase windings.

5.5 Symmetrical Components

The balanced 3-phase system of voltages and currents we have discussed so far is called a positive sequence system. The phase ordering has been a, b, c in terms of which phasor crosses the real axis first as time increases. There is another 3-phase balanced system, called a negative sequence system with the phase ordering a, c, b. In addition, there is a zero sequence system that consists of 3 phasors of equal phase and magnitude. The zero sequence system is not balanced in the sense that the phasors in this system do not sum to zero but it is symmetric. The significance of these systems of phasors is that any 3-phase collection of phasors can be written as a sum of phasors, one from each of these systems. The components of such a sum are called symmetrical components. This makes it possible to analyze an unbalanced 3-phase system by separately analyzing the symmetric sequence systems and then adding the results. This will require a systematic treatment as discussed in the following.

In a balanced 3-phase electrical system, the voltage or current phasors are of equal magnitude and separated by 120° as discussed previously. They are shown again in Figure 5.13a, in order to contrast them with the other sequence sets, and are labeled V_{a1}, V_{b1}, V_{c1}, where

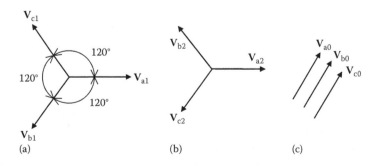

FIGURE 5.13
Symmetric systems of 3-phase phasors: (a) positive sequence, (b) negative sequence, and (c) zero sequence.

1 denotes a positive sequence set by convention. The unit vectors for this set were given in (5.12). We will use a special notation for this set of unit vectors. The unit vector 1 stays the same, but the other 2 are denoted as follows:

$$\alpha = e^{j120°} = -\frac{1}{2} + j\frac{\sqrt{3}}{2}$$

$$\alpha^2 = e^{j240°} = e^{-j120°} = -\frac{1}{2} - j\frac{\sqrt{3}}{2}$$

$$(5.22)$$

We see from this that

$$1 + \alpha + \alpha^2 = 0$$

$$\alpha^3 = 1$$

$$\alpha^4 = \alpha$$

$$\alpha^* = \alpha^2$$

$$(5.23)$$

Thus, for a balanced 3-phase positive sequence system, we have

$$\mathbf{V}_{a1} = \mathbf{V}_{a1}, \quad \mathbf{V}_{b1} = \alpha^2 \mathbf{V}_{a1}, \quad \mathbf{V}_{c1} = \alpha \mathbf{V}_{a1} \qquad (5.24)$$

Notice that $\mathbf{V}_{a1}$ need not be along the positive real axis. α^2 and α can be thought of as rotation operators that rotate the phasor they multiply by 240° and 120°, respectively. This guarantees that $\mathbf{V}_{b1}$ and $\mathbf{V}_{c1}$ are 240° and 120° from $\mathbf{V}_{a1}$ regardless of its position in the complex plane. Also, these phasors are of equal magnitude since α and its powers are of unit magnitude.

A negative sequence set of balanced phasors is one with the phase ordering a, c, b as shown in Figure 5.13b where we see that

$$\mathbf{V}_{a2} = \mathbf{V}_{a2}, \quad \mathbf{V}_{b2} = \alpha \mathbf{V}_{a2}, \quad \mathbf{V}_{c2} = \alpha^2 \mathbf{V}_{a2} \qquad (5.25)$$

where 2 refers to negative sequence quantities by convention. These are separated by 120° and have the same magnitude, which can differ from the positive sequence magnitude.

A zero sequence set of phasors is shown in Figure 5.13c. These are all in phase and have equal magnitudes, which can differ from the positive or negative sequence magnitudes. Thus,

$$\mathbf{V}_{a0} = \mathbf{V}_{b0} = \mathbf{V}_{c0} \qquad (5.26)$$

with 0 used to label zero sequence quantities. In addition, the relative phase between the a-phasor of each set can differ from zero.

We now show that it is possible to represent any unbalanced or unsymmetrical set of 3 phasors by means of these symmetric sequence sets. Let $\mathbf{V}_a$, $\mathbf{V}_b$, $\mathbf{V}_c$ be such an unbalanced set as shown in Figure 5.14. Since the positive, negative, and zero sequence symmetric sets are determined once $\mathbf{V}_{a1}$, $\mathbf{V}_{a2}$, and $\mathbf{V}_{a0}$ are specified, we need to find only these a-phase

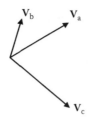

FIGURE 5.14
Unsymmetrical set of 3 phasors.

components of the symmetric sets in terms of the original phasors to prove that this representation is possible. Write

$$\begin{aligned}
\mathbf{V}_a &= \mathbf{V}_{a0} + \mathbf{V}_{a1} + \mathbf{V}_{a2} \\
\mathbf{V}_b &= \mathbf{V}_{b0} + \mathbf{V}_{b1} + \mathbf{V}_{b2} \\
\mathbf{V}_c &= \mathbf{V}_{c0} + \mathbf{V}_{c1} + \mathbf{V}_{c2}
\end{aligned} \tag{5.27}$$

Using (5.24) through (5.26), this can be written as

$$\begin{aligned}
\mathbf{V}_a &= \mathbf{V}_{a0} + \mathbf{V}_{a1} + \mathbf{V}_{a2} \\
\mathbf{V}_b &= \mathbf{V}_{a0} + \alpha^2 \mathbf{V}_{a1} + \alpha \mathbf{V}_{a2} \\
\mathbf{V}_c &= \mathbf{V}_{a0} + \alpha \mathbf{V}_{a1} + \alpha^2 \mathbf{V}_{a2}
\end{aligned} \tag{5.28}$$

In matrix notation, we have

$$\begin{pmatrix} \mathbf{V}_a \\ \mathbf{V}_b \\ \mathbf{V}_c \end{pmatrix} = \begin{pmatrix} 1 & 1 & 1 \\ 1 & \alpha^2 & \alpha \\ 1 & \alpha & \alpha^2 \end{pmatrix} \begin{pmatrix} \mathbf{V}_{a0} \\ \mathbf{V}_{a1} \\ \mathbf{V}_{a2} \end{pmatrix} \tag{5.29}$$

which is often abbreviated to

$$\mathbf{V}_{abc} = A\mathbf{V}_{012} \tag{5.30}$$

where
  $\mathbf{V}_{abc}$ and $\mathbf{V}_{012}$ are column vectors
  A is the matrix in (5.29)

$\mathbf{V}_{a0}$, $\mathbf{V}_{a1}$, $\mathbf{V}_{a2}$ can be found uniquely if A has an inverse. This can be shown to be the case and the result is

$$\begin{pmatrix} \mathbf{V}_{a0} \\ \mathbf{V}_{a1} \\ \mathbf{V}_{a2} \end{pmatrix} = \frac{1}{3} \begin{pmatrix} 1 & 1 & 1 \\ 1 & \alpha & \alpha^2 \\ 1 & \alpha^2 & \alpha \end{pmatrix} \begin{pmatrix} \mathbf{V}_a \\ \mathbf{V}_b \\ \mathbf{V}_c \end{pmatrix} \tag{5.31}$$

as can be verified by direct computation, using the identities in (5.23). Equation 5.31 can be abbreviated to

$$\mathbf{V}_{012} = A^{-1}\mathbf{V}_{abc} \tag{5.32}$$

where A^{-1} is the matrix in (5.31) including the factor of $1/3$. It is the inverse of the matrix A.

Thus, given any unsymmetrical set of phasors, the symmetric positive, negative, and zero sequence sets can be found. Conversely, given the symmetric sets or just one phasor from each symmetric set chosen customarily to be the a-phasor, the unsymmetrical set can be obtained.

Equations 5.29 through 5.32 apply equally well to currents. We can use them to compute the power in the two systems. The 3-phase power is given by

$$\begin{aligned}
P &= \mathbf{V}_a\mathbf{I}_a^* + \mathbf{V}_b\mathbf{I}_b^* + \mathbf{V}_c\mathbf{I}_c^* = \mathbf{V}_{abc}^T\mathbf{I}_{abc}^* \\
&= \left(A\mathbf{V}_{012}\right)^T\left(A\mathbf{I}_{012}\right)^* = \mathbf{V}_{012}^T A^T A^*\mathbf{I}_{012}^* \\
&= 3\mathbf{V}_{012}^T\mathbf{I}_{012}^* = 3\left(\mathbf{V}_{a0}\mathbf{I}_{a0}^* + \mathbf{V}_{a1}\mathbf{I}_{a1}^* + \mathbf{V}_{a2}\mathbf{I}_{a2}^*\right)
\end{aligned} \tag{5.33}$$

Here T denotes transpose, which converts a column vector to a row vector or vice versa. It also transforms a matrix so that its rows become columns or equivalently its columns become rows. This is needed since $\mathbf{V}_{abc}$ and $\mathbf{I}_{abc}$ are column vectors and the dot product is a matrix product between a row and column vector. We have also used $\alpha^* = \alpha^2$ and $\alpha^{2*} = \alpha$ in Equations 5.29 and 5.31 to show that $A^T A^* = 3I$, where I is the unit matrix. Thus, the 3-phase power is also equal to the sum of the powers in the 3 sequence networks. The factor of 3 enters because the last expression in (5.33) only contains the representative a-component of the sequence networks.

The virtue of using symmetrical components is that an unsymmetrical 3-phase set of phasors can be analyzed in terms of 3 symmetrical systems of phasors. Since the component systems are symmetrical, only one phase from each symmetrical system need be analyzed, in the same manner that we need only consider one phase of a conventional balanced 3-phase system. However, the circuit model applying to each sequence may differ from the normal 3-balanced phase circuit model. This latter circuit model applies generally only to the positive sequence circuit. The negative sequence circuit may differ from the positive, particularly if there are generators or motors in the system. Since generated voltages are usually of positive sequence, voltage sources are absent from the negative sequence circuit. For transformers, the impedances are independent of phase order so that the positive and negative sequence circuit impedances are identical.

The zero sequence circuit model can differ considerably from the positive or negative one. For example, a balanced set of currents can flow into the terminals of a 3-phase system, even without a path to ground, because the currents add to zero. Thus, an effective ground point can be assumed. Since zero sequence currents are of equal magnitude and phase, they do not sum to zero and therefore cannot flow into the terminals of a 3-phase system unless they there is a path to ground. Thus, for zero sequence currents to flow in a Y-connected set of windings, for example, a path to ground is necessary. Thus, in the zero sequence circuit model for a Y connection without ground, an infinite impedance to ground must be placed in the circuit. Similarly, zero sequence currents cannot flow into the terminals of a delta connection since there is no path to ground. However, such currents can circulate within the delta if they are induced there for amp-turn balance with another set of windings.

Since we alluded to the fact that the sequence impedances may differ in the 3 sequence networks, it is necessary to consider under what conditions these networks are uncoupled so that they can be analyzed separately. Let us start with the assumption that the 3-phase balanced networks for transformers are uncoupled and the same for all 3 phases. Even for a 3-phase core, the mutual reactances between windings are significant only for windings on the same leg of the core and tend to be small between windings on different legs. We also generally assume that the self-inductances are the same for windings on different legs, even though there are slight differences due to the fact that the magnetic environment of the center leg differs from that of the outer legs. This similarity also applies to leakage reactances. Consider the voltages across a circuit element having impedances of Z_a, Z_b, and Z_c in the 3 phases. The 3-phase voltages across these elements are given by

$$\begin{pmatrix} \mathbf{V}_a \\ \mathbf{V}_b \\ \mathbf{V}_c \end{pmatrix} = \begin{pmatrix} Z_a & 0 & 0 \\ 0 & Z_b & 0 \\ 0 & 0 & Z_c \end{pmatrix} \begin{pmatrix} \mathbf{I}_a \\ \mathbf{I}_b \\ \mathbf{I}_c \end{pmatrix} \tag{5.34}$$

Let Z denote the impedance matrix in (5.34). Using (5.30) and (5.32), we can convert the $\mathbf{V}$ and $\mathbf{I}$ vectors in the abc system to the 012 system to obtain

$$A^{-1}\mathbf{V}_{abc} = \mathbf{V}_{012} = A^{-1}ZA\mathbf{I}_{012} \tag{5.35}$$

If the impedances Z_a, etc., are unequal, then the matrix $A^{-1}ZA$ contains off-diagonal terms that couple the sequence networks. However, if the impedances are equal, then we obtain a diagonal matrix and the sequences are uncoupled as shown in (5.36):

$$\begin{pmatrix} \mathbf{V}_{a0} \\ \mathbf{V}_{a1} \\ \mathbf{V}_{a2} \end{pmatrix} = \begin{pmatrix} Z_a & 0 & 0 \\ 0 & Z_a & 0 \\ 0 & 0 & Z_a \end{pmatrix} \begin{pmatrix} \mathbf{I}_{a0} \\ \mathbf{I}_{a1} \\ \mathbf{I}_{a2} \end{pmatrix} \tag{5.36}$$

We see, in this case, that all the sequence impedances are the same. Therefore, this situation does not allow for the possibility that the zero sequence impedances can differ from the positive or negative sequence impedances. Let us examine the situation from the point of view of uncoupled sequence circuits, each having their own impedance. The sequence voltages across a given element are now given by

$$\begin{pmatrix} \mathbf{V}_{a0} \\ \mathbf{V}_{a1} \\ \mathbf{V}_{a2} \end{pmatrix} = \begin{pmatrix} Z_0 & 0 & 0 \\ 0 & Z_1 & 0 \\ 0 & 0 & Z_2 \end{pmatrix} \begin{pmatrix} \mathbf{I}_{a0} \\ \mathbf{I}_{a1} \\ \mathbf{I}_{a2} \end{pmatrix} \tag{5.37}$$

Letting Z represent the matrix in (5.37) and applying (5.30) and (5.32) to convert to the abc system, we get

$$A\mathbf{V}_{012} = \mathbf{V}_{abc} = AZA^{-1}\mathbf{I}_{abc} \tag{5.38}$$

For transformers, we noted previously that $Z_1 = Z_2$. For this situation, the matrix in (5.38) is given in (5.39):

$$\begin{pmatrix} \mathbf{V_a} \\ \mathbf{V_b} \\ \mathbf{V_c} \end{pmatrix} = \frac{1}{3} \begin{pmatrix} (Z_0 + 2Z_1) & (Z_0 - Z_1) & (Z_0 - Z_1) \\ (Z_0 - Z_1) & (Z_0 + 2Z_1) & (Z_0 - Z_1) \\ (Z_0 - Z_1) & (Z_0 - Z_1) & (Z_0 + 2Z_1) \end{pmatrix} \begin{pmatrix} \mathbf{I_a} \\ \mathbf{I_b} \\ \mathbf{I_c} \end{pmatrix} \tag{5.39}$$

We note that when $Z_0 = Z_1$, this becomes a diagonal matrix with all diagonal elements equal to Z_1 after multiplying by the factor of $1/3$ in front of the matrix.

Thus, when analyzing an unbalanced fault condition in transformers with zero sequence impedances that differ from the positive and negative sequence impedances, we can analyze the separate sequence circuit models using (5.37). We can then use the A matrix to convert to the abc system. On the other hand, if we wish to analyze an unbalanced fault condition from an abc circuit system approach, then we need to use the coupled set of impedances given in (5.39) when the zero and positive sequences differ [Bra82]. This would also apply to the case where all the sequence impedances were different, except that, when transforming to the abc system, the impedance matrix would be more complicated than in (5.39) unless some special conditions were placed on the sequence impedances.

6

Fault Current Analysis

6.1 Introduction

It is necessary to design transformers to withstand various possible faults, such as a short to ground of one or more phases. The high currents accompanying these faults, approximately 10–30 times normal, produce high forces and stresses in the windings and support structure. Also, depending on the fault duration, significant amounts of heat may be generated inside the unit. The design must accommodate the worst case fault that can occur from both the mechanical and thermal standpoints.

The first step in designing transformers to withstand faults is to determine the fault currents in all the windings, which is the subject of this chapter. Since this is an electrical problem, it requires a circuit model that includes leakage impedances of the transformer and also relevant system impedances. The systems are typically represented by a voltage source in series with an impedance, since we are not interested here in detailed fault currents within the system external to the transformer. The transformer circuit model considered here is that of a 2- or 3-terminal-per-phase unit with all pairs of terminal leakage reactances given either from calculations or measurement. We ignore core excitation since, for modern power transformers, its effects on the fault currents are negligible.

The transformers considered here are 3-phase units and the fault types treated are 3-phase line to ground, single-phase line to ground, line to line, and double line to ground. These are the standard fault types and are important because they are most likely to occur on actual systems. The transformer must be designed to withstand the worst of these fault types, or rather each coil must be designed to withstand the worst (highest current) fault it can experience. Note that each fault type refers to a fault on any of the single-phase terminals. For example, a 3-phase fault can occur on all the high-voltage (HV) terminals (H_1, H_2, and H_3), all the low-voltage (LV) terminals (X_1, X_2, X_3), or all the tertiary voltage terminals (Y_1, Y_2, Y_3), etc., for the other fault types. Note also that faults on a single-phase system can be considered as 3-phase faults on a 3-phase system, so these are included automatically in the analysis of faults on 3-phase systems.

Since the fault types considered include faults that produce unbalanced conditions in a 3-phase system, one of the most efficient methods of treating them is by the use of symmetrical components. In this method, an unbalanced set of voltages or currents can be represented mathematically by sets of balanced voltages or currents, called sequence voltages or currents. These latter can then be analyzed by means of sequence circuit models.

The final results are then obtained by transforming the voltages and currents from the sequence analysis into the voltages and currents of the real system. This method has been introduced in the last chapter and we will use those results here. For easy reference, we will repeat the transformation equations between the normal-phase abc system and the symmetrical component 012 system.

$$
\begin{pmatrix} \mathbf{V}_a \\ \mathbf{V}_b \\ \mathbf{V}_c \end{pmatrix} = \begin{pmatrix} 1 & 1 & 1 \\ 1 & \alpha^2 & \alpha \\ 1 & \alpha & \alpha^2 \end{pmatrix} \begin{pmatrix} \mathbf{V}_{a0} \\ \mathbf{V}_{a1} \\ \mathbf{V}_{a2} \end{pmatrix}
\tag{6.1}
$$

$$
\begin{pmatrix} \mathbf{V}_{a0} \\ \mathbf{V}_{a1} \\ \mathbf{V}_{a2} \end{pmatrix} = \frac{1}{3} \begin{pmatrix} 1 & 1 & 1 \\ 1 & \alpha & \alpha^2 \\ 1 & \alpha^2 & \alpha \end{pmatrix} \begin{pmatrix} \mathbf{V}_a \\ \mathbf{V}_b \\ \mathbf{V}_c \end{pmatrix}
\tag{6.2}
$$

It should also be noted that the circuit model calculations discussed here are for steady-state conditions, whereas actual faults would have a transient phase where the currents can exceed their steady-state values for short periods of time. These enhancement effects are included by means of an asymmetry factor. This factor takes into account the resistance and reactance present at the faulted terminal and is considered to be conservative from a design point of view. This enhancement factor was calculated in Chapter 2. The following references have been used in this chapter: [Ste62a], [Lyo37], [Blu51].

6.2 Fault Current Analysis on 3-Phase Systems

We assume that the system is balanced before the fault occurs, that is, any prefault voltages and currents are positive sequence sets. Here we consider a general electrical system as shown on the left side of Figure 6.1. The fault occurs at some location on the system where fault-phase currents I_a, I_b, and I_c flow. They are shown as leaving the system in the figure. The voltages to ground at the fault point are labeled V_a, V_b, and V_c. The system, as viewed from the fault point or terminal, is modeled by means of Thevenin's theorem. First, however, we resolve the voltages and currents into symmetrical components so that we need only analyze one phase of the positive, negative, and zero sequence sets. This is indicated on the right-hand side of Figure 6.1, where the a-phase sequence set has been singled out.

By Thevenin's theorem, each of the sequence systems can be modeled as a voltage source in series with an impedance, where the voltage source is the open-circuit voltage at the fault point and the impedance is found by shorting all voltage sources and measuring or calculating the impedance to ground at the fault terminal. Note that, in applying Thevenin's theorem, the fault point together with the ground point is the terminal that is associated with the open-circuit voltage before the fault. The load when the fault occurs is typically set to zero but may include a fault impedance. The resulting model is shown in Figure 6.2. No voltage source is included in the negative and zero sequence circuits, since the standard voltage sources in power systems are positive sequence sources.

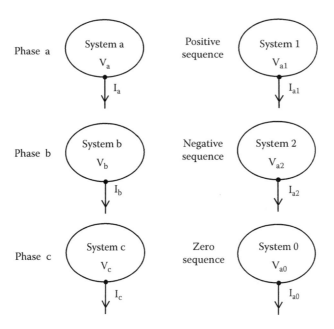

FIGURE 6.1
Fault at a point on a general electrical system.

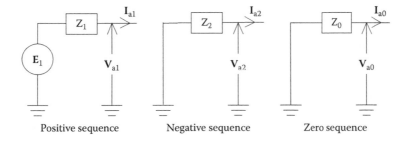

FIGURE 6.2
Thevenin equivalent sequence circuit models.

The circuit equations for Figure 6.2 are

$$\mathbf{V}_{a1} = \mathbf{E}_1 - \mathbf{I}_{a1}\mathbf{Z}_1, \quad \mathbf{V}_{a2} = -\mathbf{I}_{a2}\mathbf{Z}_2, \quad \mathbf{V}_{a0} = -\mathbf{I}_{a0}\mathbf{Z}_0 \tag{6.3}$$

Since $\mathbf{E}_1$ is the open-circuit voltage at the fault terminal, it is the voltage at the fault point before the fault occurs and can be labeled V_{pf}, where pf denotes prefault. We can omit the label 1 since it is understood to be a positive sequence voltage. We will also regard this as a reference phasor and use ordinary type for it. Thus

$$\mathbf{V}_{a1} = V_{pf} - \mathbf{I}_{a1}\mathbf{Z}_1, \quad \mathbf{V}_{a2} = -\mathbf{I}_{a2}\mathbf{Z}_2, \quad \mathbf{V}_{a0} = -\mathbf{I}_{a0}\mathbf{Z}_0 \tag{6.4}$$

If there is some impedance in the fault, this could be included in the circuit model. However, because we are interested in the worst case faults (highest fault currents), we assume that

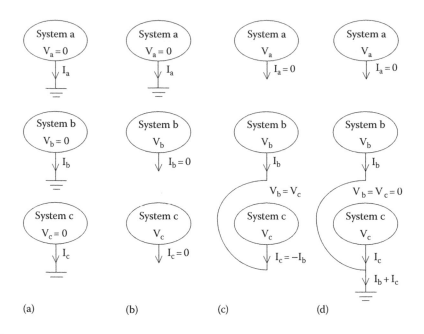

FIGURE 6.3
Standard fault types on 3-phase systems: (a) 3-phase line-to-ground fault, (b) single-phase line-to-ground fault, (c) line-to-line fault, and (d) double-line-to-ground fault.

the fault resistance is zero. If there is an impedance to ground at a neutral point in the transformer as, for example, at the junction of a Y-connected set of windings then 3× its value should be included in the single-phase zero sequence network at that point. This is because all three zero-sequence currents flow into the neutral resistor, but only one of them is represented in the zero sequence circuit.

The types of faults considered and their voltage and current constraints are shown in Figure 6.3.

6.2.1 3-Phase Line-to-Ground Fault

3-phase faults to ground are characterized by

$$\mathbf{V}_a = \mathbf{V}_b = \mathbf{V}_c = 0 \tag{6.5}$$

as shown in Figure 6.3a. From (6.5), together with (6.2), we find

$$\mathbf{V}_{a0} = \mathbf{V}_{a1} = \mathbf{V}_{a2} = 0 \tag{6.6}$$

Therefore, from (6.4) we get

$$\mathbf{I}_{a1} = \frac{\mathbf{V}_{pf}}{\mathbf{Z}_1}, \quad \mathbf{I}_{a2} = \mathbf{I}_{a0} = 0 \tag{6.7}$$

Using (6.7) and (6.1) applied to currents, we find

$$\mathbf{I}_a = \mathbf{I}_{a1}, \quad \mathbf{I}_b = \alpha^2 \mathbf{I}_{a1}, \quad \mathbf{I}_c = \alpha \mathbf{I}_{c1} \tag{6.8}$$

Thus the fault currents, as expected, form a balanced positive sequence set of magnitude V_{pf}/Z_1. This example could have been carried out without the use of symmetrical components, since the fault does not unbalance the system.

6.2.2 Single-Phase Line-to-Ground Fault

For a single-phase-to-ground fault, we assume, without loss of generality, that the a-phase is faulted. Thus we have

$$\mathbf{V}_a = 0, \quad \mathbf{I}_b = \mathbf{I}_c = 0 \tag{6.9}$$

as indicated in Figure 6.3b. From (6.2) applied to currents, we get

$$\mathbf{I}_{a0} = \mathbf{I}_{a1} = \mathbf{I}_{a2} = \frac{\mathbf{I}_a}{3} \tag{6.10}$$

From (6.9), (6.10), and (6.4), we find

$$\mathbf{V}_a = \mathbf{V}_{a0} + \mathbf{V}_{a1} + \mathbf{V}_{a2} = V_{pf} - \mathbf{I}_{a1}\left(Z_0 + Z_1 + Z_2\right) = 0$$

or

$$\mathbf{I}_{a1} = \frac{V_{pf}}{\left(Z_0 + Z_1 + Z_2\right)} = \mathbf{I}_{a2} = \mathbf{I}_{a0} \tag{6.11}$$

Using (6.9) and (6.10), we get

$$\mathbf{I}_a = \frac{3V_{pf}}{\left(Z_0 + Z_1 + Z_2\right)}, \quad \mathbf{I}_b = \mathbf{I}_c = 0 \tag{6.12}$$

However, we will keep both Z_1 and Z_2 in our formulas even though they are equal for transformers where $Z_1 = Z_2$.

Using these equations, we can also find the short-circuited phase voltages.

$$\mathbf{V}_a = 0, \quad \mathbf{V}_b = -\frac{V_{pf}}{\left(Z_0 + Z_1 + Z_2\right)}\left[\left(\frac{3}{2} + j\frac{\sqrt{3}}{2}\right)Z_0 + j\sqrt{3}Z_2\right]$$

$$\mathbf{V}_c = -\frac{V_{pf}}{\left(Z_0 + Z_1 + Z_2\right)}\left[\left(\frac{3}{2} - j\frac{\sqrt{3}}{2}\right)Z_0 - j\sqrt{3}Z_2\right] \tag{6.13}$$

We see that if all the sequence impedances were the same then, in magnitude, $V_b = V_c = V_{pf}$. However, their phases would differ.

6.2.3 Line-to-Line Fault

A line-to-line fault can, without loss of generality, be assumed to occur between lines b and c as shown in Figure 6.3c. The fault equations are

$$\mathbf{V}_b = \mathbf{V}_c, \quad \mathbf{I}_a = 0, \quad \mathbf{I}_c = -\mathbf{I}_b \tag{6.14}$$

From (6.2) applied to voltages and currents, we get

$$\mathbf{V}_{a1} = \mathbf{V}_{a2}, \quad \mathbf{I}_{a0} = 0, \quad \mathbf{I}_{a2} = -\mathbf{I}_{a1} \tag{6.15}$$

Using (6.4) and (6.15), we find

$$\mathbf{V}_{a0} = 0, \quad \mathbf{V}_{a1} - \mathbf{V}_{a2} = 0 = \mathbf{V}_{pf} - \mathbf{I}_{a1}\left(Z_1 + Z_2\right)$$

or

$$\mathbf{I}_{a1} = \frac{\mathbf{V}_{pf}}{\left(Z_1 + Z_2\right)} = -\mathbf{I}_{a2}, \quad \mathbf{I}_{a0} = 0 \tag{6.16}$$

Using (6.1) applied to currents, (6.15), and (6.16), we obtain

$$\mathbf{I}_a = 0, \quad \mathbf{I}_c = \frac{j\sqrt{3}}{\left(Z_1 + Z_2\right)} \mathbf{V}_{pf} = -\mathbf{I}_b \tag{6.17}$$

We can also find the short-circuited phase voltages by these methods.

$$\mathbf{V}_a = \mathbf{V}_{pf}\left(\frac{2Z_2}{Z_1 + Z_2}\right), \quad \mathbf{V}_b = \mathbf{V}_c = -\frac{\mathbf{V}_a}{2} \tag{6.18}$$

We notice that the fault analysis does not involve the zero-sequence circuit, that is, there are no zero-sequence currents. The fault currents flow between the b and c-phases. Also for transformers, $Z_1 = Z_2$ so $V_a = V_{pf}$.

6.2.4 Double Line-to-Ground Fault

The double line-to-ground fault, as shown in Figure 6.3d, can be regarded as involving lines b and c without loss of generality. The fault equations are

$$\mathbf{V}_b = \mathbf{V}_c = 0, \quad \mathbf{I}_a = 0 \tag{6.19}$$

From (6.19) and (6.2), we find

$$\mathbf{V}_{a0} = \mathbf{V}_{a1} = \mathbf{V}_{a2} = \frac{\mathbf{V}_a}{3}, \quad \mathbf{I}_{a0} + \mathbf{I}_{a1} + \mathbf{I}_{a2} = 0 \tag{6.20}$$

Using (6.4) and (6.20),

$$\mathbf{I}_{a0} + \mathbf{I}_{a1} + \mathbf{I}_{a2} = 0 = \frac{V_{pf}}{Z_1} - V_{a1}\left(\frac{1}{Z_0} + \frac{1}{Z_1} + \frac{1}{Z_2}\right)$$

or (6.21)

$$\mathbf{V}_{a1} = V_{pf}\left(\frac{Z_0 Z_2}{Z_0 Z_1 + Z_0 Z_2 + Z_1 Z_2}\right)$$

so that, using (6.4),

$$\mathbf{I}_{a1} = V_{pf}\left(\frac{Z_0 + Z_2}{Z_0 Z_1 + Z_0 Z_2 + Z_1 Z_2}\right)$$

$$\mathbf{I}_{a2} = -V_{pf}\left(\frac{Z_0}{Z_0 Z_1 + Z_0 Z_2 + Z_1 Z_2}\right) \tag{6.22}$$

$$\mathbf{I}_{a0} = -V_{pf}\left(\frac{Z_2}{Z_0 Z_1 + Z_0 Z_2 + Z_1 Z_2}\right)$$

Substituting into (6.1) applied to currents, we obtain

$$\mathbf{I}_a = 0, \quad \mathbf{I}_b = -V_{pf}\left[\frac{j\sqrt{3}Z_0 + \left(\frac{3}{2} + j\frac{\sqrt{3}}{2}\right)Z_2}{Z_0 Z_1 + Z_0 Z_2 + Z_1 Z_2}\right]$$

(6.23)

$$\mathbf{I}_c = V_{pf}\left[\frac{j\sqrt{3}Z_0 - \left(\frac{3}{2} - j\frac{\sqrt{3}}{2}\right)Z_2}{Z_0 Z_1 + Z_0 Z_2 + Z_1 Z_2}\right]$$

From (6.23), the fault phase voltages are given by

$$\mathbf{V}_a = V_{pf}\left(\frac{3Z_0 Z_2}{Z_0 Z_1 + Z_0 Z_2 + Z_1 Z_2}\right), \quad \mathbf{V}_b = \mathbf{V}_c = 0 \tag{6.24}$$

We see that if all the sequence impedances were equal, then $\mathbf{V}_a = V_{pf}$.

6.3 Fault Currents for Transformers with Two Terminals per Phase

A 2-terminal transformer can be modeled by a single leakage reactance that we call z_{HX} where H and X indicate high- and low-voltage terminals. All electrical quantities from this point on will be taken to mean per-unit quantities and will be written with small letters.

This will enable us to describe transformers, using a single circuit without the ideal transformer for each sequence.

The high- and low-voltage systems external to the transformer are described by system impedances z_{SH} and z_{SX} and voltage sources e_{SH} and e_{SX}. In per-unit terms, the 2-voltage sources are the same. The resulting sequence circuit models are shown in Figure 6.4. Subscripts 0, 1, and 2 are used to label the sequence circuit parameters since they can differ, although the positive and negative circuit parameters are equal for transformers but not necessarily for the systems. We have shown a fault on the X terminal in Figure 6.4. By interchanging subscripts, the H terminal faults can be obtained. In addition to the voltage source on the H system, we also show a voltage source attached to the X system. This could be a reasonable simulation of the actual system or it can simply be regarded as a device for keeping the voltage v_{a1} at the fault point at the rated per-unit value. This also allows one to consider cases where z_{SX} is small or zero in a limiting sense.

In order to use the previously developed general results, we need to compute the Thevenin impedances and prefault voltage at the fault point. From Figure 6.4, we see that

$$z_1 = \frac{z_{SX1}\left(z_{HX1} + z_{SH1}\right)}{z_{HX1} + z_{SH1} + z_{SX1}}, \quad z_2 = \frac{z_{SX2}\left(z_{HX2} + z_{SH2}\right)}{z_{HX2} + z_{SH2} + z_{SX2}}$$

$$z_0 = \frac{z_{SX0}\left(z_{HX0} + z_{SH0}\right)}{z_{HX0} + z_{SH0} + z_{SX0}}$$

(6.25)

and, since we are ignoring prefault currents,

$$v_{pf} = e_{SH1} = 1$$

(6.26)

where the prefault (pf) voltages are all positive sequence. The prefault voltage at the fault point will be taken as the rated voltage of the transformer and, in per-unit terms, is equal to 1. Figure 6.4 and the previously mentioned formulas assume that both terminals are

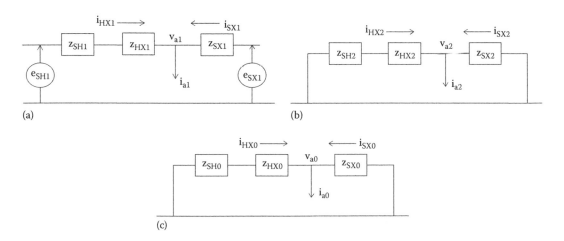

FIGURE 6.4
Sequence circuits for a fault on the low voltage, X, terminal of a 2 terminal-per-phase transformer, using per-unit quantities: (a) positive sequence, (b) negative sequence, and (c) zero sequence.

connected to the HV and LV systems. If it is desired not to consider the LV system attached beyond the fault point, then the Thevenin impedances become

$$z_1 = z_{HX1} + z_{SH1}, \quad z_2 = z_{HX2} + z_{SH2}, \quad z_0 = z_{HX0} + z_{SH0} \tag{6.27}$$

On the computer, this could be accomplished by setting z_{SX1}, z_{SX2}, and z_{SX0} to large values in (6.25), that is, open circuited. Thus, considering the LV system not attached beyond the fault point amounts to assuming that the faulted terminal is open circuited before the fault occurs. In fact the terminals associated with all three phases of the faulted terminal or terminals are open before the fault occurs. After the fault, one or more terminals are grounded or connected, depending on the fault type. The other terminals, corresponding to the other phases remain open when the system is not connected beyond the fault point.

It is sometimes desirable to let the system impedances equal zero. This increases the severity of the fault and is sometimes required for design purposes. This is not a problem mathematically for the system impedances on the nonfaulted terminal. But with zero-system impedances on the faulted terminal, Equation 6.25 shows that the Thevenin impedances would all equal zero unless one considered the system not attached beyond the fault point and used (6.27). Therefore, in this case, if one wants to consider the system attached beyond the fault point, one must give the system impedances very small values. In this way, continuity is assured when transitioning from rated system impedances to very small system impedances. Otherwise, there could be a discontinuity in the fault currents if one abruptly switches to a system connected to one not connected beyond the fault point just because the system impedances become small.

Much simplification can be achieved when the positive and negative system impedances are equal, since $z_1 = z_2$ implies $z_{HX1} = z_{HX2}$ for transformers. We will calculate the sequence currents for the general case where all the sequence impedances are different. One can then apply (6.1) expressed in terms of currents to get the phase currents for the general case. However, in this last step, we will assume the positive and negative impedances are the same. This will simplify the formulas for the phase currents and is closer to the real situation for transformers.

To obtain the currents in the transformer during the fault, according to Figure 6.4, we need to find i_{HX1}, i_{HX2}, and i_{HX0} for the standard faults. Since i_{a1}, i_{a2}, and i_{a0} have already been obtained for the standard faults, we must find the transformer currents in terms of these known fault currents. From Figure 6.4, using (6.26), we see that

$$v_{a1} = v_{pf} - i_{HX1}(z_{HX1} + z_{SH1}), \quad v_{a2} = -i_{HX2}(z_{HX2} + z_{SH2})$$
$$v_{a0} = -i_{HX0}(z_{HX0} + z_{SH0}) \tag{6.28}$$

Substituting the per-unit version of (6.4) into (6.28), we obtain

$$i_{HX1} = i_{a1}\left(\frac{z_1}{z_{HX1} + z_{SH1}}\right), \quad i_{HX2} = i_{a2}\left(\frac{z_2}{z_{HX2} + z_{SH2}}\right)$$
$$i_{HX0} = i_{a0}\left(\frac{z_0}{z_{HX0} + z_{SH0}}\right) \tag{6.29}$$

6.3.1 3-Phase Line-to-Ground Fault

For this fault case, we substitute the per-unit version of (6.7) into (6.29) to obtain

$$i_{HX1} = \frac{v_{pf}}{\left(z_{HX1} + z_{SH1}\right)}, \quad i_{HX2} = i_{HX0} = 0 \tag{6.30}$$

Then, using (6.1) applied to currents, we find

$$i_{HXa} = i_{HX1}, \quad i_{HXb} = \alpha^2 i_{HX1}, \quad i_{HXc} = \alpha i_{HX1} \tag{6.31}$$

that is, the fault currents in the transformer form a positive sequence set as expected.

6.3.2 Single-Phase Line-to-Ground Fault

For this type of fault, substitute the per-unit versions of (6.11) into (6.29), to obtain

$$i_{HX1} = \frac{v_{pf}}{\left(z_0 + z_1 + z_2\right)} \left(\frac{z_1}{\left(z_{HX1} + z_{SH1}\right)} \right)$$

$$i_{HX2} = \frac{v_{pf}}{\left(z_0 + z_1 + z_2\right)} \left(\frac{z_2}{z_{HX2} + z_{SH2}} \right) \tag{6.32}$$

$$i_{HX0} = \frac{v_{pf}}{\left(z_0 + z_1 + z_2\right)} \left(\frac{z_0}{z_{HX0} + z_{SH0}} \right)$$

If the system beyond the fault point is neglected, Equation 6.32 becomes, using (6.27),

$$i_{HX1} = \frac{v_{pf}}{\left(z_0 + z_1 + z_2\right)} = i_{HX2} = i_{HX0} \tag{6.33}$$

Substituting (6.32) into (6.1) applied to currents, we obtain the phase currents, assuming the positive and negative system impedances are equal or zero,

$$i_{HXa} = \frac{v_{pf}}{\left(z_0 + 2z_1\right)} \left[\frac{z_0}{\left(z_{HX0} + z_{SH0}\right)} + \frac{2z_1}{\left(z_{HX1} + z_{SH1}\right)} \right]$$

$$i_{HXb} = \frac{v_{pf}}{\left(z_0 + 2z_1\right)} \left[\frac{z_0}{\left(z_{HX0} + z_{SH0}\right)} - \frac{z_1}{\left(z_{HX1} + z_{SH1}\right)} \right] \tag{6.34}$$

$$i_{HXc} = \frac{v_{pf}}{\left(z_0 + 2z_1\right)} \left[\frac{z_0}{\left(z_{HX0} + z_{SH0}\right)} - \frac{z_1}{\left(z_{HX1} + z_{SH1}\right)} \right]$$

If we ignore the system beyond the fault point, Equation 6.34 becomes upon substituting from (6.27),

$$i_{HXa} = \frac{3v_{pf}}{(z_0 + 2z_1)}, \quad i_{HXb} = i_{HXc} = 0 \tag{6.35}$$

This can also be seen directly from (6.33). This makes sense because according to (6.12), we see that all the fault current flows through the transformer and none is shared with the system side of the fault point.

Note that, if the system impedance beyond the fault point is not ignored, there is fault current in phases b and c inside the transformer even though the fault is on phase a. These b and c fault currents are of lower magnitude than the phase a fault current.

6.3.3 Line-to-Line Fault

For this type of fault, we substitute the per-unit versions of (6.16) into (6.29) to obtain

$$
\begin{aligned}
i_{HX1} &= \frac{v_{pf}}{(z_{HX1} + z_{SH1})} \left(\frac{z_1}{z_1 + z_2} \right) \\
i_{HX2} &= -\frac{v_{pf}}{(z_{HX2} + z_{SH2})} \left(\frac{z_2}{z_1 + z_2} \right) \\
i_{HX0} &= 0
\end{aligned}
\tag{6.36}
$$

If we ignore the system beyond the fault point, this becomes, using (6.27),

$$i_{HX1} = \frac{v_{pf}}{(z_1 + z_2)} = -i_{HX2}, \quad i_{HX0} = 0 \tag{6.37}$$

Using (6.1) applied to currents and (6.36), we obtain for the phase currents, assuming the positive and negative system impedances are equal or zero,

$$
\begin{aligned}
i_{HXa} &= 0 \\
i_{HXb} &= -j\frac{\sqrt{3}}{2} \frac{v_{pf}}{(z_{HX1} + z_{SH1})} \\
i_{HXc} &= j\frac{\sqrt{3}}{2} \frac{v_{pf}}{(z_{HX1} + z_{SH1})}
\end{aligned}
\tag{6.38}
$$

If we ignore the system beyond the fault point, Equation 6.38 becomes

$$i_{HXa} = 0, \quad i_{HXc} = j\sqrt{3}\,\frac{v_{pf}}{2z_1} = -i_{HXb} \tag{6.39}$$

In this case, with the fault between phases b and c, phase a is unaffected. Also, as seen from (6.36) and (6.37), no zero sequence currents are involved in this type of fault.

6.3.4 Double Line-to-Ground Fault

For this fault, we substitute (6.22) expressed in per-unit terms, into (6.29),

$$
\begin{aligned}
i_{HX1} &= \frac{v_{pf}}{\left(z_{HX1} + z_{SH1}\right)} \left(\frac{z_0 z_1 + z_1 z_2}{z_0 z_1 + z_0 z_2 + z_1 z_2} \right) \\
i_{HX2} &= -\frac{v_{pf}}{\left(z_{HX2} + z_{SH2}\right)} \left(\frac{z_0 z_2}{z_0 z_1 + z_0 z_2 + z_1 z_2} \right) \\
i_{HX0} &= -\frac{v_{pf}}{\left(z_{HX0} + z_{SH0}\right)} \left(\frac{z_0 z_2}{z_0 z_1 + z_0 z_2 + z_1 z_2} \right)
\end{aligned}
\tag{6.40}
$$

If we ignore the system beyond the fault point, Equation 6.40 becomes

$$
\begin{aligned}
i_{HX1} &= v_{pf} \left(\frac{z_0 + z_2}{z_0 z_1 + z_0 z_2 + z_1 z_2} \right) \\
i_{HX2} &= -v_{pf} \left(\frac{z_0}{z_0 z_1 + z_0 z_2 + z_1 z_2} \right) \\
i_{HX0} &= -v_{pf} \left(\frac{z_2}{z_0 z_1 + z_0 z_2 + z_1 z_2} \right)
\end{aligned}
\tag{6.41}
$$

Using (6.1), applied to currents, and (6.40) we get for the phase currents, assuming equal positive and negative system impedances or zero system impedances,

$$
\begin{aligned}
i_{HXa} &= \frac{v_{pf}}{\left(2z_0 + z_1\right)} \left[\frac{z_1}{\left(z_{HX1} + z_{SH1}\right)} - \frac{z_0}{\left(z_{HX0} + z_{SH0}\right)} \right] \\
i_{HXb} &= \frac{v_{pf}}{\left(2z_0 + z_1\right)} \left[\frac{\alpha^2 z_1 - j\sqrt{3}z_0}{\left(z_{HX1} + z_{SH1}\right)} - \frac{z_0}{\left(z_{HX0} + z_{SH0}\right)} \right] \\
i_{HXc} &= \frac{v_{pf}}{\left(2z_0 + z_1\right)} \left[\frac{\alpha z_1 + j\sqrt{3}z_0}{\left(z_{HX1} + z_{SH1}\right)} - \frac{z_0}{\left(z_{HX0} + z_{SH0}\right)} \right]
\end{aligned}
\tag{6.42}
$$

If we ignore the system beyond the fault point, using (6.27) and (6.41), (6.42) becomes

$$
\begin{aligned}
i_{HXa} &= 0 \\
i_{HXb} &= -\frac{v_{pf}}{\left(2z_0 + z_1\right)} \left[\left(\frac{3}{2} + j\frac{\sqrt{3}}{2} \right) + j\sqrt{3}\frac{z_0}{z_1} \right] \\
i_{HXc} &= \frac{v_{pf}}{\left(2z_0 + z_1\right)} \left[\left(-\frac{3}{2} + j\frac{\sqrt{3}}{2} \right) + j\sqrt{3}\frac{z_0}{z_1} \right]
\end{aligned}
\tag{6.43}
$$

6.3.5 Zero-Sequence Circuits

Zero-sequence circuits require special consideration since certain transformer 3-phase connections, such as the delta connection, block the flow of zero sequence currents at the terminals and hence provide an essentially infinite impedance to their passage. This is also true of the ungrounded Y connection. The reason is that the zero sequence currents, being all in phase, require a return path in order to flow. The delta connection provides an internal path for the flow of these currents, circulating around the delta, but blocks their flow through the external lines. These considerations do not apply to positive- or negative-sequence currents which sum to zero vectorially and so require no return path.

For transformers, since the amp-turns must be balanced for each sequence, in order for zero-sequence currents to be present, they must flow in both windings. Thus in a grounded Y-Delta unit, for example, zero-sequence currents can flow within the transformer but cannot flow in the external circuit connected to the delta side. Similarly, zero-sequence currents cannot flow in either winding if one of them is an ungrounded Y.

Figure 6.5 shows some examples of zero-sequence circuit diagrams for different transformer connections. These should be compared with Figure 6.4c that applies to a grounded Y/grounded Y connection. Where a break in a line occurs, imagine that an infinite impedance is inserted. Mathematically, this is accomplished by letting the impedance approach infinity as a limiting process in the formulas or set to a very large value.

For the connection in Figure 6.5a, because the fault is on the delta side terminals, no zero sequence can flow into these terminals. The Thevenin impedance, looking in from the fault point, is $z_0 = z_{SX0}$. In this case, no zero sequence current can flow in the transformer and only flows in the external circuit on the LV side.

For the connection in Figure 6.5b, we find $z_0 = z_{HX0}z_{SX0}/(z_{HX0}+z_{SX0})$, that is, the parallel combination of z_{HX0} and z_{SX0}. In this case, zero-sequence current flows in the transformer, but no zero-sequence current flows into the system impedance on the HV side.

In Figure 6.5c, because of the ungrounded Y connection, no zero sequence current flows in the transformer. This would be true regardless of which side of the transformer has an ungrounded Y connection. In Figure 6.5d, no zero sequence current can flow into the transformer from the external circuit so effectively no zero sequence current flows in the transformer.

Another issue is the value of the zero sequence impedances themselves when they are fully in the circuit. These values tend to differ from the positive sequence impedances in transformers because the magnetic flux patterns associated with them can be quite

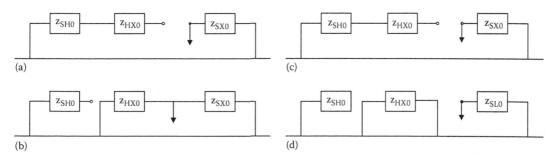

FIGURE 6.5
Some examples of zero-sequence impedance diagrams for 2-terminal transformers. The arrow indicates the fault point. (a) Y_g/delta, (b) delta/Y_g, (c) Y_g/Y_u, and (d) delta/delta. Y_g = grounded Y and Y_u = ungrounded Y.

different from the positive sequence flux pattern. This difference is taken into account by multiplying factors that multiply the positive sequence impedances to produce the zero sequence values. For 3-phase core-form transformers, these multiplying factors tend to be ≈0.85; however, they can differ for different 3-phase connections and are usually found by experimental measurements.

We should also note that if there is an impedance between the neutral and ground, for example, at a Y connected neutral, then 3 times this impedance value should be added to the zero-sequence transformer impedance. This is because identical zero-sequence current flows in the neutral from all three phases, so to account for the single-phase voltage drop across the neutral impedance, its value must be increased by a factor of 3 in the single-phase circuit diagram.

6.3.6 Numerical Example for a Single Line-to-Ground Fault

As a numerical example, consider a single line-to-ground fault on the X terminal of a 24 MVA 3-phase transformer. Assume the H and X terminals have line-to-line voltages of 112 and 20 kV respectively and that they are connected Delta and grounded-Y. Let $z_{HX1} = 10\%$, $z_{HX0} = 8\%$, $z_{SH1} = z_{SX1} = 0.01\%$, $z_{SH0} = 0\%$, and $z_{SX0} = 0.02\%$. The system impedance on the HV side was set to zero because it is a Delta winding. It is perhaps better to work on per-unit values by dividing these impedances by 100. But if we work with per-unit impedances in percentages, then we must also set $v_{pf} = 100\%$, assuming it has its rated value. This will give us fault currents in per unit (not percentages). Then we get from (6.25), assuming the system connected beyond the fault point, $z_1 = 9.99 \times 10^{-3}\%$ and $z_0 = 1.995 \times 10^{-2}\%$. Solving for the currents from (6.34), we get $i_{HXa} = 11.24$, $i_{HXb} = 3.746$, $i_{HXc} = 3.746$. These currents are per-unit values and are not percentages. For a 45 MVA 3-phase transformer with a high voltage Delta-connected winding with a line-to-line voltage of 112 kV, the base MVA per phase is 15, and the base winding voltage is 112 kV. Thus the base current is $(15/112) \times 10^3 = 133.9$ amps. Thus, the fault currents on the high voltage side of the transformer are $I_{HXa} = 1505$ amps, $I_{HXb} = 501.7$ amps, and $I_{HXc} = 501.7$ amps.

If we assume the system is not connected beyond the fault point, then we get from (6.27) $z_1 = 10.01\%$ and $z_0 = 8\%$ so that from (6.35), we get $i_{HXa} = 10.71$ and $I_{HXa} = 1434$ amps. The other phase currents are zero. This a-phase current is a bit lower than the a-phase current when assuming the system is connected beyond the fault point.

6.4 Fault Currents for Transformers with Three Terminals per Phase

A 3-terminal transformer can be represented in terms of three single winding impedances. Figure 6.6 shows the sequence circuits for such a transformer where H, X, and Y label the transformer impedances and SH, SX, and SY label the associated system impedances. Per-unit quantities are shown in the figure. The systems are represented by impedances in series with voltage sources. The positive sense of the currents is into the transformer terminals. Although the fault is shown on the X-terminal, by interchanging subscripts, the formulas that follow can apply to faults on any terminal. As before, we label the sequence impedances with subscripts 0, 1, and 2 even though the positive- and negative-sequence impedances are equal for transformers. They are not necessarily equal for the systems. The zero-sequence impedances usually differ from their positive-sequence counterparts for transformers as well as for the systems.

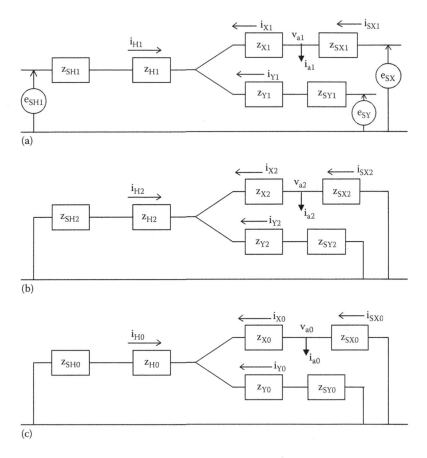

FIGURE 6.6
Sequence circuits for a fault on the X terminal of a 3-terminal per phase transformer, using per-unit quantities:
(a) positive-sequence circuit, (b) negative-sequence circuit, and (c) zero-sequence circuit.

We will calculate the sequence currents assuming possible unequal sequence impedances; however, we will calculate the phase currents assuming equal positive- and negative-system impedances. This is true for transformers and transmission lines. This would also be true for the case where small or zero-system impedances are assumed. This latter situation is often a requirement in fault analysis. Equation 6.1 could be applied to the sequence currents to get the phase currents if it is desired to obtain the fault currents for the case where all the sequence impedances are unequal.

From Figure 6.6, the Thevenin impedances, looking into the circuits from the fault point, are

$$z_1 = \frac{z_{SX1}\left(z_{X1} + w_1\right)}{z_{X1} + z_{SX1} + w_1} \quad \text{where } w_1 = \frac{\left(z_{H1} + z_{SH1}\right)\left(z_{Y1} + z_{SY1}\right)}{z_{H1} + z_{Y1} + z_{SH1} + z_{SY1}}$$

$$z_2 = \frac{z_{SX2}\left(z_{X2} + w_2\right)}{z_{X2} + z_{SX2} + w_2} \quad \text{where } w_2 = \frac{\left(z_{H2} + z_{SH2}\right)\left(z_{Y2} + z_{SY2}\right)}{z_{H2} + z_{Y2} + z_{SH2} + z_{SY2}} \tag{6.44}$$

$$z_0 = \frac{z_{SX0}\left(z_{X0} + w_0\right)}{z_{X0} + z_{SX0} + w_0} \quad \text{where } w_0 = \frac{\left(z_{H0} + z_{SH0}\right)\left(z_{Y0} + z_{SY0}\right)}{z_{H0} + z_{Y0} + z_{SH0} + z_{SY0}}$$

If the system beyond the fault point is ignored, the Thevenin impedances become

$$z_1 = z_{X1} + w_1, \quad z_2 = z_{X2} + w_2, \quad z_0 = z_{X0} + w_0 \tag{6.45}$$

where the w's remain the same. This amounts mathematically to setting z_{SX1}, z_{SX2}, and z_{SX0} equal to large values. When it is desired to set the system impedances to zero, then, as was the case for 2-terminal transformers, the system impedances must be set to small values if we wish to consider the system connected beyond the fault point.

At this point, we should consider some common transformer connections or conditions that require special consideration, including changes in some of the said impedances. For example, if one of the terminals is not loaded, then this amounts to setting the system impedance for that terminal to a large value. The large value should be infinity, but since these calculations are usually performed on a computer, we just need to make it meaningfully large compared with the other impedances. If the unloaded terminal is the faulted terminal, then this results in changing z_1 and z_2 to the values in (6.45). In other words, this is equivalent to ignoring the system beyond the fault point. If the unloaded terminal is an unfaulted terminal, then w_1 and w_2 will be changed when z_{SH1} and z_{SH2} or z_{SY1} and z_{SY2} are set to large values. This amounts to having an open circuit replace that part of the positive- or negative-system circuit associated with the open terminal. If the open terminal connection is a Y connection, then the zero-sequence system impedance should also be set to a large value and the zero-sequence circuit associated with the unloaded terminal becomes an open circuit. This will change w_0. However, if the open circuit is delta connected, as for a buried delta connection, then the system zero sequence impedance should be set to zero. This will also change w_0 but only slightly. This is because zero-sequence currents can circulate in delta-connected windings even with an open terminal situation, whereas zero-sequence currents cannot flow in Y-connected windings if the terminals are unloaded.

Working in the per-unit system, the prefault voltages are given by

$$v_{pf} = e_{SH1} = e_{SX1} = e_{SY1} = 1 \tag{6.46}$$

We are ignoring prefault currents, which can always be added later. We assume all the prefault voltages are equal to their rated values or 1 in per-unit terms. We also have

$$i_H + i_X + i_Y = 0 \tag{6.47}$$

This last equation applies to all the sequence currents. From Figure 6.6, the sequence voltages at the faulted terminal are

$$
\begin{aligned}
v_{a1} &= e_{SH1} - i_{H1}(z_{H1} + z_{SH1}) + i_{X1}z_{X1} = e_{SY1} - i_{Y1}(z_{Y1} + z_{SY1}) + i_{X1}z_{X1} \\
v_{a2} &= -i_{H2}(z_{H2} + z_{SH2}) + i_{X2}z_{X2} = -i_{Y2}(z_{Y2} + z_{SY2}) + i_{X2}z_{X2} \\
v_{a0} &= -i_{H0}(z_{H0} + z_{SH0}) + i_{X0}z_{X0} = -i_{Y0}(z_{Y0} + z_{SY0}) + i_{X0}z_{X0}
\end{aligned}
\tag{6.48}
$$

Solving (6.48), together with (6.44), (6.46), (6.47), and (6.4) expressed in per-unit terms, for the winding sequence currents in terms of the fault sequence currents, we obtain

$$
i_{H1} = i_{a1}\left(\frac{z_1}{z_{X1} + w_1}\right)\left(\frac{w_1}{z_{H1} + z_{SH1}}\right), \quad i_{H2} = i_{a2}\left(\frac{z_2}{z_{X2} + w_2}\right)\left(\frac{w_2}{z_{H2} + z_{SH2}}\right)
$$

$$
i_{H0} = i_{a0}\left(\frac{z_0}{z_{X0} + w_0}\right)\left(\frac{w_0}{z_{H0} + z_{SH0}}\right)
$$

$$
i_{X1} = -i_{a1}\left(\frac{z_1}{z_{X1} + w_1}\right), \quad i_{X2} = -i_{a2}\left(\frac{z_2}{z_{X2} + w_2}\right)
$$

$$
i_{X0} = -i_{a0}\left(\frac{z_0}{z_{X0} + w_0}\right) \tag{6.49}
$$

$$
i_{Y1} = i_{a1}\left(\frac{z_1}{z_{X1} + w_1}\right)\left(\frac{w_1}{z_{Y1} + z_{SY1}}\right), \quad i_{Y2} = i_{a2}\left(\frac{z_2}{z_{X2} + w_2}\right)\left(\frac{w_2}{z_{Y2} + z_{SY2}}\right)
$$

$$
i_{Y0} = i_{a0}\left(\frac{z_0}{z_{X0} + w_0}\right)\left(\frac{w_0}{z_{Y0} + z_{SY0}}\right)
$$

If we ignore the system beyond the fault point, using (6.45), we get for (6.49)

$$
i_{H1} = i_{a1}\left(\frac{w_1}{z_{H1} + z_{SH1}}\right), \quad i_{H2} = i_{a2}\left(\frac{w_2}{z_{H2} + z_{SH2}}\right), \quad i_{H0} = i_{a0}\left(\frac{w_0}{z_{H0} + z_{SH0}}\right)
$$

$$
i_{X1} = -i_{a1}, \quad i_{X2} = -i_{a2}, \quad i_{X0} = -i_{a0} \tag{6.50}
$$

$$
i_{Y1} = i_{a1}\left(\frac{w_1}{z_{Y1} + z_{SY1}}\right), \quad i_{Y2} = i_{a2}\left(\frac{w_2}{z_{Y2} + z_{SY2}}\right), \quad i_{Y0} = i_{a0}\left(\frac{w_0}{z_{Y0} + z_{SY0}}\right)
$$

We now use these equations, together with the fault current equations to obtain the currents in the transformer for the various types of fault. Equation 6.1, applied to currents, may be used to obtain the phase currents in terms of the sequence currents.

6.4.1 3-Phase Line-to-Ground Fault

For this type of fault, we substitute the per-unit version of (6.7) into (6.49) to obtain

$$
i_{H1} = \frac{v_{pf}}{\left(z_{X1} + w_1\right)}\left(\frac{w_1}{z_{H1} + z_{SH1}}\right), \quad i_{H2} = i_{H0} = 0
$$

$$
i_{X1} = -\frac{v_{pf}}{\left(z_{X1} + w_1\right)}, \quad i_{X2} = i_{X0} = 0 \tag{6.51}
$$

$$
i_{Y1} = \frac{v_{pf}}{\left(z_{X1} + w_1\right)}\left(\frac{w_1}{z_{Y1} + z_{SY1}}\right), \quad i_{Y2} = i_{Y0} = 0
$$

If we ignore the system beyond the fault point, we get the same sequence equations as (6.51). Thus, the system beyond the fault has no influence on 3-phase fault currents. The phase currents corresponding to (6.51) are

$$i_{Ha} = \frac{v_{pf}}{(z_{X1} + w_1)}\left(\frac{w_1}{z_{H1} + z_{SH1}}\right), \quad i_{Hb} = \alpha^2 i_{Ha}, \quad i_{Hc} = \alpha i_{Ha}$$

$$i_{Xa} = -\frac{v_{pf}}{(z_{X1} + w_1)}, \quad i_{Xb} = \alpha^2 i_{Xa}, \quad i_{Xc} = \alpha i_{Xa} \qquad (6.52)$$

$$i_{Ya} = \frac{v_{pf}}{(z_{X1} + w_1)}\left(\frac{w_1}{z_{Y1} + z_{SY1}}\right), \quad i_{Yb} = \alpha^2 i_{Ya}, \quad i_{Yc} = \alpha i_{Ya}$$

These form a positive sequence set as expected for a 3-phase fault.

6.4.2 Single-Phase Line-to-Ground Fault

For this fault, we substitute the per-unit version of (6.11) into (6.49) to get

$$i_{H1} = \frac{v_{pf}}{(z_0 + z_1 + z_2)}\left(\frac{z_1}{z_{X1} + w_1}\right)\left(\frac{w_1}{z_{H1} + z_{SH1}}\right)$$

$$i_{H2} = \frac{v_{pf}}{(z_0 + z_1 + z_2)}\left(\frac{z_2}{z_{X2} + w_2}\right)\left(\frac{w_2}{z_{H2} + z_{SH2}}\right)$$

$$i_{H0} = \frac{v_{pf}}{(z_0 + z_1 + z_2)}\left(\frac{z_0}{z_{X0} + w_0}\right)\left(\frac{w_0}{z_{H0} + z_{SH0}}\right)$$

$$i_{X1} = -\frac{v_{pf}}{(z_0 + z_1 + z_2)}\left(\frac{z_1}{z_{X1} + w_1}\right)$$

$$i_{X2} = -\frac{v_{pf}}{(z_0 + z_1 + z_2)}\left(\frac{z_2}{z_{X2} + w_2}\right) \qquad (6.53)$$

$$i_{X0} = -\frac{v_{pf}}{(z_0 + z_1 + z_2)}\left(\frac{z_0}{z_{X0} + w_0}\right)$$

$$i_{Y1} = \frac{v_{pf}}{(z_0 + z_1 + z_2)}\left(\frac{z_1}{z_{X1} + w_1}\right)\left(\frac{w_1}{z_{Y1} + z_{SY1}}\right)$$

$$i_{Y2} = \frac{v_{pf}}{(z_0 + z_1 + z_2)}\left(\frac{z_2}{z_{X2} + w_2}\right)\left(\frac{w_2}{z_{Y2} + z_{SY2}}\right)$$

$$i_{Y0} = \frac{v_{pf}}{(z_0 + z_1 + z_2)}\left(\frac{z_0}{z_{X0} + w_0}\right)\left(\frac{w_0}{z_{Y0} + z_{SY0}}\right)$$

These equations simplify by omitting the second term on the right side of each of the equations in (6.53) if the system beyond the fault point is ignored. We give them here for completeness:

$$i_{H1} = \frac{v_{pf}}{(z_0 + z_1 + z_2)}\left(\frac{w_1}{z_{H1} + z_{SH1}}\right)$$

$$i_{H2} = \frac{v_{pf}}{(z_0 + z_1 + z_2)}\left(\frac{w_2}{z_{H2} + z_{SH2}}\right)$$

$$i_{H0} = \frac{v_{pf}}{(z_0 + z_1 + z_2)}\left(\frac{w_0}{z_{H0} + z_{SH0}}\right)$$

$$i_{X1} = -\frac{v_{pf}}{(z_0 + z_1 + z_2)} = i_{X2} = i_{X0} \tag{6.54}$$

$$i_{Y1} = \frac{v_{pf}}{(z_0 + z_1 + z_2)}\left(\frac{w_1}{z_{Y1} + z_{SY1}}\right)$$

$$i_{Y2} = \frac{v_{pf}}{(z_0 + z_1 + z_2)}\left(\frac{w_2}{z_{Y2} + z_{SY2}}\right)$$

$$i_{Y0} = \frac{v_{pf}}{(z_0 + z_1 + z_2)}\left(\frac{w_0}{z_{Y0} + z_{SY0}}\right)$$

Using the per-unit version of (6.1), we can obtain the phase currents from (6.53), assuming equal positive- and negative-system impedances,

$$i_{Ha} = \frac{v_{pf}}{(z_0 + 2z_1)}\left[2\left(\frac{z_1}{z_{X1} + w_1}\right)\left(\frac{w_1}{z_{H1} + z_{SH1}}\right) + \left(\frac{z_0}{z_{X0} + w_0}\right)\left(\frac{w_0}{z_{H0} + z_{SH0}}\right)\right]$$

$$i_{Hb} = \frac{v_{pf}}{(z_0 + 2z_1)}\left[-\left(\frac{z_1}{z_{X1} + w_1}\right)\left(\frac{w_1}{z_{H1} + z_{SH1}}\right) + \left(\frac{z_0}{z_{X0} + w_0}\right)\left(\frac{w_0}{z_{H0} + z_{SH0}}\right)\right]$$

$$i_{Hc} = \frac{v_{pf}}{(z_0 + 2z_1)}\left[-\left(\frac{z_1}{z_{X1} + w_1}\right)\left(\frac{w_1}{z_{H1} + z_{SH1}}\right) + \left(\frac{z_0}{z_{X0} + w_0}\right)\left(\frac{w_0}{z_{H0} + z_{SH0}}\right)\right]$$

$$i_{Xa} = -\frac{v_{pf}}{(z_0 + 2z_1)}\left[2\left(\frac{z_1}{z_{X1} + w_1}\right) + \left(\frac{z_0}{z_{X0} + w_0}\right)\right]$$

$$i_{Xb} = -\frac{v_{pf}}{(z_0 + 2z_1)}\left[-\left(\frac{z_1}{z_{X1} + w_1}\right) + \left(\frac{z_0}{z_{X0} + w_0}\right)\right] \tag{6.55}$$

$$i_{Xc} = -\frac{v_{pf}}{(z_0 + 2z_1)}\left[-\left(\frac{z_1}{z_{X1} + w_1}\right) + \left(\frac{z_0}{z_{X0} + w_0}\right)\right]$$

$$i_{Ya} = \frac{v_{pf}}{(z_0 + 2z_1)}\left[2\left(\frac{z_1}{z_{X1} + w_1}\right)\left(\frac{w_1}{z_{Y1} + z_{SY1}}\right) + \left(\frac{z_0}{z_{X0} + w_0}\right)\left(\frac{w_0}{z_{Y0} + z_{SY0}}\right)\right]$$

$$i_{Yb} = \frac{v_{pf}}{(z_0 + 2z_1)}\left[-\left(\frac{z_1}{z_{X1} + w_1}\right)\left(\frac{w_1}{z_{Y1} + z_{SY1}}\right) + \left(\frac{z_0}{z_{X0} + w_0}\right)\left(\frac{w_0}{z_{Y0} + z_{SY0}}\right)\right]$$

$$i_{Yc} = \frac{v_{pf}}{(z_0 + 2z_1)}\left[-\left(\frac{z_1}{z_{X1} + w_1}\right)\left(\frac{w_1}{z_{Y1} + z_{SY1}}\right) + \left(\frac{z_0}{z_{X0} + w_0}\right)\left(\frac{w_0}{z_{Y0} + z_{SY0}}\right)\right]$$

We have used the fact that $\alpha^2 + \alpha + 1 = 0$, along with $w_1 = w_2$ and $z_1 = z_2$ to obtain (6.55).

For the case where the system beyond the fault point is ignored, again, restricting ourselves to the case where the positive- and negative-sequence impedances are equal, the phase currents are given by using (6.54):

$$i_{Ha} = \frac{v_{pf}}{(z_0 + 2z_1)} \left(\frac{2w_1}{z_{H1} + z_{SH1}} + \frac{w_0}{z_{H0} + z_{SH0}} \right)$$

$$i_{Hb} = \frac{v_{pf}}{(z_0 + 2z_1)} \left(-\frac{w_1}{z_{H1} + z_{SH1}} + \frac{w_0}{z_{H0} + z_{SH0}} \right)$$

$$i_{Hc} = \frac{v_{pf}}{(z_0 + 2z_1)} \left(-\frac{w_1}{z_{H1} + z_{SH1}} + \frac{w_0}{z_{H0} + z_{SH0}} \right)$$

$$i_{Xa} = -\frac{3v_{pf}}{(z_0 + 2z_1)}, \quad i_{Xb} = 0, \quad i_{Xc} = 0 \qquad (6.56)$$

$$i_{Ya} = \frac{v_{pf}}{(z_0 + 2z_1)} \left(\frac{2w_1}{z_{Y1} + z_{SY1}} + \frac{w_0}{z_{Y0} + z_{SY0}} \right)$$

$$i_{Yb} = \frac{v_{pf}}{(z_0 + 2z_1)} \left(-\frac{w_1}{z_{Y1} + z_{SY1}} + \frac{w_0}{z_{Y0} + z_{SY0}} \right)$$

$$i_{Yc} = \frac{v_{pf}}{(z_0 + 2z_1)} \left(-\frac{w_1}{z_{Y1} + z_{SY1}} + \frac{w_0}{z_{Y0} + z_{SY0}} \right)$$

6.4.3 Line-to-Line Fault

For this fault condition, substitute the per-unit version of (6.16) into (6.49), to obtain

$$i_{H1} = \frac{v_{pf}}{(z_1 + z_2)} \left(\frac{z_1}{z_{X1} + w_1} \right) \left(\frac{w_1}{z_{H1} + z_{SH1}} \right)$$

$$i_{H2} = -\frac{v_{pf}}{(z_1 + z_2)} \left(\frac{z_2}{z_{X2} + w_2} \right) \left(\frac{w_2}{z_{H2} + z_{SH2}} \right)$$

$$i_{H0} = 0$$

$$i_{X1} = -\frac{v_{pf}}{(z_1 + z_2)} \left(\frac{z_1}{z_{X1} + w_1} \right)$$

$$i_{X2} = \frac{v_{pf}}{(z_1 + z_2)} \left(\frac{z_2}{z_{X2} + w_2} \right) \qquad (6.57)$$

$$i_{X0} = 0$$

$$i_{Y1} = \frac{v_{pf}}{(z_1 + z_2)} \left(\frac{z_1}{z_{X1} + w_1} \right) \left(\frac{w_1}{z_{Y1} + z_{SY1}} \right)$$

$$i_{Y2} = -\frac{v_{pf}}{(z_1 + z_2)} \left(\frac{z_2}{z_{X2} + w_2} \right) \left(\frac{w_2}{z_{Y2} + z_{SY2}} \right)$$

$$i_{Y0} = 0$$

Ignoring the system beyond the fault point, Equation 6.57 becomes

$$i_{H1} = \frac{v_{pf}}{(z_1 + z_2)}\left(\frac{w_1}{z_{H1} + z_{SH1}}\right)$$

$$i_{H2} = -\frac{v_{pf}}{(z_1 + z_2)}\left(\frac{w_2}{z_{H2} + z_{SH2}}\right)$$

$$i_{H0} = 0$$

$$i_{X1} = -\frac{v_{pf}}{(z_1 + z_2)}, \quad i_{X2} = \frac{v_{pf}}{(z_1 + z_2)}, \quad i_{X0} = 0 \qquad (6.58)$$

$$i_{Y1} = \frac{v_{pf}}{(z_1 + z_2)}\left(\frac{w_1}{z_{Y1} + z_{SY1}}\right)$$

$$i_{Y2} = -\frac{v_{pf}}{(z_1 + z_2)}\left(\frac{w_2}{z_{Y2} + z_{SY2}}\right)$$

$$i_{Y0} = 0$$

Using the per-unit version of (6.1), we can obtain the phase currents from (6.57), assuming equal positive- and negative-system impedances:

$$i_{Ha} = 0, \quad i_{Hb} = -j\frac{\sqrt{3}}{2}\frac{v_{pf}}{(z_{X1} + w_1)}\left(\frac{w_1}{z_{H1} + z_{SH1}}\right) = -i_{Hc}$$

$$i_{Xa} = 0, \quad i_{Xb} = j\frac{\sqrt{3}}{2}\frac{v_{pf}}{(z_{X1} + w_1)} = -i_{Xc} \qquad (6.59)$$

$$i_{Ya} = 0, \quad i_{Yb} = -j\frac{\sqrt{3}}{2}\frac{v_{pf}}{(z_{X1} + w_1)}\left(\frac{w_1}{z_{Y1} + z_{SY1}}\right) = -i_{Yc}$$

For the case where we ignore the system beyond the fault point, we can obtain the phase currents from (6.58):

$$i_{Ha} = 0, \quad i_{Hb} = -j\frac{\sqrt{3}}{2}\frac{v_{pf}}{z_1}\left(\frac{w_1}{z_{H1} + z_{SH1}}\right) = -i_{Hc}$$

$$i_{Xa} = 0, \quad i_{Xb} = j\frac{\sqrt{3}}{2}\frac{v_{pf}}{z_1} = -i_{Xc} \qquad (6.60)$$

$$i_{Ya} = 0, \quad i_{Yb} = -j\frac{\sqrt{3}}{2}\frac{v_{pf}}{z_1}\left(\frac{w_1}{z_{Y1} + z_{SY1}}\right) = -i_{Yc}$$

6.4.4 Double Line-to-Ground Fault

For this fault, substitute the per-unit version of (6.22) into (6.49) to get

$$i_{H1} = v_{pf}\left(\frac{z_0 + z_2}{\Delta}\right)\left(\frac{z_1}{z_{X1} + w_1}\right)\left(\frac{w_1}{z_{H1} + z_{SH1}}\right)$$

$$i_{H2} = -v_{pf}\left(\frac{z_0}{\Delta}\right)\left(\frac{z_2}{z_{X2} + w_2}\right)\left(\frac{w_2}{z_{H2} + z_{SH2}}\right)$$

$$i_{H0} = -v_{pf}\left(\frac{z_2}{\Delta}\right)\left(\frac{z_0}{z_{X0} + w_0}\right)\left(\frac{w_0}{z_{H0} + z_{SH0}}\right)$$

$$i_{X1} = -v_{pf}\left(\frac{z_0 + z_2}{\Delta}\right)\left(\frac{z_1}{z_{X1} + w_1}\right)$$

$$i_{X2} = v_{pf}\left(\frac{z_0}{\Delta}\right)\left(\frac{z_2}{z_{X2} + w_2}\right) \tag{6.61}$$

$$i_{X0} = v_{pf}\left(\frac{z_2}{\Delta}\right)\left(\frac{z_0}{z_{X0} + w_0}\right)$$

$$i_{Y1} = v_{pf}\left(\frac{z_0 + z_2}{\Delta}\right)\left(\frac{z_1}{z_{X1} + w_1}\right)\left(\frac{w_1}{z_{Y1} + z_{SY1}}\right)$$

$$i_{Y2} = -v_{pf}\left(\frac{z_0}{\Delta}\right)\left(\frac{z_2}{z_{X2} + w_2}\right)\left(\frac{w_2}{z_{Y2} + z_{SY2}}\right)$$

$$i_{Y0} = -v_{pf}\left(\frac{z_2}{\Delta}\right)\left(\frac{z_0}{z_{X0} + w_0}\right)\left(\frac{w_0}{z_{Y0} + z_{SY0}}\right)$$

where $\Delta = z_0 z_1 + z_0 z_2 + z_1 z_2$.

For the system not attached beyond the fault point, Equation 6.61 becomes, using (6.45),

$$i_{H1} = v_{pf}\left(\frac{z_0 + z_2}{\Delta}\right)\left(\frac{w_1}{z_{H1} + z_{SH1}}\right)$$

$$i_{H2} = -v_{pf}\left(\frac{z_0}{\Delta}\right)\left(\frac{w_2}{z_{H2} + z_{SH2}}\right)$$

$$i_{H0} = -v_{pf}\left(\frac{z_2}{\Delta}\right)\left(\frac{w_0}{z_{H0} + z_{SH0}}\right)$$

$$i_{X1} = -v_{pf}\left(\frac{z_0 + z_2}{\Delta}\right), \quad i_{X2} = v_{pf}\left(\frac{z_0}{\Delta}\right), \quad i_{X0} = v_{pf}\left(\frac{z_2}{\Delta}\right) \tag{6.62}$$

$$i_{Y1} = v_{pf}\left(\frac{z_0 + z_2}{\Delta}\right)\left(\frac{w_1}{z_{Y1} + z_{SY1}}\right)$$

$$i_{Y2} = -v_{pf}\left(\frac{z_0}{\Delta}\right)\left(\frac{w_2}{z_{Y2} + z_{SY2}}\right)$$

$$i_{Y0} = -v_{pf}\left(\frac{z_2}{\Delta}\right)\left(\frac{w_0}{z_{Y0} + z_{SY0}}\right)$$

The phase currents, for the case where the system is connected beyond the fault point and assuming equal positive- and negative-sequence impedances are obtained from (6.61) and (6.1) applied to currents:

$$i_{Ha} = \frac{v_{pf}}{2z_0 + z_1}\left[\left(\frac{z_1}{z_{X1} + w_1}\right)\left(\frac{w_1}{z_{H1} + z_{SH1}}\right) - \left(\frac{z_0}{z_{X0} + w_0}\right)\left(\frac{w_0}{z_{H0} + z_{SH0}}\right)\right]$$

$$i_{Hb} = \frac{v_{pf}}{2z_0 + z_1}\left[\left(\alpha^2 - j\sqrt{3}\,\frac{z_0}{z_1}\right)\left(\frac{z_1}{z_{X1} + w_1}\right)\left(\frac{w_1}{z_{H1} + z_{SH1}}\right)\right.$$

$$\left. - \left(\frac{z_0}{z_{X0} + w_0}\right)\left(\frac{w_0}{z_{H0} + z_{SH0}}\right)\right]$$

$$i_{Hc} = \frac{v_{pf}}{2z_0 + z_1}\left[\left(\alpha + j\sqrt{3}\,\frac{z_0}{z_1}\right)\left(\frac{z_1}{z_{X1} + w_1}\right)\left(\frac{w_1}{z_{H1} + z_{SH1}}\right)\right.$$

$$\left. - \left(\frac{z_0}{z_{X0} + w_0}\right)\left(\frac{w_0}{z_{H0} + z_{SH0}}\right)\right]$$

$$i_{Xa} = -\frac{v_{pf}}{2z_0 + z_1}\left[\left(\frac{z_1}{z_{X1} + w_1}\right) - \left(\frac{z_0}{z_{X0} + w_0}\right)\right]$$

$$i_{Xb} = -\frac{v_{pf}}{2z_0 + z_1}\left[\left(\alpha^2 - j\sqrt{3}\,\frac{z_0}{z_1}\right)\left(\frac{z_1}{z_{X1} + w_1}\right) - \left(\frac{z_0}{z_{X0} + w_0}\right)\right] \qquad (6.63)$$

$$i_{Xc} = -\frac{v_{pf}}{2z_0 + z_1}\left[\left(\alpha + j\sqrt{3}\,\frac{z_0}{z_1}\right)\left(\frac{z_1}{z_{X1} + w_1}\right) - \left(\frac{z_0}{z_{X0} + w_0}\right)\right]$$

$$i_{Ya} = \frac{v_{pf}}{2z_0 + z_1}\left[\left(\frac{z_1}{z_{X1} + w_1}\right)\left(\frac{w_1}{z_{Y1} + z_{SY1}}\right) - \left(\frac{z_0}{z_{X0} + w_0}\right)\left(\frac{w_0}{z_{Y0} + z_{SY0}}\right)\right]$$

$$i_{Yb} = \frac{v_{pf}}{2z_0 + z_1}\left[\left(\alpha^2 - j\sqrt{3}\,\frac{z_0}{z_1}\right)\left(\frac{z_1}{z_{X1} + w_1}\right)\left(\frac{w_1}{z_{Y1} + z_{SY1}}\right)\right.$$

$$\left. - \left(\frac{z_0}{z_{X0} + w_0}\right)\left(\frac{w_0}{z_{Y0} + z_{SY0}}\right)\right]$$

$$i_{Yc} = \frac{v_{pf}}{2z_0 + z_1}\left[\left(\alpha + j\sqrt{3}\,\frac{z_0}{z_1}\right)\left(\frac{z_1}{z_{X1} + w_1}\right)\left(\frac{w_1}{z_{Y1} + z_{SY1}}\right)\right.$$

$$\left. - \left(\frac{z_0}{z_{X0} + w_0}\right)\left(\frac{w_0}{z_{Y0} + z_{SY0}}\right)\right]$$

For the case where the system is not connected beyond the fault point, Equation 6.63 becomes, using (6.45),

$$i_{Ha} = \frac{V_{pf}}{2z_0 + z_1}\left[\left(\frac{w_1}{z_{H1} + z_{SH1}}\right) - \left(\frac{w_0}{z_{H0} + z_{SH0}}\right)\right]$$

$$i_{Hb} = \frac{V_{pf}}{2z_0 + z_1}\left[\left(\alpha^2 - j\sqrt{3}\,\frac{z_0}{z_1}\right)\left(\frac{w_1}{z_{H1} + z_{SH1}}\right) - \left(\frac{w_0}{z_{H0} + z_{SH0}}\right)\right]$$

$$i_{Hc} = \frac{V_{pf}}{2z_0 + z_1}\left[\left(\alpha + j\sqrt{3}\,\frac{z_0}{z_1}\right)\left(\frac{w_1}{z_{H1} + z_{SH1}}\right) - \left(\frac{w_0}{z_{H0} + z_{SH0}}\right)\right]$$

$$i_{Xa} = 0$$

$$i_{Xb} = -\frac{V_{pf}}{2z_0 + z_1}\left[\left(\alpha^2 - j\sqrt{3}\,\frac{z_0}{z_1}\right) - 1\right]$$

$$i_{Xc} = -\frac{V_{pf}}{2z_0 + z_1}\left[\left(\alpha + j\sqrt{3}\,\frac{z_0}{z_1}\right) - 1\right] \qquad (6.64)$$

$$i_{Ya} = \frac{V_{pf}}{2z_0 + z_1}\left[\left(\frac{w_1}{z_{Y1} + z_{SY1}}\right) - \left(\frac{w_0}{z_{Y0} + z_{SY0}}\right)\right]$$

$$i_{Yb} = \frac{V_{pf}}{2z_0 + z_1}\left[\left(\alpha^2 - j\sqrt{3}\,\frac{z_0}{z_1}\right)\left(\frac{w_1}{z_{Y1} + z_{SY1}}\right) - \left(\frac{w_0}{z_{Y0} + z_{SY0}}\right)\right]$$

$$i_{Yc} = \frac{V_{pf}}{2z_0 + z_1}\left[\left(\alpha + j\sqrt{3}\,\frac{z_0}{z_1}\right)\left(\frac{w_1}{z_{Y1} + z_{SY1}}\right) - \left(\frac{w_0}{z_{Y0} + z_{SY0}}\right)\right]$$

6.4.5 Zero-Sequence Circuits

Figure 6.7 lists some examples of zero-sequence circuits for 3-terminal transformers. An infinite impedance is represented by a break in the circuit. When substituting into the preceding formulas, a large enough value needs to be used. Although there are many more possibilities than shown in Figure 6.7, they can serve to illustrate the method for accounting for the different 3-phase connections. The zero-sequence circuit in Figure 6.6c represents a transformer with all grounded Y terminal connections.

In Figure 6.7a, we have $z_{X0} \to \infty$, so that no zero-sequence current flows into the transformer. This can also be seen from the formulas in (6.49).

In Figure 6.7b, we see that $z_{H0} \to \infty$ since no zero-sequence current flows into an ungrounded Y-connected winding. This implies that $w_0 = z_{Y0} + z_{SY0}$, that is, the parallel combination of the H and Y impedances is replaced by the Y impedances.

In Figure 6.7c, no zero-sequence current flows into or out of the terminals of a delta winding so the delta terminal system impedance must be set to zero. This, however, allows current to circulate within the delta.

In Figure 6.7d, we see that zero-sequence current flows in all the windings but not out of the HV delta winding terminals. We just need to set $z_{SH0} = 0$ in all the formulas. This only affects w_0.

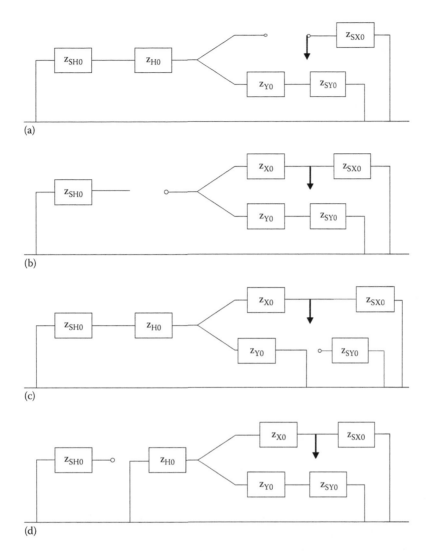

FIGURE 6.7
Some examples of zero-sequence circuit diagrams for 3-terminal transformers. The arrow indicates the fault point. (a) $Y_g/Y_u/Y_g$, (b) $Y_u/Y_g/Y_g$, (c) Y_g/Y_g/delta, and (d) delta/Y_g/Y_g. Y_g = grounded Y, Y_u = ungrounded Y.

6.4.6 Numerical Example

At this point, let us calculate the fault currents for a single line-to-ground fault on the X terminal. This is a common type of fault. Consider a Grounded-Y (H), Grounded-Y (X), buried Delta (Y) transformer. The transformer has a 3-phase MVA = 45 (15 MVA/phase). The line-to-line terminal voltages are H: 125 kV, X: 20 kV, Y: 10 kV. The base winding currents are, therefore, $I_{bH} = 15/72.17 \times 10^3 = 207.8$ amps, $I_{bX} = 15/11.55 \times 10^3 = 1299$ amps, and $I_{bY} = 15/10 \times 10^3 = 1500$ amps.

The winding-to-winding leakage reactances are given in Table 6.1.

TABLE 6.1

Leakage Reactances for a Y_g, Y_g; Delta Transformer

Winding 1	Winding 2	Pos/Neg Leakage Reactance (%)		Zero Seq Leakage Reactance (%)	
H	X	12	z_{HX}	10.2	z_{HX0}
H	Y	25	z_{HY}	21.25	z_{HY0}
X	Y	15	z_{XY}	12.75	z_{XY0}

From these, we calculate the single winding leakage reactances as

$$z_H = \frac{1}{2}\left(z_{HX} + z_{HY} - z_{XY}\right) = 11\%$$

$$z_X = \frac{1}{2}\left(z_{HX} + z_{XY} - z_{HY}\right) = 1\%$$

$$z_Y = \frac{1}{2}\left(z_{HY} + z_{XY} - z_{HX}\right) = 14\%$$

$$z_{H0} = \frac{1}{2}\left(z_{HX0} + z_{HY0} - z_{XY0}\right) = 9.35\%$$

$$z_{X0} = \frac{1}{2}\left(z_{HX0} + z_{XY0} - z_{HY0}\right) = 0.85\%$$

$$z_{Y0} = \frac{1}{2}\left(z_{HY0} + z_{XY0} - z_{HX0}\right) = 11.9\%$$

We are dropping the subscript 1 in these calculations for the positive/negative reactances. We are assuming that the zero-sequence reactances are 0.85 times the positive/negative reactances. This factor can vary with the winding connection and other factors and is usually determined by experience. The zero-sequence reactances can also be measured.

Because the Y winding is a buried delta, no positive- or negative-sequence currents can flow into the terminals of this winding. Therefore, it is necessary to set z_{SY} to a large value in the formulas. Zero sequence current can circulate around the delta, so z_{SY0} is set to zero.

We do not want to consider system impedances in our calculations. However, if we consider the system connected beyond the fault point, we need to set them to very small values as discussed previously. Considering the system not attached beyond the fault point amounts to having the faulted terminal unloaded before the fault occurs. We will do the calculation both ways.

We will assume system impedances of

$$z_{SH} = z_{SX} = 0.001\%, \quad z_{SY} = 1,000,000\%$$

$$z_{SH0} = z_{SX0} = 0.002\%, \quad z_{SY0} = 0\%$$

Again, the large value for z_{SY} and zero value for z_{SY0} are necessary because it is a buried delta. We are assuming here that the system zero sequence reactances are a factor of 2 times the system positive/negative sequence reactances. This is a reasonable assumption for transmission lines since these constitute a major part of the system [Ste62a]. Using these values, we obtain, from (6.44)

$$w_1 = 11.00088\%, \quad w_0 = 5.23663\%$$

For the system connected beyond the fault point, we get from (6.44)

$$z_1 = 0.001\%, \quad z_0 = 0.002\%$$

and for the system not connected beyond the fault point we obtain from (6.45)

$$z_1 = 12.00088\%, \quad z_0 = 6.08663\%$$

Using these values, along with the impedances in Table 6.1, we can calculate the per-unit phase and phase currents in the transformer from (6.53) through (6.56) for the two cases. Since we are working with percentages, we need to set $v_{pf} = 100\%$. These currents are given in Table 6.2 for the two cases.

The per-unit values in Table 6.2 are not percentages. They should sum to zero for each phase. These values would differ somewhat if different multiplying factors were used to get the zero-sequence winding impedances.

The currents are somewhat lower for the system not connected beyond the fault point compared with the currents for system connected beyond the fault point. In fact, the delta winding current is significantly lower. These results do not change much as the system impedances increase towards their rated values in the 1% range. Thus, the discontinuity in the currents between the two cases remains as the system sequence impedances increase. If the system is considered connected while its impedance has its rated value, it would appear awkward to consider the system disconnected when the system impedances approach zero.

TABLE 6.2

Per-Unit and Phase Currents Compared for System Connected or Not Connected beyond the Fault Point and Small System Impedances

	Phase a	Phase b	Phase c
System connected beyond fault			
Per-unit currents			
H	8.766	2.516	2.516
X	−12.380	−6.130	−6.130
Y	3.614	3.614	3.614
Phase currents			
H	1822	523	523
X	−16,082	−7,963	−7,963
Y	5422	5422	5422
System not connected beyond fault			
Per-unit currents			
H	8.508	−1.463	−1.463
X	−9.971	0	0
Y	1.463	1.463	1.463
Phase currents			
H	1768	−304	−304
X	−12,952	0	0
Y	2194	2194	2194

These currents do not include the asymmetry factor, which accounts for initial transient effects when the fault occurs. This is discussed in the next section.

6.5 Asymmetry Factor

A factor multiplying the currents calculated above is necessary to account for a transient overshoot when the fault occurs. This factor, called the asymmetry factor has been discussed in Chapter 2. It is given by

$$K = \sqrt{2}\left\{1 + \exp\left[-\left(\phi + \frac{\pi}{2}\right)\frac{r}{x}\right]\sin\phi\right\} \tag{6.65}$$

where
  x is the reactance looking into the terminal
  r is the resistance
  $\phi = \tan^{-1}(x/r)$ in radians

Usually, the system impedances are ignored when calculating these quantities so that for a 2 terminal unit,

$$\frac{x}{r} = \frac{\text{Im}(z_{HX})}{\text{Re}(z_{HX})} \tag{6.66}$$

while for a 3-terminal unit with a fault on the X terminal,

$$x = \text{Im}(z_X) + \frac{\text{Im}(z_H)\text{Im}(z_Y)}{\text{Im}(z_H) + \text{Im}(z_Y)}, \quad r = \text{Re}(z_X) + \frac{\text{Re}(z_H)\text{Re}(z_Y)}{\text{Re}(z_H) + \text{Re}(z_Y)} \tag{6.67}$$

with corresponding expressions for the other terminals should the fault be on them. When K in (6.65) multiplies the rms short-circuit current, it yields the maximum peak short-circuit current.

7

Phase-Shifting and Zigzag Transformers

7.1 Introduction

Phase-shifting transformers are used in power systems to help control power flow and line losses. They shift the input voltages and currents by an angle, which can be adjusted by means of a tap changer. They operate by adding a voltage at ±90° to the input voltage, that is, in quadrature. For 3-phase transformers, the quadrature voltage to be added to a given phase voltage can be derived by interconnecting the other phases. The many ways of doing this give rise to a large number of configurations for these transformers. We will deal with only a few common types here. Phase shifting capability can be combined with voltage magnitude control in the same transformer. This results in a more complex unit, involving two sets of tap changers, which we do not discuss here.

An example of their use can be seen by considering the feeding of a common load from two voltage sources, which could be out of phase with each other as shown in Figure 7.1. Solving for the currents, we get

$$I_1 = V_1 \frac{(Z_2 + Z_L)}{K} - V_2 \frac{Z_L}{K}, \quad I_2 = V_2 \frac{(Z_1 + Z_L)}{K} - V_1 \frac{Z_L}{K} \tag{7.1}$$

where $K = Z_1 Z_2 + Z_1 Z_L + Z_2 Z_L$. The current into the load, I_L, is

$$I_L = I_1 + I_2 = V_1 \frac{Z_2}{K} + V_2 \frac{Z_1}{K} \tag{7.2}$$

and the voltage across the load, V_L, is

$$V_L = V_1 - I_1 Z_1 = \frac{Z_L}{K} (V_1 Z_2 + V_2 Z_1) \tag{7.3}$$

Thus, the complex power delivered to the load is

$$V_L I_L^* = \frac{Z_L}{|K|^2} |V_1 Z_2 + V_2 Z_1|^2 \tag{7.4}$$

where * denotes complex conjugation.

Consider the case where $Z_1 = Z_2 = Z$, $V_1 = V$, $V_2 = Ve^{j\theta}$. Then (7.4) becomes

$$V_L I_L^* = \frac{2Z_L |Z|^2 V^2}{|K|^2} (1 + \cos\theta) \tag{7.5}$$

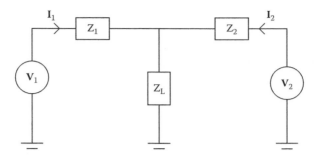

FIGURE 7.1
Two possibly out-of-phase sources feeding a common load.

Thus, it can be seen that the maximum power is transferred when $\theta = 0$. In this case, a-phase-shifting transformer could be used to adjust the phase of $\mathbf{V}_2$ so that it equals that of $\mathbf{V}_1$.

In modern power systems, which are becoming more and more interconnected, the need for these devices is growing. Other methods for introducing a quadrature voltage are being developed, for instance by means of power electronic circuits in conjunction with ac–dc converters. These can act much faster than on-load tap changers. However, at present they are more costly than phase-shifting transformers and are used primarily when response time is important. (Electronic on-load tap changers are also being developed for fast response time applications.)

In this chapter, we develop a circuit model description for three common types of phase-shifting transformer. This is useful in order to understand the regulation behavior of such devices, that is, how much the output voltage magnitude and phase change when the unit is loaded as compared with the no-load voltage output. In the process, we also find how the phase angle depends on tap position, a relationship that can be nonlinear. In addition to the positive sequence circuit model, which describes normal operation, we also determine the negative and zero sequence circuit models for use in short-circuit fault current analysis. The latter analysis is carried out for the standard types of fault for two of the phase-shifting transformers.

Very little has been published on this subject in the open literature beyond general interconnection diagrams and how they are used in specific power grids. However, two references that emphasize basic principles are [Hob39], [Cle39]. Other useful references are [Kra98], [Wes64].

7.2 Basic Principles

The basic principles have been discussed in earlier chapters, but we repeat some of them here for easier reference. We neglect exciting current and model the individual phases in terms of their leakage impedances. For a 2-winding phase, we use the circuit model shown in Figure 7.2. Z_{12} is the 2-winding leakage impedance referred to in side 1. Most of the development described here is carried out in terms of impedances in ohms. Because of differences in per-unit bases for the input and winding quantities, per-unit quantities are not as convenient in the analysis. At the appropriate place, we indicate where per-unit quantities might prove useful. The currents are assumed to flow into their respective windings.

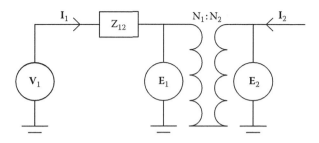

FIGURE 7.2
Model of a 2-winding transformer phase.

With N_1 equal to the number of turns on side 1, and N_2 the number of turns on side 2, the ideal transformer voltages satisfy

$$\frac{E_1}{E_2} = \frac{N_1}{N_2} \tag{7.6}$$

and the currents satisfy

$$\frac{I_1}{I_2} = -\frac{N_2}{N_1} \tag{7.7}$$

For 3 windings per phase, we use the model shown in Figure 7.3. In this case, the ideal transformer voltages satisfy

$$\frac{E_1}{N_1} = \frac{E_2}{N_2} = \frac{E_3}{N_3} \tag{7.8}$$

and the currents satisfy

$$N_1 I_1 + N_2 I_2 + N_3 I_3 = 0 \tag{7.9}$$

The single-winding impedances are given in terms of the 2-winding impedances by

$$Z_1 = \frac{1}{2}\left[Z_{12} + Z_{13} - \left(\frac{N_1}{N_2}\right)^2 Z_{23} \right]$$

$$Z_2 = \frac{1}{2}\left(\frac{N_2}{N_1}\right)^2 \left[Z_{12} + \left(\frac{N_1}{N_2}\right)^2 Z_{23} - Z_{13} \right] \tag{7.10}$$

$$Z_3 = \frac{1}{2}\left(\frac{N_3}{N_1}\right)^2 \left[Z_{13} + \left(\frac{N_1}{N_2}\right)^2 Z_{23} - Z_{12} \right]$$

Here, the 2-winding impedances are referred to the winding corresponding to the first subscript, and the second subscript refers to the winding, which would be shorted when measuring the impedance. To refer impedances to the opposite winding, use

$$Z_{ij} = \left(\frac{N_i}{N_j}\right)^2 Z_{ji} \tag{7.11}$$

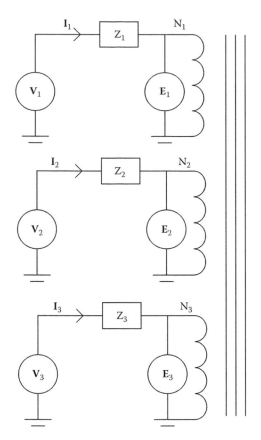

FIGURE 7.3
Model of a 3-winding transformer phase. The vertical lines connecting the circuits represent the common core.

For a 3-phase system, the positive sequence quantities correspond to the ordering of the unit phasors shown in Figure 7.4. Letting

$$\alpha = e^{j120°} = -\frac{1}{2} + j\frac{\sqrt{3}}{2} \qquad (7.12)$$

the ordering is $1, \alpha^2, \alpha$. This corresponds to the phase ordering a, b, c for the windings and 1, 2, 3 for the terminals. Negative sequence ordering is $1, \alpha, \alpha^2$, which corresponds to the phase ordering a, c, b or 1, 3, 2. This is obtained by interchanging two phases. Note that

$$\alpha^2 = e^{j240°} = e^{-j120°} = \alpha^* = -\frac{1}{2} - j\frac{\sqrt{3}}{2} \qquad (7.13)$$

Zero sequence quantities are all in phase with each other.

Our interconnections will be chosen to produce a positive phase shift at a positive tap setting. By interchanging two phases, a negative phase shift can be produced at a positive tap setting. Interchanging two phases is equivalent to inputting a negative sequence set of voltages. This implies that negative sequence circuits have the opposite phase shift to positive sequence circuits. Zero sequence circuits have zero phase shift. By positive phase shift, we mean that output voltages and currents lead input voltages and currents, that is, output quantities are rotated counterclockwise on a phasor diagram relative to input quantities.

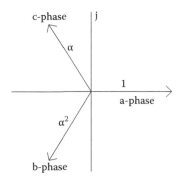

FIGURE 7.4
Positive sequence unit phasors.

7.3 Squashed Delta-Phase-Shifting Transformer

One of the simplest phase shifters to analyze is the squashed delta configuration shown in Figure 7.5. In the figure, S labels the source or input quantities and L the load or output quantities. The input and output set of voltages and corresponding currents form a balanced positive sequence set. In this study, we are taking the S currents as positive into the terminals and the L currents as positive out of the terminals. The input and output voltage phasor diagram is shown in Figure 7.5b. These are voltages to ground. The currents form a similar set but are not shown. Similarly, the internal voltages and corresponding currents form a positive sequence set as shown in Figure 7.5c for the voltages. Note that Figure 7.5b and c could be rotated relative to each other if shown on a common phasor diagram. a and a', etc., refer to windings on the same leg, with the prime labeling the tapped winding. The 2-winding impedances will be referred to the unprimed coil, that is, $Z_{aa'}$, etc. Since these impedances are all the same for the different phases, we need only this one symbol. This transformer can be designed with a single 3-phase core and is referred to generically as a single core design.

Using the 2-winding-per-phase model described in the last section, adapted to the present labeling scheme, and concentrating on one input–output pair, we can write

$$\mathbf{V}_{S1} - \mathbf{V}_{L1} = \mathbf{E}_{a'}, \quad \mathbf{V}_{S1} - \mathbf{V}_{L2} = \mathbf{I}_c Z_{aa'} + \mathbf{E}_c$$
$$\mathbf{I}_{S1} = \mathbf{I}_c + \mathbf{I}_{a'}, \quad \mathbf{I}_{L1} = \mathbf{I}_{a'} + \mathbf{I}_b \tag{7.14}$$

In these equations, the 2-winding leakage impedance is assigned to the unprimed side of the transformer. We also have the transformer relations

$$\mathbf{E}_{a'} = \frac{N_{a'}}{N_a} \mathbf{E}_a = \frac{1}{n} \mathbf{E}_a = \frac{e^{-j120°}}{n} \mathbf{E}_c$$
$$\mathbf{I}_{a'} = -\frac{N_a}{N_{a'}} \mathbf{I}_a = -n\mathbf{I}_a = -ne^{-j120°} \mathbf{I}_c \tag{7.15}$$

where we have defined $n = N_a/N_{a'}$, the turns ratio. We also have the following relationships:

$$\mathbf{V}_{S1} - \mathbf{V}_{S2} = \left(1 - \alpha^2\right)\mathbf{V}_{S1} = \sqrt{3}e^{j30°}\mathbf{V}_{S1}$$
$$\mathbf{V}_{S2} - \mathbf{V}_{L2} = \mathbf{E}_{b'} = e^{-j120°}\mathbf{E}_{a'} = \frac{e^{j120°}}{n}\mathbf{E}_c \tag{7.16}$$

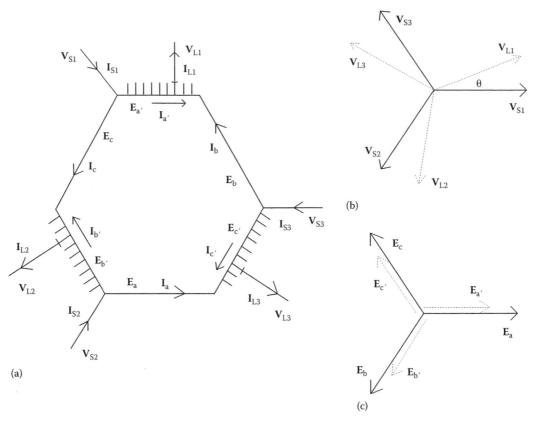

FIGURE 7.5
Squashed delta configuration with phase quantities labeled. The ideal transformer voltages, E, increase in the opposite direction to the assumed current flow. θ is a positive phase shift angle. (a) Circuit, (b) input and output phasors, and (c) internal phasors.

It is worthwhile going into the details of the solution of these equations since we will need some of the intermediate results later. We assume that $\mathbf{V}_{S1}$ and $\mathbf{I}_{S1}$ are given. From (7.14) and (7.15), we obtain

$$\mathbf{I}_c = \frac{1}{1 - ne^{-j120°}} \mathbf{I}_{S1}, \quad \mathbf{I}_{a'} = -\frac{ne^{-j120°}}{1 - ne^{-j120°}} \mathbf{I}_{S1} \tag{7.17}$$

We also have

$$\mathbf{I}_{L1} = \mathbf{I}_{a'} + \mathbf{I}_b = \mathbf{I}_{a'} + e^{j120°}\mathbf{I}_c = \left(\frac{n - e^{-j120°}}{n - e^{-j120°}} \right)\mathbf{I}_{S1} = e^{j\theta}\mathbf{I}_{S1}$$

where $\qquad\qquad\qquad\qquad\qquad\qquad\qquad\qquad\qquad\qquad\qquad\qquad\qquad$ (7.18)

$$\theta = 2\tan^{-1}\left(\frac{\sqrt{3}}{2n+1} \right)$$

Since n is determined by the tap position, this shows that θ is a nonlinear function of the tap position, assuming the taps are evenly spaced.

Adding Equations 7.16 and using (7.14),

$$\mathbf{V}_{S1} - \mathbf{V}_{L2} = \sqrt{3}e^{j30°}\mathbf{V}_{S1} + \frac{e^{j120°}}{n}\mathbf{E}_c = \mathbf{I}_c\mathbf{Z}_{aa'} + \mathbf{E}_c$$

Solving for E_c, using (7.17),

$$\mathbf{E}_c = \frac{\sqrt{3}\,ne^{j30°^*}}{n-e^{j120°}}\mathbf{V}_{S1} + \frac{ne^{j120°}\mathbf{Z}_{aa'}}{\left(n-e^{j120°}\right)^2}\mathbf{I}_{S1} \tag{7.19}$$

Combining with the first of Equations 7.14 and 7.15, we obtain

$$\mathbf{V}_{L1} = \left[\mathbf{V}_{S1} - \left(\frac{\mathbf{Z}_{aa'}}{n^2+n+1}\right)\mathbf{I}_{S1}\right]e^{j\theta} \tag{7.20}$$

where θ is given in (7.18). Although the current is shifted by θ in all cases, in general the voltage is shifted by θ only under no-load conditions.

The circuit model suggested by (7.20) is shown in Figure 7.6. This applies to all three phases with appropriate labeling. The equivalent impedance shown in Figure 7.6 is given by

$$\mathbf{Z}_{eq} = \frac{\mathbf{Z}_{aa'}}{n^2+n+1} \tag{7.21}$$

Figure 7.6 is a positive sequence circuit model. The negative sequence model is obtained simply by changing θ to $-\theta$ with Z_{eq} unchanged.

While the input power per phase is $P_{in} = \mathbf{V}_{S1}\mathbf{I}_{S1}^*$, the transformed or winding power per phase, P_{wdg}, is somewhat less. Ignoring impedance drops, we have from (7.17) and (7.19)

$$P_{wdg} = \mathbf{E}_c\mathbf{I}_c^* = j\left(\frac{n\sqrt{3}}{n^2+n+1}\right)P_{in} \tag{7.22}$$

where the factor of j shows that the transformed power is in quadrature with the input power. As a numerical example, let $\theta = 30°$. Then, from (7.18) and (7.22),

$$\tan\left(\frac{\theta}{2}\right) = \frac{\sqrt{3}}{2n+1} \quad \Rightarrow \quad n = 2.732 \quad \Rightarrow \quad P_{wdg} = j0.4227P_{in} \tag{7.23}$$

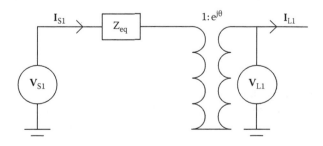

FIGURE 7.6
Circuit model of one phase of a-phase-shifting transformer. This is for positive sequence. For negative sequence, change θ to $-\theta$.

We can express (7.20) in per-unit terms; however, because of the difference between the input power and voltage base and that of the windings, we must be careful to specify the base used. As usual, the power base is taken to be the terminal power base. This is the terminal power per phase. Thus the rated input power base is $P_{b,in}$. The rated input voltage, $V_{b,in}$, and the rated input current, which can be derived from the power and voltage base, $I_{b,in} = P_{b,in}/V_{b,in}$. Similarly, the input impedance base can be derived from the power and voltage base, $Z_{b,in} = V_{b,in}/I_{b,in} = (V_{b,in})^2/P_{b,in}$. Because the transformation ratio is 1:1 in terms of magnitude, the output base values are the same as the input base values. Thus (7.20) can be written as

$$\frac{\mathbf{V}_{L1}}{V_{b,in}} = \left[\frac{\mathbf{V}_{S1}}{V_{b,in}} - Z_{eq}\frac{\mathbf{I}_{S1}}{I_{b,in}}\left(\frac{I_{b,in}}{V_{b,in}} \right) \right]e^{j\theta} = \left[\frac{\mathbf{V}_{S1}}{V_{b,in}} - \frac{Z_{eq}}{Z_{b,in}}\frac{\mathbf{I}_{S1}}{I_{b,in}} \right]e^{j\theta}$$

or

$$\mathbf{v}_{L1} = \left[\mathbf{v}_{S1} - z_{eq}\mathbf{i}_{S1} \right]e^{j\theta} \tag{7.24}$$

where small letters are used to denote per-unit quantities.

Assuming the rated input power per phase is the common power base, the winding impedance and input impedance bases are proportional to the square of the respective voltages magnitudes. Using (7.19), and ignoring impedance drops,

$$\frac{Z_{b,wdg}}{Z_{b,in}} = \left| \frac{\mathbf{E}_c}{\mathbf{V}_{S1}} \right|^2 = \frac{3n^2}{n^2 + n + 1} \tag{7.25}$$

Thus, z_{eq} can be expressed as, using (7.21),

$$z_{eq} = \frac{Z_{eq}}{Z_{b,in}} = \frac{1}{n^2 + n + 1}\left(\frac{Z_{b,wdg}}{Z_{b,in}} \right)\left(\frac{Z_{aa'}}{Z_{b,wdg}} \right) = \frac{3n^2}{\left(n^2 + n + 1\right)^2}z_{aa'} \tag{7.26}$$

where
   z_{eq} is on an input base
   $z_{aa'}$ is on a winding base

Note that the winding base depends on the turns ratio as indicated in (7.25). Thus, this base is perhaps most useful for design purposes when n refers to the maximum phase angle. At the other extreme, with $\theta = 0$, we have $n = \infty$, so that $Z_{eq} = z_{eq} = 0$. In this case, the input is directly connected to the output, bypassing the coils.

7.3.1 Zero Sequence Circuit Model

The zero sequence circuit model may be derived with reference to Figure 7.5, but assuming all quantities are zero sequence. Thus, Figure 7.5b and c should be replaced by diagrams with all phasors in parallel. Using a zero subscript for zero sequence quantities, we can write

$$\mathbf{V}_{S1,0} - \mathbf{V}_{L1,0} = \mathbf{E}_{a',0}, \quad \mathbf{V}_{S1,0} - \mathbf{V}_{L2,0} = \mathbf{I}_{c,0}Z_{aa',0} + \mathbf{E}_{c,0}$$
$$\mathbf{I}_{S1,0} = \mathbf{I}_{c,0} + \mathbf{I}_{a',0}, \quad \mathbf{I}_{L1,0} = \mathbf{I}_{a',0} + \mathbf{I}_{b,0} \tag{7.27}$$

and

$$E_{a',0} = \frac{N_{a'}}{N_a} E_{a,0} = \frac{1}{n} E_{a,0} = \frac{1}{n} E_{c,0}$$

$$I_{a',0} = -\frac{N_a}{N_{a'}} I_{a,0} = -nI_{a,0} = -nI_{c,0}$$

(7.28)

Since $I_{b,0} = I_{c,0}$, Equation 7.27 shows that

$$I_{L1,0} = I_{S1,0}$$

(7.29)

so the current is not phase shifted. We also have

$$V_{S1,0} - V_{S2,0} = 0, \quad V_{S2,0} - V_{L2,0} = E_{b',0} = \frac{1}{n} E_{b,0} = \frac{1}{n} E_{c,0}$$

(7.30)

Solving the above zero sequence equations, we obtain

$$I_{c,0} = \frac{I_{S1,0}}{1-n}, \quad E_{c,0} = \frac{n}{1-n} I_{c,0} Z_{aa',0} = \frac{n}{(1-n)^2} I_{S1,0} Z_{aa',0}$$

(7.31)

and

$$V_{L1,0} = V_{S1,0} - \frac{Z_{aa',0}}{(1-n)^2} I_{S1,0}$$

(7.32)

The circuit model for this last equation is shown in Figure 7.7, where we have defined

$$Z_{eq,0} = \frac{Z_{aa',0}}{(1-n)^2}$$

(7.33)

This has a different dependence on n from the positive sequence circuit. It becomes infinite when n = 1. This is reasonable, because from (7.28) $I_{a',0} = -I_{c,0}$ when n = 1 so that from (7.27) $I_{S1,0} = 0$. Thus, no zero sequence current can flow into the squashed delta transformer when n = 1. Internal current can, however, circulate around the delta. From (7.18), $\theta = 60°$ when n = 1.

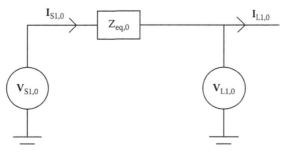

FIGURE 7.7
Zero sequence circuit model of one phase of a squashed delta-phase-shifting transformer.

Equation 7.32 can be written in per-unit terms referring to the same input base as used for positive sequence.

$$\mathbf{v}_{L1,0} = \mathbf{v}_{S1,0} - z_{eq,0}\mathbf{i}_{S1,0} \tag{7.34}$$

Using the positive sequence winding base for the 2-winding zero sequence impedance, $Z_{aa',0}$ we can write

$$z_{eq,0} = \frac{3n^2}{\left(n^2+n+1\right)\left(1-n\right)^2} Z_{aa',0} \tag{7.35}$$

We postpone a discussion of regulation effects and short-circuit current calculations until other types of phase-shifting transformers are treated. This is because the positive, negative, and zero sequence circuit diagrams for all the cases treated will have the same appearance, although Z_{eq} or z_{eq} and θ will differ among the various types.

7.4 Standard Delta-Phase-Shifting Transformer

As opposed to the squashed delta design, the standard delta design utilizes a regular delta winding but is still a single 3-phase core design. The connection diagram is given in Figure 7.8. The two tap windings are on the same core phase as the corresponding parallel delta winding in the figure. The taps are symmetrically placed with respect to the point of contact at the delta vertex. This assures that there is no change in current or no-load voltage magnitude from input to output. It also means that $\mathbf{E}_{a'} = \mathbf{E}_{a''}$, etc., for the other phases. Each phase consists of three windings, the two tap windings and the delta winding opposite and parallel in the figure, where primes and double primes are used to distinguish them. Thus a 3-winding-per-phase model is needed. Z_a, $Z_{a'}$, $Z_{a''}$ will be used to label the single-winding impedances for phase a. Since all the phases have equivalent impedances by symmetry, these same designations will be used for the other phase impedances as well. The same remarks apply to the phasor diagrams as for the squashed delta case. Although not shown, the current phasor diagrams have the same sequence order as their corresponding voltage phasor diagrams; however, the two diagrams could be rotated relative to each other.

In Figure 7.8a, $\mathbf{V}_1$, $\mathbf{V}_2$, $\mathbf{V}_3$ designate the phasor voltages to ground at the delta vertices. Thus we have, using $\mathbf{E}_{a'} = \mathbf{E}_{a''}$, etc. and $\mathbf{I}_{a'} = \mathbf{I}_{S1}$, $\mathbf{I}_{a''} = \mathbf{I}_{L1}$, etc.,

$$\mathbf{V}_{S1} - \mathbf{V}_1 = \mathbf{I}_{S1}Z_{a'} + \mathbf{E}_{a'}, \quad \mathbf{V}_1 - \mathbf{V}_{L1} = \mathbf{I}_{L1}Z_{a''} + \mathbf{E}_{a'}$$

$$\mathbf{V}_2 - \mathbf{V}_3 = \left(\alpha^2 - \alpha\right)\mathbf{V}_1 = -j\sqrt{3}\mathbf{V}_1 = \mathbf{I}_a Z_a + \mathbf{E}_a \tag{7.36}$$

$$\mathbf{I}_{S1} - \mathbf{I}_{L1} = \mathbf{I}_c - \mathbf{I}_b = \left(\alpha - \alpha^2\right)\mathbf{I}_a = j\sqrt{3}\mathbf{I}_a$$

In addition, we have the transformer relations

$$\frac{\mathbf{E}_a}{\mathbf{E}_{a'}} = \frac{N_a}{N_{a'}} = n$$

$$N_a\mathbf{I}_a + N_{a'}\mathbf{I}_{S1} + N_{a'}\mathbf{I}_{L1} = 0 \quad \text{or} \quad n\mathbf{I}_a + \mathbf{I}_{S1} + \mathbf{I}_{L1} = 0 \tag{7.37}$$

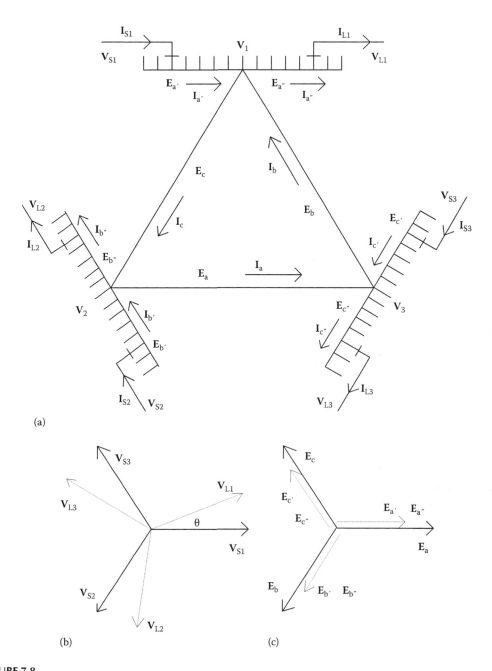

FIGURE 7.8
Standard delta configuration with phase quantities labeled. The ideal transformer voltages, **E**, increase in the opposite direction of the assumed current flow. (a) Circuit, (b) input and output phasors, and (c) internal phasors.

where we have defined the turns ratio n as the ratio between the turns in one of the delta windings to the turns in one of the tap windings. Both tap windings have the same number of turns. Solving for I_a in the last equation in (7.36) and inserting into the last equation in (7.37), we obtain

$$\mathbf{I}_{L1} = e^{j\theta}\mathbf{I}_{S1} \quad \text{with} \quad \theta = 2\tan^{-1}\left(\frac{\sqrt{3}}{n}\right) \tag{7.38}$$

We also have

$$\mathbf{I}_a = -\frac{2}{n\left(1-j\dfrac{\sqrt{3}}{n}\right)}\mathbf{I}_{S1} \tag{7.39}$$

From the these equations, we obtain

$$\mathbf{E}_a = -\frac{j\sqrt{3}}{\left(1-j\dfrac{\sqrt{3}}{n}\right)}\mathbf{V}_{S1} + \frac{j\sqrt{3}}{\left(1-j\dfrac{\sqrt{3}}{n}\right)}\left[Z_{a'} - \frac{2jZ_a}{\sqrt{3}n\left(1-j\dfrac{\sqrt{3}}{n}\right)}\right]\mathbf{I}_{S1} \tag{7.40}$$

and

$$\mathbf{V}_{L1} = \left[\mathbf{V}_{S1} - \left(Z_{a'} + Z_{a''} + \frac{4Z_a}{n^2+3}\right)\mathbf{I}_{S1}\right]e^{j\theta} \tag{7.41}$$

with θ as given in (7.38). θ is the no-load phase angle shift. This conforms to the circuit model shown in Figure 7.6 with

$$Z_{eq} = Z_{a'} + Z_{a''} + \frac{4Z_a}{n^2+3} \tag{7.42}$$

or in terms of 2-winding impedances, using (7.10),

$$Z_{eq} = Z_{a'a''} + \left(\frac{2}{n^2+3}\right)\left(Z_{aa'} + Z_{aa''} - n^2Z_{a'a''}\right) \tag{7.43}$$

This is the positive sequence circuit. As before, the negative sequence circuit is found by changing θ to $-\theta$ without change in Z_{eq}.

Let us again determine the winding power per phase, P_{wdg}, in terms of the input power per phase, $P_{in} = \mathbf{V}_{S1}\mathbf{I}_{S1}^*$. Ignoring impedance drops, we get

$$P_{wdg} = \mathbf{E}_a\mathbf{I}_a^* = j\left(\frac{2\sqrt{3}n}{n^2+3}\right)P_{in} \tag{7.44}$$

where j indicates that the transformed power is in quadrature with the input power. Using trigonometric identities, it can be shown that

$$P_{wdg} = j\sin\theta P_{in} \tag{7.45}$$

with θ as given in (7.38). Thus for $\theta = 30°$, $|P_{wdg}| = 0.5|P_{in}|$.

In per-unit terms, based on input quantities, Equation 7.41 can be cast in the form of (7.24). For this, we need to determine the relation between the winding bases and the input base for impedances. The 2-winding impedances given earlier are referred to either the a or a' winding. Their bases are related to the input base by keeping the power base the same as the input power per phase:

$$\frac{Z_{b,a\,wdg}}{Z_{b,in}} = \left|\frac{E_a}{V_{S1}}\right|^2 = \left(\frac{3n^2}{n^2+3}\right), \quad \frac{Z_{b,a'wdg}}{Z_{b,in}} = \left|\frac{E_{a'}}{V_{S1}}\right|^2 = \left(\frac{3}{n^2+3}\right) \tag{7.46}$$

Thus, in per-unit terms, Z_{eq} in (7.43) becomes

$$z_{eq} = \left(\frac{3n^2}{n^2+3}\right)\left[\frac{z_{a'a''}}{n^2} + \frac{2\left(z_{aa'} + z_{aa''} - z_{a'a''}\right)}{\left(n^2+3\right)}\right] \tag{7.47}$$

Again, this is primarily useful in design when n refers to the maximum phase angle. At zero phase shift where $n = \infty$, Equation 7.47 indicates that $z_{eq} = 0$. At this tap position, the tap windings are effectively out of the circuit and the input is directly connected to the output.

7.4.1 Zero Sequence Circuit Model

The zero sequence circuit model can be derived with reference to Figure 7.8 but with all phasors taken to be zero sequence. Rewriting (7.36) and (7.37) with this in mind and appending a zero subscript, we get

$$\mathbf{V}_{S1,0} - \mathbf{V}_{1,0} = \mathbf{I}_{S1,0}\mathbf{Z}_{a',0} + \mathbf{E}_{a',0}, \quad \mathbf{V}_{1,0} - \mathbf{V}_{L1,0} = \mathbf{I}_{L1,0}\mathbf{Z}_{a'',0} + \mathbf{E}_{a',0}$$
$$\mathbf{V}_{2,0} - \mathbf{V}_{3,0} = 0 = \mathbf{I}_{a,0}\mathbf{Z}_{a,0} + \mathbf{E}_{a,0}, \quad \mathbf{I}_{S1,0} - \mathbf{I}_{L1,0} = \mathbf{I}_{c,0} - \mathbf{I}_{b,0} = 0 \tag{7.48}$$
$$\frac{\mathbf{E}_{a,0}}{\mathbf{E}_{a',0}} = \frac{N_a}{N_{a'}} = n, \quad n\mathbf{I}_{a,0} + \mathbf{I}_{S1,0} + \mathbf{I}_{L1,0} = 0$$

Solving, we find

$$\mathbf{I}_{L1,0} = \mathbf{I}_{S1,0} \tag{7.49}$$

so that the zero sequence current undergoes no phase shift. We also have

$$\mathbf{E}_{a,0} = -\mathbf{I}_{a,0}\mathbf{Z}_{a,0}, \quad \mathbf{I}_{a,0} = -\frac{2}{n}\mathbf{I}_{S1,0} \tag{7.50}$$

and

$$\mathbf{V}_{L1,0} = \mathbf{V}_{S1,0} - \left(\mathbf{Z}_{a',0} + \mathbf{Z}_{a'',0} + \frac{4}{n^2}\mathbf{Z}_{a,0}\right)\mathbf{I}_{S1,0} \tag{7.51}$$

Thus the voltage undergoes no phase shift at no-load. The circuit model of Figure 7.7 applies with

$$Z_{eq} = Z_{a',0} + Z_{a'',0} + \frac{4}{n^2}Z_{a,0} \tag{7.52}$$

or, in terms of 2 winding impedances,

$$Z_{eq,0} = Z_{a'a'',0} + \frac{2}{n^2}\left(Z_{aa',0} + Z_{aa'',0} - n^2 Z_{a'a'',0}\right) \tag{7.53}$$

On a per-unit basis, using rated input quantities, Equation 7.51 has the same appearance as (7.34). The equivalent per-unit impedance is given by

$$z_{eq,0} = \left(\frac{3}{n^2 + 3}\right)\left[2\left(z_{aa',0} + z_{aa'',0}\right) - z_{a'a'',0}\right] \tag{7.54}$$

At zero phase shift, the output is directly connected to the input as was the case for positive sequence.

7.5 2-Core Phase-Shifting Transformer

For large power applications, phase shifters are often designed as two units, the series unit and the excitor unit, each having its own 3-phase core and associated coils. Depending on size, the two units can be inside the same tank or be housed in separate tanks. This construction is largely dictated by tap changer limitations. A commonly used circuit diagram is shown in Figure 7.9. The phasor diagrams refer to positive sequence quantities and although the phase ordering is consistent among the diagrams, their relative orientation is not specified. Currents have the same phase ordering as their associated voltages. A 3-winding model is needed for the series unit, while a 2-winding model applies to the excitor unit. We have used two different labeling schemes for the series and excitor units. Letter subscripts are used for series quantities (a, a', a''), and number subscripts for excitor quantities (1, 1'). Primes and double primes are used to distinguish different windings associated with the same phase. Note that the input–output coils in the figure are part of the series unit but are attached to the excitor unit at their midpoints. The input and output voltages are voltages to ground as before. We assume the input voltage and current phasors are given.

Following an analysis similar to that of the last section for the series unit, we can write, using $\mathbf{E}_{a'} = \mathbf{E}_{a''}$, etc.,

$$\mathbf{V}_{S1} - \mathbf{V}_1 = \mathbf{I}_{S1}Z_{a'} + \mathbf{E}_{a'}, \quad \mathbf{V}_1 - \mathbf{V}_{L1} = \mathbf{I}_{L1}Z_{a''} + \mathbf{E}_{a'}$$

$$\mathbf{E}_{2'} - \mathbf{E}_{3'} = \left(\alpha^2 - \alpha\right)\mathbf{E}_{1'} = -j\sqrt{3}\mathbf{E}_{1'} = \mathbf{I}_a Z_a + \mathbf{E}_a \tag{7.55}$$

$$\mathbf{V}_1 = \mathbf{I}_1 Z_{11'} + \mathbf{E}_1, \quad \mathbf{I}_1 = \mathbf{I}_{S1} - \mathbf{I}_{L1}, \quad \mathbf{I}_{2'} = \mathbf{I}_c - \mathbf{I}_a$$

We also have the transformer relations,

$$\frac{\mathbf{E}_a}{\mathbf{E}_{a'}} = \frac{N_a}{N_{a'}} = n_s, \quad \frac{\mathbf{E}_1}{\mathbf{E}_{1'}} = \frac{N_1}{N_{1'}} = n_e, \quad \frac{\mathbf{I}_{1'}}{\mathbf{I}_1} = -\frac{N_1}{N_{1'}} = -n_e$$

$$N_a \mathbf{I}_a + N_{a'} \mathbf{I}_{S1} + N_{a'} \mathbf{I}_{L1} = 0 \quad \text{or} \quad n_s \mathbf{I}_a + \mathbf{I}_{S1} + \mathbf{I}_{L1} = 0$$

where $\qquad\qquad\qquad\qquad\qquad\qquad\qquad\qquad\qquad\qquad\qquad\qquad$ (7.56)

$$n_s = \frac{N_a}{N_{a'}} \quad \text{and} \quad N_{a'} = N_{a''}$$

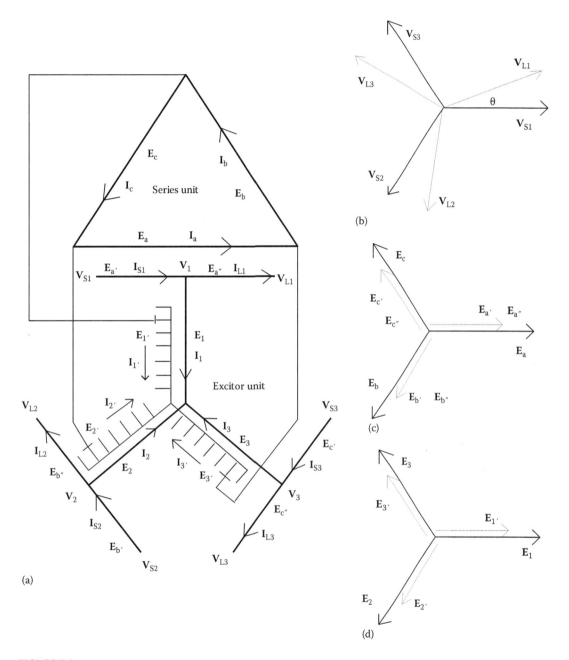

FIGURE 7.9
Circuit diagram of a 2-core phase-shifting transformer. The ideal transformer voltages, the E's, increase in the opposite direction to the assumed current flow. (a) Circuit, (b) input and output phasors, (c) internal phasors for series, and (d) internal phasors for excitor.

where we have defined the turns ratio of the series unit, n_s, as the ratio of the turns in a coil of the delta to the turns in the first or second half of the input–output winding, that is, from the input to the midpoint or from the output to the midpoint. We have also defined the excitor winding ratio, n_e, as the ratio of the turns in the winding connected to the midpoint of the input–output winding to the turns in the tapped winding. This latter ratio will depend on the tap position. From (7.55), (7.56), and the phasor diagrams, we obtain

$$\mathbf{I}_{2'} = e^{-j120°}\mathbf{I}_{1'} = -n_e e^{-j120°}\mathbf{I}_1 = \mathbf{I}_c - \mathbf{I}_a = (\alpha - 1)\mathbf{I}_a = -\sqrt{3}e^{-j30°}\mathbf{I}_a$$

which implies   (7.57)

$$\mathbf{I}_1 = j\frac{\sqrt{3}}{n_e}\mathbf{I}_a$$

Substituting $\mathbf{I}_1$ from (7.55) and $\mathbf{I}_a$ from (7.56) into (7.57), we get

$$\mathbf{I}_{L1} = \left[\frac{1+j\left(\dfrac{\sqrt{3}}{n_e n_s}\right)}{1-j\left(\dfrac{\sqrt{3}}{n_e n_s}\right)}\right]\mathbf{I}_{S1} = e^{j\theta}\mathbf{I}_{S1} \quad \text{where } \theta = 2\tan^{-1}\left(\frac{\sqrt{3}}{n_e n_s}\right) \quad (7.58)$$

Thus, in terms of the known input current, we can write, using (7.55) and (7.56),

$$\mathbf{I}_1 = -\frac{2j\left(\dfrac{\sqrt{3}}{n_e n_s}\right)}{\left[1-j\left(\dfrac{\sqrt{3}}{n_e n_s}\right)\right]}\mathbf{I}_{S1}, \quad \mathbf{I}_a = -\frac{2}{n_s\left[1-j\left(\dfrac{\sqrt{3}}{n_e n_s}\right)\right]}\mathbf{I}_{S1} \quad (7.59)$$

From the initial set of equations, we also obtain

$$\mathbf{E}_1 = j\frac{n_e}{\sqrt{3}}\left(\mathbf{I}_a \mathbf{Z}_a + \mathbf{E}_a\right) \quad (7.60)$$

Substituting (7.57) and (7.60) into the $\mathbf{V}_1$ equation in (7.55), we find

$$\mathbf{V}_1 = \left(\mathbf{Z}_{11'} + \frac{n_e^2}{3}\mathbf{Z}_a\right)\mathbf{I}_1 + j\frac{n_e}{\sqrt{3}}\mathbf{E}_a \quad (7.61)$$

Substituting into the first equation in (7.55) and using (7.56) and the first equation of (7.59),

$$\mathbf{E}_a = -\frac{j\left(\dfrac{\sqrt{3}}{n_e}\right)}{\left[1-j\left(\dfrac{\sqrt{3}}{n_e n_s}\right)\right]}\mathbf{V}_{S1} + \left\{\frac{\dfrac{6n_s}{(n_e n_s)^2}}{\left[1-j\left(\dfrac{\sqrt{3}}{n_e n_s}\right)\right]^2}\left(\mathbf{Z}_{11'} + \frac{n_e^2}{3}\mathbf{Z}_a\right) + \frac{j\dfrac{\sqrt{3}}{n_e}}{\left[1-j\left(\dfrac{\sqrt{3}}{n_e n_s}\right)\right]}\mathbf{Z}_{a'}\right\}\mathbf{I}_{S1} \quad (7.62)$$

Adding the first two equations in (7.55) and using (7.62), we get

$$\mathbf{V}_{L1} = \left\{ \mathbf{V}_{S1} - \left[Z_{a'} + Z_{a''} + \frac{12}{\left(n_e n_s\right)^2 + 3} \left(Z_{11'} + \frac{n_e^2}{3} Z_a \right) \right] \mathbf{I}_{S1} \right\} e^{j\theta} \tag{7.63}$$

where θ is given in (7.58). This can be represented with the same circuit model as Figure 7.6, with

$$Z_{eq} = Z_{a'} + Z_{a''} + \left[\frac{12}{\left(n_e n_s\right)^2 + 3} \right] \left(Z_{11'} + \frac{n_e^2}{3} Z_a \right) \tag{7.64}$$

or, in terms of 2-winding impedances,

$$Z_{eq} = Z_{a'a''} + \left(\frac{12}{\left(n_e n_s\right)^2 + 3} \right) \left[Z_{11'} + \frac{n_e^2}{6} \left(Z_{aa'} + Z_{aa''} - n_s^2 Z_{a'a''} \right) \right] \tag{7.65}$$

Using (7.59) and (7.60) and ignoring impedance drops, we can show that the winding power per phase is the same for the series and excitor units, that is,

$$P_{wdg} = \mathbf{E}_1 \mathbf{I}_1^* = \mathbf{E}_a \mathbf{I}_a^* = j \left[\frac{2\sqrt{3} n_e n_s}{\left(n_e n_s\right)^2 + 3} \right] P_{in} \tag{7.66}$$

where $P_{in} = \mathbf{V}_{S1} \mathbf{I}_{S1}^*$. In terms of the phase shift θ given in (7.58), this last equation can be written

$$P_{wdg} = j \sin \theta P_{in} \tag{7.67}$$

as was the case for the standard delta-phase shifter.

In per-unit terms, keeping the power base constant at the rated input power, Equation 7.63 can be cast in the form of (7.24). To do this, we need to find the ratio of the impedance bases for both the series and excitor windings to the input or terminal impedance base. Using these formulas, we find

$$\frac{Z_{b,awdg}}{Z_{b,in}} = \left| \frac{\mathbf{E}_a}{\mathbf{V}_{S1}} \right|^2 = \left[\frac{3 n_s^2}{\left(n_e n_s\right)^2 + 3} \right]$$

$$\frac{Z_{b,a'wdg}}{Z_{b,in}} = \left| \frac{\mathbf{E}_{a'}}{\mathbf{V}_{S1}} \right|^2 = \left[\frac{3}{\left(n_e n_s\right)^2 + 3} \right] \tag{7.68}$$

$$\frac{Z_{b,1wdg}}{Z_{b,in}} = \left| \frac{\mathbf{E}_1}{\mathbf{V}_{S1}} \right|^2 = \left[\frac{\left(n_e n_s\right)^2}{\left(n_e n_s\right)^2 + 3} \right]$$

Using these relations, the per-unit equivalent impedance can be written:

$$z_{eq} = \frac{3}{\left[\left(n_e n_s\right)^2 + 3 \right]} \left\{ z_{a'a''} + \frac{4\left(n_e n_s\right)^2}{\left[\left(n_e n_s\right)^2 + 3 \right]} \left[z_{11'} + \frac{1}{2} \left(z_{aa'} + z_{aa''} - z_{a'a''} \right) \right] \right\} \tag{7.69}$$

The negative sequence circuit has the same equivalent impedance, but a-phase angle shift in the opposite direction to the positive sequence circuit.

7.5.1 Zero Sequence Circuit Model

The zero sequence circuit model is derived with reference to Figure 7.9 by assuming all quantities are in zero sequence. Thus we simply rewrite the basic equations, appending a zero subscript,

$$
\begin{aligned}
&\mathbf{V}_{S1,0} - \mathbf{V}_{1,0} = \mathbf{I}_{S1,0}Z_{a',0} + \mathbf{E}_{a',0}, \quad \mathbf{V}_{1,0} - \mathbf{V}_{L1,0} = \mathbf{I}_{L1,0}Z_{a'',0} + \mathbf{E}_{a',0} \\
&\mathbf{E}_{2',0} - \mathbf{E}_{3',0} = 0 = \mathbf{I}_{a,0}Z_{a,0} + \mathbf{E}_{a,0} \\
&\mathbf{V}_{1,0} = \mathbf{I}_{1,0}Z_{11',0} + \mathbf{E}_{1,0}, \quad \mathbf{I}_{1,0} = \mathbf{I}_{S1,0} - \mathbf{I}_{L1,0}, \quad \mathbf{I}_{2',0} = \mathbf{I}_{c,0} - \mathbf{I}_{a,0} = 0
\end{aligned}
\tag{7.70}
$$

and

$$
\frac{\mathbf{E}_{a,0}}{\mathbf{E}_{a',0}} = \frac{N_a}{N_{a'}} = n_s, \quad \frac{\mathbf{E}_{1,0}}{\mathbf{E}_{1',0}} = \frac{N_1}{N_{1'}} = n_e, \quad \frac{\mathbf{I}_{1',0}}{\mathbf{I}_{1,0}} = -\frac{N_1}{N_{1'}} = -n_e
\tag{7.71}
$$
$$
N_a\mathbf{I}_{a,0} + N_{a'}\mathbf{I}_{S1,0} + N_{a'}\mathbf{I}_{L1,0} = 0 \quad \text{or} \quad n_s\mathbf{I}_{a,0} + \mathbf{I}_{S1,0} + \mathbf{I}_{L1,0} = 0
$$

Solving, we find,

$$
\mathbf{I}_{1,0} = 0, \quad \mathbf{I}_{S1,0} = \mathbf{I}_{L1,0}, \quad \mathbf{I}_{a,0} = -\frac{2}{n_s}\mathbf{I}_{S1,0}
\tag{7.72}
$$
$$
\mathbf{E}_{a,0} = \frac{2}{n_s}\mathbf{I}_{S1,0}Z_{a,0}, \quad \mathbf{V}_{1,0} = \mathbf{E}_{1,0}
$$

Notice that, even if both Y windings of the excitor were grounded, no zero sequence current flows into the excitor because the secondary current from the tap winding would have to flow into the closed delta of the series unit, and this is not possible for zero sequence currents.

From the said formulas, we obtain

$$
\mathbf{V}_{L1,0} = \mathbf{V}_{S1,0} - \left(Z_{a',0} + Z_{a'',0} + \frac{4}{n_s^2} Z_{a,0} \right)\mathbf{I}_{S1,0}
\tag{7.73}
$$

Thus, we see that the zero sequence current and no-load voltage have no phase angle shift and the circuit of Figure 7.7 applies with

$$
Z_{eq,0} = Z_{a',0} + Z_{a'',0} + \frac{4}{n_s^2} Z_{a,0}
\tag{7.74}
$$

or, in terms of 2-winding impedances,

$$
Z_{eq,0} = Z_{a'a'',0} + \frac{2}{n_s^2}\left(Z_{aa',0} + Z_{aa'',0} - n_s^2 Z_{a'a'',0} \right)
\tag{7.75}
$$

Using the same basis as for positive sequence, the per-unit version of this is

$$
z_{eq,0} = \frac{3}{\left[(n_e n_s)^2 + 3 \right]}\left[2\left(z_{aa',0} + z_{aa'',0} \right) - z_{a'a'',0} \right]
\tag{7.76}
$$

7.6 Regulation Effects

Because all the phase-shifting transformers examined, have the same basic positive sequence (as well as negative and zero sequence) circuit, Figure 7.6, with different expressions for Z_{eq} or z_{eq} and θ, the effect of a load on the output can be studied in common. The relevant circuit model is shown in Figure 7.10.

Since all phases are identical, we drop the phase subscript for the impedances. We use per-unit quantities for this development since a common base employing rated input quantities can be used for the various quantities in the figure. The ideal transformer is included in the figure to account for the phase shift. From the figure, we see that

$$\mathbf{v}_L = \left(\mathbf{v}_S - z_{eq}\mathbf{i}_S\right)e^{j\theta} = \mathbf{i}_L z_L = e^{j\theta}\mathbf{i}_S z_L \tag{7.77}$$

where z_L is the per-unit load impedance. Solving for $\mathbf{i}_S$, we find

$$\mathbf{i}_S = \frac{\mathbf{v}_S}{\left(z_{eq} + z_L\right)} \tag{7.78}$$

so that

$$\mathbf{v}_L = \mathbf{v}_S\left[1 - \frac{z_{eq}}{\left(z_{eq} + z_L\right)}\right]e^{j\theta} = \mathbf{v}_S\left[\frac{1}{1 + \left(z_{eq}/z_L\right)}\right]e^{j\theta} \tag{7.79}$$

Thus, any-phase angle or magnitude shift from no-load conditions is due to a nonzero z_{eq}/z_L. Since z_{eq} is almost entirely inductive, a purely inductive or capacitive load will not affect θ but will result only in a magnitude change in the voltage. On the other hand, a resistive or complex load will lead to both magnitude and phase angle shifts.

Under no-load, or NL, conditions ($z_L = \infty$), Equation 7.79 shows that

$$\mathbf{v}_{L,NL} = \mathbf{v}_S e^{j\theta} \tag{7.80}$$

Thus taking ratios, we see that

$$\frac{\mathbf{v}_L}{\mathbf{v}_{L,NL}} = \frac{1}{\left(1 + \left(z_{eq}/z_L\right)\right)} \tag{7.81}$$

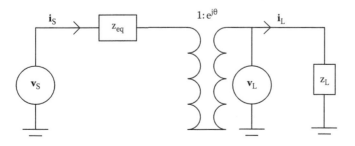

FIGURE 7.10
One phase of a-phase-shifting transformer under load, using per-unit quantities.

Thus the presence of a load will lower the magnitude and shift the phase, depending on the phase of the previously mentioned ratio of impedances. This analysis is similar to that given previously in Chapter 3, except there, we ignored the shift in phase. Because these transformers are designed to shift the phase by a given amount under no-load conditions, it is important to find the additional shift in this phase caused by transformer loading. Let $z_{eq}/z_L = pe^{j\beta}$. Then (7.81) becomes

$$\frac{v_L}{v_{L,NL}} = \frac{1}{\sqrt{1 + 2p\cos\beta + p^2}} e^{-j\tan^{-1}(p\sin\beta/(1+p\cos\beta))} \qquad (7.82)$$

As an example, let z_{eq} be a 10% inductive leakage impedance and z_L be a 100% resistive load. Then $z_{eq}/z_L = 0.1e^{j90°}$ and $v_L/v_{L,NL} = 0.995\angle{-5.71°}$, where $\angle$ indicates the angular dependence of the complex number. This angle of nearly 6° will subtract from the phase angle of the transformer output, so the behavior of the phase under load is quite important. In the said example, the magnitude will only be lowered by 0.5%. Since the transformer impedance varies with the tap setting (phase shift), the angle shift due to loading will also vary with the tap setting.

7.7 Fault Current Analysis

The phase-shifting transformers discussed here are all 2-terminal transformers, which can be characterized by a single reactance, Z_{eq} or $Z_{eq,0}$, for positive/negative and zero sequence circuits respectively. We will assume the positive and negative sequence reactances are the same and not distinguish them by subscripts. Thus, the theory developed in Chapter 6 can be applied to them with the recognition that negative sequence currents have the opposite phase shift from positive sequence currents. We will retain the source, S, and load, L, notation used previously instead of H and X. We assume the fault is on the L or output terminal. The fault currents at the fault are the same for the various fault types as given in Chapter 6.

The faults of interest are

1. 3-phase line-to-ground fault
2. Single-phase line-to-ground fault
3. Line-to-line fault
4. Double line-to-ground fault

The fault currents for these faults are summarized next. We will use per-unit notation.
For fault type (1), we have

$$i_{a1} = \frac{v_{pf}}{z_1}, \quad i_{a2} = i_{a0} = 0 \qquad (7.83)$$

For fault type (2), we assume that the a-phase line is shorted, so that

$$i_{a1} = \frac{v_{pf}}{(z_0 + z_1 + z_2)} = i_{a2} = i_{a0} \qquad (7.84)$$

For fault type (3), assuming the b and c lines are shorted together,

$$i_{a1} = \frac{v_{pf}}{(z_1 + z_2)} = -i_{a2}, \quad i_{a0} = 0 \tag{7.85}$$

For fault type (4), assuming the b and c lines are shorted to ground,

$$i_{a1} = v_{pf}\left(\frac{z_0 + z_2}{z_0 z_1 + z_0 z_2 + z_1 z_2}\right)$$

$$i_{a2} = -v_{pf}\left(\frac{z_0}{z_0 z_1 + z_0 z_2 + z_1 z_2}\right) \tag{7.86}$$

$$i_{a0} = -v_{pf}\left(\frac{z_2}{z_0 z_1 + z_0 z_2 + z_1 z_2}\right)$$

We have, assuming equal positive and negative reactances for the transformer and the systems, for the Thevenin impedances

$$z_1 = z_2 = \frac{z_{SL}(z_{eq} + z_{SS})}{z_{eq} + z_{SS} + z_{SL}}, \quad z_0 = \frac{z_{SL,0}(z_{eq,0} + z_{SS,0})}{z_{eq,0} + z_{SS,0} + z_{SL,0}} \tag{7.87}$$

The per-unit currents feeding the fault from the transformer side are given by

$$i_{eq1} = i_{a1}\left(\frac{z_1}{z_{eq} + z_{SS}}\right), \quad i_{eq2} = i_{a2}\left(\frac{z_1}{z_{eq} + z_{SS}}\right)$$

$$i_{eq0} = i_{a0}\left(\frac{z_0}{z_{eq,0} + z_{SS,0}}\right) \tag{7.88}$$

Figure 7.11 shows the terminal circuit diagrams for the three sequences, which covers all the fault types. We assume for simplicity that no current flows in the prefault condition. These can always be added later.

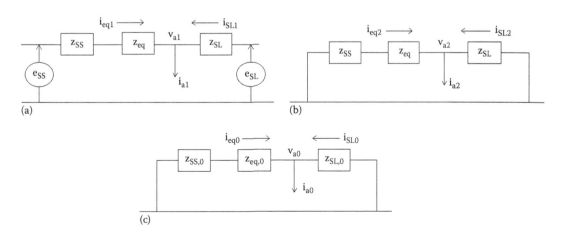

FIGURE 7.11
Sequence circuits showing system impedances and transformer equivalent impedances. The fault is on the load-side terminal: (a) positive sequence, (b) negative sequence, and (c) zero sequence.

As shown in the figure, the per-unit currents flowing out of the transformer into the fault are on the load side of the transformer and are given by (7.88) in terms of the fault sequence currents. Thus, in terms of finding the fault currents in the individual windings, we need to find their expressions in terms of the load currents. However, in a-phase-shifting transformer, positive and negative sequence currents experience opposite phase shifts. This makes the analysis of fault currents inside these transformers quite different from fault currents in a standard 2-winding transformer.

We have already considered currents in the windings of a-phase-shifting transformer when we derived their equivalent impedance for positive, negative, and zero sequence circuits. We make use of these results now, together with the fact that negative sequence currents inside a-phase-shifting transformer can be obtained from positive sequence currents by changing their phase shift to its negative. Note that i_{eq} equals i_{S1} in the previous formulas in per-unit terms. Although we have singled out the a-phase in the analysis, the equations could refer to any phase in our previous results for the different phase-shifting transformers, provided we maintain the correct phase ordering. We now apply the fault formulas to our previously studied phase-shifting transformers. Since we are using per-unit notation, we need to express the formulas for the currents in per-unit terms. Because of the length of the formulas, we only present results for the winding sequence currents.

7.7.1 Squashed Delta Fault Currents

From (7.17), we see that

$$\mathbf{I}_{a'} = \frac{n}{n - e^{j120°}} \mathbf{I}_{S1} = \frac{n}{n - e^{j120°}} e^{-j\theta} \mathbf{I}_{L1} \tag{7.89}$$

where

$$\theta = 2\tan^{-1}\left(\frac{\sqrt{3}}{2n+1}\right) \tag{7.90}$$

Using (7.90), (7.89) can be manipulated to the form

$$\mathbf{I}_{a'} = \frac{n}{\sqrt{n^2 + n + 1}} e^{-j(\theta/2)} \mathbf{I}_{L1} \tag{7.91}$$

To convert this to per-unit form, we need the ratio of the current bases. These are the reciprocal of the ratio of the voltage bases, assuming the same power level for both bases. This is given by

$$\frac{I_{b,L1}}{I_{b,a'}} = \frac{V_{b,a'}}{V_{b,L1}} = \frac{\sqrt{3}}{\sqrt{n^2 + n + 1}} \tag{7.92}$$

Using this, we can convert (7.91) to per-unit form

$$i_{a'} = \frac{n\sqrt{3}}{n^2 + n + 1} e^{-j(\theta/2)} i_{L1} \tag{7.93}$$

Since this is a 2-winding transformer $\mathbf{i}_a = -\mathbf{i}_{a'}$.

For the zero sequence currents, we have from (7.28) and (7.31)

$$\mathbf{I}_{a',0} = -n\mathbf{I}_{c,0} = \frac{n}{n-1}\mathbf{I}_{L1} \tag{7.94}$$

The zero sequence current is the same in all the phases. Using the base transformation given in (7.92), this can be converted to per-unit form

$$\mathbf{i}_{a',0} = \frac{n\sqrt{3}}{(n-1)\sqrt{n^2+n+1}}\mathbf{i}_{L1} \tag{7.95}$$

We also have $\mathbf{i}_{a0} = -\mathbf{i}_{a'0}$.

Using (7.88) and the fact that negative sequence currents have the opposite phase shift from positive sequence currents, we can find the sequence currents in the windings

$$
\begin{aligned}
\mathbf{i}_{a',1} &= \mathbf{i}_{a1}e^{-j(\theta/2)}\left(\frac{n\sqrt{3}}{n^2+n+1}\right)\left(\frac{z_1}{z_{eq}+z_{SS}}\right) \\
\mathbf{i}_{a',2} &= \mathbf{i}_{a2}e^{j(\theta/2)}\left(\frac{n\sqrt{3}}{n^2+n+1}\right)\left(\frac{z_1}{z_{eq}+z_{SS}}\right) \\
\mathbf{i}_{a',0} &= \mathbf{i}_{a0}\left(\frac{n\sqrt{3}}{(n-1)\sqrt{n^2+n+1}}\right)\left(\frac{z_0}{z_{eq,0}+z_{SS,0}}\right)
\end{aligned}
\tag{7.96}
$$

Here $\mathbf{i}_{a1}$, $\mathbf{i}_{a2}$, and $\mathbf{i}_{a0}$ are the fault sequence currents for the various fault types as given in Chapter 6. These can then be transformed to the phase or winding currents by using (6.1).

7.7.2 Standard Delta Fault Currents

This transformer has 3 windings per phase. We have, from (7.39)

$$\mathbf{I}_a = -\frac{2}{n\left(1-j\dfrac{\sqrt{3}}{n}\right)}\mathbf{I}_{S1} = -\frac{2}{n\left(1-j\dfrac{\sqrt{3}}{n}\right)}e^{-j\theta}\mathbf{I}_{L1} = -\frac{2}{\sqrt{n^2+3}}e^{-j(\theta/2)}\mathbf{I}_{L1} \tag{7.97}$$

where

$$\mathbf{I}_{L1} = e^{j\theta}\mathbf{I}_{S1} \quad \text{and} \quad \theta = 2\tan^{-1}\left(\frac{\sqrt{3}}{n}\right) \tag{7.98}$$

Converting to a per-unit system, we find for the ratio of current bases

$$\frac{I_{b,L1}}{I_{b,a}} = \frac{E_a}{V_{L1}} = \frac{n\sqrt{3}}{\sqrt{n^2+3}} \tag{7.99}$$

Using this, Equation 7.97 becomes in per-unit terms,

$$\mathbf{i}_a = -\frac{2n\sqrt{3}}{n^2+3}e^{-j(\theta/2)}\mathbf{i}_{L1} \tag{7.100}$$

We also have

$$\mathbf{I}_{a'} = e^{-j\theta}\mathbf{I}_{L1}, \quad \mathbf{I}_{a''} = \mathbf{I}_{L1} \tag{7.101}$$

The ratio of current bases is the same for the currents in (7.101) and is given by

$$\frac{I_{b,L1}}{I_{b,a'}} = \frac{E_{a'}}{V_{L1}} = \frac{\sqrt{3}}{\sqrt{n^2+3}} \tag{7.102}$$

Using this, Equation 7.101 becomes in per-unit terms,

$$i_{a'} = \frac{\sqrt{3}}{\sqrt{n^2+3}} e^{-j\theta} i_{L1}, \quad i_{a''} = \frac{\sqrt{3}}{\sqrt{n^2+3}} i_{L1} \tag{7.103}$$

Using trigonometric identities and the formula for θ in (7.98), we can show that $i_a + i_{a'} + i_{a''} = 0$

For the zero sequence currents, we get from (7.49) and (7.50)

$$\mathbf{I}_{a,0} = -\frac{2}{n}\mathbf{I}_{L1,0}, \quad \mathbf{I}_{a',0} = \mathbf{I}_{L1,0}, \quad \mathbf{I}_{a'',0} = \mathbf{I}_{L1,0} \tag{7.104}$$

Putting these on a per-unit basis, using the same current bases as for the positive sequence case,

$$i_{a,0} = -\frac{2\sqrt{3}}{\sqrt{n^2+3}} i_{L1,0}, \quad i_{a',0} = \frac{\sqrt{3}}{\sqrt{n^2+3}} i_{L1,0}, \quad i_{a'',0} = \frac{\sqrt{3}}{\sqrt{n^2+3}} i_{L1,0} \tag{7.105}$$

These zero sequence per-unit currents also sum to zero as expected.

Using (7.88) and noting that i_{eq} corresponds to i_{L1}, we get for the winding sequence currents

$$
\begin{aligned}
&i_{a,1} = -i_{a1} \frac{2n\sqrt{3}}{n^2+3} e^{-j(\theta/2)} \left(\frac{z_1}{z_{eq} + z_{SS}} \right), \quad i_{a,2} = -i_{a2} \frac{2n\sqrt{3}}{n^2+3} e^{j(\theta/2)} \left(\frac{z_1}{z_{eq} + z_{SS}} \right) \\
&i_{a,0} = -i_{a0} \frac{2\sqrt{3}}{\sqrt{n^2+3}} \left(\frac{z_0}{z_{eq,0} + z_{SS,0}} \right) \\
&i_{a',1} = i_{a1} \frac{\sqrt{3}}{\sqrt{n^2+3}} e^{-j\theta} \left(\frac{z_1}{z_{eq} + z_{SS}} \right), \quad i_{a',2} = i_{a2} \frac{\sqrt{3}}{\sqrt{n^2+3}} e^{j\theta} \left(\frac{z_1}{z_{eq} + z_{SS}} \right) \\
&i_{a',0} = i_{a0} \frac{\sqrt{3}}{\sqrt{n^2+3}} \left(\frac{z_0}{z_{eq,0} + z_{SS,0}} \right) \\
&i_{a'',1} = i_{a1} \frac{\sqrt{3}}{\sqrt{n^2+3}} \left(\frac{z_1}{z_{eq} + z_{SS}} \right), \quad i_{a'',2} = i_{a2} \frac{\sqrt{3}}{\sqrt{n^2+3}} \left(\frac{z_1}{z_{eq} + z_{SS}} \right) \\
&i_{a',0} = i_{a0} \frac{\sqrt{3}}{\sqrt{n^2+3}} \left(\frac{z_0}{z_{eq,0} + z_{SS,0}} \right)
\end{aligned} \tag{7.106}
$$

The per-unit currents for the various fault types can be inserted for i_{a1}, i_{a2}, and i_{a0} as given in Chapter 6. One can then get the winding-per-unit phase currents via (6.1) for the various fault types.

7.7.3 2-Core Phase-Shifting Transformer Fault Currents

We will simply quote the results for this transformer based on the formulas in Section 7.5. This transformer has 5 windings labeled 1, 1′, a, a′, a″. The per-unit winding currents, expressed in terms of the output L1 current are given by

$$i_1 = -j\frac{2\sqrt{3}n_en_s}{\left(n_en_s\right)^2+3}e^{-j(\theta/2)}i_{L1} = -i_{1'}, \quad i_a = -\frac{2\sqrt{3}n_en_s}{\left(n_en_s\right)^2+3}e^{-j(\theta/2)}i_{L1}$$

$$i_{a'} = \frac{\sqrt{3}}{\sqrt{\left(n_en_s\right)^2+3}}e^{-j\theta}i_{L1}, \quad i_{a''} = \frac{\sqrt{3}}{\sqrt{\left(n_en_s\right)^2+3}}i_{L1}$$

$$i_{1,0} = i_{1',0} = 0, \quad i_{a,0} = -\frac{2\sqrt{3}}{\sqrt{\left(n_en_s\right)^2+3}}i_{L1,0}$$

$$i_{a',0} = \frac{\sqrt{3}}{\sqrt{\left(n_en_s\right)^2+3}}i_{L1,0} = i_{a'',0}$$

(7.107)

The winding sequence currents are given by

$$i_{1,1} = -j\frac{2\sqrt{3}n_en_s}{\left(n_en_s\right)^2+3}e^{-j(\theta/2)}i_{L1,1}, \quad i_{1,2} = j\frac{2\sqrt{3}n_en_s}{\left(n_en_s\right)^2+3}e^{j(\theta/2)}i_{L1,2}$$

$$i_{1,0} = i_{1',0} = 0$$

$$i_{a,1} = -\frac{2\sqrt{3}n_en_s}{\left(n_en_s\right)^2+3}e^{-j(\theta/2)}i_{L1,1}, \quad i_{a,2} = -\frac{2\sqrt{3}n_en_s}{\left(n_en_s\right)^2+3}e^{j(\theta/2)}i_{L1,2}$$

$$i_{a,0} = -\frac{2\sqrt{3}}{\sqrt{\left(n_en_s\right)^2+3}}i_{L1,0}$$

(7.108)

$$i_{a',1} = \frac{\sqrt{3}}{\sqrt{\left(n_en_s\right)^2+3}}e^{-j\theta}i_{L1,1}, \quad i_{a',2} = \frac{\sqrt{3}}{\sqrt{\left(n_en_s\right)^2+3}}e^{j\theta}i_{L1,2}$$

$$i_{a',0} = \frac{\sqrt{3}}{\sqrt{\left(n_en_s\right)^2+3}}i_{L1,0}$$

$$i_{a'',1} = \frac{\sqrt{3}}{\sqrt{\left(n_en_s\right)^2+3}}i_{L1,1} = i_{a'',2}, \quad i_{a'',0} = \frac{\sqrt{3}}{\sqrt{\left(n_en_s\right)^2+3}}i_{L1,0}$$

Substituting the i_{eq} currents from (7.88) for i_{L1} in the formulas of (7.108), we can get the sequence currents in all the windings. Since j in the formula is a rotation by 90° in the complex plane, in the negative sequence current formula it was replaced by −j for a 90° rotation in the opposite direction.

7.8 Zigzag Transformer

A Delta HV (H) winding and zigzag LV (X) winding transformer is analyzed. The complex currents are calculated as well as the terminal positive/negative and zero sequence impedances. Although explicit expressions for these impedances are given in terms of 2-winding impedances, they can also be obtained directly from a program, which calculates the magnetic field associated with complex currents. These impedances are then used to obtain fault currents for a short-circuit analysis.

The 3-phase connections for the delta primary and zigzag secondary are given schematically in Figure 7.12. Parallel lines denote windings on the same phase. The delta winding phases are denoted a, b, and c. The zigzag windings consist of windings from 2 phases connected in series. Thus, the zigzag X1 terminal is attached to a series combination of windings from the b and c-phases. The orientation of the zigzag windings in Figure 7.12 is such that the terminal voltage of the X terminals to ground is in phase with the terminal voltage of the delta terminals V to ground. This requires the two separate windings of the zigzag to have the same number of turns. A 180° phase shift between X and V could be achieved by winding the X windings in the opposite sense to the delta windings.

Figure 7.13 shows the positive sequence phasors associated with the winding and terminal voltages of the delta winding. $\mathbf{E}_a$, $\mathbf{E}_b$, $\mathbf{E}_c$ are a positive sequence set of winding voltages and $\mathbf{V}_1$, $\mathbf{V}_2$, $\mathbf{V}_3$ are a positive sequence set of terminal-to-ground voltages. The E's are no-load voltages due to core excitation and the diagram in Figure 7.13 does not take into account the voltage drops caused by the currents and impedances. For these, we use a 3-winding circuit model. The single-winding impedances will be the same for all three phases. Concentrating on the a-phase, the winding voltages are given by

$$
\begin{aligned}
\mathbf{V}_a &= \mathbf{E}_a + \mathbf{I}_a \mathbf{Z}_a \quad (\text{Delta HV}) \\
\mathbf{V}_{a'} &= \mathbf{E}_{a'} + \mathbf{I}_{a'} \mathbf{Z}_{a'} \quad (\text{Zig-Y LV}) \\
\mathbf{V}_{a''} &= \mathbf{E}_{a''} + \mathbf{I}_{a''} \mathbf{Z}_{a''} \quad (\text{Zag-Y LV})
\end{aligned}
\tag{7.109}
$$

where the currents are taken as positive into the winding terminals. The E's are proportional to the winding turns and they increase in the opposite sense to the current direction.

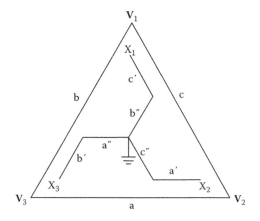

FIGURE 7.12
3-Phase schematic connection diagram of the delta zigzag transformer.

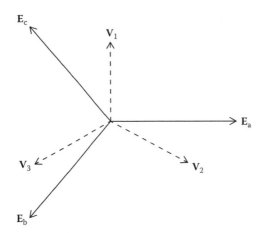

FIGURE 7.13
Phasor diagram of delta terminal and winding voltages.

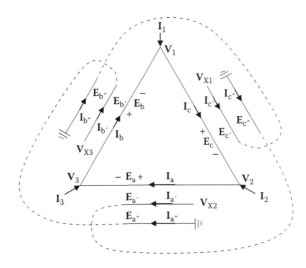

FIGURE 7.14
3-phase detailed connection diagram of the zigzag transformer with the electrical quantities labeled.

The winding connection diagram for all three phases is shown in Figure 7.14. The subscripts a, a′, a″ refer to the delta, zig, and zag windings, respectively, of the a-phase. The b and c subscripts have the same significance for the b and c-phases. The zig and zag windings are connected in such a way that their phase voltages are subtractive. This gives the maximum net voltage across both windings.

7.8.1 Calculation of Electrical Characteristics

Like the voltages, the currents I_1, I_2, I_3 and I_a, I_b, I_c, as shown in Figure 7.14, form a 3-phase set. From the figure, we have

$$I_1 = I_c - I_b$$
$$I_2 = I_a - I_c \qquad (7.110)$$
$$I_3 = I_b - I_a$$

We also have, from Figure 7.14,

$$\mathbf{V}_1 - \mathbf{V}_2 = \mathbf{E}_c + \mathbf{I}_c Z_a$$
$$\mathbf{V}_3 - \mathbf{V}_1 = \mathbf{E}_b + \mathbf{I}_b Z_a \tag{7.111}$$
$$\mathbf{V}_2 - \mathbf{V}_3 = \mathbf{E}_a + \mathbf{I}_a Z_a$$

(Note that $Z_a = Z_b = Z_c$ and similarly for $Z_{a'}$ and $Z_{a''}$, so we will only use the a, a', a'' subscripts on the Z's.)

For the X1 terminal, we see from Figure 7.14 that

$$\mathbf{V}_{X1} = \mathbf{E}_{c'} + \mathbf{I}_{c'} Z_{a'} - \mathbf{E}_{b''} - \mathbf{I}_{b''} Z_{a''} \tag{7.112}$$

Letting N_1 = the number of turns in the HV delta winding and N_2 = the number of turns in the zig winding = the number of turns in the zag winding, we have

$$\frac{\mathbf{E}_a}{\mathbf{E}_{a'}} = \frac{N_1}{N_2} = n$$
$$\frac{\mathbf{E}_{a'}}{\mathbf{E}_{a''}} = 1 \tag{7.113}$$

where we have defined n as the ratio of turns in the delta winding to turns in the zig or zag winding. Because of the phase relationship among a positive sequence set, we can write

$$\mathbf{E}_{c'} - \mathbf{E}_{b''} = \mathbf{E}_{a'}\left(e^{j120°} - e^{-j120°}\right) = j\sqrt{3}\mathbf{E}_{a'} = j\frac{\sqrt{3}}{n}\mathbf{E}_a \tag{7.114}$$

From (7.111), we have

$$\mathbf{E}_a = \left(\mathbf{V}_2 - \mathbf{V}_3\right) - \mathbf{I}_a Z_a = \left(e^{-j120°} - e^{j120°}\right)\mathbf{V}_1 - \mathbf{I}_a Z_a = -j\sqrt{3}\mathbf{V}_1 - \mathbf{I}_a Z_a \tag{7.115}$$

Substituting from (7.114) and (7.115) into (7.112), we get

$$\mathbf{V}_{X1} = \frac{3}{n}\mathbf{V}_1 - j\frac{\sqrt{3}}{n}\mathbf{I}_a Z_a + \mathbf{I}_{c'} Z_{a'} - \mathbf{I}_{b''} Z_{a''} \tag{7.116}$$

By current continuity in the zig and zag windings, we must have

$$\mathbf{I}_{c'} = -\mathbf{I}_{b''}$$
$$\mathbf{I}_{a'} = -\mathbf{I}_{c''} \tag{7.117}$$
$$\mathbf{I}_{b'} = -\mathbf{I}_{a''}$$

so that (7.116) can be written:

$$\mathbf{V}_{X1} = \frac{3}{n}\mathbf{V}_1 - j\frac{\sqrt{3}}{n}\mathbf{I}_a Z_a + \mathbf{I}_{c'}\left(Z_{a'} + Z_{a''}\right) \tag{7.118}$$

Since the currents form a balanced set, we have

$$\mathbf{I}_{c'} = e^{j120°}\mathbf{I}_{a'} \tag{7.119}$$

From amp-turn balance, we have

$$\begin{aligned} N_1\mathbf{I}_a + N_2\mathbf{I}_{a'} + N_2\mathbf{I}_{a''} &= 0 \\ n\mathbf{I}_a + \mathbf{I}_{a'} + \mathbf{I}_{a''} &= 0 \end{aligned} \tag{7.120}$$

From (7.117), we have

$$\mathbf{I}_{a''} = -\mathbf{I}_{b'} = -e^{-j120°}\mathbf{I}_{a'} \tag{7.121}$$

Thus from (7.119) through (7.121), we get

$$\mathbf{I}_{c'} = -j\frac{n}{\sqrt{3}}\mathbf{I}_a \tag{7.122}$$

Using this last result, Equation 7.118 becomes

$$\mathbf{V}_{X1} = \frac{3}{n}\mathbf{V}_1 - j\frac{\sqrt{3}}{n}\mathbf{I}_a\left[Z_a + \frac{n^2}{3}\left(Z_{a'} + Z_{a''}\right)\right] \tag{7.123}$$

From (7.110), we have

$$\mathbf{I}_1 = \left(e^{j120°} - e^{-j120°}\right)\mathbf{I}_a = j\sqrt{3}\mathbf{I}_a \tag{7.124}$$

Substituting into (7.123), we get

$$\mathbf{V}_{X1} = \frac{3}{n}\mathbf{V}_1 - \frac{\mathbf{I}_1}{n}\left[Z_a + \frac{n^2}{3}\left(Z_{a'} + Z_{a''}\right)\right] \tag{7.125}$$

Thus, under no-load conditions, we see from (7.125) that the terminal voltages are related by

$$\mathbf{V}_{X1} = \frac{3}{n}\mathbf{V}_1 \tag{7.126}$$

This shows that the H and X no-load terminal voltages are in phase, as desired. The current into the X1 terminal is, from (7.122) and (7.124),

$$\mathbf{I}_{X1} = \mathbf{I}_{c'} = -\frac{n}{3}\mathbf{I}_1 \tag{7.127}$$

Thus we see that the power into the X1 terminal, ignoring the impedance drop, is

$$\mathbf{V}_{X1}\mathbf{I}_{X1}^* = -\mathbf{V}_1\mathbf{I}_1^* \tag{7.128}$$

where* indicates complex conjugation. This says that the power is flowing out of the X1 terminal and that it equals the power into the $\mathbf{V}_1$ terminal. We also have, from (7.115) and (7.124), ignoring the impedance drop,

$$\mathbf{E}_a\mathbf{I}_a^* = \mathbf{V}_1\mathbf{I}_1^* \tag{7.129}$$

so that the delta winding power equals the input power. The a-phase winding currents can be obtained from (7.120) and (7.121).

$$\mathbf{I}_{a'} = n\mathbf{I}_a\left(-\frac{1}{2} + \frac{j}{2\sqrt{3}}\right)$$

$$\mathbf{I}_{a''} = n\mathbf{I}_a\left(-\frac{1}{2} - \frac{j}{2\sqrt{3}}\right) \tag{7.130}$$

Multiplying (7.125) by n/3, we get

$$\frac{n}{3}\mathbf{V}_{X1} = \mathbf{V}_1 - \frac{\mathbf{I}_1}{3}\left[Z_a + \frac{n^2}{3}(Z_{a'} + Z_{a''})\right] \tag{7.131}$$

Thus, as seen from the high-voltage side, the effective impedance, Z_{eff}, is given by

$$Z_{eff} = \frac{1}{3}\left[Z_a + \frac{n^2}{3}(Z_{a'} + Z_{a''})\right] \tag{7.132}$$

7.8.2 Per-Unit Formulas

At this point, it is desirable to put everything on a per-unit basis. We will use the input power per phase as the power base. This also equals, by (7.129), the delta winding power. Thus $P_b = V_1I_1 = E_aI_a$. The asterisks are not needed since these quantities are real. Thus V_1 and I_1 are the voltage and current bases for the H terminal. We label these $V_{b,1}$ and $I_{b,1}$. In terms of these, the X1 terminal base voltage and current are, from (7.126) and (7.127),

$$V_{b,X1} = \frac{3}{n}V_{b,1}$$

$$I_{b,X1} = \frac{n}{3}I_{b,1} \tag{7.133}$$

Dividing (7.131) by $V_{b,1}$ and using (7.133), we find

$$\frac{n}{3}\left(\frac{\mathbf{V}_{X1}}{V_{b,1}}\right) = \frac{\mathbf{V}_1}{V_{b,1}} - \frac{\mathbf{I}_1}{V_{b,1}}Z_{eff} \quad \Rightarrow \quad \mathbf{v}_{X1} - \mathbf{i}_1\left(\frac{I_{b,1}}{V_{b,1}}\right)Z_{eff}$$

$$\mathbf{v}_{X1} = \mathbf{v}_1 - \mathbf{i}_1\frac{Z_{eff}}{Z_{b,1}} \quad \Rightarrow \quad \mathbf{v}_{X1} = \mathbf{v}_1 - \mathbf{i}_1z_{eff} \tag{7.134}$$

where

$$Z_{b,1} = \frac{V_{b,1}}{I_{b,1}} = \frac{V_{b,1}^2}{P_b} \tag{7.135}$$

Since the H winding power equals the terminal power, we can resort to a winding base to calculate z_{eff}. We have, from (7.115) and (7.124), ignoring impedance drops,

$$V_{b,a} = \sqrt{3}V_{b,1}$$

$$I_{b,a} = \frac{1}{\sqrt{3}}I_{b,1} \tag{7.136}$$

$$Z_{b,a} = \frac{V_{b,a}}{I_{b,a}} = \frac{V_{b,a}^2}{P_b} = 3Z_{b,1}$$

We also have for the a′ and a″ windings, using (7.113) and keeping the power base the same,

$$V_{b,a'} = V_{b,a''} = \frac{V_{b,a}}{n}$$

$$Z_{b,a'} = Z_{b,a''} = \frac{V_{b,a'}^2}{P_b} = \frac{Z_{b,a}}{n^2} \tag{7.137}$$

From (7.134) and (7.132), we can write

$$z_{\text{eff}} = \frac{Z_{\text{eff}}}{Z_{b,1}} = \frac{1}{3}\left[\left(\frac{Z_{b,a}}{Z_{b,1}}\right)\frac{Z_a}{Z_{b,a}} + \frac{n^2}{3}\left(\frac{Z_{b,a}}{Z_{b,1}}\right)\left(\frac{Z_{b,a'}}{Z_{b,a}}\right)\frac{(Z_{a'} + Z_{a''})}{Z_{b,a'}} \right] \tag{7.138}$$

Using (7.136) and (7.137), we can put each winding on its own base to get

$$z_{\text{eff}} = z_a + \frac{1}{3}\left(z_{a'} + z_{a''}\right) \tag{7.139}$$

In terms of 2-winding-per-unit impedances, we have the standard formulas

$$z_a = \frac{z_{aa'} + z_{aa''} - z_{a'a''}}{2}$$

$$z_{a'} = \frac{z_{aa'} + z_{a'a''} - z_{aa''}}{2} \tag{7.140}$$

$$z_{a''} = \frac{z_{aa''} + z_{a'a''} - z_{aa'}}{2}$$

Substituting these into (7.139), we obtain

$$z_{\text{eff}} = \frac{z_{aa'} + z_{aa''}}{2} - \frac{z_{a'a''}}{6} \tag{7.141}$$

Thus, the effective terminal impedance can be derived from the 2-winding impedances.

7.8.3 Zero Sequence Impedance

We can go through a similar analysis to get the zero sequence impedance, using a zero subscript to label zero sequence quantities. Thus, instead of (7.112), we have

$$\mathbf{V}_{X1,0} = \mathbf{E}_{c',0} + \mathbf{I}_{c',0}Z_{a',0} - \mathbf{E}_{b'',0} - \mathbf{I}_{b'',0}Z_{a'',0} \tag{7.142}$$

But now $\mathbf{E}_{c',0}$ and $\mathbf{E}_{b'',0}$ are in phase and have the same magnitude and therefore cancel in (7.142). In addition, using (7.117), which also applies to zero sequence currents, Equation 7.142 becomes

$$\mathbf{V}_{X1,0} = \mathbf{I}_{c',0}\left(Z_{a',0} + Z_{a'',0}\right) \tag{7.143}$$

From (7.117) and the equality of the zero phase currents in the three phases, we have

$$\mathbf{I}_{a',0} = \mathbf{I}_{b',0} = -\mathbf{I}_{a'',0} \tag{7.144}$$

Hence, from (7.120) applied to zero phase currents, we have

$$\mathbf{I}_{a,0} = 0 \tag{7.145}$$

Thus no zero sequence current flows in the delta HV winding. Since $\mathbf{I}_{X1,0} = \mathbf{I}_{c',0}$, Equation 7.143 becomes

$$\mathbf{V}_{X1,0} = \mathbf{I}_{X1,0}\left(Z_{a',0} + Z_{a'',0}\right) \tag{7.146}$$

Reverting to per-unit quantities, we divide (7.146) by $V_{b,X1}$ to get

$$\mathbf{v}_{X1,0} = \mathbf{i}_{X1,0}z_{eff,0} \tag{7.147}$$

where

$$z_{eff,0} = \frac{\left(Z_{a',0} + Z_{a'',0}\right)}{Z_{b,X1}} \tag{7.148}$$

and

$$Z_{b,X1} = \frac{V_{b,X1}^2}{P_b} = \left(\frac{V_{b,X1}}{V_{b,a'}}\right)^2 \frac{V_{b,a'}^2}{P_b} = 3Z_{b,a'} \tag{7.149}$$

Substituting into (7.148) and using (7.140), we get

$$z_{eff,0} = \frac{1}{3}\left(z_{a',0} + z_{a'',0}\right) = \frac{1}{3}z_{a'a'',0} \tag{7.150}$$

Thus, the effective zero sequence impedance can be found from the per-unit 2-winding zero sequence impedance between the zig and zag windings. Normally, the positive sequence impedance is calculated, and this is adjusted via an empirical factor to get the zero sequence impedance.

7.8.4 Fault Current Analysis

The phase sequence diagrams are shown in Figure 7.15. We will calculate the sequence currents in the 3 windings in terms of the fault sequence currents, which will depend on the fault type. The phase currents can then be obtained by the transformation given in (6.1). We consider equal positive and negative sequence reactances for the transformer and the systems and will not distinguish them by subscripts. Also we will consider a fault on the X terminal, the fault being fed from the HV line.

The Thevenin impedances, in terms of those given in Figure 7.15 are

$$z_1 = z_2 = \frac{z_{SX}\left(z_{eff} + z_{SH}\right)}{z_{eff} + z_{SH} + z_{SX}}, \quad z_0 = \frac{z_{SX,0}\left(z_{eff,0} + z_{SH,0}\right)}{z_{eff,0} + z_{SH,0} + z_{SX,0}} \tag{7.151}$$

$z_{SH,0}$ must be set equal to zero in the formulas. The per-unit sequence currents feeding the fault from the transformer side are given by

$$i_{X,1} = i_{a1}\left(\frac{z_1}{z_{eff} + z_{SH}}\right), \quad i_{X,2} = i_{a2}\left(\frac{z_1}{z_{eff} + z_{SH}}\right)$$

$$i_{X,0} = i_{a0}\left(\frac{z_0}{z_{eff,0} + z_{SH,0}}\right) \tag{7.152}$$

We need to find the per-unit fault sequence currents in the windings in terms of the fault sequence currents out of the X1 terminal. 1 preceded by a comma in (7.152) refers to positive sequence, etc., but X1 refers to X terminal number 1. From (7.124) and (7.127) and (7.130), we find

$$\mathbf{I}_a = j\frac{\sqrt{3}}{n}\mathbf{I}_{X1}, \quad \mathbf{I}_{a'} = n\mathbf{I}_a\left(-\frac{1}{2} + j\frac{1}{2\sqrt{3}}\right), \quad \mathbf{I}_{a''} = n\mathbf{I}_a\left(-\frac{1}{2} - j\frac{1}{2\sqrt{3}}\right) \tag{7.153}$$

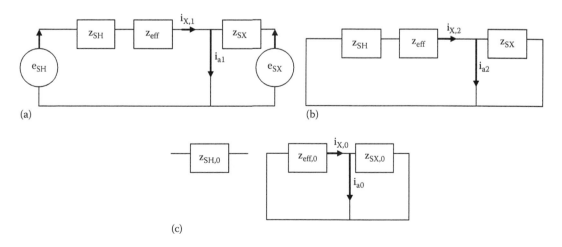

(a)   (b)

(c)

FIGURE 7.15
Sequence circuits for the zigzag transformer: (a) positive sequence, (b) negative sequence, and (c) zero sequence.

From (7.133), (7.136), and (7.137), the current bases are related by

$$\frac{I_{b,X1}}{I_{b,a}} = \frac{n}{\sqrt{3}}, \quad \frac{I_{b,X1}}{I_{b,a'}} = \frac{I_{b,X1}}{I_{b,a''}} = \frac{1}{\sqrt{3}} \tag{7.154}$$

Using these base current relations, Equation 7.153 can be written in per-unit terms

$$\mathbf{i}_a = j\mathbf{i}_{X1}, \quad \mathbf{i}_{a'} = j\mathbf{i}_{X1}\left(-\frac{1}{2} + j\frac{1}{2\sqrt{3}}\right), \quad \mathbf{i}_{a''} = j\mathbf{i}_{X1}\left(-\frac{1}{2} - j\frac{1}{2\sqrt{3}}\right) \tag{7.155}$$

The 3 per-unit currents sum to zero as expected.

Replacing X1 by X,1; X,2; and X,0 for the sequence currents obtained from (7.152) for the different fault types, one can obtain the fault sequence currents in the various windings from (7.155). From these, one can then obtain the phase currents from (6.1).

8

Multiterminal 3-Phase Transformer Model*

8.1 Introduction

It has been shown in earlier chapters that it is possible to model 2- and 3-terminal transformers using leakage inductances or reactances associated with the leakage flux. The 2-winding leakage inductance is the self-inductance of the two windings, which takes into account that the sum of the amp-turns is zero. It is easily measured by applying a voltage to the first winding, shorting the second winding, and measuring the reactance, while keeping any other windings open circuited. It can also be calculated with finite element codes or analytic methods.

For a 2-winding transformer, all that is needed to model the circuit is an ideal transformer in series with the leakage inductance. For a 3-winding transformer, it is possible to combine the three pairs of leakage inductances in such a way that, together with an ideal transformer, a simple circuit without mutual couplings, called a T-equivalent circuit, results. For a 4-winding transformer, a similar process leads to a much more complicated circuit, again without mutual couplings [Blu51]. There does not appear to be a circuit model without mutual couplings for a transformer with any number of terminals or windings. However, there are methods that use only the 2-winding leakage inductances to model the transformer [Bra82], [deL92], [Mom02], [Shi63], [Aco89]. These have mutual couplings and assume an ideal or separately treated iron core. We explore such a method here and show how it lends itself naturally to matrix methods to obtain complicated connections among the windings on the same or different legs. The single-winding resistances are also included in the model for greater generality.

The basic equations are derived here from first principles in order to highlight the assumptions in the model and to establish terminology. Most of the examples given are for 2- or 3-terminal transformers in order to show that the formulas do reduce to the familiar formulas derived by previous methods. We restrict our discussion to sinusoidal conditions at power frequencies. The theory is developed initially for balanced conditions. It is then extended to account for unbalanced conditions, including short circuits. This latter development requires inclusion of the core in order to get a solvable system. Both methods can be useful, depending on the type of problem to be solved.

For higher-frequency applications, capacitances must be included [deL92a]. The inductances and, to a lesser extent, the leakage inductances also vary with frequency. The resistances have a strong frequency dependence, and it also becomes necessary to consider mutual resistances at higher frequencies [Mom02]. However, here, we will only deal with the low-frequency inductive and resistive aspects of the windings.

---

* This chapter closely follows our published papers, [Del06], *IEEE Trans. Power Deliv.*, 21(3), July 2006, 1300–1308. © 2006 IEEE, and [Del08], *IEEE Trans. Power Deliv.*, 23(3), July 2008, 1439–1447. © 2008 IEEE.

8.2 Theory

8.2.1 Two-Winding Leakage Inductance

Since we are concerned with transformer operation at a fixed frequency and sinusoidal conditions, we will most often be dealing with reactances in our circuit models. We use X to denote a reactance, where $X = j\omega L$, with L = inductance, j the imaginary unit, and $\omega = 2\pi f$, where f is the frequency. We also use $X = j\omega M$ to refer to mutual reactance. A subscript pair on X will distinguish self from mutual reactances. These quantities are constants, as they are associated with the leakage flux.

The general circuit equations for a 2-winding transformer can be written, ignoring resistance, as

$$\begin{aligned}
\mathbf{V}_1 &= X_{11}\mathbf{I}_1 + X_{12}\mathbf{I}_2 + \mathbf{E}N_1 \\
\mathbf{V}_2 &= X_{21}\mathbf{I}_1 + X_{22}\mathbf{I}_2 + \mathbf{E}N_2
\end{aligned} \tag{8.1}$$

Since the X's are linear circuit elements, $X_{12} = X_{21}$. $\mathbf{E}$ is the volts/turn and N the number of turns, so the terms $\mathbf{E}N$ refer to the voltages induced by the ideal transformer. V are the voltages and I the currents. We are using phasor notation here. The currents are directed into the winding at the positive voltage end. Since the amp-turns sum to zero, we have

$$N_1\mathbf{I}_1 + N_2\mathbf{I}_2 = 0 \tag{8.2}$$

Substituting for $\mathbf{I}_2$ from (8.2) and $\mathbf{E}$ from the second equation in (8.1), we get

$$\mathbf{V}_1 = \frac{N_1}{N_2}\mathbf{V}_2 + \left[X_{11} - 2\frac{N_1}{N_2}X_{12} + \left(\frac{N_1}{N_2}\right)^2 X_{22} \right]\mathbf{I}_1 \tag{8.3}$$

The leakage reactance is

$$Z_{12} = \left.\frac{\mathbf{V}_1}{\mathbf{I}_1}\right|_{V_2=0} \tag{8.4}$$

The term in brackets in (8.4) is the leakage reactance between windings 1 and 2 in that order, denoted by Z_{12}:

$$Z_{12} = X_{11} - 2\frac{N_1}{N_2}X_{12} + \left(\frac{N_1}{N_2}\right)^2 X_{22} \tag{8.5}$$

The order is important, and we have

$$Z_{21} = \left(\frac{N_2}{N_1}\right)^2 Z_{12} \tag{8.6}$$

We also note from (8.5) that $Z_{11} = 0$, that is, when the subscripts are equal, the 2-winding leakage reactance vanishes. The circuit model for a 2-winding transformer would consist of the leakage reactance in series with an ideal transformer.

8.2.2 Multi-Winding Transformer

We use the same approach to model a multi-winding transformer. However, we will include the winding resistances for greater generality. The circuit equations are

$$\mathbf{V}_j = R_j \mathbf{I}_j + \sum_{k=1}^{n} X_{jk} \mathbf{I}_k + E\mathbf{N}_j, \quad j=1,\ldots,n \tag{8.7}$$

together with

$$\sum_{k=1}^{n} N_k \mathbf{I}_k = 0 \tag{8.8}$$

Here, n is the number of windings and the other symbols are the same as before.

Using (8.8), we can eliminate one voltage equation in (8.7), as was done for the 2-winding transformer, where the winding 2 equation was chosen for elimination. We choose the winding 1 equation here for elimination. In the following discussion, this will be taken as the primary winding. Substituting for $\mathbf{I}_1$ from (8.8) and $\mathbf{E}$ from the winding 1 equation of (8.7), we obtain, for the remaining equations,

$$\mathbf{V}_j = \frac{N_j}{N_1} \mathbf{V}_1 + R_j \mathbf{I}_j + \sum_{k=2}^{n} \left(\frac{N_j N_k}{N_1^2} R_1 + D_{1jk} \right) \mathbf{I}_k, \quad j=2,\ldots,n \tag{8.9}$$

where

$$D_{1jk} = X_{jk} - \frac{N_k}{N_1} X_{j1} + \frac{N_j N_k}{N_1^2} X_{11} - \frac{N_j}{N_1} X_{1k} \tag{8.10}$$

Rewrite this as

$$D_{1jk} = \frac{1}{2} \left(2 X_{jk} - 2 \frac{N_k}{N_1} X_{j1} + 2 \frac{N_j N_k}{N_1^2} X_{11} - 2 \frac{N_j}{N_1} X_{1k} \right) \tag{8.11}$$

Add and subtract $(N_j/N_k)X_{kk} + (N_k/N_j)X_{jj}$ from the term in parentheses in (8.11). After some rearrangement, we get

$$D_{1jk} = \frac{1}{2} \Bigg\{ \frac{N_j N_k}{N_1^2} \left[X_{11} - 2 \frac{N_1}{N_j} X_{1j} + \left(\frac{N_1}{N_j} \right)^2 X_{jj} \right]$$

$$+ \frac{N_j N_k}{N_1^2} \left[X_{11} - 2 \frac{N_1}{N_k} X_{1k} + \left(\frac{N_1}{N_k} \right)^2 X_{kk} \right]$$

$$- \frac{N_k}{N_j} \left[X_{jj} - 2 \frac{N_j}{N_k} X_{jk} + \left(\frac{N_j}{N_k} \right)^2 X_{kk} \right] \Bigg\} \tag{8.12}$$

Using the definition of 2-winding leakage reactance given in (8.5), we can rewrite (8.12) as

$$
\begin{aligned}
D_{1jk} &= \frac{1}{2}\left\{ \frac{N_j N_k}{N_1^2}\left(Z_{1j}+Z_{1k}\right) - \frac{N_k}{N_j}Z_{jk}\right\} \\
&= \frac{N_j N_k}{N_1^2}\frac{1}{2}\left[Z_{1j}+Z_{1k} - \left(\frac{N_1}{N_j}\right)^2 Z_{jk}\right] \\
&= \frac{N_j N_k}{N_1^2}Z_{1jk}
\end{aligned}
\tag{8.13}
$$

where we have defined the quantity

$$
Z_{1jk} = \frac{1}{2}\left[Z_{1j}+Z_{1k} - \left(\frac{N_1}{N_j}\right)^2 Z_{jk}\right]
\tag{8.14}
$$

This is just the single-winding leakage impedance of winding 1 as part of a 3-winding system, represented by a T-equivalent circuit. It is usually denoted by Z_1 when there are only two other windings involved, which are understood to be 2 and 3. We need the additional subscript labeling here since there can be more than three windings. Substituting (8.13) into (8.9), the circuit equations become

$$
V_j = \frac{N_j}{N_1}V_1 + R_j I_j + \sum_{k=2}^{n}\frac{N_j N_k}{N_1^2}\left(R_1 + Z_{1jk}\right)I_k, \quad \text{for } j=2,\dots,n
\tag{8.15}
$$

The circuit equations (8.15) contain only 2-winding leakage reactances via the term Z_{1jk}. The first term, containing V_1, expresses the voltage transfer across the ideal transformer. Winding 1, with voltage V_1, is on one side of the ideal transformer and its voltage is assumed known. All the other windings are on the other side of the ideal transformer. They are coupled to each other via the cross terms in (8.15). The presence of winding 1 is reflected in these terms. It follows from (8.6), with j, k substituted for 1, 2, that

$$
Z_{1jk} = Z_{1kj}
\tag{8.16}
$$

that is, Z_{1jk} is symmetric in j and k. We should also note, from the discussion following (8.6), that

$$
Z_{1jj} = Z_{1j}
\tag{8.17}
$$

that is, when the second and third indices are equal, Z_{1jj} reduces to the 2-winding leakage reactance.

To simplify the notation further, we define a new quantity:

$$
W_{1jk} = \frac{N_j N_k}{N_1^2}\left(R_1 + Z_{1jk}\right)
\tag{8.18}
$$

so that (8.15) becomes

$$\mathbf{V}_j = \frac{N_j}{N_1}\mathbf{V}_1 + R_j\mathbf{I}_j + \sum_{k=2}^{n} W_{1jk}\mathbf{I}_k, \quad \text{for } j = 2,\dots,n \tag{8.19}$$

W_{1jk} is, in general, a complex quantity since it contains resistive and reactive elements. From (8.16), we see that

$$W_{1jk} = W_{1kj} \tag{8.20}$$

so that it is symmetric in j and k. We also see, from (8.17), that when j = k,

$$W_{1jj} = \left(\frac{N_j}{N_1}\right)^2 (R_1 + Z_{1j}) \tag{8.21}$$

A circuit diagram, illustrating the meaning of (8.19), is shown in Figure 8.1 for a 4-winding and 4-terminal transformer. All the impedances are on the secondary side of the transformer.

Although the Z's were treated as inductive elements, the derivations given earlier would be unchanged with the Z's containing mutual resistances. This could be useful in high-frequency work [Mom02].

It is often simpler to work in the per-unit system. Using lowercase letters for per-unit quantities, Equation 8.19 becomes in this system

$$\mathbf{v}_j = \mathbf{v}_1 + r_j\mathbf{i}_j + \sum_{k=2}^{n} w_{1jk}\mathbf{i}_k, \quad \text{for } j = 2,\dots,n \tag{8.22}$$

where

$$w_{1jk} = r_1 + z_{1jk}$$
$$z_{1jk} = \frac{1}{2}(z_{1j} + z_{1k} - z_{jk}) \tag{8.23}$$

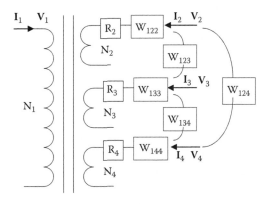

FIGURE 8.1
Circuit model of a 4-winding transformer using only 2-winding leakage reactances and an ideal transformer. (From Del Vecchio, R.M., *IEEE Trans. Power Deliv.*, 21(3), 1300–1308, July 2006. © 2006 IEEE.)

We prefer to derive formulas in the regular system of units, since it is relatively easy to convert them to the per-unit system afterward. Much of the literature is written in the per-unit system, which we will use later to perform calculations.

8.2.3 Transformer Loading

It is convenient in the following applications to consider winding 1 as the primary winding with known voltage. Windings j = 2, ..., n will be regarded as secondary windings. When the secondaries are loaded, the voltages V_j are applied across a load Z_{Lj}. This can be, and often is, complex, containing resistive and reactive terms. The load current flows out of the secondary terminal and into the load. Since our convention has been that the winding currents flow into the terminal, the load current is the negative of the winding current used in the formulas given earlier. Therefore, we have

$$V_j = -Z_{Lj}I_j \qquad (8.24)$$

Substituting into (8.19), the circuit equations with load become

$$\left(R_j + Z_{Lj}\right)I_j + \sum_{k=2}^{n} W_{1jk}I_k = -\frac{N_j}{N_1}V_1, \quad \text{for } j = 2,\ldots,n \qquad (8.25)$$

This can be written as a matrix equation (complex and symmetric) and solved by standard methods. V_1, the N's, the resistances, the 2-winding leakage reactances, and the load impedances are assumed known. Therefore, the I's can be solved for.

8.3 Transformers with Winding Connections within a Phase

Thus far, we have been discussing a single-phase transformer or one phase of a 3-phase transformer with uncoupled phases. Furthermore, the windings have been without internal interconnections. We now apply the theory to more complicated transformers, where the windings can be interconnected within a single phase. In general, a multiterminal transformer with n windings may have less than n − 1 secondary terminals. Since the number of equations to solve equals the number of secondary terminals, we may have fewer than n − 1 equations to solve.

8.3.1 Two Secondary Windings in Series

This is a common situation. For example, a tap winding is often connected to a secondary winding. There are also cases where two series-connected secondary windings are used. Let windings p and q be the series-connected secondaries. Let s denote the terminal to which one of these windings is attached. We have

$$\begin{aligned} V_s &= V_p + V_q \\ I_s &= I_p = I_q \end{aligned} \qquad (8.26)$$

Adding the j = p and j = q equations in (8.19) yields for the resulting system

$$\mathbf{V}_s = \left(\frac{N_p + N_q}{N_1}\right)\mathbf{V}_1 + \left(R_p + R_q\right)\mathbf{I}_s$$

$$+ \left(W_{1pp} + 2W_{1pq} + W_{1qq}\right)\mathbf{I}_s + \sum_{k=2,k\neq p,k\neq q}^{n}\left(W_{1pk} + W_{1qk}\right)\mathbf{I}_k \qquad (8.27)$$

$$\mathbf{V}_j = \frac{N_j}{N_1}\mathbf{V}_1 + R_j\mathbf{I}_j + \left(W_{1jp} + W_{1jq}\right)\mathbf{I}_s + \sum_{k=2,k\neq p,k\neq q}^{n} W_{1jk}\mathbf{I}_k \quad \text{for } j = 2,\ldots,n, j \neq p, j \neq q$$

Thus, the equations for windings p and q are eliminated and an equation for terminal s is added, resulting in one less equation.

As an example, consider a 3-winding system with windings 2 and 3 in series. Then (8.27) reduces to one equation, the s terminal equation, which becomes

$$\mathbf{V}_s = \left(\frac{N_2 + N_3}{N_1}\right)\mathbf{V}_1 + \left(R_2 + R_3 + W_{122} + 2W_{123} + W_{133}\right)\mathbf{I}_s \qquad (8.28)$$

If we omit the resistance terms from the W's in this equation, we obtain an effective reactance on the secondary side, Z_{eff}:

$$Z_{eff} = \frac{N_2\left(N_2 + N_3\right)}{N_1^2}Z_{12} + \frac{N_3\left(N_2 + N_3\right)}{N_1^2}Z_{13} - \frac{N_3}{N_2}Z_{23} \qquad (8.29)$$

This is the effective leakage reactance of this 2-terminal transformer on the secondary side. We obtained a formula for the reactance for this connection in Chapter 4. However, there the joined windings were windings 1 and 2. Here the joined windings are windings 2 and 3. By relabeling the indices and expressing the single-winding reactances in terms of 2-winding reactances, the previous formula reduces to (8.29).

8.3.2 Primary Winding in Series with a Secondary Winding

This is also a fairly common situation. We could have two high voltage primary windings connected in series or a high voltage winding connected in series with a tap winding. This requires special consideration, because we have singled out the primary winding in the treatment mentioned earlier. Let winding p be in series with winding 1. We have

$$\mathbf{V}_s = \mathbf{V}_1 + \mathbf{V}_p$$

$$\mathbf{I}_1 = \mathbf{I}_p = -\frac{1}{\left(N_1 + N_p\right)}\sum_{k=2,k\neq p}^{n} N_k\mathbf{I}_k \qquad (8.30)$$

We have used the amp-turn balance condition here. Substituting for $\mathbf{I}_p$ in (8.19), we get

$$
\mathbf{V}_j = \frac{N_j}{N_1}\mathbf{V}_1 + R_j\mathbf{I}_j + \sum_{k=2,k\neq p}^{n}\left[W_{1jk} - \left(\frac{N_k}{N_1 + N_p}\right)W_{1jp}\right]\mathbf{I}_k \quad \text{for } j = 2,\dots,n, j \neq p
$$

$$
\mathbf{V}_p = \frac{N_p}{N_1}\mathbf{V}_1 + \sum_{k=2,k\neq p}^{n}\left[W_{1pk} - \left(\frac{N_k}{N_1 + N_p}\right)(R_p + W_{1pp})\right]\mathbf{I}_k
$$

$$(8.31)$$

We therefore get for $\mathbf{V}_s$

$$
\mathbf{V}_s = \left(\frac{N_1 + N_p}{N_1}\right)\mathbf{V}_1 + \sum_{k=2,k\neq p}^{n}\left[W_{1pk} - \left(\frac{N_k}{N_1 + N_p}\right)(R_p + W_{1pp})\right]\mathbf{I}_k \tag{8.32}
$$

Solving (8.32) for $\mathbf{V}_1$ and substituting into the $\mathbf{V}_j$ equations,

$$
\mathbf{V}_j = \left(\frac{N_j}{N_1 + N_p}\right)\mathbf{V}_s + R_j\mathbf{I}_j + \sum_{k=2,k\neq p}^{n}\left[W_{1jk} - \left(\frac{N_k}{N_1 + N_p}\right)W_{1jp}\right.
$$

$$
\left. + \frac{N_jN_k}{(N_1 + N_p)^2}(R_p + W_{1pp}) - \left(\frac{N_j}{N_1 + N_p}\right)W_{1pk}\right]\mathbf{I}_k \quad \text{for } j = 2,\dots,n, j \neq p \tag{8.33}
$$

This last formula shows that the two windings in series are placed on one side of the ideal transformer and that their series voltage, $\mathbf{V}_s$, is assumed known. The presence of both windings is reflected in the terms in (8.33) on the secondary side.

As an example, consider a 3-winding transformer with windings 1 and 2 in series. There is just one equation (8.33) for winding 3. Dropping the resistance terms in (8.33), we obtain an effective leakage reactance of

$$
Z_{\text{eff}} = \left(\frac{N_3}{N_1 + N_2}\right)^2\left[-\frac{N_2}{N_1}Z_{12} + \left(\frac{N_1 + N_2}{N_1}\right)Z_{13} + \left(\frac{N_1 + N_2}{N_2}\right)Z_{23}\right] \tag{8.34}
$$

This is similar to the previous example except that the joined windings are windings 1 and 2. The formula in Chapter 4 applies directly to this connection without relabeling the indices, except that the reactance must be transformed to the secondary side. This transference is done by the $(N_3/(N_1 + N_2))^2$ term in (8.34).

8.3.3 Autotransformer

An autotransformer connection is shown in Figure 8.2. We are letting winding 1 be the series or HV winding and winding p be the common or LV winding. The current at the

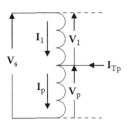

FIGURE 8.2
Autotransformer connection. (From Del Vecchio, R.M., *IEEE Trans. Power Deliv.*, 21(3), 1300–1308, July 2006. © 2006 IEEE.)

common terminal is positive into the terminal by the convention used here and is labeled I_{Tp}. The voltage at the HV terminal is labeled V_s.

From the diagram, we can see that

$$V_s = V_1 + V_p$$

$$I_{Tp} = I_p - I_1$$

$$= \left(\frac{N_1 + N_p}{N_1}\right) I_p + \frac{1}{N_1} \sum_{k=2, k \neq p}^{n} N_k I_k \tag{8.35}$$

where we have used amp-turn balance. Proceeding similarly as in the previous section, we find for the secondary terminal equations

$$V_j = \left(\frac{N_j}{N_1 + N_p}\right) V_s + R_j I_j$$

$$+ \left(\frac{N_1}{N_1 + N_p}\right)\left[W_{1jp} - \left(\frac{N_j}{N_1 + N_p}\right)\left(R_p + W_{1pp}\right)\right] I_{Tp}$$

$$+ \sum_{k=2, k \neq p}^{n} \left\{ W_{1jk} - \left(\frac{N_k}{N_1 + N_p}\right)\left[W_{1jp} + \frac{N_j}{N_k} W_{1pk} - \left(\frac{N_j}{N_1 + N_p}\right)\left(R_p + W_{1pp}\right)\right]\right\} I_k \tag{8.36}$$

for $j = 2, \ldots, n, j \neq p$

$$V_p = \left(\frac{N_p}{N_1 + N_p}\right) V_s + \left(\frac{N_1}{N_1 + N_p}\right)^2 \left(R_p + W_{1pp}\right) I_{Tp}$$

$$+ \left(\frac{N_1}{N_1 + N_p}\right) \sum_{k=2, k \neq p}^{n} \left[W_{1pk} - \left(\frac{N_k}{N_1 + N_p}\right)\left(R_p + W_{1pp}\right)\right] I_k \tag{8.37}$$

The primary terminal voltage, V_s, is assumed to be given. It is on one side of the ideal transformer. The secondaries, with their impedances and couplings, are on the other side of the ideal transformer. All the direct and coupling terms are expressed in terms of winding resistances and leakage reactances.

8.4 Multiphase Transformers

At this point, some simplification in notation is useful. We will recast (8.19), using vector and matrix notation. Vectors will be denoted by boldfaced type and matrices by italicized type. Since the voltage and current vectors are vectors of phasors, this is consistent with our phasor notation; however, the elements of these vectors, although phasors, will be denoted in ordinary type. Other vectors such as the turns vector are not vectors of phasors but will still be denoted with boldfaced type. The vector notation will supersede our previous phasor notation here. Let $\mathbf{V}$, $\mathbf{I}$, and $\mathbf{N}$ be vectors of winding secondary voltages, currents, and turns. The turns will be divided by N_1:

$$\mathbf{V} = \left(V_2, V_3, \ldots, V_n \right)^T$$
$$\mathbf{I} = \left(I_2, I_3, \ldots, I_n \right)^T \tag{8.38}$$
$$\mathbf{N} = \left(\frac{N_2}{N_1}, \frac{N_3}{N_1}, \ldots, \frac{N_n}{N_1} \right)^T$$

Here superscript T denotes transpose, which converts these row vectors to column vectors. Let R denote the diagonal matrix of secondary winding resistances:

$$R = \begin{pmatrix} R_2 & 0 & 0 & 0 \\ 0 & R_3 & 0 & 0 \\ 0 & 0 & \ddots & 0 \\ 0 & 0 & 0 & R_n \end{pmatrix} \tag{8.39}$$

Let W denote the matrix, W_{1jk}, with rows labeled j and columns k, for j, k = 2, …, n. Then (8.19) becomes

$$\mathbf{V} = \mathbf{N}V_1 + \left(R + W \right)\mathbf{I} \tag{8.40}$$

This equation refers to a single phase. For a 3-phase system, label the phases a, b, and c. The unit phasors, associated with these respective phases, are, as before, denoted by 1, α^2, and α. They are 120° apart in the complex plane and, for convenience, are given here and diagramed in Figure 8.3:

$$1$$
$$\alpha^2 = -\frac{1}{2} - j\frac{\sqrt{3}}{2} \tag{8.41}$$
$$\alpha = -\frac{1}{2} + j\frac{\sqrt{3}}{2}$$

where j is the imaginary unit.

Using a, b, and c as subscripts to label the phase quantities, we need to expand (8.40) for a 3-phase system:

$$\begin{pmatrix} \mathbf{V}_a \\ \mathbf{V}_b \\ \mathbf{V}_c \end{pmatrix} = \begin{pmatrix} \mathbf{N}V_{1a} \\ \mathbf{N}V_{1b} \\ \mathbf{N}V_{1c} \end{pmatrix} + \begin{pmatrix} R+W & 0 & 0 \\ 0 & R+W & 0 \\ 0 & 0 & R+W \end{pmatrix} \begin{pmatrix} \mathbf{I}_a \\ \mathbf{I}_b \\ \mathbf{I}_c \end{pmatrix} \tag{8.42}$$

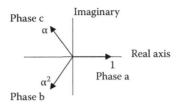

FIGURE 8.3
Unit phasors of a 3-phase system. (From Del Vecchio, R.M., *IEEE Trans. Power Deliv.*, 21(3), 1300–1308, July 2006. © 2006 IEEE.)

The vectors of each phase are now considered as phasors. In particular, $V_{1a} = V_1$, $V_{1b} = V_1\alpha^2$, and $V_{1c} = V_1\alpha$ since V_1 is the driving voltage, assumed known. However V_a, V_b, V_c or I_a, I_b, I_c cannot be expressed so simply because of the complex nature of the matrix in (8.42). The V's and I's will still form a balanced 3-phase system, but there could be an overall phase shift from the input V_1 system. The extended vectors shown earlier are of dimension $3(n-1)$ for an n-winding system. Equation 8.42 represents a system of uncoupled phases. We now explore several possibilities for coupling the phases and for which this matrix formalism is very useful. For our further discussion, the vectors $\mathbf{V}$, $\mathbf{NV}_1$, and $\mathbf{I}$ will refer to the 3-phase system and the matrix $R + W$ will be assumed to have the expanded form in (8.42). We now need to treat $\mathbf{NV}_1$ as a 3-phase vector entity, since $\mathbf{V}_1$ contains phase information. This is why $\mathbf{V}_1$ is in boldfaced type here. Thus, in our abbreviated notation, Equation 8.42 is expressed as

$$\mathbf{V} = \mathbf{NV}_1 + (R+W)\mathbf{I} \tag{8.43}$$

Writing (8.43) out for two secondary windings, it becomes

$$
\begin{pmatrix} V_{a2} \\ V_{a3} \\ V_{b2} \\ V_{b3} \\ V_{c2} \\ V_{c3} \end{pmatrix} = \begin{pmatrix} (N_2/N_1)V_{1a} \\ (N_3/N_1)V_{1a} \\ (N_2/N_1)V_{1b} \\ (N_3/N_1)V_{1b} \\ (N_2/N_1)V_{1c} \\ (N_3/N_1)V_{1c} \end{pmatrix}
$$

$$
+ \begin{pmatrix} R_2 + W_{122} & W_{123} & 0 & 0 & 0 & 0 \\ W_{132} & R_3 + W_{133} & 0 & 0 & 0 & 0 \\ 0 & 0 & R_2 + W_{122} & W_{123} & 0 & 0 \\ 0 & 0 & W_{132} & R_3 + W_{133} & 0 & 0 \\ 0 & 0 & 0 & 0 & R_2 + W_{122} & W_{123} \\ 0 & 0 & 0 & 0 & W_{132} & R_3 + W_{133} \end{pmatrix} \begin{pmatrix} I_{a2} \\ I_{a3} \\ I_{b2} \\ I_{b3} \\ I_{c2} \\ I_{c3} \end{pmatrix}
$$

Equation 8.43 is basically the same as (8.40), except that the vector quantities refer to all three phases and $\mathbf{V}_1$ is a phasor vector. We will normally be dealing with 3-phase quantities, unless otherwise noted, with $\mathbf{V}_1$ representing a balanced 3-phase system of phasors.

8.4.1 Delta Connection

The delta connection diagram for a set of windings from three phases is shown in Figure 8.4. This is actually a −30° delta since the terminal voltages are shifted by −30° relative to the phase voltages.

The voltages at the terminals are labeled V_{T1}, etc. The terminal voltages are assumed to be relative to ground potential. Since there is no ground point or reference voltage in the delta connection, one is assumed at the middle of the delta as shown in Figure 8.4b. Relative to this, the voltage relationships can be expressed in matrix form as

$$\begin{pmatrix} V_{T1} \\ V_{T2} \\ V_{T3} \end{pmatrix} = \begin{pmatrix} 1/3 & 0 & -1/3 \\ -1/3 & 1/3 & 0 \\ 0 & -1/3 & 1/3 \end{pmatrix} \begin{pmatrix} V_a \\ V_b \\ V_c \end{pmatrix}$$
$$\begin{pmatrix} V_a \\ V_b \\ V_c \end{pmatrix} = \begin{pmatrix} 1 & -1 & 0 \\ 0 & 1 & -1 \\ -1 & 0 & 1 \end{pmatrix} \begin{pmatrix} V_{T1} \\ V_{T2} \\ V_{T3} \end{pmatrix} \tag{8.44}$$

V_{T1}, etc., are phasors. By convention, we are assuming that positive currents flow into the terminals and, within the winding, opposite to the direction of voltage increase. The voltage representation in the first equation of (8.44) assumes balanced 3-phase delta voltages. It is not the inverse of the matrix in the second equation in (8.44), which applies to balanced or unbalanced voltages.

Labeling the terminal currents, I_{T1}, etc., the winding and terminal currents are related by

$$\begin{pmatrix} I_{T1} \\ I_{T2} \\ I_{T3} \end{pmatrix} = \begin{pmatrix} 1 & 0 & -1 \\ -1 & 1 & 0 \\ 0 & -1 & 1 \end{pmatrix} \begin{pmatrix} I_a \\ I_b \\ I_c \end{pmatrix}$$
$$\begin{pmatrix} I_a \\ I_b \\ I_c \end{pmatrix} = \begin{pmatrix} 1/3 & -1/3 & 0 \\ 0 & 1/3 & -1/3 \\ -1/3 & 0 & 1/3 \end{pmatrix} \begin{pmatrix} I_{T1} \\ I_{T2} \\ I_{T3} \end{pmatrix} \tag{8.45}$$

The second current equation in (8.45) assumes balanced 3-phase currents in the delta. The two matrices in (8.45) are not inverses. The first equation in (8.45) applies to balanced or unbalanced delta currents.

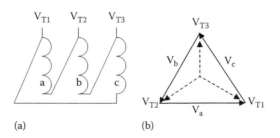

(a)   (b)

FIGURE 8.4
Delta connection: (a) winding phase and (b) phasor relationships. (From Del Vecchio, R.M., *IEEE Trans. Power Deliv.*, 21(3), 1300–1308, July 2006. © 2006 IEEE.)

Assume winding p is delta connected. Let M_V be a unit diagonal matrix of dimension $3(n - 1)$ for an n-winding system. Modify the p winding rows of M_V for all three phases by inserting the entries from the first matrix in (8.44) in the appropriate slots, for example, place 1/3 in the p winding diagonal position for all three phases, place a $-1/3$ in the a-phase row and in the c-phase column for winding p, etc. For example, if the p winding were winding 3 of a transformer with two secondary windings, 2 and 3 (winding 1 is the primary winding that has been eliminated from the matrix), we would have

$$M_V = \begin{pmatrix} 1 & 0 & 0 & 0 & 0 & 0 \\ 0 & 1/3 & 0 & 0 & 0 & -1/3 \\ 0 & 0 & 1 & 0 & 0 & 0 \\ 0 & -1/3 & 0 & 1/3 & 0 & 0 \\ 0 & 0 & 0 & 0 & 1 & 0 \\ 0 & 0 & 0 & -1/3 & 0 & 1/3 \end{pmatrix} \tag{8.46}$$

Let M_I be a similar matrix corresponding to the second matrix of (8.45) for the currents, for example, for winding 3 of a transformer with two secondaries:

$$M_I = \begin{pmatrix} 1 & 0 & 0 & 0 & 0 & 0 \\ 0 & 1/3 & 0 & -1/3 & 0 & 0 \\ 0 & 0 & 1 & 0 & 0 & 0 \\ 0 & 0 & 0 & 1/3 & 0 & -1/3 \\ 0 & 0 & 0 & 0 & 1 & 0 \\ 0 & -1/3 & 0 & 0 & 0 & 1/3 \end{pmatrix} \tag{8.47}$$

Thus, we have, assuming the expanded 3-phase notation,

$$\mathbf{V}_T = M_V \mathbf{V}, \quad \mathbf{I} = M_I \mathbf{I}_T \tag{8.48}$$

Here T as a subscript is labeling the terminal quantities. In the case of a delta connection, the number of terminals is the same as the number of windings. Applying these to (8.43), we get

$$\mathbf{V}_T = M_V (\mathbf{N}\mathbf{V}_1) + M_V (R + W) M_I \mathbf{I}_T \tag{8.49}$$

This equation now applies to terminal quantities. We can impose a load on these, as was done in Section 8.2.3, and solve the resulting equations to get the terminal currents. The terminal voltages can then be found from (8.49) or from the load voltage drop equation (8.24). The phase voltages and currents can then be found by applying the appropriate matrix in (8.44) and (8.45) in expanded form.

8.4.2 Zigzag Connection

The zigzag connection is often used for grounding purposes. It affords a low impedance to zero sequence currents but a high impedance to balanced currents. The connection involves

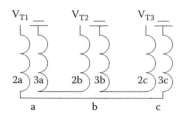

FIGURE 8.5
Zigzag winding connection. (From Del Vecchio, R.M., *IEEE Trans. Power Deliv.*, 21(3), 1300–1308, July 2006. © 2006 IEEE.)

interconnecting two secondary windings from different phases. The diagram is shown in Figure 8.5 where the secondary windings are labeled 2a, 2b, and 2c and 3a, 3b, and 3c.

The winding voltages are assumed to increase upward in the diagram. Therefore, the windings are connected so that their phasor voltages subtract. The currents in the connected windings are of opposite sign based on our sign convention. It is also helpful to view this from the phasor diagram viewpoint as shown in Figure 8.6.

Assuming the transformer has only two secondary windings, the matrices connecting the terminal to the winding voltages and currents, M_V and M_I, as illustrated in (8.48), are

$$M_V = \begin{pmatrix} 1 & 0 & 0 & 0 & 0 & -1 \\ 0 & -1 & 1 & 0 & 0 & 0 \\ 0 & 0 & 0 & -1 & 1 & 0 \end{pmatrix}$$

$$M_I = \begin{pmatrix} 1 & 0 & 0 \\ 0 & -1 & 0 \\ 0 & 1 & 0 \\ 0 & 0 & -1 \\ 0 & 0 & 1 \\ -1 & 0 & 0 \end{pmatrix} \tag{8.50}$$

These are no longer square matrices. The column dimension of M_V is 6 since it operates on the voltages in the original configuration consisting of three phases with two secondary

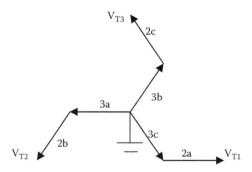

FIGURE 8.6
Zigzag phasor relationships. (From Del Vecchio, R.M., *IEEE Trans. Power Deliv.*, 21(3), 1300–1308, July 2006. © 2006 IEEE.)

terminal windings per phase. The row dimension is 3 since there are three phases with only one terminal per phase after the zigzag connection is made. The opposite is true for M_I since this operates on the terminal currents after the zigzag connection is made to produce the currents in the two separate windings. We should note that windings 2 and 3 should have the same number of turns in this connection. Let $N_z = N_2 = N_3$ be their common number of turns. Using this and the (8.50) equations in (8.49) and dropping the resistance matrix, we find

$$\begin{pmatrix} V_{T1} \\ V_{T2} \\ V_{T3} \end{pmatrix} = \frac{N_z}{N_1} \sqrt{3} V_1 \begin{pmatrix} e^{-j30°} \\ e^{-j150°} \\ e^{j90°} \end{pmatrix}$$
$$+ \begin{pmatrix} W_{122} + W_{133} & -W_{123} & -W_{123} \\ -W_{123} & W_{122} + W_{133} & -W_{123} \\ -W_{123} & -W_{123} & W_{122} + W_{133} \end{pmatrix} \begin{pmatrix} I_{T1} \\ I_{T2} \\ I_{T3} \end{pmatrix} \tag{8.51}$$

If we make use of the fact that the I_T's form a balanced set, that is, $I_{T1} + I_{T2} + I_{T3} = 0$, we find that this reduces to a set of three uncoupled equations. The first of these is

$$V_{T1} = \frac{N_z}{N_1} \sqrt{3} V_1 e^{-j30°} + \left(W_{122} + W_{133} + W_{123} \right) I_{T1} \tag{8.52}$$

Using the definition of the W's and ignoring the resistance terms, the effective impedance term multiplying I_{T1} is

$$Z_{eff} = \frac{3}{2} \left(\frac{N_z}{N_1} \right)^2 \left(Z_{12} + Z_{13} \right) - \frac{1}{2} Z_{23} \tag{8.53}$$

This agrees with the formula found in Chapter 7 after that is expressed in terms of 2-winding reactances and transferred to the zigzag side of the ideal transformer.

After applying a load as in Section 8.2.3, the resulting equations may be solved for the terminal currents. Then the version of (8.51), which includes the resistance terms and possibly other secondary windings, in conjunction with the load voltage drop equation (8.24), can be used to find the terminal voltages. Applying M_I to the terminal current vector, we can find the currents in the windings. Then, (8.42) can be used to find the voltages across the individual windings.

8.5 Generalizing the Model

The procedure outlined earlier for the multiphase transformers can be generalized. For example, the addition of two windings in series from the same phase can be accomplished by applying the appropriate matrices M_V and M_I to (8.40). In this case, we do not need to work with the expanded 3-phase version of Equation 8.43 since the phases are uncoupled.

For two secondary windings, labeled 2 and 3, with the terminal T_1 at the positive end of winding 2, the appropriate matrix transformations are

$$V_{T1} = \begin{pmatrix} 1 & 1 \end{pmatrix} \begin{pmatrix} V_2 \\ V_3 \end{pmatrix}, \quad \begin{pmatrix} I_2 \\ I_3 \end{pmatrix} = \begin{pmatrix} 1 \\ 1 \end{pmatrix} (I_{T1})$$

where
$$M_V = \begin{pmatrix} 1 & 1 \end{pmatrix}, \quad M_I = \begin{pmatrix} 1 \\ 1 \end{pmatrix}$$

(8.54)

Naturally, if the two secondaries are part of a larger system of secondaries, the matrices M_V and M_I must be expanded with unit diagonal elements for the other windings and the 1's in (8.54) need to be inserted into the appropriate slots.

In general, we can consider a series of winding interconnections for the secondaries, each described by matrices that we label 1, 2, …, s, corresponding to the order in which they are applied, where s is the number of separate connections. This results in the series

$$\mathbf{V}_{T1} = M_{V1}\mathbf{V}, \quad \mathbf{V}_{T2} = M_{V2}\mathbf{V}_{T1}, \ldots, \mathbf{V}_{Ts} = M_{Vs}\mathbf{V}_{Ts-1}$$
$$\Rightarrow \quad \mathbf{V}_{Ts} = M_{Vs}M_{Vs-1}, \ldots, M_{V1}\mathbf{V}$$
$$\mathbf{I} = M_{I1}\mathbf{I}_{T1}, \quad \mathbf{I}_{T1} = M_{I2}\mathbf{I}_{T2}, \ldots, \mathbf{I}_{Ts-1} = M_{Is}\mathbf{I}_{Ts}$$
$$\Rightarrow \quad \mathbf{I} = M_{I1}M_{I2}, \ldots, M_{Is}\mathbf{I}_{Ts}$$

(8.55)

In this case, the resulting matrices

$$M_V = M_{Vs}M_{Vs-1}, \ldots, M_{V2}M_{V1}$$
$$M_I = M_{I1}M_{I2}, \ldots, M_{Is-1}M_{Is}$$

(8.56)

should be used in (8.49). Notice that the M_V steps are applied in reverse order of the M_I steps. Here, as previously, the matrices and Equation 8.49 could refer to a single phase if there are no phase-to-phase interconnections or to a system including all three phases if there are phase-to-phase interconnections.

This procedure will lead to a set of terminal equations to which the loads may be attached to get an equation for the terminal currents. Let the loading condition be given, in vector matrix form, by

$$\mathbf{V}_L = Z_L\mathbf{I}_L$$

(8.57)

where
   $\mathbf{V}_L$ is a vector of load voltages at the various terminals
   $\mathbf{I}_L$ is a vector of load currents out of the various terminals

In general, $\mathbf{V}_L$ and $\mathbf{I}_L$ can be 3-phase vectors. The load impedance matrix, Z_L, can be in general complex and may not necessarily be diagonal.

By our convention, $\mathbf{I}_L = -\mathbf{I}_T$, and setting $\mathbf{V}_T = \mathbf{V}_L$ in (8.49) with M_V and M_I as in (8.56) for the particular connections involved, the terminal equations to solve are

$$\left[Z_L + M_V (R + W) M_I \right] \mathbf{I}_T = -M_V (\mathbf{NV}_1)$$

(8.58)

Once the terminal currents are found, the winding currents can be found by applying M_I to the terminal current vector. The terminal voltages can be found from (8.57) and the winding voltages by substituting the winding currents in (8.43) or (8.40) for a single-phase system.

By singling out a particular winding, for example, the high voltage winding, as winding 1, any interconnections involving this winding must be handled separately from the procedure mentioned earlier, as was done previously for the case of a winding in series with winding 1 or in the case of an autotransformer. However, after this winding's interconnections are accounted for, the procedure mentioned earlier can be applied to the modified equations to account for any additional interconnections among the secondaries.

We should note that M_I is the transpose of M_V. This was seen in (8.50) for the zigzag connection, in (8.46) and (8.47) for a delta connection, as well as in (8.54). This follows from our convention regarding the current direction in a winding. Therefore, if $R + W$ is symmetric, then $M_V(R + W)M_I$ will also be symmetric.

8.6 Regulation and Terminal Impedances

Regulation effects can be obtained by comparing the output terminal voltages with the no-load terminal voltages. The output terminal voltages can be found from (8.57) once the load currents are solved for. The no-load terminal voltages are given by the negative of the right-hand side of (8.58), which amounts to setting the currents to zero in (8.49), that is, $\mathbf{V}_{T,\text{no-load}} = M_V(\mathbf{NV}_1)$. One can obtain the voltage magnitude drop and angle shift from no-load to load conditions from these voltage vectors.

As an example, consider a Y–Y autotransformer with X-line taps and a delta-connected tertiary. This is a 3-phase 60 MVA unit. The HV winding also has no-load taps within the winding. The 2-winding leakage reactances when the HV winding no-load taps are in the center position are, in per-unit %, given in Table 8.1.

Single-winding resistances in per-unit percent were also used in the calculation and are shown in Table 8.2.

Regulation effects at different power factor loads with the on-load taps out, expressed as a percent voltage drop from no-load conditions, are given in Table 8.3, where calculations

TABLE 8.1

Winding to Winding (Wdg 1 to Wdg 2) Leakage Reactances

2-Winding Leakage Reactances		
Wdg 1	**Wdg 2**	**p.u. %**
HV	LV	23.12
HV	Taps	44.22
HV	TV	45.71
LV	Taps	16.53
LV	TV	19.44
Taps	TV	11.77

Source: Del Vecchio, R.M., *IEEE Trans. Power Deliv.*, 21(3), 1300–1308, July 2006. © 2006 IEEE.

TABLE 8.2

Single Winding Resistances

Winding	HV	LV	Taps	TV
	Winding Resistances (p.u. %)			
Resistance	0.391	0.254	1.3	0.62

Source: Del Vecchio, R.M., *IEEE Trans. Power Deliv.*, 21(3), 1300–1308, July 2006. © 2006 IEEE.

TABLE 8.3

Regulation as a Percent of Voltage Drop from No-Load Conditions

	Regulation as % Drop			
Power factor %	100	90	80	70
Calculation	0.47	3.55	4.63	5.36
Test	0.5	3.71	4.88	5.68

Source: Del Vecchio, R.M., *IEEE Trans. Power Deliv.*, 21(3), 1300–1308, July 2006. © 2006 IEEE.

are compared with test results. The calculated results were obtained by the methods described here, using the per-unit reactances and resistances given in Tables 8.1 and 8.2. These per-unit quantities were converted to ohmic quantities for the calculation.

These regulation effects can also be obtained at other tap positions but were not measured. However, we can use them to get the terminal to terminal reactances at various tap positions, which were measured.

To obtain the terminal to terminal impedances, Z_{tt}, the regulation, R, can be expressed as a voltage ratio, using the simplified model of Figure 3.12, where $Z_{2,eq}$ is replaced by Z_{tt}:

$$R = \frac{V_{Term,load}}{V_{Term,no-load}} = \frac{1}{\left(1 + \left(Z_{tt}/Z_L\right)\right)} \tag{8.59}$$

Here Z_L is the load impedance, assumed known, which can be expressed in terms of its real and imaginary parts. Z_{tt} contains unknown real and imaginary parts, so there are two unknowns. The no-load terminal voltages are known and the load terminal voltages are obtained by solving (8.58) and then using (8.57). The ratio of these two quantities, R, contains magnitude and phase information. The magnitude is used to obtain the regulation expressed as a percentage drop as given in Table 8.3. The phase shift from the no-load condition is the phase of R. These two calculated quantities allow us to solve (8.59) for the two unknowns in Z_{tt}. The imaginary part of Z_{tt} is the terminal to terminal reactance. We compare this with test values of the HV to LV reactance in Table 8.4. Note that the 2-winding leakage reactances between the HV and the other windings change slightly at the different no-load tap settings, and these were used in the calculations in Table 8.4 for the All in and All out no-load tap positions.

The HV to TV terminal to terminal reactances can be found by the same methods. They are compared with test results in Table 8.5. They were only measured with the load taps out.

This type of calculated information can be valuable at the design stage to determine whether regulation and impedance requirements will be met.

TABLE 8.4

HV to LV Percent Leakage Reactances at Different Tap
Positions from Calculation and Test

	HV–LV Terminal Reactances (p.u. %)		
Load/No-Load Taps	All In	Center	All Out
Boost	8.46 C	7.94 C	7.68 C
	8.53 T	8.08 T	7.74 T
Out	8.11 C	7.56 C	7.28 C
	8.04 T	7.56 T	7.20 T
Buck	8.05 C	7.48 C	7.16 C
	8.07 T	7.56 T	7.16 T

Source: Del Vecchio, R.M., *IEEE Trans. Power Deliv.*, 21(3),
1300–1308, July 2006. © 2006 IEEE.
C, calc; T, test.

TABLE 8.5

HV to TV Percent Leakage Reactances at Several Tap
Positions from Calculation and Test

	HV–TV Reactances (p.u. %)		
Load/No-Load Taps	All In	Center	All Out
Out	29.35 C	28.81 C	28.54 C
	29.54 T	29.05 T	28.58 T

Source: Del Vecchio, R.M., *IEEE Trans. Power Deliv.*, 21(3),
1300–1308, July 2006. © 2006 IEEE.
C, calc; T, test.

8.7 Multiterminal Transformer Model for Balanced and Unbalanced Load Conditions

Thus far, the multiterminal model developed required balanced loading when windings, such as a delta winding, are present. The model also singled out the input (HV) winding for special treatment. This leads to complications when the input winding is interconnected with other windings. In this section, both of these issues are addressed so that a more general treatment emerges. This generalization can accommodate balanced or unbalanced loading, including unbalanced short circuits.

The starting point for the circuit model is the calculation (or measurement) of the 2-winding leakage reactances between all pairs of windings. Both the positive and zero sequence reactances are needed, although the latter are often estimated from the positive sequence reactances. This contrasts with the previous development where only the positive sequence reactances were needed. We assume that the positive and negative sequence reactances are equal, whereas the zero sequence reactances can differ. The 2-winding leakage reactances are combined to form a reduced reactance matrix. This matrix is called reduced because it excludes one of the windings. We also want to make this distinction because we will later be adding the excluded winding back so all the windings can be treated on the same footing.

This reduced matrix is the same matrix used earlier for the balanced model but is given a slightly different notation here. This matrix is then inverted to get an admittance matrix. An amp-turn balance condition is imposed on the admittance matrix, which extends it to a matrix involving all the windings. This full admittance matrix is singular.

Although one can work in the admittance representation, it is often convenient to solve problems in the impedance representation. In order to invert the full admittance matrix, the core excitation data are added. Since the resulting matrix is close to being singular, the inversion must be carried out with high accuracy.

Superposed on this overall scheme, we will be making winding interconnections as needed via voltage and current transfer matrices. We will then show how these transfer matrices can be used to obtain winding currents and voltages, after the terminal loading problem has been solved.

Our main concern here is with constant frequency steady-state a.c. applications. We will therefore work with reactances rather than inductances. We will include single-winding resistances and ignore mutual resistances, which may be present at high frequencies [Mom02].

8.7.1 Theory

The starting point for this development is the formula relating voltages and currents developed earlier, where we singled out winding 1 (usually the HV winding) for exclusion. Including the single-winding resistances, the basic formula is given by (8.15), which we repeat here:

$$V_j = \frac{N_j}{N_1} V_1 + R_j I_j + \sum_{k=2}^{n} \frac{N_j N_k}{N_1^2} \left(R_1 + Z_{1jk} \right) I_k \quad \text{for } j = 2, \ldots, n \tag{8.60}$$

where n is the number of windings. The V_j's, I_j's, R_j's, and N_j's are the single-winding voltages, currents, resistances, and turns. Z_{1jk} is a leakage reactance matrix element, derived from 2-winding positive sequence leakage reactances. It is given by

$$Z_{1jk} = \frac{1}{2} \left[Z_{1j} + Z_{1k} - \left(\frac{N_1}{N_j} \right)^2 Z_{jk} \right] \quad j, k = 2, \ldots, n \tag{8.61}$$

where the Z_{ij} are the 2-winding positive sequence leakage reactances.

It is convenient to introduce separate matrices for the real and imaginary parts of (8.60). Since the R's are real and the Z's are imaginary, we define the reduced matrices as

$$R_{jk}^r = R_j \delta_{jk} + \frac{N_j N_k}{N_1^2} R_1$$

$$W_{jk}^r = \frac{N_j N_k}{N_1^2} Z_{1jk} \quad j, k = 2, \ldots, n \tag{8.62}$$

where $\delta_{jk} = 1$ if $j = k$ and 0 otherwise. The superscript r indicates a reduced quantity. Previously, W, without the superscript r, included the resistance of winding 1. Define the reduced vectors as

$$\mathbf{V}^r = \begin{pmatrix} V_2 \\ V_3 \\ \vdots \\ V_n \end{pmatrix}, \quad \mathbf{I}^r = \begin{pmatrix} I_2 \\ I_3 \\ \vdots \\ I_n \end{pmatrix}, \quad \mathbf{N}^r = \begin{pmatrix} N_2 \\ N_3 \\ \vdots \\ N_n \end{pmatrix} \tag{8.63}$$

Using this notation, Equation 8.60 becomes

$$\mathbf{V}^{\mathrm{r}} - \frac{V_1}{N_1}\mathbf{N}^{\mathrm{r}} = \left(R^{\mathrm{r}} + jW^{\mathrm{r}}\right)\mathbf{I}^{\mathrm{r}} \tag{8.64}$$

As before, italicized quantities represent matrices and boldfaced quantities vectors. We have inserted the imaginary unit j to take account of the complex nature of the reduced reactance matrix and therefore consider W^{r} to be real.

Extending this to three phases, labeled a, b, and c, we have

$$\begin{pmatrix} \mathbf{V}_a^r - \dfrac{V_{1a}}{N_1}\mathbf{N}^r \\ \mathbf{V}_b^r - \dfrac{V_{1b}}{N_1}\mathbf{N}^r \\ \mathbf{V}_c^r - \dfrac{V_{1c}}{N_1}\mathbf{N}^r \end{pmatrix} = \left[\begin{pmatrix} R^r & 0 & 0 \\ 0 & R^r & 0 \\ 0 & 0 & R^r \end{pmatrix} + j \begin{pmatrix} W^r & 0 & 0 \\ 0 & W^r & 0 \\ 0 & 0 & W^r \end{pmatrix} \right] \begin{pmatrix} \mathbf{I}_a^r \\ \mathbf{I}_a^r \\ \mathbf{I}_a^r \end{pmatrix} \tag{8.65}$$

The 0's in (8.65) are zero matrices of the same size as the W^{r}'s and R^{r}'s.

At this point, since we are interested in unbalanced conditions in general, we need to include the zero sequence 2-winding leakage reactances, $Z_{0,ij}$. Using these in an expression similar to (8.61), we obtain the analogous quantities, $Z_{0,1jk}$, and the analogous matrix, W_0^{r}, defined by an expression similar to W^{r} in (8.62). Following [Bra82], we need to modify the 3-phase W^{r} matrix in (8.65). We do this by replacing the W^{r} matrices along the diagonal in (8.65) by

$$W_{\mathrm{S}}^{\mathrm{r}} = \frac{1}{3}\left(W_0^{\mathrm{r}} + 2W^{\mathrm{r}}\right) \tag{8.66}$$

We also replace the off-diagonal 0 matrices in (8.65) by

$$W_{\mathrm{M}}^{\mathrm{r}} = \frac{1}{3}\left(W_0^{\mathrm{r}} - W^{\mathrm{r}}\right) \tag{8.67}$$

The R^{r} matrices remain unchanged in this procedure. Thus, the 3-phase W^{r} matrix in (8.65) becomes

$$W^{3\mathrm{r}} = \begin{pmatrix} W_{\mathrm{S}}^{\mathrm{r}} & W_{\mathrm{M}}^{\mathrm{r}} & W_{\mathrm{M}}^{\mathrm{r}} \\ W_{\mathrm{M}}^{\mathrm{r}} & W_{\mathrm{S}}^{\mathrm{r}} & W_{\mathrm{M}}^{\mathrm{r}} \\ W_{\mathrm{M}}^{\mathrm{r}} & W_{\mathrm{M}}^{\mathrm{r}} & W_{\mathrm{S}}^{\mathrm{r}} \end{pmatrix} \tag{8.68}$$

This method is slightly different from that in [Bra82] in the way the matrix (8.68) is organized. For balanced conditions, set $W_0^{\mathrm{r}} = W^{\mathrm{r}}$ and the matrix in (8.68) reduces to that in (8.65). Labeling the quantities in (8.65) with a 3 superscript to indicate that three phases are involved, we rewrite it in simplified notation as

$$\mathbf{V}^{3\mathrm{r}} - \frac{V_1^3}{N_1}\mathbf{N}^{3\mathrm{r}} = \left(R^{3\mathrm{r}} + jW^{3\mathrm{r}}\right)\mathbf{I}^{3\mathrm{r}} \tag{8.69}$$

The relabeled vectors and matrices correspond to the expanded versions in (8.65), with $W^{3\mathrm{r}}$ given by (8.68).

8.7.2 Admittance Representation

In order to include all the windings on a similar footing, it is necessary to invert (8.69). The matrices will then become admittance matrices. This requires the inversion of a complex matrix. This can be done, working with real matrix inverses only, via the formula

$$\left(M_{\text{Re}} + jM_{\text{Im}}\right)^{-1} = \left(M_{\text{Re}} + M_{\text{Im}}M_{\text{Re}}^{-1}M_{\text{Im}}\right)^{-1} - j\left(M_{\text{Im}} + M_{\text{Re}}M_{\text{Im}}^{-1}M_{\text{Re}}\right)^{-1} \tag{8.70}$$

Here M_{Re} and M_{Im} are the real and imaginary components of a general complex matrix and both are assumed invertible. If either component matrix is zero, this reduces to the usual matrix inverse. If software is available for dealing with complex matrix algebra and inverses, then this procedure is unnecessary.

Complex matrices and vectors can be handled by treating them as enlarged real matrices and vectors according to

$$M = \begin{pmatrix} M_{\text{Re}} & -M_{\text{Im}} \\ M_{\text{Im}} & M_{\text{Re}} \end{pmatrix}, \quad \mathbf{V} = \begin{pmatrix} \mathbf{V}_{\text{Re}} \\ \mathbf{V}_{\text{Im}} \end{pmatrix} \tag{8.71}$$

where
   M_{Re} is the real part
   M_{Im} the imaginary part of the complex matrix and similarly for the vector $\mathbf{V}$

Using these, matrix algebra can be carried out as if they were real matrices and vectors. The real and imaginary parts of vectors can then be extracted as needed, the real part at the top and the imaginary part at the bottom of the column vector.

Inverting (8.69), we get

$$\mathbf{I}^{3r} = Y^{3r}\left(\mathbf{V}^{3r} - \frac{V_1^3}{N_1}\mathbf{N}^{3r}\right) \tag{8.72}$$

where Y^{3r} is the complex inverse of $R^{3r} + jW^{3r}$ in (8.69).

In order to include winding 1 on an equal footing with the other windings, we use amp-turn balance to get its current. Thus,

$$I_1 = -\frac{N_2}{N_1}I_2 - \frac{N_3}{N_1}I_3 - \cdots - \frac{N_n}{N_1}I_n \tag{8.73}$$

Using the following matrix, we can transform the reduced current vector in (8.63) to a full current vector, $\mathbf{I}$:

$$\begin{pmatrix} I_1 \\ I_2 \\ I_3 \\ \vdots \\ I_n \end{pmatrix} = \begin{pmatrix} -\dfrac{N_2}{N_1} & -\dfrac{N_3}{N_1} & \cdots & -\dfrac{N_n}{N_1} \\ 1 & 0 & \cdots & 0 \\ 0 & 1 & \cdots & 0 \\ \vdots & \vdots & \ddots & 0 \\ 0 & 0 & 0 & 1 \end{pmatrix} \begin{pmatrix} I_2 \\ I_3 \\ \vdots \\ I_n \end{pmatrix}$$

or (8.74)

$$\mathbf{I} = A\mathbf{I}^r$$

where A is an amp-turn balance matrix for one phase. This matrix has n rows and n − 1 columns. By tripling it along the diagonal, we get a matrix that applies to all three phases. Calling this matrix A^3 for a 3-phase amp-turn balance matrix, we can rewrite (8.74) as

$$\mathbf{I}^3 = A^3\mathbf{I}^{3r} \tag{8.75}$$

where $\mathbf{I}^3$ stands for the full 3-phase current vector of dimension 3n.

Using the transpose of the matrix in (8.74), we can express one phase of the reduced voltage vector in (8.72) in terms of the full voltage vector:

$$
\begin{pmatrix}
V_2 - \dfrac{N_2}{N_1} V_1 \\[2mm]
V_3 - \dfrac{N_3}{N_1} V_1 \\[2mm]
\vdots \\[2mm]
V_n - \dfrac{N_n}{N_1} V_1
\end{pmatrix}
=
\begin{pmatrix}
-\dfrac{N_2}{N_1} & 1 & 0 & \cdots & 0 \\[2mm]
-\dfrac{N_3}{N_1} & 0 & 1 & \cdots & 0 \\[2mm]
\vdots & \vdots & \vdots & \ddots & \vdots \\[2mm]
-\dfrac{N_n}{N_1} & 0 & 0 & \cdots & 1
\end{pmatrix}
\begin{pmatrix}
V_1 \\[2mm]
V_2 \\[2mm]
V_3 \\[2mm]
\vdots \\[2mm]
V_n
\end{pmatrix}
$$

or

$$\mathbf{V}^r - \dfrac{V_1}{N_1}\mathbf{N}^r = A^T\mathbf{V} \tag{8.76}$$

where the T superscript stands for transpose. By tripling the matrix in (8.76) along the diagonal so that it applies to all three phases and calling the resulting matrix A^{3T}, we can write

$$\left(\mathbf{V}^{3r} - \dfrac{V_1^3}{N_1}\mathbf{N}^{3r}\right) = A^{3T}\mathbf{V}^3 \tag{8.77}$$

Here, $\mathbf{V}^3$ is the voltage vector for all the windings and all three phases of dimension 3n. Using these transformations, Equation 8.72 can be put in a form that includes all the windings:

$$\mathbf{I}^3 = A^3 Y^{3r} A^{3T}\mathbf{V}^3 \tag{8.78}$$

Letting $Y^3 = A^3 Y^{3r} A^{3T}$, this equation can be written more succinctly as

$$\mathbf{I}^3 = Y^3\mathbf{V}^3 \tag{8.79}$$

where Y^3 is a 3n × 3n complex matrix.

At this point, we will drop the superscript 3 since all the matrix equations we deal with from here on will involve all three phases, unless otherwise noted. Thus, Equation 8.79 becomes simply

$$\mathbf{I} = Y\mathbf{V} \tag{8.80}$$

8.7.2.1 Delta Winding Connection

As mentioned in the introduction, the matrix Y is singular. It is usual to add the core excitation shunt admittances to Y^3 to render it invertible as discussed in [Bra82]. However, at this point, it is natural to make the delta connections so that balanced loading need

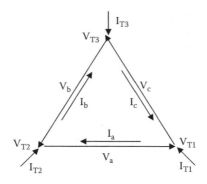

FIGURE 8.7
Delta winding connection with the phase and terminal quantities labeled. Subscript T labels terminal quantities and small letters the phase quantities. (From Del Vecchio, R.M., *IEEE Trans. Power Deliv.*, 23(3), 1439–1447, July 2008. © 2008 IEEE.)

not be assumed. Figure 8.7 shows a delta connection with the labeling convention assumed for the phase and terminal voltages and currents.

The terminal currents are positive into the terminals. The phase currents are opposite to the direction of phase voltage increase. With these conventions, the terminal and phase voltages and currents are related for the delta connection by

$$\begin{pmatrix} V_a \\ V_b \\ V_c \end{pmatrix} = \begin{pmatrix} 1 & -1 & 0 \\ 0 & 1 & -1 \\ -1 & 0 & 1 \end{pmatrix} \begin{pmatrix} V_{T1} \\ V_{T2} \\ V_{T3} \end{pmatrix} \tag{8.81}$$

$$\begin{pmatrix} I_{T1} \\ I_{T2} \\ I_{T3} \end{pmatrix} = \begin{pmatrix} 1 & 0 & -1 \\ -1 & 1 & 0 \\ 0 & -1 & 1 \end{pmatrix} \begin{pmatrix} I_a \\ I_b \\ I_c \end{pmatrix} \tag{8.82}$$

These equations were given before in (8.44) and (8.45). However, they were not used for the balanced terminal model. They are needed here in order to allow unbalanced conditions. Notice that the matrices in (8.81) and (8.82) are transposes of each other. This transformation needs to be embedded in a general transformation matrix that includes all the other windings and all three phases. The other winding currents and voltages are left unchanged so they are transformed by a unit matrix. We thus embed the delta transformations in a unit matrix for the other windings. For example, if we have a 2-winding transformer and winding 2 is a delta winding, Equation 8.82 is embedded as

$$\begin{pmatrix} I_{1,T1} \\ I_{2,T1} \\ I_{1,T2} \\ I_{2,T2} \\ I_{1,T3} \\ I_{2,T3} \end{pmatrix} = \begin{pmatrix} 1 & 0 & 0 & 0 & 0 & 0 \\ 0 & 1 & 0 & 0 & 0 & -1 \\ 0 & 0 & 1 & 0 & 0 & 0 \\ 0 & -1 & 0 & 1 & 0 & 0 \\ 0 & 0 & 0 & 0 & 1 & 0 \\ 0 & 0 & 0 & -1 & 0 & 1 \end{pmatrix} \begin{pmatrix} I_{1,a} \\ I_{2,a} \\ I_{1,b} \\ I_{2,b} \\ I_{1,c} \\ I_{2,c} \end{pmatrix} \tag{8.83}$$

Here all the windings in one phase are listed consecutively, followed by the next phase, etc. The corresponding voltage transformation in (8.81) is embedded by the transpose of the matrix in (8.83).

Calling the matrix in (8.83) M_y, where y indicates admittance representation, Equation 8.83 and the corresponding voltage transformation can be written to include all the windings as

$$
\begin{aligned}
\mathbf{I}_T &= M_y \mathbf{I} \\
\mathbf{V} &= M_y^T \mathbf{V}_T
\end{aligned}
\tag{8.84}
$$

If there is more than one delta winding, say, K, label the transforming matrices, $M_{y1}, M_{y2}, \ldots, M_{yK}$. Then, the transformed currents and voltages become

$$
\begin{aligned}
\mathbf{I}_{T1} &= M_{y1}\mathbf{I}, \quad \mathbf{I}_{T2} = M_{y2}\mathbf{I}_{T1}, \ldots, \mathbf{I}_{TK} = M_{yK}\mathbf{I}_{TK-1} \\
\mathbf{V} &= M_{y1}^T \mathbf{V}_{T1}, \quad \mathbf{V}_{T1} = M_{y2}^T\mathbf{V}_{T2}, \ldots, \mathbf{V}_{TK-1} = M_{yK}^T\mathbf{V}_{TK}
\end{aligned}
\tag{8.85}
$$

Combining these, we get

$$
\begin{aligned}
\mathbf{I}_{TK} &= M_{yK}\cdots M_{y2}M_{y1}\mathbf{I} = M_Y\mathbf{I} \\
\mathbf{V} &= M_{y1}^T M_{y2}^T \cdots M_{yK}^T \mathbf{V}_{TK} = M_Y^T\mathbf{V}_{TK}
\end{aligned}
\tag{8.86}
$$

where we have defined the product matrix M_Y. Then (8.80) becomes, after the delta transformations,

$$
\mathbf{I}_{TK} = M_Y Y M_Y^T \mathbf{V}_{TK}
\tag{8.87}
$$

Calling the resulting matrix in (8.87) Y_{mod} to indicate that Y has been modified to include the deltas, we have simply

$$
\mathbf{I}_{TK} = Y_{mod}\mathbf{V}_{TK}
\tag{8.88}
$$

The delta connection treated here is a $-30°$ delta since the terminal voltages are shifted by $-30°$ relative to the phase voltages. There is also a $+30°$ delta connection that can be treated by a similar procedure.

If there are squashed delta windings, they should also be handled in the admittance representation in order to allow unbalanced loading. This connection is discussed in Chapter 7. The transformations are a bit more complicated than the delta transformations and will not be dealt with here. We should also note that parallel winding connections should also be handled in this admittance representation.

8.7.3 Impedance Representation

Although one can work in the admittance representation, it is often convenient to make other types of winding connections and solve terminal equations in the impedance representation. It should be noted that, even with the matrix transformation in (8.87), the resulting matrix is still singular. Therefore, to invert it, some modifications must be made. This is done, as in [Bra82], by adding the core excitation characteristics.

Let Y_{exc-1} and Y_{exc-0} be the positive and zero sequence core excitation shunt admittances. These are used to construct diagonal and off-diagonal admittances for unbalanced conditions similar to Equations 8.66 through 8.68:

$$Y_S = -j\frac{1}{3}\left(Y_{exc-0} + 2Y_{exc-1}\right)$$

$$(8.89)$$

$$Y_M = -j\frac{1}{3}\left(Y_{exc-0} - Y_{exc-1}\right)$$

These can be added to a single winding or $1/n$ times these to all the windings. It is added to the imaginary part of Y_{mod}.

If the core losses are known, these can be used to obtain a core conductance, which also should be added to a single winding or $1/n$ times it to all the windings. This is a real quantity and should be added to the diagonal terms in the real part of Y_{mod}.

Once this is done, the matrix Y_{mod} becomes invertible. Again, if this is a complex matrix, complex inversion must be used. Note that if Y_{mod} has real and imaginary parts, both must be modified by the procedure mentioned earlier in order to get an invertible matrix. Thus, core losses are needed if winding resistances are included. Since the resulting matrix is close to being singular, a double-precision inversion may be needed.

Inverting (8.88) with these modifications, we get

$$\mathbf{V}_{TK} = Z_{mod}\mathbf{I}_{TK} \tag{8.90}$$

where Z_{mod} is the inverse of the core modified version of Y_{mod} and is an impedance matrix. It is in this representation that most of the other winding connections can be made more naturally without sacrificing unbalanced loading. These will be discussed in the next sections.

8.7.3.1 Ungrounded Y Connection

The ungrounded Y connection can be generalized to include Y connections with an impedance connecting the neutral point to ground as shown in Figure 8.8. For a true ungrounded Y connection, this impedance is set to a large enough value to keep the neutral current close to zero. If it is too large, numerical instabilities may occur when the problem is solved.

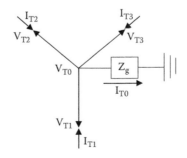

FIGURE 8.8
Y connection grounded through an impedance. (From Del Vecchio, R.M., *IEEE Trans. Power Deliv.*, 23(3), 1439–1447, July 2008. © 2008 IEEE.)

Another equation must be added to (8.90) to account for the additional branch to ground. It now becomes

$$\begin{pmatrix} \mathbf{V}_{TK} \\ \mathbf{V}_{T0} \end{pmatrix} = \begin{pmatrix} Z_{mod} & 0 \\ 0 & Z_g \end{pmatrix} \begin{pmatrix} \mathbf{I}_{TK} \\ \mathbf{I}_{T0} \end{pmatrix} \tag{8.91}$$

The vectors in (8.90) have been enlarged to include one more component and are of dimension $3n + 1$. The matrix in (8.90) has also been enlarged to a $3n + 1 \times 3n + 1$ matrix. The current equation

$$\mathbf{I}_{T0} = \mathbf{I}_{T1} + \mathbf{I}_{T2} + \mathbf{I}_{T3} \tag{8.92}$$

can be added via the matrix

$$\begin{pmatrix} \mathbf{I}_{T1} \\ \mathbf{I}_{T2} \\ \mathbf{I}_{T3} \\ \mathbf{I}_{T0} \end{pmatrix} = \begin{pmatrix} 1 & 0 & 0 \\ 0 & 1 & 0 \\ 0 & 0 & 1 \\ 1 & 1 & 1 \end{pmatrix} \begin{pmatrix} \mathbf{I}_{S1} \\ \mathbf{I}_{S2} \\ \mathbf{I}_{S3} \end{pmatrix} \tag{8.93}$$

We have relabeled the new terminal currents with S instead of T to distinguish the different terminal situations. As before, this must be embedded in a larger unit matrix to include all the other windings. The transformed voltages are given by the transpose of the matrix in (8.93):

$$\begin{pmatrix} \mathbf{V}_{S1} \\ \mathbf{V}_{S2} \\ \mathbf{V}_{S3} \end{pmatrix} = \begin{pmatrix} 1 & 0 & 0 & 1 \\ 0 & 1 & 0 & 1 \\ 0 & 0 & 1 & 1 \end{pmatrix} \begin{pmatrix} \mathbf{V}_{T1} \\ \mathbf{V}_{T2} \\ \mathbf{V}_{T3} \\ \mathbf{V}_{T0} \end{pmatrix} \tag{8.94}$$

Here we see that the new terminal voltages are given in terms of the old by $\mathbf{V}_{S1} = \mathbf{V}_{T1} + \mathbf{V}_{T0}$, etc. We note that the matrices in (8.93) and (8.94) are not square matrices. In both cases, the new terminal voltage and current vectors have one less component than the original vectors for the grounded Y connection.

We need to embed (8.94) in a larger matrix that leaves the other winding terminal voltages unchanged and similarly for the currents in (8.93). Calling this enlarged matrix in (8.94) M_{zu}, with z standing for impedance representation and u for ungrounded Y, we can rewrite (8.93) and (8.94) to include all the windings as

$$\mathbf{V}_S = M_{zu} \begin{pmatrix} \mathbf{V}_{TK} \\ \mathbf{V}_{T0} \end{pmatrix}$$

$$\begin{pmatrix} \mathbf{I}_{TK} \\ \mathbf{I}_{T0} \end{pmatrix} = M_{zu}^T \mathbf{I}_S \tag{8.95}$$

where the T superscript indicates transpose. Then (8.91) is transformed into

$$\mathbf{V}_S = M_{zu} \begin{pmatrix} Z_{mod} & 0 \\ 0 & Z_g \end{pmatrix} M_{zu}^T \mathbf{I}_S \tag{8.96}$$

The terminal voltage and current vectors in (8.96) include all the terminals and all three phases but exclude the neutral terminals. We are using the S label for the terminals in the transformed situation.

The procedure mentioned earlier can be generalized for other ungrounded (or grounded through an impedance) Y-connected windings. Simply enlarge the matrix in (8.91) to include other grounding impedances along the diagonal, labeled Z_{g1}, Z_{g2}, etc., and enlarge the matrices M_{zu} and M_{zu}^T appropriately. The resulting matrix as in (8.96) will be a 3n × 3n matrix, having the same terminals as before the grounding impedances were added. It is convenient to refer to the enlarged matrix in (8.96) as $Z_{\text{mod-g}}$ so that the equation can be written as

$$\mathbf{V_S} = M_{zu} Z_{\text{mod-g}} M_{zu}^T \mathbf{I_S} \tag{8.97}$$

If there are no ungrounded or impedance grounded Y's, $Z_{\text{mod-g}}$ becomes Z_{mod} and (8.97) reduces to (8.90).

8.7.3.2 Series-Connected Windings from the Same Phase

This is a common connection, particularly for tap windings. The windings can be connected so that the voltages add or subtract. For simplicity, consider connecting windings 1 and 2, with 1 retaining its terminal status. The case of voltage addition is shown in Figure 8.9. Here the label p, Sq refers to winding p in phase q.

The new terminal quantities will be labeled with U. They are $V_{1,U1} = V_{1,S1} + V_{2,S1}$ and $I_{1,S1} = I_{1,U1}$, $I_{2,S1} = I_{1,U1}$. These can be expressed in matrix form, including the other phases, as

$$
\begin{pmatrix} V_{1,U1} \\ V_{1,U2} \\ V_{1,U3} \end{pmatrix} = \begin{pmatrix} 1 & 1 & 0 & 0 & 0 & 0 \\ 0 & 0 & 1 & 1 & 0 & 0 \\ 0 & 0 & 0 & 0 & 1 & 1 \end{pmatrix} \begin{pmatrix} V_{1,S1} \\ V_{2,S1} \\ V_{1,S2} \\ V_{2,S2} \\ V_{1,S3} \\ V_{2,S3} \end{pmatrix} \tag{8.98}
$$

The currents are transformed by the transpose of the matrix in (8.98) and are

$$
\begin{pmatrix} I_{1,S1} \\ I_{2,S1} \\ I_{1,S2} \\ I_{2,S2} \\ I_{1,S3} \\ I_{2,S3} \end{pmatrix} = \begin{pmatrix} 1 & 0 & 0 \\ 1 & 0 & 0 \\ 0 & 1 & 0 \\ 0 & 1 & 0 \\ 0 & 0 & 1 \\ 0 & 0 & 1 \end{pmatrix} \begin{pmatrix} I_{1,U1} \\ I_{1,U2} \\ I_{1,U3} \end{pmatrix} \tag{8.99}
$$

These matrices are not square. They reduce the size of the 3-phase voltage and current vectors by 3. We need to embed the matrices in (8.98) and (8.99) into enlarged matrices that

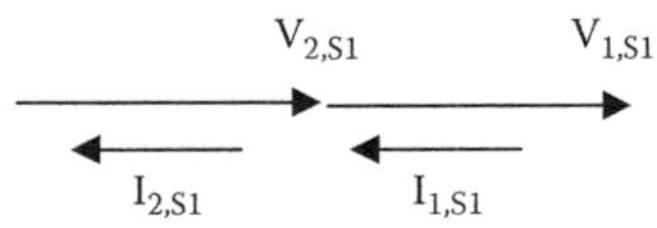

FIGURE 8.9
Voltage addition of two windings of the same phase. (From Del Vecchio, R.M., *IEEE Trans. Power Deliv.*, 21(3), 1300–1308, July 2006. © 2006 IEEE.)

include the other windings without changing their terminal voltages and currents, that is, a unit matrix. If different winding numbers are connected in series, then the patterns in these matrices must be inserted in the appropriate positions of the final full winding matrix. Call the enlarged matrix, corresponding to the voltage transformation matrix in (8.98), M_{zs}, where z refers to impedance representation and s to series connection. Then we have

$$\mathbf{V}_U = M_{zs}\mathbf{V}_S$$
$$\mathbf{I}_S = M_{zs}^T\mathbf{I}_U \tag{8.100}$$

This transforms (8.97) into

$$\mathbf{V}_U = M_{zs}M_{zu}Z_{\text{mod-g}}M_{zu}^T M_{zs}^T\mathbf{I}_U \tag{8.101}$$

This has fewer terminals than before the series addition.

If there are ungrounded Y connections, from (8.95) and (8.100), we see that

$$\mathbf{V}_U = M_{zs}M_{zu}\begin{pmatrix}\mathbf{V}_{TK}\\\mathbf{V}_{T0}\end{pmatrix}$$
$$\begin{pmatrix}\mathbf{I}_{TK}\\\mathbf{I}_{T0}\end{pmatrix} = M_{zu}^T M_{zs}^T\mathbf{I}_U \tag{8.102}$$

If there are more series-connected windings, then these are handled in the same way, and we would end up with additional transformation matrices, for example, M_{zs1}, M_{zs2}, etc. Similarly, we could add more than 2 windings together by adding the additional windings to the previous ones by similar matrices operating on the previous terminal quantities.

We see the general pattern here. Voltages and currents are transformed by matrices that are typically transposes of each other and nonsquare. These also transform the terminal equations as in (8.101).

From here on, we will only calculate the transformation matrices M for the terminal voltages and currents and assume that they are used as in the examples given earlier. To avoid notational difficulties, we will assume that the starting terminal quantities are labeled S and the transformed ones U. The S terminal quantities will be, in general, the result of multiple previous transformations.

8.7.3.3 Zigzag Connection

This is really a series connection involving two windings from different phases and is treated in a similar manner as in the previous section. The connection is shown in Figure 8.10,

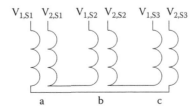

FIGURE 8.10
Zigzag winding connection. The phases are labeled a, b, and c. (From Del Vecchio, R.M., *IEEE Trans. Power Deliv.*, 23(3), 1439–1447, July 2008. © 2008 IEEE.)

where we are assuming that windings 1 and 2 are involved in the connection, with winding 1 being the final terminal winding. This connection was given in Figure 8.5, but here the labeling is different and the HV or input winding can be part of the zigzag connection.

The connection is one of voltage subtraction and results in the elimination of the terminal voltages and currents associated with winding 2. The transformed voltages are given by

$$\begin{pmatrix} V_{1,U1} \\ V_{1,U2} \\ V_{1,U3} \end{pmatrix} = \begin{pmatrix} 1 & 0 & 0 & 0 & 0 & -1 \\ 0 & -1 & 1 & 0 & 0 & 0 \\ 0 & 0 & 0 & -1 & 1 & 0 \end{pmatrix} \begin{pmatrix} V_{1,S1} \\ V_{2,S1} \\ V_{1,S2} \\ V_{2,S2} \\ V_{1,S3} \\ V_{2,S3} \end{pmatrix} \tag{8.103}$$

and the transformed currents by the transpose

$$\begin{pmatrix} I_{1,S1} \\ I_{2,S1} \\ I_{1,S2} \\ I_{2,S2} \\ I_{1,S3} \\ I_{2,S3} \end{pmatrix} = \begin{pmatrix} 1 & 0 & 0 \\ 0 & -1 & 0 \\ 0 & 1 & 0 \\ 0 & 0 & -1 \\ 0 & 0 & 1 \\ -1 & 0 & 0 \end{pmatrix} \begin{pmatrix} I_{1,U1} \\ I_{1,U2} \\ I_{1,U3} \end{pmatrix} \tag{8.104}$$

Since the positive current direction is into the terminals, we see that the current in winding 2 is the negative of the current into winding 1 to which it is connected. These matrices must be extended to include all the windings and are used to transform the terminal equation as discussed previously. This will result in a terminal equation with three less terminals. Label the full voltage transformation matrix, corresponding to that in (8.104), M_{zz}, where the second z stands for zigzag connection.

8.7.3.4 Autoconnection

This connection involves two windings from the same phase as shown in Figure 8.11. We will assume, again without loss of generality, that windings 1 and 2 are involved in this connection.

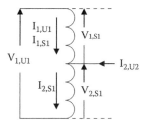

FIGURE 8.11
Autotransformer connection. (From Del Vecchio, R.M., *IEEE Trans. Power Deliv.*, 23(3), 1439–1447, July 2008. © 2008 IEEE.)

For simplicity, we will only consider the transformations for one phase as shown in the figure. The other phases will simply be repeats along the diagonal of the complete matrix. The voltage terminal transformation is

$$\begin{pmatrix} V_{1,U1} \\ V_{2,U1} \end{pmatrix} = \begin{pmatrix} 1 & 1 \\ 0 & 1 \end{pmatrix} \begin{pmatrix} V_{1,S1} \\ V_{2,S1} \end{pmatrix} \tag{8.105}$$

and the currents are transformed by the transpose matrix

$$\begin{pmatrix} I_{1,S1} \\ I_{2,S1} \end{pmatrix} = \begin{pmatrix} 1 & 0 \\ 1 & 1 \end{pmatrix} \begin{pmatrix} I_{1,U1} \\ I_{2,U1} \end{pmatrix} \tag{8.106}$$

This leaves the number of terminals unchanged. Label the full voltage transformation, including all the windings and phases, M_{za}, with a standing for auto.

Notice how much simpler the autotransformation is when all the windings are treated on the same footing as compared with the autoconnection discussed for the balanced model when the HV or input winding is the series winding in this connection.

8.7.3.5 Three Windings Joined

Here the three windings are connected at a common point. This is an important connection for phase shifting transformers or autotransformers with X-line taps. The three windings can be from different phases and the voltage connections can be additive or subtractive in general. This is a transformation that can be performed in two steps. In the first step, join two of the windings but retain both terminals, as is done in the autoconnection outlined earlier. In the second step, join the third winding to one of the terminals of the first 2 joined windings as in a series connection. This can be generalized if the windings are on different phases.

8.7.4 Terminal Loading

The various winding connections have been represented by matrices, which transform the terminal voltages and currents. These matrices are typically transposes of each other. In the impedance representation, we have labeled these with a z for impedance and another letter to indicate the type of connection. Here, to generalize this, we will label these z1, z2, etc., with the number indicating the order in which they are applied. We will also label the terminal vectors S1, S2, S3, etc., with the number corresponding to the matrix transformation order. For simplicity, we will ignore ungrounded Y connections, but their inclusion is straightforward as discussed in Section 8.7.3.1. Thus, we will start with (8.90).

In the impedance representation, the multiple voltage and current transformations result in

$$\begin{aligned} \mathbf{V}_{S1} = M_{z1}\mathbf{V}_{TK}, \quad \mathbf{V}_{S2} = M_{z2}\mathbf{V}_{S1},\ldots, \mathbf{V}_{SL} = M_{zL}\mathbf{V}_{SL-1}, \\ \mathbf{I}_{TK} = M_{z1}^{T}\mathbf{I}_{S1}, \quad \mathbf{I}_{S1} = M_{z2}^{T}\mathbf{I}_{S2},\ldots, \mathbf{I}_{SL-1} = M_{zL}^{T}\mathbf{I}_{SL} \end{aligned} \tag{8.107}$$

We are considering a total of L transformations of various types. When applied successively, they result in

$$\begin{aligned} \mathbf{V}_{SL} = M_{zL}\cdots M_{z2}M_{z1}\mathbf{V}_{TK} = M_{Z}\mathbf{V}_{TK} \\ \mathbf{I}_{TK} = M_{z1}^{T}M_{z2}^{T}\cdots M_{zL}^{T}\mathbf{I}_{SL} = M_{Z}^{T}\mathbf{I}_{SL} \end{aligned} \tag{8.108}$$

Here we have defined the total transformation matrix M_Z. The terminal equation (8.90) that we started with is transformed into

$$\mathbf{V}_{SL} = M_Z Z_{mod} M_Z^T \mathbf{I}_{SL} \tag{8.109}$$

If ungrounded Y's are present, then $Z_{mod\text{-}g}$ should replace Z_{mod} in (8.109) and $\mathbf{V}_{TK}$ and $\mathbf{I}_{TK}$ should be expanded to include $\mathbf{V}_{T0}$ and $\mathbf{I}_{T0}$ as in (8.95) or additional ones if more than one ungrounded Y is present.

We are assuming that terminal 1 is the input terminal having a given voltage although, in general, other terminals could be specified for this role. The noninput terminals are loaded with some impedance. In general, we can write

$$\mathbf{V}_{SL} = -Z_{Ld} \mathbf{I}_{SL} + \mathbf{V}_{Ld} \tag{8.110}$$

Here Z_{Ld} is a complex load matrix that is typically diagonal. The minus sign is necessary because the terminal currents are assumed to flow into the transformer while the load currents flow out of it. $\mathbf{V}_{Ld}$ is an applied voltage load vector that typically includes only a nonzero terminal voltage for the input terminal. However, there may be situations where it may be nonzero for other terminals.

As an example, if there are three terminals for a single phase, which we will label 1, 2, and 3, and terminal 1 was driven by a voltage V_o, Equation 8.110 would take the form, for one phase,

$$\begin{pmatrix} V_1 \\ V_2 \\ V_3 \end{pmatrix} = - \begin{pmatrix} 0 & 0 & 0 \\ 0 & Z_{Ld,2} & 0 \\ 0 & 0 & Z_{Ld,3} \end{pmatrix} \begin{pmatrix} I_1 \\ I_2 \\ I_3 \end{pmatrix} + \begin{pmatrix} V_o \\ 0 \\ 0 \end{pmatrix} \tag{8.111}$$

This must be enlarged to include the other phases. The diagonal matrix entries are the complex load impedances for terminals 2 and 3. The voltage V_o will change for the different phases if, for example, the input is a 3-phase balanced set of voltages. The Z_{Ld}'s will be the same for the different phases under balanced conditions but will be different for unbalanced loading. In fact, for this, we would need to include the zero sequence load impedances, if they differ from the positive sequence impedances, by the procedure given in (8.66) through (8.68).

Combining (8.110) and (8.109), we get

$$\left(M_Z Z_{mod} M_Z^T + Z_{Ld} \right) \mathbf{I}_{SL} = \mathbf{V}_{Ld} \tag{8.112}$$

This is the terminal equation that must be solved. If a real matrix solver is used, then this must be cast in the form of (8.71) to separate the real and imaginary parts.

8.7.5 Solution Process

8.7.5.1 Terminal Currents and Voltages

Equation 8.112 is, in general, complex so complex solution techniques must be employed, unless the methods discussed in Section 8.7.2 are used. Most scientific or math libraries have matrix equation solvers. A double-precision one is recommended.

Once the terminal currents, $\mathbf{I}_{SL}$, are found, the terminal voltages, $\mathbf{V}_{SL}$, can be found from (8.110) or (8.109). From these quantities, one can also obtain the power into and out of the various terminals. In addition, regulation effects and transformer terminal impedances can be obtained by the methods outlined for balanced loading.

8.7.5.2 Winding Currents and Voltages

Equation 8.108 can be used to obtain $\mathbf{I}_{TK}$ from the terminal currents $\mathbf{I}_{SL}$. If ungrounded Y's are present, then the $\mathbf{I}_{T0}$'s are also obtained. Then (8.90) or (8.91), depending on whether ungrounded Y's are present, can be used to determine $\mathbf{V}_{TK}$ from $\mathbf{I}_{TK}$. Equation 8.86 can then be used to obtain $\mathbf{V}$ from $\mathbf{V}_{TK}$, where $\mathbf{V}$ contains the winding voltages under load. Finally, Equation 8.80 can be used to get the winding currents $\mathbf{I}$ from $\mathbf{V}$. This is summarized in the following steps:

$$
\begin{aligned}
\mathbf{I}_{TK} &= M_Z^T \mathbf{I}_{SL} \\
\mathbf{V}_{TK} &= Z_{mod} \mathbf{I}_{TK} \\
\mathbf{V} &= M_Y^T \mathbf{V}_{TK} \\
\mathbf{I} &= Y\mathbf{V}
\end{aligned}
\tag{8.113}
$$

These are slightly modified if ungrounded Y's are present. In this case, the neutral currents and voltages are also obtained.

8.7.6 Unbalanced Loading Examples

The examples given here will be for a single line to ground fault. If the unfaulted terminals are open, they are given a very high load impedance. For unfaulted loaded terminals, we assume that the load impedances are very low, about 0.01%, but that the load voltages are kept at their rated values. In general, a full or partial load could be added to the unfaulted terminals. For the faulted terminal, we assume that the load impedance and voltage are zero. Generally, the a-phase terminal is faulted. The b- and c-phase terminals corresponding to the faulted terminal are given very low impedances, and their voltages are set to their rated values. These terminal conditions correspond to those used in the sequence analysis calculations for the case where the system is connected beyond the fault point. If the system were not connected beyond the fault point, then the b- and c-phase terminals corresponding to the faulted terminal would be given very high impedances to signify an unloaded condition. The results obtained by the methods discussed here will be compared with the results of the sequence analysis method.

It is also convenient to use per-unit impedance and admittance quantities as inputs for the computer program that performs the calculations given earlier. However, they are converted to Ohms or Mhos before performing the analysis. We have also kept the exciting admittances low, typically about 0.05%, in order to minimize the influence of the core for better comparison with the sequence results. We also ignore resistance effects in this analysis. The fault currents are rms amps with no asymmetry factor correction.

8.7.6.1 Autotransformer with Buried Delta Tertiary and Fault on LV Terminal

The autotransformer is a 3-phase 50 MVA unit, Y connected with a buried delta. The line–line voltages in kV are HV 154.63, LV 69.28, and TV 7.2. The per-unit winding to winding leakage

TABLE 8.6

Winding to Winding Leakage Reactances in Per-Unit % for the 50 MVA Autotransformer

| Wdg1/Wdg2 | Positive Sequence | | Zero Sequence | |
	LV	TV	LV	TV
HV	24.518	40.428	20.842	30.508
LV		11.090		9.430

Source: Del Vecchio, R.M., *IEEE Trans. Power Deliv.*, 23(3), 1439–1447, July 2008. © 2008 IEEE.

TABLE 8.7

Comparison of Terminal and Winding Fault Currents for the 50 MVA Autotransformer

| | Present Method Phase | | | Sequence Analysis Phase | | |
	a	b	c	a	b	c
Terminal						
HV	2650	154	154	2654	154	154
LV	7401	1830	1830	7412	1833	1833
TV	0.27	0.27	0	0	0	0
Winding						
Series	2650	154	154	2654	154	154
Common	4754	1678	1678	4759	1679	1679
TV	8274	8264	8264	8274	8274	8274

Source: Del Vecchio, R.M., *IEEE Trans. Power Deliv.*, 23(3), 1439–1447, July 2008. © 2008 IEEE.

reactances are given in Table 8.6. Note that these are winding to winding leakage reactances and not terminal to terminal leakage reactances, since the method presented here is based on winding to winding leakage reactances. The relationship between winding to winding and terminal to terminal leakage reactances for autotransformers was given in Chapter 4.

The single line to ground fault is on phase a of the LV terminal. The terminal and winding currents by the present method are compared with the sequence analysis results in Table 8.7. Only the current magnitudes are given.

The neutral current is 8104 amps by the present method and 8117 amps by the sequence analysis method.

8.7.6.2 Power Transformer with Fault on Delta Tertiary

This is a 3-phase 16.8 MVA power transformer with line–line voltages in kV: HV 138, LV 13.8, and TV 7.22. The HV and LV windings are Y connected and the TV is delta connected. The leakage reactances are given in Table 8.8.

The single line to ground fault is on phase a of the loaded delta tertiary. There are no line-to-line loads present. The comparison of the fault currents by the present and sequence analysis methods is shown in Table 8.9. The current magnitudes only are given.

The terminal currents sum to zero for the delta. There are no neutral currents for either analysis method, since this would be a zero sequence current that cannot flow out of the delta. If the system were not connected beyond the fault point, there would be no short circuit currents in any of the windings, as was discussed in Chapter 6.

TABLE 8.8

Winding to Winding Leakage Reactances in Per-Unit % for the 16.8 MVA Power Transformer

Wdg1/Wdg2	Positive Sequence		Zero Sequence	
	LV	TV	LV	TV
HV	10.00	17.40	8.50	13.92
LV		7.10		6.39

Source: Del Vecchio, R.M., *IEEE Trans. Power Deliv.*, 23(3), 1439–1447, July 2008. © 2008 IEEE.

TABLE 8.9

Comparison of Terminal and Winding Fault Currents for the 16.8 MVA Power Transformer

	Present Method Phase			Sequence Analysis Phase		
	a	b	c	a	b	c
Terminal						
HV	8.7	0	8.7	8.6	0	8.6
LV	5,804	0.23	5804	5803	0	5803
TV	12,628	6314	6314	12627	6313	6313
Winding						
HV	8.6	0	8.6	8.6	0	8.6
LV	5,803	0	5803	5803	0	5803
TV	6,314	0	6314	6313	0	6313

Source: Del Vecchio, R.M., *IEEE Trans. Power Deliv.*, 23(3), 1439–1447, July 2008. © 2008 IEEE.

8.7.6.3 Power Transformer with Fault on Ungrounded Y Secondary

This is a 3-phase 40 MVA power transformer with line–line voltages in kV: HV 242.93, LV 34.5, and TV 13.2. The HV is a grounded Y, the LV an ungrounded Y, and the TV a loaded delta. The leakage reactances are given in Table 8.10.

The single line to ground fault is on phase a of the ungrounded Y secondary. The terminal and winding current magnitudes are given in Table 8.11 for both methods. There are no neutral currents in either method of analysis. If the system were not connected beyond the fault point, there would be no fault currents in the transformer, as discussed in Chapter 6 and as shown by this method.

TABLE 8.10

Winding to Winding Leakage Reactances in Per-Unit % for the 40 MVA Power Transformer

Wdg1/Wdg2	Positive Sequence		Zero Sequence	
	LV	TV	LV	TV
HV	8.45	12.62	7.18	10.10
LV		3.3		3.00

Source: Del Vecchio, R.M., *IEEE Trans. Power Deliv.*, 23(3), 1439–1447, July 2008. © 2008 IEEE.

TABLE 8.11

Comparison of Terminal and Winding Fault Currents for the 40 MVA Power Transformer

	Present Method Phase			Sequence Analysis Phase		
	a	b	c	a	b	c
Terminal						
HV	854	427	427	852	426	426
LV	20,281	10,140	10,140	20,162	10,081	10,081
TV	32,300	32,300	0.2	32,063	32,063	0
Winding						
HV	852	426	426	852	426	426
LV	20,167	10,084	10,084	20,162	10,081	10,081
TV	21,382	10,691	10,691	21,375	10,688	10,688

Source: Del Vecchio, R.M., *IEEE Trans. Power Deliv.*, 23(3), 1439–1447, July 2008. © 2008 IEEE.

8.7.7 Balanced Loading Example

8.7.7.1 Standard Delta Phase Shifting Transformer

This type of transformer has been discussed in Chapter 7 so there are analytic expressions for the phase shift, terminal and winding currents, and effective leakage reactance for comparison with the values determined by this method. The connection is shown in Figure 8.12.

The secondary windings are loaded with their rated terminal resistance on all three phases. The delta is a +30° delta. The current transformation corresponding to the +30° delta is

$$\begin{pmatrix} I_{T1} \\ I_{T2} \\ I_{T3} \end{pmatrix} = \begin{pmatrix} 1 & -1 & 0 \\ 0 & 1 & -1 \\ -1 & 0 & 1 \end{pmatrix} \begin{pmatrix} I_a \\ I_b \\ I_c \end{pmatrix} \tag{8.114}$$

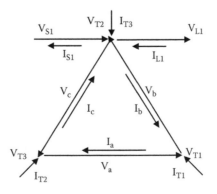

FIGURE 8.12

Standard delta phase shifter. S_1 and L_1 label the input and output terminals of the A phase. The other input and output terminals are not shown. The delta is a +30° delta. (From Del Vecchio, R.M., *IEEE Trans. Power Deliv.*, 23(3), 1439–1447, July 2008. © 2008 IEEE.)

The voltages are transformed by the transpose of this matrix but with the voltage vectors interchanged right to left. The −30° delta transformation was given in (8.82). Note that the S1 winding is connected negatively to the delta terminal, while the L1 winding is connected positively.

The transformer is a 3-phase 75 MVA unit. Labeling the input and output windings S and L, respectively, and the delta winding T, the winding phase voltages are in kV: S 8.0, L 8.0, and T 38.0. The positive sequence leakage reactances are, in per-unit %, $Z_{SL} = 20$, $Z_{ST} = 10$, and $Z_{LT} = 10$. The zero sequence reactances are the same, since balanced loading is assumed.

The parameter n is the ratio of delta winding turns to turns in the S or L winding, which have equal numbers of turns. This turns ratio is 4.76 for this unit. The phase shift is given by

$$\theta = 2\tan^{-1}\left(\frac{\sqrt{3}}{n}\right) \tag{8.115}$$

For the value of n given, this is 40°. The present method gives the same value for no-load conditions but 38.66° for load conditions.

The magnitude of the terminal voltage is given by

$$\left|V_{L1}\right| = \sqrt{\frac{n^2+3}{3n^2}}\left|V_a\right| \tag{8.116}$$

This is equal to 23.396 kV for this unit. The present method gives this value for no-load conditions but a value of 23.389 for load conditions. The angle and voltage magnitude shift under load conditions are what are termed regulation effects.

Based on the input power and terminal voltage, the terminal current S1 is 1068.6 amps. The current in the delta is given by

$$\left|I_a\right| = \frac{2}{\sqrt{n^2+3}}\left|I_{S1}\right| \tag{8.117}$$

This gives 421.9 amps compared with a value of 422.6 amps by the present method.

The winding power is smaller than the input power:

$$P_{wdg} = \sin\theta P_{in} \tag{8.118}$$

For P_{in} of 75 MVA, P_{wdg} is 48.21 MVA. The present method gives 48.26 MVA.

The effective per-unit leakage reactance, as given in Chapter 7, is

$$z_{eff} = \left(\frac{3n^2}{n^2+3}\right)\left[\frac{z_{SL}}{n^2} + \frac{2\left(z_{TS}+z_{TL}-z_{SL}\right)}{n^2+3}\right] \tag{8.119}$$

This gives 2.34%. Using the technique discussed in Section 8.6, we obtain 2.324% for this quantity by the present method of analysis.

8.7.8 Discussion

The multiterminal circuit analysis method presents a systematic way of modeling transformers, especially the ones with complicated winding interconnections. The inputs required are (1) the winding to winding positive and zero sequence leakage reactances, with only the positive ones required for balanced conditions, (2) winding turns, (3) the core excitation conditions for unbalanced loading, (4) winding resistances and core losses if desired, (5) terminal quantities such as MVA and voltages, and (6) the winding interconnections.

The method then permits the calculation of terminal voltages and currents and winding voltages and currents under load, regardless of whether the loading is balanced or unbalanced. This is particularly important for unbalanced faults since the winding currents are needed for subsequent force and stress analysis. For balanced conditions, the method can yield terminal to terminal leakage reactances and regulation effects. For the terminal to terminal per-unit leakage reactances, only two terminals should be loaded, one primary and one secondary, at its rated load. The other windings should be loaded with a very large impedance to simulate an unloaded condition.

The main purpose of the examples given earlier is to show that the method agrees with other types of analysis, where comparisons can be made. There is no claim that the methods proposed here are more accurate than other methods of analysis when they are feasible to perform. Rather, we expect that these methods are more general and can be applied to more complicated winding configurations than are possible or reasonable with more standard methods such as sequence analysis methods.

The terminal equation (8.112) could be integrated into a larger system for power flow analysis, for example. As part of a larger network, only the terminal quantities of the transformer are of interest. However, once the network equations are solved, one can obtain the internal winding currents by the methods described earlier, and these can be used, for example, for loss or short circuit analysis.

If transient analysis is desired, then inductance matrices can be used from the start instead of reactance matrices. The procedures described here can be carried out with inductance matrices, eliminating the imaginary unit j. These matrices would then be incorporated into a differential matrix equation in time. Nonlinear core characteristics can be added as discussed in [Bra82]. These would mainly be used in transient analysis and are most easily inserted in the admittance representation. However, Reference [Bra82] also discusses a method for inserting nonlinear core characteristics in the impedance representation.

We should also note that two core transformers can be modeled by the methods mentioned earlier. In this case, it is necessary to add any connections between windings on different cores. Voltage and current transfer matrices can be constructed to do this as shown in the next section.

8.8 2-Core Analysis*

When the second 3-phase core is added, both cores are analyzed individually, as discussed up to this point. Initially, the two cores are assumed to be uncoupled before interconnections between the cores are made. It is necessary to introduce additional labels to

---

* Much of the section is excerpted from [Ahu14] and reprinted with permission from Analysis of 2 core transformer designs, Cigre session paper no. A2_210_2014, Cigre 2014 General Session, Session 45, August 24–30, 2014. © 2014 Cigre.

distinguish the individual core matrices and vectors. Use c1 and c2 for this purpose and C to denote the combined voltages, current, and matrices. Thus, starting out,

$$\mathbf{V}_C = \begin{pmatrix} \mathbf{V}_{c1} \\ \mathbf{V}_{c2} \end{pmatrix}, \quad \mathbf{I}_C = \begin{pmatrix} \mathbf{I}_{c1} \\ \mathbf{I}_{c2} \end{pmatrix}, \quad \begin{pmatrix} \mathbf{V}_{c1} \\ \mathbf{V}_{c2} \end{pmatrix} = \begin{pmatrix} Z_{mod,c1} & 0 \\ 0 & Z_{mod,c2} \end{pmatrix} \begin{pmatrix} \mathbf{I}_{c1} \\ \mathbf{I}_{c2} \end{pmatrix}$$

or in matrix notation, $\quad$ (8.120)

$$\mathbf{V}_C = Z_C \mathbf{I}_C$$

where

$\mathbf{V}_C$ and $\mathbf{I}_C$ are two core vectors

Z_C a 2-core matrix

8.8.1 2-Core Parallel Connection

If there is a parallel connection between windings on the two cores, this should be done first and performed in the admittance representation. Thus, we need to invert the final equation in (8.120). Since Z_C is invertible, we get

$$\mathbf{I}_C = Y_C \mathbf{V}_C \quad (8.121)$$

As an example, assume each core has two terminals and that terminal 1 of the first core is in parallel with terminal 1 of the second core. The second terminals on both cores are unaffected by the transformation. Label the initial voltages and currents with the core and terminal number, for example, c1,1 for core 1 and terminal 1, and the final voltages and currents with the core number and terminal number preceded by t, for example, c1, t1. Considering only a single phase from each core, the transformation is given by

$$\begin{pmatrix} I_{c1,t1} \\ I_{c1,t2} \\ I_{c2,t2} \end{pmatrix} = \begin{pmatrix} 1 & 0 & 1 & 0 \\ 0 & 1 & 0 & 0 \\ 0 & 0 & 0 & 1 \end{pmatrix} \begin{pmatrix} I_{c1,1} \\ I_{c1,2} \\ I_{c2,1} \\ I_{c2,2} \end{pmatrix}, \quad \begin{pmatrix} V_{c1,1} \\ V_{c1,2} \\ V_{c2,1} \\ V_{c2,2} \end{pmatrix} = \begin{pmatrix} 1 & 0 & 0 \\ 0 & 1 & 0 \\ 1 & 0 & 0 \\ 0 & 0 & 1 \end{pmatrix} \begin{pmatrix} V_{c1,t1} \\ V_{c1,t2} \\ V_{c2,t2} \end{pmatrix}$$

or in matrix notation, $\quad$ (8.122)

$$\mathbf{I}_{C,T} = M_{C,Y} \mathbf{I}_C, \quad \mathbf{V}_C = M_{C,Y}^T \mathbf{V}_{C,T}$$

In general, this would be expanded to include all three phases on each core. In this connection, the voltages on core 1 are left intact but the voltages on core 2 would be changed in general. The terminal current, $I_{c1,t1}$, is the sum of the currents on the two parallel-connected terminals as the transformation indicates. The transformation removes any redundant terminals from the second core.

Applying (8.122) to (8.121), we get

$$\mathbf{I}_{C,T} = M_{C,Y} Y_C M_{C,Y}^T \mathbf{V}_{C,T} = Y_{C,mod} \mathbf{V}_{C,T} \quad (8.123)$$

Since this transformation is in the admittance representation, the terminal voltages after the transformation are not determined. However, these must be determined for calculations performed later when we transition to the impedance representation. The terminal voltage transformation that accomplishes this, in the opposite direction to that in (8.122), is

$$
\begin{pmatrix} V_{c1,t1} \\ V_{c1,t2} \\ V_{c2,t2} \end{pmatrix} = \begin{pmatrix} 1 & 0 & 0 & 0 \\ 0 & 1 & 0 & 0 \\ X & 0 & 0 & 0 \end{pmatrix} \begin{pmatrix} V_{c1,1} \\ V_{c1,2} \\ V_{c2,1} \\ V_{c2,2} \end{pmatrix}, \quad \mathbf{V}_{C,T} = M_{C,V}\mathbf{V}_C \tag{8.124}
$$

Here $X = |V_{c2,2}|/|V_{c2,1}|$. This is the voltage magnitude ratio of the voltage of terminal 2 on the second core to the voltage of terminal 1 on the first core. This is necessary to scale up or down the voltages on the second core terminals that do not participate in the parallel connection. This is done because the parallel connection sets the voltage of terminal 1 on the second core to equal the voltage of terminal 1 on the first core to which it is paralleled, and this may be different from the original voltage on terminal 1 of the second core. A similar ratio should be applied to all the other windings on core 2, which do not participate in the parallel connection, to scale them up or down as well. If the parallel-connected windings are at the same voltage before the parallel connection, then $X = 1$.

8.8.2 2-Core Series Connection

In order to make series connections, Equation 8.123 must be inverted to get an impedance representation. This requires that the core excitations be added, as was the case for the single-core inversion. Having done this, we obtain the inverse matrix to $Y_{C,\text{mod}}$, call it Z_C,

$$
\mathbf{V}_{C,T} = Z_C\mathbf{I}_{C,T} \tag{8.125}
$$

As before, we must save $M_{C,Y}^T$ and Y_C in order to recover the starting voltages and currents.

We use a similar notation as that used for the parallel connection, that is, a c label is added to distinguish between windings on either core. Let these two series-connected windings be winding number 1 on the two cores and let both cores have a second unconnected winding 2. The transformation for a single phase is

$$
\begin{pmatrix} V_{c1,u1} \\ V_{c1,u2} \\ V_{c2,u2} \end{pmatrix} = \begin{pmatrix} 1 & 0 & 1 & 0 \\ 0 & 1 & 0 & 0 \\ 0 & 0 & 0 & 1 \end{pmatrix} \begin{pmatrix} V_{c1,tf1} \\ V_{c1,tf2} \\ V_{c2,tf1} \\ V_{c2,tf2} \end{pmatrix}, \quad \begin{pmatrix} I_{c1,tf1} \\ I_{c1,tf2} \\ I_{c2,tf1} \\ I_{c2,tf2} \end{pmatrix} = \begin{pmatrix} 1 & 0 & 0 \\ 0 & 1 & 0 \\ 1 & 0 & 0 \\ 0 & 0 & 1 \end{pmatrix} \begin{pmatrix} I_{c1,u1} \\ I_{c1,u2} \\ I_{c2,u2} \end{pmatrix}
$$

or in matrix notation, \hfill (8.126)

$$
\mathbf{V}_{C,U} = M_{C,Z}\mathbf{V}_{C,T}, \quad \mathbf{I}_{C,T} = M_{C,Z}^T\mathbf{I}_{C,U}
$$

This transformation is applied to (8.125). A sequence of such transformations can be performed in this representation, and the resulting equation is

$$
\mathbf{V}_{C,U} = M_{C,Z}Z_CM_{C,Z}^T\mathbf{I}_{C,U}, \quad \mathbf{V}_{C,U} = Z_{C,\text{mod}}\mathbf{I}_{C,U} \tag{8.127}
$$

where the M_C's represent a possible sequence of such transformations and $Z_{C,\text{mod}}$ is the modified 2-core impedance matrix. The $M_{C,Z}^T$ and Z_C matrices need to be saved for later use.

8.8.3 Terminal Loading

Starting with (8.127), apply a load to the secondary terminals and a voltage to the input terminal and solve for the terminal currents. Typically, the terminals are loaded with some possibly complex impedance, which, when all the terminals are included, can be expressed as a complex impedance matrix, Z_L. For open terminals, a very large value for the load should be used. Any unloaded terminals such as the input terminal can be loaded with a given voltage, which takes the form of a load voltage vector $\mathbf{V}_L$. Loaded terminals would normally have their $\mathbf{V}_L$ entries set to 0. Thus, the terminal voltage vector becomes

$$\mathbf{V}_{C,U} = Z_L \mathbf{I}_L + \mathbf{V}_L \tag{8.128}$$

We use the convention where the terminal currents are into the terminals whereas the load currents are out of the terminals so that $\mathbf{I}_L = -\mathbf{I}_{C,U}$. Substituting into (8.127), we obtain

$$\left(Z_{C,\text{mod}} + Z_L \right) \mathbf{I}_{C,U} = \mathbf{V}_L \tag{8.129}$$

We can use complex matrix procedures to solve (8.129) for the terminal currents. Taking their negatives for use in (8.128), we can determine the load voltages. One of the reasons for keeping track of the terminal voltages all along is so that they can be used to obtain the unloaded terminal voltages for comparison with the loaded terminal voltages in order to calculate regulation effects.

Using previously saved matrices, the procedure for obtaining winding currents can be summarized:

$$\mathbf{I}_{C,T} = M_{C,Z}^T \mathbf{I}_{C,U}, \quad \mathbf{V}_{C,T} = Z_C \mathbf{I}_{C,T}, \quad \mathbf{V}_C = M_{C,Y}^T \mathbf{V}_{C,T}, \quad \mathbf{I}_C = Y_C \mathbf{V}_C \tag{8.130}$$

From $\mathbf{I}_C$, via (8.120), we can recover $\mathbf{I}_{c1}$ and $\mathbf{I}_{c2}$, which are the individual core current vectors. At this point, we can proceed with the determination of the winding currents for each core separately as discussed in previous sections.

8.8.4 Example of a 2-Core Phase Shifting Transformer

A 2-core phase shifting transformer was analyzed in Section 7.5, using phasor methods. This consists of a series unit with three windings per phase and an exciting unit consisting of two windings per phase. The same configuration, notation, and numerical values used in the phasor analysis are also used in this multiterminal calculation so a direct comparison can be made.

The input and output voltages and currents are labeled S and L, respectively, with a subscripted number indicating the phase. They are voltages to ground. The exciting unit's main winding is connected to the input and output windings of the series unit. The voltage to ground of the connecting point is labeled V with the subscripted number indicating the phase involved. The delta series unit winding is connected at its terminals to a particular tap on the tap winding. The phase shift will vary with the tap position. The winding currents are taken as positive into the terminal, and the load currents are taken as positive out of the load terminal.

8.8.4.1 Normal Loading

Assuming the turns ratio between the delta series unit winding and the input or output winding be n_s and the turns ratio between the main exciting winding and the tap winding be n_e, the phase shift is

$$\frac{E_a}{E_{a'}} = \frac{E_a}{E_{a''}} = n_s, \quad \frac{E_1}{E_{1'}} = n_e, \quad \theta = 2\tan^{-1}\left(\frac{\sqrt{3}}{n_e n_s}\right) \tag{8.131}$$

Let the winding to winding per-unit leakage reactances be $z_{aa'} = z_{aa''}$ and $z_{a'a''}$ for the series unit and $z_{11'}$ for the exciting unit, both on the same power base. Then from our previous analysis, the input to output per-unit terminal reactance is given by

$$z_{eq} = \frac{3}{\left[(n_e n_s)^2 + 3\right]}\left\{z_{a'a''} + \frac{4(n_e n_s)^2}{\left[(n_e n_s)^2 + 3\right]}\left[z_{11'} + \frac{1}{2}\left(z_{aa'} + z_{aa''} - z_{a'a''}\right)\right]\right\} \tag{8.132}$$

Other quantities of interest are the current in the delta, I_a, and the current in the main exciting winding, I_1. Their magnitudes are given in terms of the input current I_{S1} as

$$I_a = \frac{2}{n_s\sqrt{1 + 3/(n_e n_s)^2}} I_{S1}, \quad I_1 = \frac{2\sqrt{3}/(n_e n_s)}{\sqrt{1 + 3/(n_e n_s)^2}} I_{S1} \tag{8.133}$$

Also the power into the series or exciting transformer is given in terms of the input power, P_{in}, as $P_{series} = P_{exciting} = j\sin\theta\, P_{in}$, where j is the imaginary unit indicating a reactive power. It is also of interest to determine the regulation effects. This is given by

$$\text{Regulation} = \frac{1}{1 + z_{eq}/z_L} \tag{8.134}$$

where z_L is the per-unit load impedance. For a real 100% load and an imaginary z_{eq}, we can extract the magnitude and angle shift away from the no-load conditions from (8.134).

A phase shifting transformer of 100 MVA 3-phase rating was modeled with the 2-core multiterminal computer program. The winding turns were $N_a = 500$, $N_{a'} = N_{a''} = 200$, $N_1 = 393$, and $N_{1'} = 100$. The turns ratios were $n_s = 2.5$ and $n_e = 3.93$. The N_1 turns and therefore the turns ratio in the exciting unit were adjusted to get the desired phase shift. The per-unit percentage winding to winding reactances were $z_{aa'} = z_{aa''} = 22$, $z_{a'a''} = 50$, and $z_{11'} = 8$. Winding resistances were set to 0 in the calculations. The comparison of the analytical formula results with the multiterminal model results is shown in Table 8.12. The input current, I_{S1}, is calculated from the input power and the volts/turn of the series transformer, whose value is 100. We get $I_{S1} = 289.4$ amps. The magnitude shift was negligible so it is not given in the table. It was necessary to adjust the volts/turn of the exciting unit to make the voltages of the parallel-connected terminals equal so that regulation effects and terminal to terminal reactances are correct. In addition, it is also necessary to have both the series and exciting units on the same power base.

TABLE 8.12

Comparison of the Analytical with Multiterminal Results for Normal Loading

	θ (°)[a]	z_{eq} (%)	P_{series} (MVA)	I_a (amps)	I_1 (amps)	θ Shift (°)[b]
Analytical	20	2.093	34.202	228.03	100.52	−1.199
Multiterminal	20	2.093	34.206	228.01	100.52	−1.199

Source: Reprinted with permission from Ahuja, R. and Del Vecchio, R.M., Analysis of 2 core transformer designs, Cigre session paper no. A2_210_2014, Cigre 2014 General Session, Session 45, August 24–30, 2014. © 2014 Cigre.

[a] Angles are no load.
[b] Angle shift results from loading.

8.8.4.2 Single Line-to-Ground Fault

A single line to ground fault on the output terminal was analyzed in Section 7.7.3 using phasors and with the multiterminal computer program. For this, the zero sequence per-unit percent reactances must be determined. We used $z_{aa',0} = z_{aa'',0} = 18.7$, $z_{a'a'',0} = 42.5$, and $z_{11',0} = 8$. The 0 subscript denotes zero sequence. From these, the zero sequence terminal to terminal reactance, $z_{eq,0}$, is given by, using the same power base as for z_{eq},

$$z_{eq,0} = \frac{3}{\left[(n_e n_s)^2 + 3\right]}\left[2(z_{aa',0} + z_{aa'',0}) - z_{a'a'',0}\right] \tag{8.135}$$

The exciting unit's reactance is not included in (8.135) because zero sequence current cannot flow out of the delta winding in the series unit. We can also calculate this zero sequence reactance using the multiterminal program by running it with zero sequence voltages and reactances.

The analytical calculation of the fault currents in the various windings requires the use of sequence analysis. The positive and negative sequence quantities are labeled 1 and 2, and we assume the positive and negative sequence reactances are equal, $z_{eq,1} = z_{eq,2}$. For a single line to ground fault on the A phase, the per-unit sequence fault currents are given by

$$i_{L,1} = i_{L,2} = i_{L,0} = \frac{1}{z_{eq,0} + z_{eq,1} + z_{eq,2}} \qquad i_L = i_{L,1} + i_{L,2} + i_{L,0} \tag{8.136}$$

where i_L is the resulting phase A per-unit fault current. For the other winding currents, which are typically shifted in phase relative to the fault current, it should be noted that the sign of the phase should be reversed for the negative sequence currents. Thus, for the delta winding, a, on the series unit,

$$i_{a,1} = -\frac{2\sqrt{3}n_e n_s}{(n_e n_s)^2 + 3}e^{-j(\theta/2)}i_{L,1}, \quad i_{a,2} = -\frac{2\sqrt{3}n_e n_s}{(n_e n_s)^2 + 3}e^{j(\theta/2)}i_{L,2},$$

$$i_{a,0} = -\frac{2\sqrt{3}}{(n_e n_s)^2 + 3}i_{L,0} \tag{8.137}$$

To get the per-unit current in the A phase for winding a, we need to sum these three sequence currents.

TABLE 8.13

Comparison of the Analytical and Multiterminal Results for a Single Line
to Ground Fault

	$z_{eq,0}$ (%)	Fault Current, I_L (amps)	I_a (amps)	I_1 (amps)
Analytical	0.974	16,826	13,191	676.5
Multiterminal	0.974	16,828	13,192	676.6

Source: Reprinted with permission from Ahuja, R. and Del Vecchio, R.M., Analysis of
2 core transformer designs, Cigre session paper no. A2_210_2014, Cigre 2014
General Session, Session 45, August 24–30, 2014. © 2014 Cigre.

The per-unit sequence fault currents in winding 1 of the exciting unit are given by

$$i_{1,1} = -j\frac{2\sqrt{3}n_e n_s}{\left(n_e n_s\right)^2 + 3} e^{-j(\theta/2)} i_{L,1}, \quad i_{1,2} = j\frac{2\sqrt{3}n_e n_s}{\left(n_e n_s\right)^2 + 3} e^{j(\theta/2)} i_{L,2}, \quad i_{1,0} = 0 \tag{8.138}$$

These phase currents sum to get the phase A per-unit current, and we see that the sum will be purely imaginary relative to the fault current.

Inserting the values for i_{L1}, i_{L2}, and i_{L0} from (8.136) into (8.137) and (8.138) and summing the 3 per-unit phase currents, we obtain the per-unit currents in windings a and 1. The base fault load current is 289.4 amps. The winding a base voltage is 50 kV. So, using the same power base, its base current is 666.67 amps. The volts/turn of the exciting unit is 288.68 so the base voltage for winding 1 is 113.42 kV. Using the same power base, its base current is 293.89 amps Table 8.13. Inserting these values into the last four formulas, we obtain values for comparison with the output of the multiterminal program.

8.8.5 Discussion

The extension of the single-core multiterminal winding analysis to two cores is straightforward except for the parallel or terminals equal connection. This requires the adjustment of all the winding voltages and phases on the second core to be consistent with the parallel-connected voltage and phase of the winding on the first core. For purposes of obtaining the correct regulation effects and terminal to terminal impedances, it also requires that the two transformers be on the same power base and that the second transformer's volts/turn be adjusted so that the parallel-connected terminal voltages are equal.

9

Rabins' Method for Calculating Leakage Fields, Inductances, and Forces in Iron Core Transformers, Including Air Core Methods

9.1 Introduction

Modern general-purpose computer programs are available for calculating the magnetic field inside the complex geometry of a transformer. These numerical methods generally employ finite elements or boundary elements. Geometric details such as the tank wall and clamping structure can be included. While 3D programs are available, 2D programs using an axisymmetric geometry are adequate for many purposes. Although inputting the geometry, the Ampere-turns in the winding sections and the boundary conditions can be tedious, parametric procedures are often available for simplifying this task. Along with the magnetic field, associated quantities such as inductances and forces can be calculated by these methods. In addition, eddy currents in structural parts and their accompanying losses can be obtained with the appropriate a.c. solver.

In spite of these modern advances in computational methods, older procedures can often be profitably employed to obtain quantities of interest very quickly and with a minimum of input. One of these is Rabins' method, which uses an idealized transformer geometry [Rab56]. This simplified geometry permits analytic formulas to be developed for the magnetic field and other useful quantities. The geometry consists of a single leg of a 1- or possibly 3-phase transformer. The leg consists of a core and surrounding coils, which are assumed to be axisymmetric, along with yokes, which are assumed to be of infinite radial extent at the top and bottom of the leg. The entire axisymmetric geometry is of infinite extent radially. Thus there are no tank walls or clamping structures in the geometry. In addition, the core and yokes are assumed to be infinitely permeable.

In spite of these simplifications, Rabins' method does a good job of calculating the magnetic field in the immediate vicinity of the windings. Thus forces and inductances, which depend largely on the field near the windings, are also accurately obtained. This can be shown by direct comparison with a finite element solution applied to a more complex geometry, including the tank wall and clamps. Although the finite element procedure can obtain losses in structural parts, Rabins' method is not suited for this. However, because the magnetic field near or inside the windings is obtained accurately, eddy current losses in the windings as well as their spatial distribution can be accurately obtained from formulas based on this leakage field.

In the following, we present Rabins' method and show how it can be used to obtain the leakage magnetic field, forces, leakage inductances, and mutual inductances between winding sections. The leakage inductances and mutual inductances can also be used in detailed circuit models of transformers such as are needed in impulse calculations.

9.2 Theory

A cylindrical coil or section of a coil surrounding a core leg with top and bottom yokes is modeled in Figure 9.1. The yokes are really the top boundary of the geometry, which extends infinitely far radially. The method assumes d.c. conditions so that eddy current effects are not modeled directly. The current density in the coil is assumed to be piecewise constant axially and uniform radially as shown in Figure 9.2. As indicated, there can be regions within the coil where the current density drops to zero. Since there are no nonlinear effects within the geometry modeled, the fields from several coils can be added vectorially so that it is only necessary to model one coil at a time.

Maxwell's equations for the magnetic field, applied to the geometry outside the core and yokes and assuming static conditions are, in SI units,

$$\nabla \times \mathbf{H} = \mathbf{J} \tag{9.1}$$

$$\nabla \cdot \mathbf{B} = 0 \tag{9.2}$$

where
$\mathbf{H}$ is the magnetic field
$\mathbf{B}$ is the induction
$\mathbf{J}$ is the current density

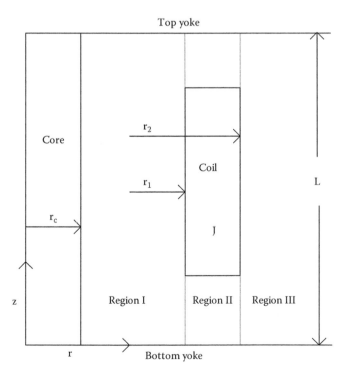

FIGURE 9.1
Geometry of iron core, yokes, and coil or coil section.

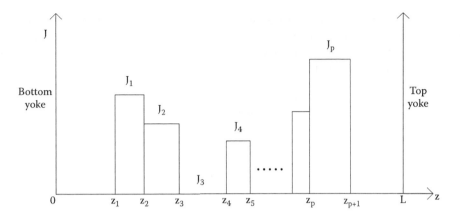

FIGURE 9.2
Axial distribution of current density in the coil.

Define a vector potential **A** by

$$\mathbf{B} = \nabla \times \mathbf{A} \tag{9.3}$$

Using this formula for **B**, Equation 9.2 is automatically satisfied. We also have

$$\mathbf{B} = \mu_o \mathbf{H} \tag{9.4}$$

in the region of interest where μ_o is the permeability of vacuum, oil, or air, which are all nearly the same. Substituting (9.3) and (9.4) into (9.1), we obtain

$$\nabla \times (\nabla \times \mathbf{A}) = \nabla(\nabla \cdot \mathbf{A}) - \nabla^2 \mathbf{A} = \mu_o \mathbf{J} \tag{9.5}$$

The vector potential is not completely defined by (9.3). It contains some arbitrariness, which can be removed by setting

$$\nabla \cdot \mathbf{A} = 0 \tag{9.6}$$

Using this, (9.5) becomes

$$\nabla^2 \mathbf{A} = -\mu_o \mathbf{J} \tag{9.7}$$

The current density vector is azimuthal so that

$$\mathbf{J} = J_\varphi \mathbf{a}_\varphi \tag{9.8}$$

where $\mathbf{a}_\varphi$ is the unit vector in the azimuthal direction. Because of the axisymmetric geometry, all the field quantities are independent of the azimuthal angle φ. With these assumptions, (9.7) becomes in cylindrical coordinates

$$\frac{\partial^2 A_\varphi}{\partial r^2} + \frac{1}{r}\frac{\partial A_\varphi}{\partial r} - \frac{A_\varphi}{r^2} + \frac{\partial^2 A_\varphi}{\partial z^2} = -\mu_o J_\varphi \tag{9.9}$$

Thus, **A** and **J** have only a φ component and we drop this subscript in the following formulas for simplicity.

Write the current density as a Fourier series in terms of a fundamental spatial period of length L, the yoke-to-yoke distance, or window height. In practice, L can be some multiple of the yoke-to-yoke distance to approximate the fact that the yoke does not extend uniformly in the azimuthal direction.

$$J = J_0 + \sum_{n=1}^{\infty} J_n \cos\left(\frac{n\pi z}{L}\right) \tag{9.10}$$

where

$$J_0 = \frac{1}{L}\int_0^L J\,dz, \quad J_n = \frac{2}{L}\int_0^L J\cos\left(\frac{n\pi z}{L}\right)dz \tag{9.11}$$

For the current density described in Figure 9.2 for example, we have

$$J = \begin{cases} 0, & 0 \le z \le z_1 \\ J_1, & z_1 \le z \le z_2 \\ \quad \vdots \\ J_i, & z_i \le z \le z_{i+1} \\ \quad \vdots \\ J_p, & z_p \le z \le z_{p+1} \\ 0, & z_{p+1} \le z \le L \end{cases} \tag{9.12}$$

Using (9.12), the integrals in (9.11) can be evaluated to get

$$J_0 = \frac{1}{L}\sum_{i=1}^{p} J_i\left(z_{i+1} - z_i\right)$$

$$J_n = \frac{2}{n\pi}\sum_{i=1}^{p} J_i\left[\sin\left(\frac{n\pi z_{i+1}}{L}\right) - \sin\left(\frac{n\pi z_i}{L}\right)\right] \tag{9.13}$$

Thus for one section with constant current density J, we would have

$$J_0 = J\frac{\left(z_2 - z_1\right)}{L}$$

$$J_n = \frac{2J}{n\pi}\left[\sin\left(\frac{n\pi z_2}{L}\right) - \sin\left(\frac{n\pi z_1}{L}\right)\right] \tag{9.14}$$

As shown in Figure 9.1, we have divided the solution space into three regions:

Region I $r_c \le r \le r_1,$ $0 \le z \le L$
Region II $r_1 \le r \le r_2,$ $0 \le z \le L$
Region III $r_2 \le r \le \infty,$ $0 \le z \le L$

In Regions I and III, the current density is 0 so Equation 9.9 becomes, dropping the subscript φ

$$\frac{\partial^2 A}{\partial r^2} + \frac{1}{r}\frac{\partial A}{\partial r} - \frac{A}{r^2} + \frac{\partial^2 A}{\partial z^2} = 0 \tag{9.15}$$

This is a homogeneous partial differential equation. We look for a solution of the form

$$A(r,z) = R(r)Z(z) \tag{9.16}$$

Substituting this into (9.15) and dividing by RZ, we get

$$\frac{1}{R}\frac{\partial^2 R}{\partial r^2} + \frac{1}{rR}\frac{\partial R}{\partial r} - \frac{1}{r^2} + \frac{1}{Z}\frac{\partial^2 Z}{\partial z^2} = 0 \tag{9.17}$$

This equation contains terms, which are only a function of r and terms, which are only a function of z whose sum is a constant = 0. Therefore, each set of terms must separately equal a constant whose sum is zero. Let the constant be m², a positive number. Then

$$\frac{1}{R}\frac{\partial^2 R}{\partial r^2} + \frac{1}{rR}\frac{\partial R}{\partial r} - \frac{1}{r^2} = m^2, \quad \frac{1}{Z}\frac{\partial^2 Z}{\partial z^2} = -m^2 \tag{9.18}$$

Rearranging terms, we get

$$r^2\frac{\partial^2 R}{\partial r^2} + r\frac{\partial R}{\partial r} - (m^2 r^2 + 1)R = 0, \quad \frac{\partial^2 Z}{\partial z^2} + m^2 Z = 0 \tag{9.19}$$

We need to consider separately the cases m = 0 and m > 0.

For m = 0, the solution to the z equation, which satisfies the boundary conditions at the top and bottom yokes is a constant independent of z. The m = 0, r equation in (9.19) becomes

$$r^2\frac{\partial^2 R}{\partial r^2} + r\frac{\partial R}{\partial r} - R = 0 \tag{9.20}$$

The solution to this equation is

$$R_0 = Sr + \frac{T}{r} \tag{9.21}$$

where S and T are constants to be determined by the boundary conditions and the solution is labeled with a subscript 0.

For m > 0, the solution to the z equation in (9.19) can be written:

$$Z = Z_m \cos(mz + \varphi_m) \tag{9.22}$$

where Z_m and φ_m are constants to be determined by the boundary conditions. Since we assume that the yoke material has infinite permeability, this requires the B-field to be perpendicular to the yoke surfaces. To determine this boundary condition, we use (9.3) expressed in cylindrical coordinates:

$$\mathbf{B} = -\frac{\partial A}{\partial z}\mathbf{a}_r + \left(\frac{\partial A}{\partial r} + \frac{A}{r}\right)\mathbf{k} \tag{9.23}$$

where $\mathbf{a}_r$ and $\mathbf{k}$ are unit vectors in the r and z directions. For $\mathbf{B}$ to be perpendicular to the upper and lower yokes, we must have the $\mathbf{a}_r$ component of the B-field vanish at these yoke positions. Thus from (9.16),

$$\frac{\partial A}{\partial z} = 0 \quad \Rightarrow \quad \frac{\partial Z}{\partial z} = 0 \quad \text{at } z = 0, L \tag{9.24}$$

Using this, Equation 9.22 must have the form

$$Z = Z_n \cos(mz), \quad m = \frac{n\pi}{L}, \quad n = 1, 2, \dots \tag{9.25}$$

since its derivative yields a sine function, which is zero at these m values. We use n rather than m as a label. m also depends on n, but we omit this dependence for simplicity.

For m > 0, the radial equation in (9.19) can be written with the substitution x = mr:

$$x^2 \frac{\partial^2 R}{\partial x^2} + x \frac{\partial R}{\partial x} - R(x^2 + 1) = 0 \tag{9.26}$$

The solution to this equation is

$$R_n = C_n I_1(x) + D_n K_1(x) \tag{9.27}$$

where I_1 and K_1 are modified Bessel functions of the first and second kind, respectively, of order 1 [Dwi61]. We have also labeled the constants C and D with the subscript n since the solution depends on n through m that occurs in x.

In general, the solution to (9.15) is expressible as a sum of these individual solutions each, except the m = 0 term, with a product of an R and Z term,

$$A = Sr + \frac{T}{r} + \sum_{n=1}^{\infty} \left[C_n I_1(mr) + D_n K_1(mr) \right] \cos(mz), \quad m = \frac{n\pi}{L} \tag{9.28}$$

Here, the Z solution constants have been absorbed in the overall constants shown. This solution satisfies the boundary condition at z = 0, L. Because we also assume an infinitely permeable core, the B-field must be normal to the core surface. Thus from (9.23), we require that the $\mathbf{k}$-component vanish at the core surface (Region I):

$$\frac{\partial A}{\partial r} + \frac{A}{r} = 0 \quad \text{at } r = r_c \tag{9.29}$$

Substituting (9.28) into this last equation, we get

$$2S + \sum_{n=1}^{\infty} \left[mC_n \frac{\partial I_1}{\partial x} + mD_n \frac{\partial K_1}{\partial x} + m \frac{(C_n I_1 + D_n K_1)}{x} \right] \cos(mz) = 0 \quad \text{at } x = mr_c \tag{9.30}$$

Although (9.30) would seem to require that S = 0, it will turn out that when all the windings are considered with their Ampere-turns, which sum to zero, we can satisfy this boundary condition with S ≠ 0 for each winding. Assuming the S term will ultimately vanish, the remaining terms in (9.30) must vanish for all z at x = mr$_c$. These remaining terms constitute a Fourier expansion in cos(nπz/L). In order for these terms to vanish for all z, the individual Fourier coefficients must vanish. This requires

$$mC_n\left(\frac{\partial I_1}{\partial x} + \frac{I_1}{x}\right) + mD_n\left(\frac{\partial K_1}{\partial x} + \frac{K_1}{x}\right) = mC_n I_0 - mD_n K_0 = 0 \tag{9.31}$$

where we have used modified Bessel function identities to get the first equality in (9.31) [Rab56]. I_0 and K_0 are modified Bessel functions of order 0. Thus we have from (9.31)

$$D_n = C_n \frac{I_0(mr_c)}{K_0(mr_c)} \tag{9.32}$$

Labeling the unknown constants with the region number as superscript, (9.28) becomes

$$A^I = S^I r + \sum_{n=1}^{\infty} C_n^I\left[I_1(mr) + \frac{I_0(mr_c)}{K_0(mr_c)}K_1(mr)\right]\cos(mz), \quad m = \frac{n\pi}{L} \tag{9.33}$$

We have dropped the T/r term since it is not needed to satisfy the boundary condition in this region and becomes increasingly large as the core radius approaches zero. With I superscript labeling the solution constants in this region, this amounts to setting $T^I = 0$.

In Region III, we require that A be finite as r → ∞. Since I_1 → ∞ as r → ∞ and the Sr term also → ∞, we have from (9.28), using the appropriate region label,

$$A^{III} = \frac{T^{III}}{r} + \sum_{n=1}^{\infty} D_n^{III} K_1(mr)\cos(mz) \tag{9.34}$$

With III superscript labeling the solution constants in this region, this amounts to setting $S^{III} = 0$ and $C_n^{III} = 0$.

In Region II, we must keep the current density term in (9.9). Substituting the Fourier series (9.10) into (9.9) and dropping the φ subscript, we have in Region II

$$\frac{\partial^2 A}{\partial r^2} + \frac{1}{r}\frac{\partial A}{\partial r} - \frac{A}{r^2} + \frac{\partial^2 A}{\partial z^2} = -\mu_o\left[J_0 + \sum_{n=1}^{\infty} J_n \cos(mz)\right] \tag{9.35}$$

We look for a solution to this equation in the form of a series expansion,

$$A = \sum_{n=0}^{\infty} R_n(r)\cos(mz), \quad m = \frac{n\pi}{L} \tag{9.36}$$

Substituting into (9.35), we get

$$\sum_{n=0}^{\infty}\left(\frac{\partial^2 R_n}{\partial r^2} + \frac{1}{r}\frac{\partial R_n}{\partial r} - \frac{R_n}{r^2}\right)\cos(mz) - m^2\sum_{n=1}^{\infty} R_n \cos(mz)$$

$$= -\mu_o\left[J_0 + \sum_{n=1}^{\infty} J_n \cos(mz)\right] \tag{9.37}$$

Since the cosine functions are orthogonal, we can equate corresponding coefficients on both sides of this equation. We obtain

$$\frac{\partial^2 R_0}{\partial r^2} + \frac{1}{r}\frac{\partial R_0}{\partial r} - \frac{R_0}{r^2} = -\mu_o J_0$$

$$\frac{\partial^2 R_n}{\partial r^2} + \frac{1}{r}\frac{\partial R_n}{\partial r} - \frac{R_n}{r^2} - m^2 R_n = -\mu_o J_n \tag{9.38}$$

The solution to the n = 0 equation can be written in terms of a solution of the homogeneous equation, which was found earlier, plus a particular solution:

$$R_0 = Sr + \frac{T}{r} - \frac{\mu_o J_0 r^2}{3} \tag{9.39}$$

Multiplying the second equation in (9.38) by r^2, we see that the homogeneous part to the left of the equal sign, as given in the following formula, is the same as that in (9.19) or (9.26), although the latter are expressed in terms of x.

$$r^2 \frac{\partial^2 R_n}{\partial r^2} + r\frac{\partial R_n}{\partial r} - \left(m^2 r^2 + 1\right) R_n = -\mu_o J_n r^2$$

Its solution was given in (9.27) so that, combined with a particular solution, the solution to the n > 0 equation in (9.38) is:

$$R_n = C_n I_1(mr) + D_n K_1(mr) - \frac{\pi \mu_o J_n}{2m^2} L_1(mr) \tag{9.40}$$

where the last term is the particular solution to the inhomogeneous equation. L_1 is a modified Struve function of order 1 [Abr72]. Thus the solution (9.36) in Region II is given explicitly as

$$A^{II} = S^{II} r + \frac{T^{II}}{r} - \frac{\mu_o J_0 r^2}{3}$$

$$+ \sum_{n=1}^{\infty} \left[C_n^{II} I_1(mr) + D_n^{II} K_1(mr) - \frac{\pi \mu_o J_n}{2m^2} L_1(mr) \right] \cos(mz) \tag{9.41}$$

where we have labeled the constants with the region number II as superscripts.

   This equation already satisfies the boundary conditions at z = 0 and L as do the other region equations. The unknown constants must be determined by satisfying the boundary conditions at r = r_1, r_2 (see Figure 9.1). We require that the vector potential be continuous across the interfaces. Otherwise, the B-field given by (9.23) would contain infinities. Thus at r = r_1, using (9.33) and (9.41),

$$S^I r_1 + \sum_{n=1}^{\infty} C_n^I \left[I_1(mr_1) + \frac{I_0(mr_c)}{K_0(mr_c)} K_1(mr_1) \right] \cos(mz) = S^{II} r_1 + \frac{T^{II}}{r_1} - \frac{\mu_o J_0 r_1^2}{3}$$

$$+ \sum_{n=1}^{\infty} \left[C_n^{II} I_1(mr_1) + D_n^{II} K_1(mr_1) - \frac{\pi \mu_o J_n}{2m^2} L_1(mr_1) \right] \cos(mz) \tag{9.42}$$

Since this must be satisfied for all z, we require

$$S^I r_1 = S^{II} r_1 + \frac{T^{II}}{r_1} - \frac{\mu_o J_o r_1^2}{3}$$

$$C_n^I \left[I_1(x_1) + \frac{I_0(x_c)}{K_0(x_c)} K_1(x_1) \right] = C_n^{II} I_1(x_1) + D_n^{II} K_1(x_1) - \frac{\pi \mu_o J_n}{2m^2} L_1(x_1)$$

(9.43)

where $x_1 = mr_1$, $x_c = mr_c$.

At $r = r_2$, we obtain similarly, using (9.34) and (9.41),

$$\frac{T^{III}}{r_2} = S^{II} r_2 + \frac{T^{II}}{r_2} - \frac{\mu_o J_o r_2^2}{3}$$

$$D_n^{III} K_1(x_2) = C_n^{II} I_1(x_2) + D_n^{II} K_1(x_2) - \frac{\pi \mu_o J_n}{2m^2} L_1(x_2)$$

(9.44)

where $x_2 = mr_2$.

In addition to the continuity of **A** at the interfaces between regions, we also require, according to Maxwell's equations, that the normal component of **B** and the tangential component of **H** be continuous across these interfaces. According to (9.23), the normal **B** components (a_r components) are already continuous across these interfaces since all regional solutions have the same z dependence and because the A's are continuous. Since **B** is proportional to **H** in all regions of interest here, we require that the tangential **B** components be continuous. Thus from (9.23) we require that, using $x = mr$,

$$\frac{\partial A}{\partial r} + \frac{A}{r} = \frac{1}{r} \frac{\partial}{\partial r}(rA) = \frac{m}{x} \frac{\partial}{\partial x}(xA)$$

(9.45)

be continuous at $r = r_1, r_2$ ($x = x_1, x_2$). Using this, we obtain the additional conditions on the unknown constants:

$$2S^I = 2S^{II} - \mu_o J_o r_1, \quad 2S^{II} - \mu_o J_o r_2 = 0$$

$$C_n^I \left\{ \frac{\partial}{\partial x} \left[x I_1(x) \right] + \frac{I_0(x_c)}{K_0(x_c)} \frac{\partial}{\partial x} \left[x K_1(x) \right] \right\} = C_n^{II} \frac{\partial}{\partial x} \left[x I_1(x) \right] + D_n^{II} \frac{\partial}{\partial x} \left[x K_1(x) \right]$$

$$- \frac{\pi \mu_o J_n}{2m^2} \frac{\partial}{\partial x} \left[x L_1(x) \right] \quad \text{at } x = x_1 = mr_1$$

$$C_n^{II} \frac{\partial}{\partial x} \left[x I_1(x) \right] + D_n^{II} \frac{\partial}{\partial x} \left[x K_1(x) \right] - \frac{\pi \mu_o J_n}{2m^2} \frac{\partial}{\partial x} \left[x L_1(x) \right] = D_n^{III} \frac{\partial}{\partial x} \left[x K_1(x) \right] \quad \text{at } x = x_2 = mr_2$$

(9.46)

Using the identities [Rab56], [Abr72],

$$\frac{\partial}{\partial x} \left[x I_1(x) \right] = x I_0(x), \quad \frac{\partial}{\partial x} \left[x K_1(x) \right] = -x K_0(x), \quad \frac{\partial}{\partial x} \left[x L_1(x) \right] = x L_0(x)$$

(9.47)

We obtain for the last two equations in (9.46)

$$C_n^I \left[I_0(x_1) - \frac{I_0(x_c)}{K_0(x_c)} K_0(x_1) \right] = C_n^{II} I_0(x_1) - D_n^{II} K_0(x_1) - \frac{\pi\mu_o J_n}{2m^2} L_0(x_1) C_n^{II} I_0(x_2)$$

$$- D_n^{II} K_0(x_2) - \frac{\pi\mu_o J_n}{2m^2} L_0(x_2) = -D_n^{III} K_0(x_2) \qquad (9.48)$$

Solving for the S and T constants from (9.43), (9.44), and (9.46), we obtain

$$S^I = \frac{\mu_o J_0(r_2 - r_1)}{2}, \quad S^{II} = \frac{\mu_o J_0 r_2}{2}$$

$$T^{II} = -\frac{\mu_o J_0 r_1^3}{6}, \quad T^{III} = \frac{\mu_o J_0(r_2^3 - r_1^3)}{6} \qquad (9.49)$$

Solving for the C_n and D_n constants from (9.43), (9.44), and (9.46), we get

$$C_n^{II} = \frac{\pi\mu_o J_n}{2m^2} \left[\frac{K_0(x_2) L_1(x_2) + K_1(x_2) L_0(x_2)}{K_0(x_2) I_1(x_2) + K_1(x_2) I_0(x_2)} \right]$$

$$C_n^{I} = C_n^{II} - \frac{\pi\mu_o J_n}{2m^2} \left[\frac{K_0(x_1) L_1(x_1) + K_1(x_1) L_0(x_1)}{K_0(x_1) I_1(x_1) + K_1(x_1) I_0(x_1)} \right]$$

$$D_n^{II} = \frac{I_0(x_c)}{K_0(x_c)} C_n^{I} + \frac{\pi\mu_o J_n}{2m^2} \left[\frac{I_0(x_1) L_1(x_1) - I_1(x_1) L_0(x_1)}{K_0(x_1) I_1(x_1) + K_1(x_1) I_0(x_1)} \right] \qquad (9.50)$$

$$D_n^{III} = D_n^{II} - \frac{\pi\mu_o J_n}{2m^2} \left[\frac{I_0(x_2) L_1(x_2) - I_1(x_2) L_0(x_2)}{K_0(x_2) I_1(x_2) + K_1(x_2) I_0(x_2)} \right]$$

Using the identities [Rab56],

$$I_0(x) K_1(x) + I_1(x) K_0(x) = \frac{1}{x}$$

$$x \left[L_0(x) K_1(x) + L_1(x) K_0(x) \right] = \frac{2}{\pi} \int_0^x t K_1(t) \, dt \qquad (9.51)$$

$$x \left[L_0(x) I_1(x) - L_1(x) I_0(x) \right] = \frac{2}{\pi} \int_0^x t I_1(t) \, dt$$

we can transform (9.50) into the form

$$C_n^{II} = \frac{\mu_o J_n}{m^2} \int_0^{x_2} t K_1(t) dt$$

$$C_n^{I} = \frac{\mu_o J_n}{m^2} \int_{x_1}^{x_2} t K_1(t) dt$$

$$D_n^{II} = \frac{\mu_o J_n}{m^2} \left[\frac{I_0(x_c)}{K_0(x_c)} \int_{x_1}^{x_2} t K_1(t) dt - \int_0^{x_1} t I_1(t) dt \right] \qquad (9.52)$$

$$D_n^{III} = \frac{\mu_o J_n}{m^2} \left[\frac{I_0(x_c)}{K_0(x_c)} \int_{x_1}^{x_2} t K_1(t) dt + \int_{x_1}^{x_2} t I_1(t) dt \right]$$

Summarizing and simplifying the notation slightly, the solutions in the three regions are given by

$$A^I = \frac{\mu_o J_0 (r_2 - r_1)}{2} r + \mu_o \sum_{n=1}^{\infty} \frac{J_n}{m^2} \left[C_n I_1(x) + D_n K_1(x) \right] \cos(mz)$$

$$A^{II} = \mu_o J_0 \left(\frac{r_2 r}{2} - \frac{r_1^3}{6r} - \frac{r^2}{3} \right) + \mu_o \sum_{n=1}^{\infty} \frac{J_n}{m^2} \left[E_n I_1(x) + F_n K_1(x) - \frac{\pi}{2} L_1(x) \right] \cos(mz) \qquad (9.53)$$

$$A^{III} = \frac{\mu_o J_0 (r_2^3 - r_1^3)}{6r} + \mu_o \sum_{n=1}^{\infty} \frac{J_n}{m^2} G_n K_1(x) \cos(mz)$$

where $m = n\pi/L$, $x = mr$, and

$$C_n = \int_{x_1}^{x_2} t K_1(t) dt, \quad D_n = \frac{I_0(x_c)}{K_0(x_c)} C_n$$

$$E_n = \int_0^{x_2} t K_1(t) dt, \quad F_n = \frac{I_0(x_c)}{K_0(x_c)} \int_{x_1}^{x_2} t K_1(t) dt - \int_0^{x_1} t I_1(t) dt \qquad (9.54)$$

$$G_n = \frac{I_0(x_c)}{K_0(x_c)} \int_{x_1}^{x_2} t K_1(t) dt + \int_{x_1}^{x_2} t I_1(t) dt$$

Here $x_1 = mr_1$ and $x_2 = mr_2$.

Note that, using (9.23), the axial component of the induction vector at the core radius is

$$B_z(r_c, z) = \mu_o J_0 (r_2 - r_1) \qquad (9.55)$$

This is μ_o times the average current density times the radial build of the winding. If we multiply and divide (9.55) by the winding height L, we see that B_z is proportional to the amp-turns in the winding with proportionality constant μ_o/L. Since the axial height L is the same for all the windings, when we add this axial component for all the windings, we will get zero, assuming Ampere-turn balance. Thus, the flux will enter the core radially as required.

Although the modified Bessel functions are generally available in mathematical computer libraries, the modified Struve functions are not so easily obtained. We therefore indicate here some methods of obtaining these and the integrals in (9.54) in terms of readily available functions or easily evaluated integrals. From [Abr72], the modified Struve functions are given in integral form as

$$
\begin{aligned}
L_0(x) &= \frac{2}{\pi} \int_0^{\pi/2} \sinh(x\cos\theta)\,d\theta \\
L_1(x) &= \frac{2x}{\pi} \int_0^{\pi/2} \sinh(x\cos\theta)\sin^2\theta\,d\theta
\end{aligned}
\tag{9.56}
$$

It can be seen from (9.56) that these functions become asymptotically large as $x \to \infty$. Such large values of x would occur when evaluating the higher harmonics in (9.53). We indicate here some way of avoiding these large values by looking for a means of cancellation. [Abr72] gives integral expressions for the modified Bessel functions. In particular, we have

$$
\begin{aligned}
I_0(x) &= \frac{1}{\pi} \int_0^{\pi} e^{\pm x\cos\theta}\,d\theta = \frac{1}{\pi} \int_0^{\pi/2} \left(e^{x\cos\theta} + e^{-x\cos\theta}\right) d\theta \\
I_1(x) &= \frac{x}{\pi} \int_0^{\pi} e^{\pm x\cos\theta}\sin^2\theta\,d\theta = \frac{x}{\pi} \int_0^{\pi/2} \left(e^{x\cos\theta} + e^{-x\cos\theta}\right)\sin^2\theta\,d\theta
\end{aligned}
\tag{9.57}
$$

These also become asymptotically large as x increases. However, by defining the difference functions

$$
M_0(x) = I_0(x) - L_0(x), \quad M_1(x) = I_1(x) - L_1(x)
\tag{9.58}
$$

we can show that

$$
\begin{aligned}
M_0(x) &= \frac{2}{\pi} \int_0^{\pi/2} e^{-x\cos\theta}\,d\theta \\
M_1(x) &= \frac{2x}{\pi} \int_0^{\pi/2} e^{-x\cos\theta}\sin^2\theta\,d\theta = \frac{2}{\pi}\left[1 - \int_0^{\pi/2} e^{-x\cos\theta}\cos\theta\,d\theta\right]
\end{aligned}
\tag{9.59}
$$

The last equality in (9.59) is obtained by integration by parts. The integrals in (9.59) are well behaved as x increases and can be evaluated numerically. In subsequent developments, we will also need the integral of M_0. This is obtained from the first equation in (9.59) as

$$\int_0^x M_0(t)dt = \frac{2}{\pi}\int_0^{\pi/2}\frac{\left(1-e^{-x\cos\theta}\right)}{\cos\theta}d\theta$$

(9.60)

The integrand approaches x as $\theta \to \pi/2$ and so the integral is also well behaved. For high x values, asymptotic series can be found in [Abr72] for some of these functions. This reference also lists a table of the difference functions defined in (9.58) and the integral given in (9.60).

Let us now write other expressions of interest in terms of the M functions and modified Bessel functions. Using (9.51) and (9.58), we can write

$$\int_0^x tI_1(t)dt = \frac{\pi}{2}x\left[M_1(x)I_0(x)-M_0(x)I_1(x)\right]$$

$$\int_0^x tK_1(t)dt = \frac{\pi}{2}\left\{1-x\left[M_1(x)K_0(x)+M_0(x)K_1(x)\right]\right\}$$

(9.61)

We will also need a similar integral for the modified Struve function. We need the following identities given in [Rab56].

$$\frac{d}{dt}L_0(t) = \frac{2}{\pi}+L_1(t), \quad \frac{d}{dt}I_0(t) = I_1(t)$$

(9.62)

Multiplying these by t and integrating from 0 to x, we get

$$xL_0(x)-\int_0^x L_0(t)dt = \frac{x^2}{\pi}+\int_0^x tL_1(t)dt$$

$$xI_0(x)-\int_0^x I_0(t)dt = \int_0^x tI_1(t)dt$$

(9.63)

Substituting for L_0 in terms of the M_0 function and rearranging, we get

$$\int_0^x tL_1(t)dt = -xM_0(x)-\frac{x^2}{\pi}+\int_0^x M_0(t)dt+\int_0^x tI_1(t)dt$$

(9.64)

Thus, all the functions needed to determine the vector potential are obtainable in terms of the modified Bessel functions and the M functions. We will need (9.64) in obtaining derived quantities in the following sections.

The determination of the vector potential for the other coils or coil sections proceeds identically to the previous development. Because Maxwell's equations are linear in the fields and potentials in the region outside the core and yokes, we can simply add the potentials or fields from the various coils vectorially at each point to get the net potential or field. We need to be aware of the fact that a given point may not be in the same region number (I, II, or III) for the different coils.

9.3 Rabins' Formula for Leakage Reactance

The leakage inductance for a 2-winding transformer can be obtained from the total magnetic energy as discussed in Chapter 4 when there is amp-turn balance. We use (4.9) for the magnetic energy, W, which is repeated here

$$W = \frac{1}{2} \int_{\substack{\text{volume} \\ \text{of circuits}}} \mathbf{A} \cdot \mathbf{J} dV \tag{9.65}$$

This, together with (4.23), can be used to obtain the leakage inductance. Here $\mathbf{A}$ is the total vector potential due to both windings. Since $\mathbf{A}$ and $\mathbf{J}$ are both azimuthally directed, the dot product becomes simply the ordinary product. We also drop the implied φ subscript on J and A.

9.3.1 Rabins' Method Applied to Calculate the Leakage Reactance between Two Windings Which Occupy Different Radial Positions

We assume that the two windings occupy different radial positions as shown in Figure 9.3.

Writing A_1 and A_2 for the vector potential due to coils 1 and 2 and J_1, J_2 for their current densities, Equation 9.65 becomes

$$2W = \int_{V_1} (A_1 + A_2) J_1 dV_1 + \int_{V_2} (A_1 + A_2) J_2 dV_2$$

$$= \int_{V_1} A_1 J_1 dV_1 + \int_{V_1} A_2 J_1 dV_1 + \int_{V_2} A_1 J_2 dV_2 + \int_{V_2} A_2 J_2 dV_2 \tag{9.66}$$

For the $A_1 J_1$ or $A_2 J_2$ integrals, we must use the Region II solution corresponding to each coil. Assuming coil 1 is the inner coil, the $A_2 J_1$ integral requires that we use the Region I solution for A_2 while the $A_1 J_2$ integral requires that we use the Region III solution for A_1.

From (9.53) and (9.10), using a second subscript to distinguish the coil 1 quantities from the coil 2 ones,

$$\int_{V_1} A_1 J_1 dV_1 = 2\pi \int_{r_1}^{r_2} r dr \int_0^L dz \left[J_{0,1} + \sum_{p=1}^{\infty} J_{p,1} \cos\left(\frac{p\pi z}{L}\right) \right] \left\{ \mu_o J_{0,1} \left(\frac{r_2 r}{2} - \frac{r_1^3}{6r} - \frac{r^2}{3} \right) \right.$$

$$\left. + \mu_o \sum_{n=1}^{\infty} \frac{J_{n,1}}{m^2} \left[E_{n,1} I_1(mr) + F_{n,1} K_1(mr) - \frac{\pi}{2} L_1(mr) \right] \cos\left(\frac{n\pi z}{L}\right) \right\} \tag{9.67}$$

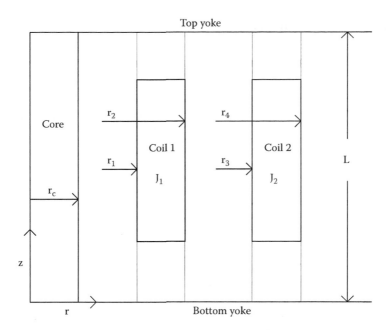

FIGURE 9.3
Geometry for 2-coil leakage inductance calculation.

Note that the subscript 1 on the I and K modified Bessel functions are unrelated to the labeling for winding 1. Also note that $E_{n,1}$ and $F_{n,1}$ are the same as E_n and F_n in (9.54) and similarly for the other coefficients. However, later when we consider winding 2, $E_{n,2}$ and $F_{n,2}$, etc., will require that the integration limits in (9.54) be changed from x_1, x_2 to x_3, x_4. Performing the z-integral first, we notice that product terms that contain a single cosine term vanish, that is,

$$\int_0^L \cos\left(\frac{n\pi z}{L}\right)dz = \frac{L}{n\pi}\sin\left(\frac{n\pi z}{L}\right)\Bigg|_0^L = 0 \tag{9.68}$$

Product terms that contain two cosine factors have the value

$$\int_0^L \cos\left(\frac{n\pi z}{L}\right)\cos\left(\frac{p\pi z}{L}\right)dz = \begin{cases} 0, & n \neq p \\ L/2, & n = p \end{cases} \tag{9.69}$$

After performing the z-integral, Equation 9.67 becomes

$$\int_{V_1} A_1 J_1 dV_1 = 2\pi\int_{r_1}^{r_2} r dr\left\{\mu_o L J_{0,1}^2\left(\frac{r_2 r}{2} - \frac{r_1^3}{6r} - \frac{r^2}{3}\right)\right.$$

$$\left. + \frac{\mu_o L}{2}\sum_{n=1}^{\infty}\frac{J_{n,1}^2}{m^2}\left[E_{n,1}I_1(mr) + F_{n,1}K_1(mr) - \frac{\pi}{2}L_1(mr)\right]\right\} \tag{9.70}$$

Performing the r-integration, we get

$$
\int_{V_1} A_1 J_1 dV_1 = 2\pi\mu_o L \left\{ J_{0,1}^2 \left(\frac{r_1^4}{4} + \frac{r_2^4}{12} - \frac{r_2 r_1^3}{3} \right) \right.
$$

$$
\left. + \frac{1}{2} \sum_{n=1}^{\infty} \frac{J_{n,1}^2}{m^2} \left[E_{n,1} \int_{r_1}^{r_2} r I_1(mr) dr + F_{n,1} \int_{r_1}^{r_2} r K_1(mr) dr - \frac{\pi}{2} \int_{r_1}^{r_2} r L_1(mr) dr \right] \right\}
$$

$$
= 2\pi\mu_o L \left\{ J_{0,1}^2 \left(\frac{r_1^4}{4} + \frac{r_2^4}{12} - \frac{r_2 r_1^3}{3} \right) \right.
$$

$$
\left. + \frac{1}{2} \sum_{n=1}^{\infty} \frac{J_{n,1}^2}{m^4} \left[E_{n,1} \int_{x_1}^{x_2} x I_1(x) dx + F_{n,1} \int_{x_1}^{x_2} x K_1(x) dx - \frac{\pi}{2} \int_{x_1}^{x_2} x L_1(x) dx \right] \right\} \qquad (9.71)
$$

In the second equality in (9.71), the r integrals have been converted to integrals in $x = mr$. Formulas for the x-integrals above have been given previously in terms of known or calculable quantities.

The $A_2 J_2$ term in (9.66) is given by (9.71) with a 2 subscript and with the r-integration from r_3 to r_4 (see Figure 9.3). Thus we have

$$
\int_{V_2} A_2 J_2 dV_2 = 2\pi\mu_o L \left\{ J_{0,2}^2 \left(\frac{r_3^4}{4} + \frac{r_4^4}{12} - \frac{r_4 r_3^3}{3} \right) \right.
$$

$$
\left. + \frac{1}{2} \sum_{n=1}^{\infty} \frac{J_{n,2}^2}{m^4} \left[E_{n,2} \int_{x_3}^{x_4} x I_1(x) dx + F_{n,2} \int_{x_3}^{x_4} x K_1(x) dx - \frac{\pi}{2} \int_{x_3}^{x_4} x L_1(x) dx \right] \right\} \qquad (9.72)
$$

The $A_2 J_1$ integral in (9.66) is given by

$$
\int_{V_1} A_2 J_1 dV_1 = 2\pi \int_{r_1}^{r_2} r dr \int_0^L dz \left[J_{0,1} + \sum_{p=1}^{\infty} J_{p,1} \cos\left(\frac{p\pi z}{L} \right) \right] \left\{ \frac{\mu_o J_{0,2}}{2} (r_4 - r_3) r \right.
$$

$$
\left. + \mu_o \sum_{n=1}^{\infty} \frac{J_{n,2}}{m^2} \left[C_{n,2} I_1(mr) + D_{n,2} K_1(mr) \right] \cos\left(\frac{n\pi z}{L} \right) \right\} \qquad (9.73)
$$

Using the previous method for carrying out the cos integrals, we obtain

$$
\int_{V_1} A_2 J_1 dV_1 = 2\pi\mu_o L \left\{ \frac{J_{0,1} J_{0,2}}{6} (r_4 - r_3)(r_2^3 - r_1^3) \right.
$$

$$
\left. + \frac{1}{2} \sum_{n=1}^{\infty} \frac{J_{n,1} J_{n,2}}{m^4} \left[C_{n,2} \int_{x_1}^{x_2} x I_1(x) dx + D_{n,2} \int_{x_1}^{x_2} x K_1(x) dx \right] \right\} \qquad (9.74)
$$

The $A_1 J_2$ integral in (9.66) is given similarly by

$$
\int_{V_2} A_1 J_2 dV_2 = 2\pi\mu_o L \left\{ \frac{J_{0,1} J_{0,2}}{6} (r_4 - r_3)(r_2^3 - r_1^3) + \frac{1}{2} \sum_{n=1}^{\infty} \frac{J_{n,1} J_{n,2}}{m^4} G_{n,1} \int_{x_3}^{x_4} x K_1(x) dx \right\} \qquad (9.75)
$$

In spite of their different appearances, Equations 9.74 and 9.75 are the same. This can be seen from the definition of the coefficients in (9.54), remembering that the 1 subscript on the coefficients is associated with x_1, x_2, and the 2 subscript with x_3, x_4. This is an example of a reciprocity relation that holds for linear systems.

Summing up the terms in (9.66), we get after some algebraic manipulation

$$
\begin{aligned}
2W = \frac{\pi\mu_0 L}{6} &\left\{ J_{0,1}^2 \left(r_2 - r_1\right)^2 \left[\left(r_1 + r_2\right)^2 + 2r_1^2\right] + J_{0,2}^2 \left(r_4 - r_3\right)^2 \left[\left(r_3 + r_4\right)^2 + 2r_3^2\right] \right. \\
&+ 4J_{0,1}J_{0,2}\left(r_2 - r_1\right)\left(r_4 - r_3\right)\left(r_1^2 + r_1 r_2 + r_2^2\right) \bigg\} \\
+ \pi\mu_0 L \sum_{n=1}^{\infty} &\left\{ \frac{J_{n,1}^2}{m^4}\left[E_{n,1}\int_{x_1}^{x_2}xI_1(x)dx + F_{n,1}\int_{x_1}^{x_2}xK_1(x)dx - \frac{\pi}{2}\int_{x_1}^{x_2}xL_1(x)dx \right] \right. \\
&+ \frac{J_{n,2}^2}{m^4}\left[E_{n,2}\int_{x_3}^{x_4}xI_1(x)dx + F_{n,2}\int_{x_3}^{x_4}xK_1(x)dx - \frac{\pi}{2}\int_{x_3}^{x_4}xL_1(x)dx \right] \\
&+ 2\frac{J_{n,1}J_{n,2}}{m^4} G_{n,1}\int_{x_3}^{x_4}xK_1(x)dx \bigg\}
\end{aligned}
\tag{9.76}
$$

When the coefficients E_n and F_n, etc., have a 1 subscript, the expressions given by (9.54) apply with x_1, x_2 for the integration limits. However, when the second subscript is a 2, then x_3, x_4 must be substituted for x_1, x_2 in the formulas.

The leakage inductance, referred to as coil 1 as given in Chapter 4, (4.23), is given by $2W/I_1^2$. For I_1^2, we can write $(NI)^2/N^2$. This makes the formula work for either winding as the reference winding since $(NI)^2$ is the same for both windings, and whatever winding's N is chosen will make that the reference winding. Using (9.76), this is

$$
\begin{aligned}
L_{leak} = \frac{\pi\mu_0 L N^2}{6(NI)^2} &\left\{ J_{0,1}^2 \left(r_2 - r_1\right)^2 \left[\left(r_1 + r_2\right)^2 + 2r_1^2\right] \right. \\
&+ J_{0,2}^2 \left(r_4 - r_3\right)^2 \left[\left(r_3 + r_4\right)^2 + 2r_3^2\right] + 4J_{0,1}J_{0,2}\left(r_2 - r_1\right)\left(r_4 - r_3\right)\left(r_1^2 + r_1 r_2 + r_2^2\right) \bigg\} \\
+ \frac{\pi\mu_0 L N^2}{(NI)^2} \sum_{n=1}^{\infty} &\left\{ \frac{J_{n,1}^2}{m^4}\left[E_{n,1}\int_{x_1}^{x_2}xI_1(x)dx + F_{n,1}\int_{x_1}^{x_2}xK_1(x)dx - \frac{\pi}{2}\int_{x_1}^{x_2}xL_1(x)dx \right] \right. \\
&+ \frac{J_{n,2}^2}{m^4}\left[E_{n,2}\int_{x_3}^{x_4}xI_1(x)dx + F_{n,2}\int_{x_3}^{x_4}xK_1(x)dx - \frac{\pi}{2}\int_{x_3}^{x_4}xL_1(x)dx \right] \\
&+ 2\frac{J_{n,1}J_{n,2}}{m^4} G_{n,1}\int_{x_3}^{x_4}xK_1(x)dx \bigg\}
\end{aligned}
\tag{9.77}
$$

To get the leakage reactance multiply L_{leak} by $\omega = 2\pi f$.

9.3.2 Rabins' Method Applied to Calculate the Leakage Reactance between Two Axially Stacked Windings

There are sometimes transformer designs where the two coils are stacked axially and therefore occupy the same radial position. In this case, the coil radial builds are usually the same.

Often, the coils are duplicates. Rabins' method can also be used to find the leakage inductance for this situation. The current densities for each coil will be nonzero within different axial regions, but the Fourier decomposition should reflect this. In this case, only the region II solution is needed for both coils and (9.66) still holds. In this case, $r_3 = r_1$ and $r_4 = r_2$. Evaluating the terms in (9.66), we find that the A_1J_1 and A_2J_2 terms are the same as before except that we must replace r_3, r_4, x_3, x_4 by r_1, r_2, x_1, x_2.

The A_2J_1 integral is

$$\int_{V_1} A_2 J_1 dV_1 = 2\pi \int_{r_1}^{r_2} r dr \int_0^L dz \left[J_{0,1} + \sum_{p=1}^{\infty} J_{p,1} \cos\left(\frac{p\pi z}{L}\right) \right] \left\{ \mu_o J_{0,2} \left(\frac{r_2 r}{2} - \frac{r_1^3}{6r} - \frac{r^2}{3} \right) \right.$$

$$\left. + \mu_o \sum_{n=1}^{\infty} \frac{J_{n,2}}{m^2} \left[E_{n,1} I_1(mr) + F_{n,1} K_1(mr) - \frac{\pi}{2} L_1(mr) \right] \cos\left(\frac{n\pi z}{L}\right) \right\} \quad (9.78)$$

We can use a 1 subscript on E_n and F_n in (9.78) since the radial extent of the two windings is the same. However, the 2 subscript is needed on J_n since the current distributions are different. Carrying out the z and r integrations, we get

$$\int_{V_1} A_2 J_1 dV_1 = 2\pi \mu_o L \left\{ J_{0,1} J_{0,2} \left(\frac{r_1^4}{4} + \frac{r_2^4}{12} - \frac{r_2 r_1^3}{3} \right) \right.$$

$$\left. + \frac{1}{2} \sum_{n=1}^{\infty} \frac{J_{n,1} J_{n,2}}{m^4} \left[E_{n,1} \int_{x_1}^{x_2} x I_1(x) dx + F_{n,1} \int_{x_1}^{x_2} x K_1(x) dx - \frac{\pi}{2} \int_{x_1}^{x_2} x L_1(x) dx \right] \right\} \quad (9.79)$$

This expression is also equal to the A_1J_2 term. Hence we get, combining all the terms in (9.66),

$$2W = \frac{\pi \mu_o L}{6} \left\{ \left(J_{0,1}^2 + J_{0,2}^2 + 2 J_{0,1} J_{0,2} \right) (r_2 - r_1)^2 \left[(r_1 + r_2)^2 + 2r_1^2 \right] \right\}$$

$$+ \pi \mu_o L \sum_{n=1}^{\infty} \left\{ \frac{\left(J_{n,1}^2 + J_{n,2}^2 + 2 J_{n,1} J_{n,2} \right)}{m^4} \right.$$

$$\left. \times \left[E_{n,1} \int_{x_1}^{x_2} x I_1(x) dx + F_{n,1} \int_{x_1}^{x_2} x K_1(x) dx - \frac{\pi}{2} \int_{x_1}^{x_2} x L_1(x) dx \right] \right\} \quad (9.80)$$

The leakage inductance between these two coils is then given by

$$L_{\substack{\text{leak, axially} \\ \text{stacked coils}}} = \frac{\pi \mu_o L N^2}{6(NI)^2} \left\{ \left(J_{0,1}^2 + J_{0,2}^2 + 2 J_{0,1} J_{0,2} \right) (r_2 - r_1)^2 \left[(r_1 + r_2)^2 + 2r_1^2 \right] \right\}$$

$$+ \frac{\pi \mu_o L N^2}{(NI)^2} \sum_{n=1}^{\infty} \left\{ \frac{\left(J_{n,1}^2 + J_{n,2}^2 + 2 J_{n,1} J_{n,2} \right)}{m^4} \right.$$

$$\left. \times \left[E_{n,1} \int_{x_1}^{x_2} x I_1(x) dx + F_{n,1} \int_{x_1}^{x_2} x K_1(x) dx - \frac{\pi}{2} \int_{x_1}^{x_2} x L_1(x) dx \right] \right\} \quad (9.81)$$

We note that $J_{0,1}$ and $J_{0,2}$ are the average current densities in windings 1 and 2 and $(r_2 - r_1)L$ is the cross-sectional area of either winding. Thus, since the amp-turns are balanced between the 2 windings, we have

$$N_1I_1 = J_{0,1}(r_2 - r_1)L = -N_2I_2 = -J_{0,2}(r_2 - r_1)L \qquad (9.82)$$

Using this, we see that the first set of terms in (9.81) vanishes. Thus we get for the leakage inductance

$$L_{\substack{\text{leak, axially} \\ \text{displaced coils}}} = \frac{\pi\mu_o LN^2}{(NI)^2} \sum_{n=1}^{\infty} \left\{ \frac{\left(J_{n,1}^2 + J_{n,2}^2 + 2J_{n,1}J_{n,2}\right)}{m^4} \right.$$

$$\left. \times \left[E_{n,1} \int_{x_1}^{x_2} xI_1(x)dx + F_{n,1} \int_{x_1}^{x_2} xK_1(x)dx - \frac{\pi}{2} \int_{x_1}^{x_2} xL_1(x)dx \right] \right\} \qquad (9.83)$$

The leakage reactance is obtained by multiplying L_{leak} by $\omega = 2\pi f$. We should note that both coils extend the full axial height L. However, the top coil would have a large section below it where the bottom winding is located with 0 current density, and similarly, the bottom coil would have a large section above it where the top winding is located with 0 current density.

9.3.3 Rabins' Method Applied to Calculate the Leakage Reactance for a Collection of Windings

Rabins' method can be used to find the leakage reactance for a collection of any number of coils. The amp-turns of the coils must sum to zero. Then the leakage induction, assuming coil 1 is the reference winding, can be obtained from

$$W = \frac{1}{2} \int_{\substack{\text{volume} \\ \text{of circuits}}} \mathbf{A} \cdot \mathbf{J} dV = \frac{1}{2} L_{\text{leak},1} I_1^2 \qquad (9.84)$$

In this case, formula (9.66) for the magnetic energy of N_w windings becomes

$$2W = \int_{V_1} \left(\sum_{j=1}^{N_w} A_j \right) J_1 dV_1 + \int_{V_2} \left(\sum_{j=1}^{N_w} A_j \right) J_2 dV_2 + \cdots$$

$$= \sum_{k=1}^{N_w} \int_{V_k} \left(\sum_{j=1}^{N_w} A_j \right) J_k dV_k = \sum_{k=1}^{N_w} \sum_{j=1}^{N_w} \int_{V_k} A_j J_k dV_k \qquad (9.85)$$

Care must be taken to use the solution of the vector potential for a given winding in the appropriate region. The double sum in (9.85) can be simplified by noting that

$$\int_{V_k} A_j J_k dV_k = \int_{V_j} A_k J_j dV_j \qquad (9.86)$$

The leakage inductance relative to winding 1 is then given by

$$L_{leak,1} = \frac{1}{I_1^2} \left(\sum_{j=1}^{N_w} \int_{V_k} A_j J_j \, dV_k + 2 \sum_{j<k=1}^{N_w} \int_{V_k} A_j J_k \, dV_k \right) \tag{9.87}$$

The leakage reactance is obtained by multiplying $L_{leak,1}$ by $\omega = 2\pi f$.

9.4 Rabins' Method Applied to Calculate the Self-Inductance of and Mutual Inductance between Coil Sections

Here, we are considering a collection of coils subdivided into coil sections. A coil section for our purposes consists of a contiguous part of a coil, which has a uniform current density. In this case, Equation 9.14 holds for the Fourier coefficients. In all other respects, it is treated like a full coil of height L. Later, we will apply these results to obtain a detailed model of the transformer for impulse simulation studies.

For these calculations, we use the vector potential solution for each winding section obtained by Rabins' method. However, this requires amp-turn balance in order to satisfy the B-field boundary condition at the core surface. Although taken individually, the self-inductance calculation of a coil section cannot satisfy this boundary condition; when considering a pair of coil sections, amp-turn balance can be achieved so that the self- and mutual inductances associated with this pair of coil sections are valid within the context of Rabins' method. By this, we mean that the self-inductances, mutual inductance, and leakage inductance of the coil section pair satisfy the relation between these quantities given in (4.23).

These Rabins' inductances tend to be about an order of magnitude higher than the corresponding air core inductances and mutual inductances, which are often used in impulse simulation studies. We find that the Rabins' method inductances tend to give good agreement with impulse data obtained via RSO methods.

Wilcox et al., have obtained analytic expressions for the mutual and self-impedances and inductances of coils on a ferromagnetic core [Wil88], [Wil89]. These could also be used in simulation studies. However, the self-inductance of a single coil on a high-permeability ferromagnetic core increases with the core permeability and can become quite large as is shown by finite element methods. Self- and mutual inductances determined in this manner tend to be about an order of magnitude higher than those obtained by Rabins' method. Since the permeability is not that well known for a transformer core with joints, etc., this can present difficulties when deciding which value to use.

Having obtained Rabins' solution for the vector potential, it is relatively straightforward to apply this solution to the calculation of self- and mutual inductances. We begin with the mutual inductance calculation. The mutual induction between two coil sections is given by formula (4.26) in Chapter 4, which we repeat here:

$$M_{12} = \frac{1}{I_1 I_2} \int_{V_1} \mathbf{A}_2 \cdot \mathbf{J}_1 \, dV_1 \tag{9.88}$$

We treat the coil sections as if they were full coils of height L, albeit with much of the coil height having 0 current density in applying Rabins' method.

If the sections are on the same coil or two coils axially displaced, then A_2 is the solution in Region II. The A_2J_1 integral has already been done in this case and is given in (9.79). Thus we have, after some algebraic manipulation of the J_0 radial terms,

$$
\begin{aligned}
M_{12,\text{ same radial}\atop\text{position}} &= \frac{\pi\mu_o L}{6 I_1 I_2} J_{0,1} J_{0,2} (r_2 - r_1)^2 \left[(r_2 + r_1)^2 + 2r_1^2 \right] \\
&+ \frac{\pi\mu_o L}{I_1 I_2} \sum_{n=1}^{\infty} \frac{J_{n,1} J_{n,2}}{m^4} \left[E_{n,1} \int_{x_1}^{x_2} x I_1(x)dx + F_{n,1} \int_{x_1}^{x_2} x K_1(x)dx - \frac{\pi}{2} \int_{x_1}^{x_2} x L_1(x)dx \right]
\end{aligned}
\tag{9.89}
$$

Since

$$
N_1 I_1 = J_{0,1}(r_2 - r_1)L, \quad N_2 I_2 = J_{0,2}(r_2 - r_1)L
\tag{9.90}
$$

we can rewrite (9.89)

$$
\begin{aligned}
M_{12,\text{ same radial}\atop\text{position}} &= \frac{\pi\mu_o N_1 N_2}{6L} \left[(r_2 + r_1)^2 + 2r_1^2 \right] + \frac{\pi\mu_o L N_1 N_2}{(N_1 I_1)(N_2 I_2)} \\
&\times \sum_{n=1}^{\infty} \frac{J_{n,1} J_{n,2}}{m^4} \left[E_{n,1} \int_{x_1}^{x_2} x I_1(x)dx + F_{n,1} \int_{x_1}^{x_2} x K_1(x)dx - \frac{\pi}{2} \int_{x_1}^{x_2} x L_1(x)dx \right]
\end{aligned}
\tag{9.91}
$$

When the coil sections are part of different coils radially displaced, then we can assume that 1 is the inner coil and 2 is the outer coil. The result will be independent of this assumption. Thus we need the Region I solution for coil 2. The A_2J_1 integral has already been done and the result is given in (9.74). Thus we have

$$
\begin{aligned}
M_{12,\text{ different radial}\atop\text{positions}} &= \frac{\pi\mu_o L J_{0,1} J_{0,2}}{3 I_1 I_2} (r_4 - r_3)(r_2^3 - r_1^3) \\
&+ \frac{\pi\mu_o L}{I_1 I_2} \sum_{n=1}^{\infty} \frac{J_{n,1} J_{n,2}}{m^4} \left[C_{n,2} \int_{x_1}^{x_2} x I_1(x)dx + D_{n,2} \int_{x_1}^{x_2} x K_1(x)dx \right]
\end{aligned}
\tag{9.92}
$$

In this case, the amp-turns can be expressed as

$$
N_1 I_1 = J_{0,1}(r_2 - r_1)L, \quad N_2 I_2 = J_{0,2}(r_4 - r_3)L
\tag{9.93}
$$

Using this, we can rewrite (9.92):

$$
\begin{aligned}
M_{12,\text{ different radial}\atop\text{positions}} &= \frac{\pi\mu_o N_1 N_2}{3L} (r_1^2 + r_1 r_2 + r_2^2) \\
&+ \frac{\pi\mu_o L N_1 N_2}{(N_1 I_1)(N_2 I_2)} \sum_{n=1}^{\infty} \frac{J_{n,1} J_{n,2}}{m^4} \left[C_{n,2} \int_{x_1}^{x_2} x I_1(x)dx + D_{n,2} \int_{x_1}^{x_2} x K_1(x)dx \right]
\end{aligned}
\tag{9.94}
$$

If the coil sections are oppositely wound, then the mutual inductances should be multiplied by −1. This can be effected by replacing N by −N for an oppositely wound coil section.

For a single coil, say coil 1, the self-inductance can be obtained from

$$W = \frac{1}{2}\int_{V_1}\mathbf{A}_1\cdot\mathbf{J}_1\,dV = \frac{1}{2}L_1I_1^2 \quad\Rightarrow\quad L_1 = \frac{1}{I_1^2}\int_{V_1}\mathbf{A}_1\cdot\mathbf{J}_1\,dV \tag{9.95}$$

This was discussed in Chapter 4. In this case, we use the solution for A_1 in Region II since that is where the current density is nonzero. The integral in (9.95) has already been carried out in (9.89), so we have

$$L_1 = \frac{\pi\mu_o L}{6I_1^2}J_{0,1}^2\left(r_2 - r_1\right)^2\left[\left(r_2 + r_1\right)^2 + 2r_1^2\right]$$

$$+ \frac{\pi\mu_o L}{I_1^2}\sum_{n=1}^{\infty}\frac{J_{n,1}^2}{m^4}\left[E_{n,1}\int_{x_1}^{x_2}xI_1(x)\,dx + F_{n,1}\int_{x_1}^{x_2}xK_1(x)\,dx - \frac{\pi}{2}\int_{x_1}^{x_2}xL_1(x)\,dx\right] \tag{9.96}$$

Since

$$N_1I_1 = J_{0,1}\left(r_2 - r_1\right)L \tag{9.97}$$

we can rewrite (9.96):

$$L_1 = \frac{\pi\mu_o N_1^2}{6L}\left[\left(r_2 + r_1\right)^2 + 2r_1^2\right]$$

$$+ \frac{\pi\mu_o L N_1^2}{\left(N_1I_1\right)^2}\sum_{n=1}^{\infty}\frac{J_{n,1}^2}{m^4}\left[E_{n,1}\int_{x_1}^{x_2}xI_1(x)\,dx + F_{n,1}\int_{x_1}^{x_2}xK_1(x)\,dx - \frac{\pi}{2}\int_{x_1}^{x_2}xL_1(x)\,dx\right] \tag{9.98}$$

9.5 Determining the B-field

It is important to determine the B-field within the windings if one is interested in obtaining the electromagnetic forces acting within the windings as a result of a fault and associated high winding currents. The B-field present for rated currents can also be useful in estimating the losses in various parts of the transformer, including the stray flux eddy losses in the windings.

Before calculating the B-field, a plot of the flux lines can be obtained from the vector potential directly. This type of plot gives a visual sense of the B-field. In the r–z geometry, the appropriate flux lines are lines of constant rA_φ, that is, the r position vector × the azimuthal vector potential. The B-field points along these lines of constant rA_φ. In addition, the flux per radian between two lines equals the difference in the values of rA_φ on the two lines. Therefore, if lines of equally spaced values of rA_φ are plotted, the B-field is strongest where the lines are closest together since the average B-field is the flux per unit area or distance between the flux lines in this case. Hence, these give a good sense of what the flux is doing in the transformer. A flux plot, using Rabins' method, is given in Figure 9.4 for a 2-winding transformer.

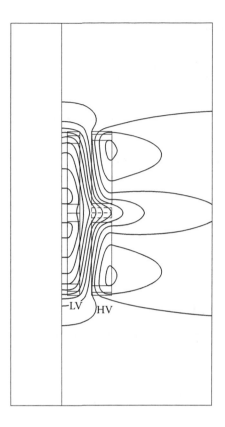

FIGURE 9.4
Flux plot for a 2-winding transformer based on Rabins' method. The amp-turns are not uniform along the windings, resulting in the radially directed flux, especially at the winding center where the center taps are out. The vertical and radial dimensions are not necessarily to the same scale.

The B-field or induction vector is given by (9.23). We need to evaluate it in the three regions using (9.53). The radial component is, using the region label as superscript,

$$B_r^I = \mu_o \sum_{n=1}^{\infty} \frac{J_n}{m} \left[C_n I_1(x) + D_n K_1(x) \right] \sin(mz)$$

$$B_r^{II} = \mu_o \sum_{n=1}^{\infty} \frac{J_n}{m} \left[E_n I_1(x) + F_n K_1(x) - \frac{\pi}{2} L_1(x) \right] \sin(mz) \qquad (9.99)$$

$$B_r^{III} = \mu_o \sum_{n=1}^{\infty} \frac{J_n}{m} G_n K_1(x) \sin(mz)$$

From (9.45),

$$B_z = \frac{1}{r} \frac{\partial}{\partial r}(rA) = \frac{m}{x} \frac{\partial}{\partial x}(xA) \qquad (9.100)$$

the axial component, using (9.47), is

$$B_z^I = \mu_0 J_0 (r_2 - r_1) + \mu_0 \sum_{n=1}^{\infty} \frac{J_n}{m} \left[C_n I_0(x) - D_n K_0(x) \right] \cos(mz)$$

$$B_z^{II} = \mu_0 J_0 (r_2 - r) + \mu_0 \sum_{n=1}^{\infty} \frac{J_n}{m} \left[E_n I_0(x) - F_n K_0(x) - \frac{\pi}{2} L_0(x) \right] \cos(mz) \qquad (9.101)$$

$$B_z^{III} = -\mu_0 \sum \frac{J_n}{m} G_n K_0(x) \cos(mz)$$

To find the net B-field associated with a collection of coils, we simply add their components at the point of interest, keeping in mind what region the point is in relative to each coil.

9.6 Determining the Winding Forces

The force density vector, **F**, in SI units (Newtons/m³), is given by

$$\mathbf{F} = \mathbf{J} \times \mathbf{B} \qquad (9.102)$$

Since **J** is azimuthal and **B** has only r and z components, this reduces to

$$\mathbf{F} = J B_z \mathbf{a}_r - J B_r \mathbf{k} \qquad (9.103)$$

where
  $\mathbf{a}_r$ is the unit vector in the radial direction
  $\mathbf{k}$ is the unit axial vector

We have omitted the φ subscript on J. Thus, the radial forces are due to the axial field component and vice versa.

In (9.103), the B-field values are the resultant from all the coils and J is the current density at the point in question. This force density is nonzero only over those parts of the winding that carry current. In order to obtain net forces over all or part of a winding, it is necessary to integrate (9.103) over the winding or winding part. For this purpose, the winding can be subdivided into as fine a mesh as desired and **F** computed at the centroids of these subdivisions, and the resulting values times the mesh volume element added.

One force that is useful to know is the compressive force, which acts axially at each axial position along the winding. We assume that the windings are constrained at the two ends by pressure rings of some type. We ignore the gravitational force here. Starting from the bottom, we integrate the axial forces upwards along the winding, stopping when the sum of the downward-acting forces reach a maximum. This is the net downward force on the bottom pressure ring, which is countered by an equal upward force on the winding exerted by the pressure ring. We do the same thing, starting from the top of the winding, integrating the axial forces until the largest upwards acting summed force is reached. This then constitutes the net upward force acting on the top pressure ring, which is countered by an equal downward force exerted by the pressure ring. Including the reaction forces of the

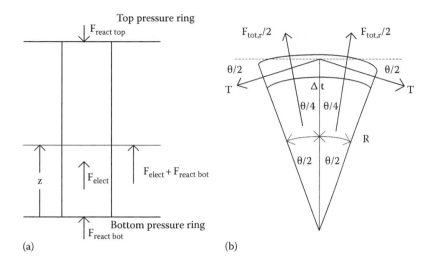

FIGURE 9.5
Forces on the winding needed for stress analysis. (a) Compressive force at each axial position and (b) hoop forces (tensile shown) from the radial forces.

pressure rings, we then integrate upwards say, starting at the bottom of the winding and at each vertical position, the force calculated will be the compressive force at that position since there will be an equal and opposite force acting from above. This is illustrated in Figure 9.5a.

Another force of interest for coil design is the radial force, which results in hoop stress, tensile or compressive, on the winding. Because of the cylindrical symmetry, the radial forces vectorially add to zero. We must therefore handle them a bit differently if we are to arrive at a useful resultant for hoop stress calculations. In Figure 9.5b, we isolate a small portion of the winding and show the forces acting in a horizontal section. We show the radial forces acting outward and the tensile forces, T, applied by the missing part of the winding, which are necessary to maintain equilibrium. If the radial forces acted inward, then the rest of the winding would need to apply compressive forces to maintain equilibrium and all the force arrows in the figure would be reversed. The calculation would, however, proceed similarly.

In Figure 9.5b, θ is assumed to be a very small angle. The force balance in the upward direction in the figure is

$$2\frac{F_{tot,r}}{2}\cos\left(\frac{\theta}{4}\right) = 2T\sin\left(\frac{\theta}{2}\right) \tag{9.104}$$

where
  $F_{tot,r}$ is the sum of the radial electromagnetic forces acting on the winding section shown
  T is the total tensile force acting on the winding cross-section

For small θ, this reduces to

$$F_{tot,r} = T\theta \tag{9.105}$$

In radians, $\theta = \Delta t/R$, where Δt is the circumferential length of the winding section shown and R its average radius. The average tensile stress is $\sigma = T/A$, where A is the cross-sectional area of the winding section. The volume of the winding section is $\Delta V = A\Delta t$. Using these relations and (9.105), we can obtain the tensile (or compressive) stress as

$$\sigma = R\frac{F_{tot,r}}{\Delta V} \tag{9.106}$$

Thus, the average tensile stress in the winding is the average radius times the average radial force density, assuming the radial forces act outward. For inward radial forces, σ is negative, that is, compressive. The average radial force density at a particular axial position z can be found by computing the radial component of (9.103) at several radial positions in the winding, keeping z fixed and taking their average. One should weight these by the volume element, which is proportional to r.

Other forces of interest can be found from the known force densities.

9.7 Numerical Considerations

First of all, the infinite sums in the various expressions given in this chapter have to be reduced to finite sums without significant loss of accuracy. Since the nth term is reduced by some power of n, they tend to get smaller and smaller the larger the value of n. We have found that 200 terms provides sufficient accuracy. The sums should also be carried out in double precision.

We have used a mathematical library for the modified Bessel functions of the first and second kinds of order zero and one. The library gives two forms for these functions, one for the function itself and another expresses the function as an exponential times a remainder. For instance,

$$\begin{gathered} I_0(x) = e^x \times \text{Remainder}_{I_0}, \quad I_1(x) = e^x \times \text{Remainder}_{I_1} \\ K_0(x) = e^{-x} \times \text{Remainder}_{K_0}, \quad K_1(x) = e^{-x} \times \text{Remainder}_{K_1} \end{gathered} \tag{9.107}$$

These exponential forms should be used. It turns out that when these functions occur in the equations of this chapter, the exponentials can be combined and the exponents usually cancel to a large extent.

The modified Struve function does not appear in our mathematical library, so we obtained it via the difference functions defined in (9.58) and their integral expressions given in (9.59) and (9.60). These integrals can be done with high accuracy and they have finite values of reasonably small magnitudes for the various values of x. These expressions then can be used in (9.61) and (9.64) to obtain the integrals of the modified Bessel and Struve functions.

9.8 Air Core Inductance

Often, an air core inductor is used in a transformer. This is sometimes used to increase the LV to tertiary winding impedance for better short circuit withstand capability. These air core inductors can be layered windings, consisting of one or more concentric coils.

Here, we consider the calculation of this inductance in the general case where one or more coils of the same or different sizes are present. We assume that each coil has a uniform turn distribution over its cross-section. We also assume that the coil cross-section is rectangular.

In order to obtain the inductance of a collection of coils, the mutual inductance between the coils must be calculated. This is because the self-inductance of a collection of N_c coils is given by

$$L = \sum_{i=1}^{N_c} L_i + 2 \sum_{i<j=1}^{N_c} M_{ij} \tag{9.108}$$

where
  L_i is the self-inductance owf coil i
  M_{ij} is the mutual inductance between coils i and j

The second sum is over all possible pairs of coils. The mutual inductances can be positive or negative depending on the winding directions of the coils. Equation 9.108 follows from the magnetic energy expression for a collection of coils in terms of their self- and mutual inductances given in Chapter 4 and the fact that, for an inductor, all the coils must carry the same current. Note that (9.108) would apply if the L_i's and M_{ij}'s referred to subdivisions of a single or multiple coils.

The starting point is the mutual inductance between two circular coaxial thin wires of radii r_1 and r_2 separated axially by a distance d as shown in Figure 9.6.

The mutual inductance M_{12} between these wires is given by [Smy68]

$$M_{12} = \frac{2\mu_o}{k} \sqrt{r_1 r_2} \left[\left(1 - \frac{k^2}{2} \right) K(k) - E(k) \right]$$

where

$$k^2 = \frac{4 r_1 r_2}{\left[(r_1 + r_2)^2 + d^2 \right]}$$

(9.109)

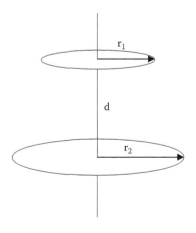

FIGURE 9.6
Coaxial thin circular wire geometry.

Here, K(k) and E(k) are complete elliptic integrals of the first and second kind, respectively. These are generally available in mathematical libraries. SI units are used and $\mu_o = 4\pi \times 10^{-7}$ H/m in this system.

Lyle's method is used for obtaining the mutual inductance between coils of rectangular cross-section. This is described in Grover [Gro73]. In its simplest form, each coil is replaced by two thin wire circular coils that are strategically placed. The position of the thin wire coils depends on whether the coil's radial build b is larger or smaller than the coil's axial height h. If h > b, the two thin wire coils each have a radius r_1 given by

$$r_1 = r_{ave}\left[1 + \frac{1}{24}\left(\frac{b}{r_{ave}}\right)^2\right] \tag{9.110}$$

where r_{ave} is the average radius of the rectangular cross-section coil. The two substitute coils are separated axially from the radial centerline of the main coil by a distance β, as shown in Figure 9.7, where

$$\beta^2 = \frac{h^2 - b^2}{12} \tag{9.111}$$

The radial centerline is a distance h/2 from the bottom of the main coil. Each sub thin wire coil is assigned ½ the turns of the original coil.

If h < b, then the substitute thin wire coils are both positioned along the radial centerline of the main coil at radial distances $r_2 + \delta$ and $r_2 - \delta$, where

$$r_2 = r_{ave}\left[1 + \frac{1}{24}\left(\frac{h}{r_{ave}}\right)^2\right], \quad \delta^2 = \frac{b^2 - h^2}{12} \tag{9.112}$$

Also, each sub thin wire coil is given ½ the turns of the main coil.

Both of these cases are illustrated in Figure 9.7.

Notice that if b = h, the two cases become identical, with $r_1 = r_2$ and β = δ. In this case, the two thin circular wires coincide. This single wire would be given the full number of turns in the coil.

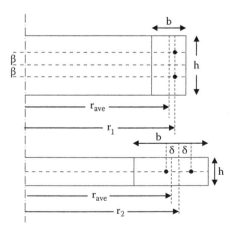

FIGURE 9.7
Geometry of Lyle's method.

Since the two wires are now a proxy for the coil they represent, the mutual inductance between this coil and another also represented by two wires is obtained by summing over all possible pairs of one wire from one coil and one wire from the second coil. For one wire pair, since the coils are assumed coaxial, (9.109) is used to obtain their mutual inductance. If the coils have N_1 and N_2 turns respectively, each wire is assigned ½ the turns of its coil. Thus, if coil 1 is represented by wires 1 and 2 and coil 2 by wires 3 and 4, the mutual inductance M between these coils is given by

$$M = \frac{N_1 N_2}{4} \left(M_{13} + M_{14} + M_{23} + M_{24} \right) \tag{9.113}$$

Since we also need the self-inductance of each coil, we use an expression given in [Gro73] which applies to circular coils having a square cross-section. Letting c = the length of the side of the square and r_{ave} = the average coil radius as before, the self-inductance is given by

$$L = \mu_o r_{ave} N^2 \times \left\{ \frac{1}{2} \left[1 + \frac{1}{6} \left(\frac{c}{2r_{ave}} \right)^2 \right] \ln \left[\frac{8}{\left(c/2r_{ave} \right)^2} \right] + 0.2041 \left(\frac{c}{2r_{ave}} \right)^2 - 0.84834 \right\} \tag{9.114}$$

where the SI system is used with $\mu_o = 4\pi \times 10^{-7}$ H/m and N is the number of turns in the square. This expression applies when $c/2r_{ave} \leq 0.2$. Since most of the coils of interest will not have a square cross-section, it will be necessary to subdivide the coils into squares, which can be made small enough so that this criterion is met. At the same time, these squares can be used for Lyle's method to get mutual inductances. In this case, the two proxy wires coincide.

Thus, the strategy is to subdivide each coil into small enough squares so that (9.114) applies for calculating its self-inductance. This may not be exactly possible, depending on the coil dimensions, but can be approximated to any desired accuracy. Each square is also represented by two coincident wires, each carrying ½ the turns in the square. This is equivalent to a single wire with the full number of turns in the square. These wires are used to obtain the mutual inductances between this square and all the other square subdivisions in the system. The self-inductance of the system of coils is given by (9.108), where N_c is now the total number of squares in the system. If one of the original coils has N_t turns and is subdivided into N_s squares, each square has N_t/N_s turns and each of the subcoils used for calculating mutual inductances has ½(N_t/N_s) turns. Since the two mutual inductances between the two wires representing the squares are the same, only one needs to be calculated, and the turns associated with it should be N_t/N_s. Note also that r_{ave}, used in the previously mentioned formulas, is the average radius of the square in the subdivision.

Although a lot of terms must be calculated, symmetry can be used to reduce the number of terms. The sums can be performed without too much difficulty by computer. This method has been checked by comparison with known formulas for the self-inductance of special coils and also via finite element calculations and is quite accurate.

For the self-inductance of a single coil, the said method of subdividing into squares, which will all be the same size for a given coil, and performing all the self- and mutual inductance calculations among the squares is useful. However, when more than one coil is involved in the inductance calculation, the self-inductance of each coil can be calculated by the square subdivision method with each coil usually having a different square size. However, the mutual inductance between a pair of coils in the group can perhaps best be obtained using Lyle's method based on a different type of subdivision for each coil.

These latter subdivisions would not necessarily be square and would be chosen so that they conform to one of the types shown in Figure 9.7. They would possibly be coarser than the square subdivisions used for the self-inductance of the coil, resulting in fewer terms to sum. These subdivisions, which will be called Lyle subdivisions, could be chosen to more exactly fill the coil's cross-section. For a given coil, they could be chosen to all have the same dimensions or not. For obtaining the mutual inductance between a pair of coils, where each coil has its own set of Lyle subdivisions, the mutual inductances between all pairs, consisting of one wire from each coil, are summed. The number of proxy wires in each coil will be two times the number of Lyle subdivisions if these subdivisions are not square.

10

Mechanical Design

10.1 Introduction

Transformers must be designed to withstand the large forces that occur during fault conditions. Fault currents for the standard fault types such as single line to ground, line to line, double line to ground, and all three lines to ground must be calculated. Since these faults can occur during any part of the a.c. cycle, the worst-case transient overcurrent must be used to determine the forces. This can be calculated and is specified in the standards as an asymmetry factor, which multiplies the rms steady-state currents. The asymmetry is given by [IEE93]

$$K = \sqrt{2}\left[1 + e^{-(\phi+(\pi/2))(r/x)}\sin\phi\right] \tag{10.1}$$

where
$\phi = \tan^{-1}(x/r)$
x/r is the ratio of the effective a.c. reactance to resistance

They are part of the total impedance, which limits the fault current in the transformer when the short circuit occurs. Using this factor, the resulting currents are used to obtain the magnetic field (leakage field) surrounding the coils and, in turn, the resulting forces on the windings. Analytic methods such as Rabins' method [Rab56] as discussed in Chapter 9 or finite element methods can be used to calculate the magnetic field. An example of such a leakage field is shown in Figure 10.1, which was generated by the finite element program Maxwell Version 12® [Ansoft]. Since this uses its 2D option, the figure is cylindrically symmetrical about the core centerline. Only the bottom half of the windings and core are shown because of assumed symmetry about a horizontal center plane. Although details such as clamps and shields can be included in the calculation using a finite element approach, they are not part of Rabins' analytic method, which assumes a simpler idealized geometry. However, calculations show that the magnetic field in the windings and hence the forces are nearly identical in the two cases.

The force density (force/unit volume), $\mathbf{f}$, generated in the windings by the magnetic induction, $\mathbf{B}$, is given by the Lorentz force law:

$$\mathbf{f} = \mathbf{J} \times \mathbf{B} \tag{10.2}$$

where $\mathbf{J}$ is the current density and SI units are used. These force densities can be integrated to get total forces, forces/unit length, or pressures depending on the type of

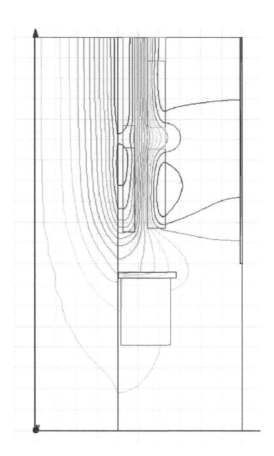

FIGURE 10.1
Plot of transformer leakage flux. Only the bottom half is shown. The figure is assumed to be cylindrically symmetrical about the core centerline. A bottom clamp with shunts on top and tank shunts on the right are shown.

integration performed. The resulting values or the maxima of these values can then be used to obtain stresses or maximum stresses in the winding materials.

The procedure described earlier is static in that the field and force calculations assume steady-state d.c. conditions. The currents, however, were multiplied by an asymmetry factor to account for a transient overshoot. Because the fault currents are applied suddenly, the resulting forces or pressures are also suddenly applied. This could result in transient mechanical effects such as the excitation of mechanical resonances, which could produce higher forces for a brief period than those obtained by a steady-state calculation. A few studies have been done on these transient mechanical effects in transformers [Hir71], [Bos72], [Pat80], [Tho79]. Although these studies are approximate, where they have indicated a large effect, due primarily to the excitation of a mechanical resonance, we use an appropriate enhancement factor to multiply the steady-state force. In other cases, these studies have indicated little or no transient effects so that no correction is necessary.

Because controlled short circuit tests are rarely performed on large transformers to validate a transformer's mechanical design, there are few empirical studies that directly test the theoretical calculations. To some extent, our confidence in these calculations comes from tests on smaller units and the fact that, in recent years, very few field failures have been directly attributable to mechanical causes.

10.2 Force Calculations

The electromagnetic forces acting on the coils are obtained by means of an analytic or finite element magnetic field calculation in conjunction with the Lorentz force law. In our analytic calculation, using Rabins' method [Rab56], each coil carrying current is subdivided into 4 radial sections and 100 axial sections as shown in Figure 10.2a. In the case of a finite element method, the coil cross section is subdivided into an irregular net of triangles as shown in Figure 10.2b. The principles used to determine the forces or pressures needed in the stress calculation are similar for the two cases, so we will focus on the analytic method.

The current density is uniform in each block in Figure 10.2a, but it can vary from block to block axially due to tapping out sections of the coil or thinning of sections of the winding to achieve a better amp-turn balance with tapped out sections in adjacent winding. These sections of reduced or zero current density produce radially bulging flux lines, as are evident in Figure 10.1. From the Lorentz force law, since the current is azimuthally directed, radially directed flux produces axial forces and axially directed flux produces radial forces. There are no azimuthal forces since the force density vector in (10.1) vanishes in this case.

The radial and axial forces are computed for each block in Figure 10.2a. Because of the cylindrical symmetry, each block in the figure is really a ring. The radial force can be thought of as a pressure acting inward or outward on the ring depending on its sign. If these forces are summed over the four radial blocks and then divided by the area of the

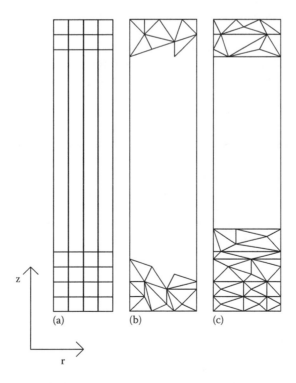

FIGURE 10.2
Coil subdivisions in (a) analytic and (b) finite element calculations of the magnetic field. In (c), a modified finite element mesh is shown, which is more useful for subsequent stress analysis.

cylindrical surface at the average radius of the winding, we obtain an effective total radial pressure acting on the coil at the axial position of the block. The maximum of these pressures in absolute value for the 100 axial positions is the worst-case radial pressure and is used in subsequent stress analysis. Note that these forces are radially directed so that integrating them vectorially over 360° would produce zero. This integration must therefore be done without regard to their vectorial nature.

The axially directed force summed over the four radial blocks is also needed. The maximum of the absolute value of this force for the 100 axial positions is a worst-case force used in the stress analysis. These axial forces are also summed, starting at the bottom of the coil. The total of these axial forces (summed over all the blocks of the coil) is a net upward or downward force depending on its sign. This force is countered by an equal and opposite force exerted by the pressure ring. If this net force acts downward, then the bottom pressure ring exerts an upward equal force on the coil. If it acts upward, then the top pressure ring exerts an equal downward force. In the former case, the top ring exerts no force on the coil, and in the latter case, the bottom ring exerts no force on the coil. We are ignoring the gravitational forces in comparison with the electromagnetic forces here. Starting with the force exerted upward, if any, by the bottom pressure ring, if we keep adding to this the forces produced by the horizontal layers of 4 blocks starting from the bottom row, we will arrive at a maximum net upward force at some vertical position along the winding. This force is called the maximum compressive force and is a worst-case force used in the stress analysis.

The total axial forces on the windings are added together if they are upward (positive). Similarly, windings having downward (negative) total axial forces contribute to a total downward force. These total upward or downward forces due to all the windings should be equal in absolute value since the net electromagnetic force acting axially must be zero. In practice, there are slight differences in these calculated quantities due to rounding errors. The net upward or downward axial force due to all the windings is called the total end thrust and is used in sizing the pressure rings. If the windings are symmetric about a horizontal center plane, the total axial force on each winding is nearly zero, so there is little or no end thrust. However, when one or more windings are offset vertically from the others, even slightly, net axial forces develop on each winding, which push some windings up and some down. It is good practice to include some offset, say, 6.35–12.7 mm, in the calculations to take into account possible misalignment in the transformer's construction.

As can be seen from the way forces needed for the stress analysis were extracted from the block forces in Figure 10.2a, for finite element calculations, the mesh shown in Figure 10.2c would be more useful for obtaining the needed forces than the mesh in Figure 10.2b. Here the coil is subdivided into a series of axial blocks, which can be of different heights, before the triangular mesh is generated. Then the forces can be summed for all the triangles within each block to correspond to the forces obtained from the 4 radial blocks at the same axial height in Figure 10.2a.

10.3 Stress Analysis

Now, we will relate the forces discussed in Section 10.2 to the stresses in the coil in order to determine whether the coil can withstand them without permanently deforming or buckling. The coils have a rather complex structure as indicated in Figure 10.3, which shows a disk-type coil. Helical windings are disk windings with only 1 turn per section and thus

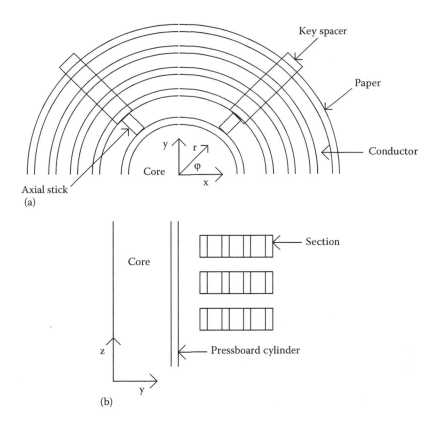

FIGURE 10.3
Details of a typical disk winding: (a) horizontal view (x, y) plane and (b) vertical view (y, z) plane.

are a special case of Figure 10.3. The sections are separated vertically by means of key spacers made of pressboard. These are spaced around the coil so as to allow cooling oil to flow between them. The coils are supported radially by means of axial sticks and pressboard cylinders or barriers. These brace the coil on the inside against an inner coil or against the core. There are similar support structures on the outside of the coil, except for the outermost coil. Spaces between the sticks allow cooling oil to flow vertically. Because of the dissimilar materials used, pressboard, paper, and copper for the conductors, and because of the many openings for the cooling oil, the stress analysis would be very complicated unless suitable approximations are made.

Although we have shown distinct separate axial sections in Figure 10.3, in reality, the wires must maintain electrical continuity from section to section so that the coil has a helical (springlike) structure. For the stress analysis, it is generally assumed that the helical pitch is small enough that the coil can be regarded as having distinct horizontal sections as shown in the figure. Moreover, these sections are assumed to close on themselves, forming rings.

Another approximation concerns the cable that comprises the winding turns. Figure 10.4 shows the two most common types. Magnet wire consists of a single strand of typically copper surrounded by a paper covering and is treated almost without approximation in the stress analysis. Continuously transposed cable (CTC) consists of multiple enamel-coated copper strands arranged in a nearly rectangular pattern shown. Not shown are how the transpositions are made, which rotate the different strands so that they each occupy all

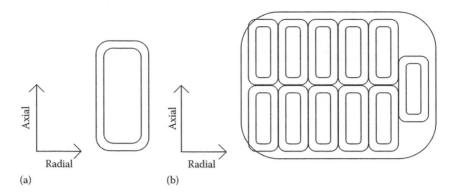

FIGURE 10.4
Types of wire or cable used in transformer coils: (a) magnet wire and (b) transposed cable.

the positions shown as one moves along the wire. The transpositions give some rigidity to the collection of strands. In addition, use is often made of bonded cable in which all the strands are bonded together by means of an epoxy coating over the enamel, which is subjected to a heat treatment. In this case, the cable can be treated as a rigid structure, although there is some question as to how to evaluate its material properties. With or without bonding, one has to make some approximations as to how to model the cable for stress analysis purposes. Without bonding, we assume for radial force considerations that the cable has a radial thickness equivalent to 2 radial strands. With bonding, we assume a radial thickness equivalent to 80% of the actual radial build. These assumed thicknesses apply when the particular stress calculation critically depends on them.

We will first look at the stresses produced by the axial forces and then consider the stresses due to the radial forces. We need to examine worst-case stresses in the copper winding turns, in the key spacers, in the pressure rings, and in the tie plates, which are attached to the top and bottom clamps.

10.3.1 Compressive Stress in the Key Spacers

The maximum axial compressive force is used to determine the worst-case compressive stress in the key spacers. This force, F_c, obtained for each coil from the magnetic field analysis program, is divided by the area of the key spacers covering one 360-degree disk section to obtain the key spacer compressive stress σ_{ks}:

$$\sigma_{ks} = \frac{F_c}{N_{ks}W_{ks}B} \tag{10.3}$$

where
  N_{ks} is the number of key spacers around one 360-degree section
  W_{ks} is the width of a key spacer
  B is the radial build of the coil

We use key spacers made of precompressed pressboard, which can withstand a maximum compressive stress of 310 MPa (MPa = Mega Pascal = N/mm^2). Therefore, the stress calculated by (10.3) should not exceed this number and is usually restricted to be much less than this.

10.3.2 Axial Bending Stress per Strand

The maximum axial force over all the vertical subdivisions is computed for each coil. Let N_v equal the number of vertical subdivisions. Call this maximum force F_a. In order to compute the bending stress, we need to know the force/unit length along an individual strand, since these forces are continuously distributed along the strand. The number of strands in the entire coil, N_s, is given by

$$N_s = N_t N_h N_w N_{st} \tag{10.4}$$

where
 N_t is the number of turns/leg
 N_h is the number of cables/turn radially
 N_w is the number of cables/turn axially
 N_{st} is the number of strands/cable

We are allowing for the fact that each turn can consist of several cables in parallel, some radially and some axially positioned, each having N_{st} strands. If the coil consists of 2 separate windings stacked axially (center fed) each having N_e electrical turns, then $N_t = 2N_e$. Since the section with the maximum axial force is only $1/N_v$ of the coil, it only has $1/N_v$ of the above number of strands. Hence, the maximum force/unit length on a single strand, q_{st}, is given by

$$q_{st} = \frac{N_v F_a}{N_s \pi D_m} \tag{10.5}$$

where D_m is the mean diameter of the coil.

The problem can be analyzed as a uniformly loaded rectangular beam with built-in ends as shown in Figure 10.5a. There are six unknowns, the horizontal and vertical components of the reaction forces and the bending moments at the two built-in ends, but only three equations of statics making this a statically indeterminate problem. The two horizontal reaction forces are equal and opposite and produce a tensile stress in the beam, which is small for small deflections and will be ignored. The vertical reaction forces are equal and share the downward load equally. They are therefore given by $R_{1,up} = R_{2,up} = qL/2$, where $q = q_{st}$ is the downward force/unit length along the beam and L the beam's length. By symmetry, the bending moments M_1 and M_2 shown in Figure 10.5a

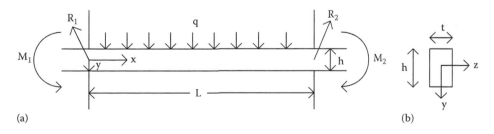

FIGURE 10.5
Uniformly loaded beam with built-in ends: (a) side view of beam and (b) cross-sectional view of beam.

are equal and produce a constant bending moment along the beam. In addition, a bending moment as a function of position is also present, resulting in a total bending moment at position x of

$$M(x) = \frac{qLx}{2} - \frac{qx^2}{2} - M_1 \tag{10.6}$$

using the sign convention of [Tim55].

To solve for the unknown M_1, we need to solve the equation of the beam's deflection curve [Tim55]:

$$\frac{d^2y}{dx^2} = -\frac{M(x)}{EI_z} \tag{10.7}$$

where

E is Young's modulus for the beam material
I_z is the area moment of inertia about the z-axis
y is the downward deflection of the beam

Substituting (10.6) into (10.7), we have

$$\frac{d^2y}{dx^2} = \frac{1}{EI_z}\left(M_1 - \frac{qLx}{2} + \frac{qx^2}{2}\right) \tag{10.8}$$

Integrating once, we get

$$\frac{dy}{dx} = \frac{1}{EI_z}\left(M_1x - \frac{qLx^2}{4} + \frac{qx^3}{6}\right) + C \tag{10.9}$$

where C is a constant of integration. Since the beam is rigidly clamped at the ends, the slope dy/dx = 0 at x = 0 and x = L. Setting dy/dx = 0 at x = 0 in (10.9) yields C = 0. Setting dy/dx = 0 at x = L, we find that

$$M_1 = \frac{qL^2}{12} \tag{10.10}$$

Substituting these values into (10.9), we obtain

$$\frac{dy}{dx} = \frac{q}{EI_z}\left(\frac{L^2x}{12} - \frac{Lx^2}{4} + \frac{x^3}{6}\right) \tag{10.11}$$

Integrating again, we obtain

$$y = \frac{q}{EI_z}\left(\frac{L^2x^2}{24} - \frac{Lx^3}{12} + \frac{x^4}{24}\right) = \frac{qx^2(L-x)^2}{24EI_z} \tag{10.12}$$

where the constant of integration was set to zero since y = 0 at x = 0 and x = L.

Inserting (10.10) into (10.6), the bending moment as a function of position along the beam is

$$M(x) = \frac{q}{2}\left[x(L-x) - \frac{L^2}{6}\right] \tag{10.13}$$

The maximum positive value occurs at $x = L/2$ and is $M_{max} = qL^2/24$. The minimum value occurs at $x = 0$ or $x = L$ and is $M_{min} = -qL^2/12$. Since the minimum bending moment is larger in absolute value, we use it in the formula to obtain the stress due to bending in the beam [Tim55]

$$\sigma_x = \frac{My}{I_z} \tag{10.14}$$

where y is measured downward from the centroid of the beam cross section as shown in Figure 10.5b. For a given x, σ_x is a maximum or minimum when $y = \pm h/2$, where h is the beam height in the bending direction. If σ_x is positive, the stress is tensile, and if negative, compressive. Inserting M_{min} into (10.14), taking $y = -h/2$, and using

$$I_z = \frac{th^3}{12} \tag{10.15}$$

for the area moment of inertia for a rectangular cross section with respect to the z-axis through the centroid (see Figure 10.5b), we obtain

$$\sigma_{x,max} = \frac{q}{2t}\left(\frac{L}{h}\right)^2 \tag{10.16}$$

Here, t is the thickness of the beam perpendicular to the bending plane. This is a tensile stress and occurs at the top of the beam at the supports. There is a compressive stress of equal magnitude at the bottom of the beam at the supports.

Inserting the actual load (10.5) into (10.16), we get for the maximum axial bending stress

$$\sigma_{max,axial} = \frac{N_v F_a}{2N_s \pi D_m t}\left(\frac{L}{h}\right)^2 \tag{10.17}$$

The span length, L, can be determined from the number of key spacers, their width, and the mean circumference:

$$L = \frac{\pi D_m - N_{ks} W_{ks}}{N_{ks}} = \frac{\pi D_m}{N_{ks}} - W_{ks} \tag{10.18}$$

The strand height h and thickness t apply to a single strand, whether as part of a cable having many strands or as a single strand in a magnet wire. If the cable is bonded, the maximum axial bending stress in (10.17) is divided by 3. This is simply an empirical correction to take into account the greater rigidity of bonded cable.

We can also get the maximum downward deflection of the beam from (10.12), which occurs at $x = L/2$. Inserting (10.15) for I_z into (10.12), we get for the maximum deflection, y_{max},

$$y_{max} = \frac{q}{32E}\left(\frac{L^4}{th^3}\right) = \frac{N_v F_a}{32N_s \pi D_m E}\left(\frac{L^4}{th^3}\right) \tag{10.19}$$

where we have used (10.5) to get the second equality.

10.3.3 Tilting Strength

The axial compressive force that is applied to the key spacers can cause the individual strands of the conductors, which are pressed under the key spacers to tilt if the force is large enough. Figure 10.6 shows an idealized geometry of this situation: an individual strand that has the form of a closed ring acted on by a uniform axial compressive pressure, P_c. We assume initially that the strand has rounded ends that do not dig into the adjacent key spacers to prevent or oppose the tilting shown. (There can be several layers of strands in the axial direction separated by paper, which plays the same role as the key spacers in the figure.)

Analyzing a small section in the azimuthal direction of length $\Delta\ell$, the applied pressure exerts a torque, τ_c, given by

$$\tau_c = P_c\left(t\Delta\ell\right)h\sin\theta \tag{10.20}$$

where
  t is the radial thickness of the strand
  $t\Delta\ell$ is the area on which the pressure P_c acts

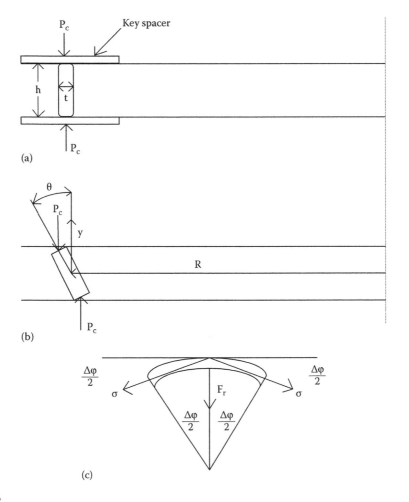

FIGURE 10.6
Geometry of strand tilting due to the axial compressive force. The y direction is perpendicular to the paper. (a) Unstressed geometry, (b) geometry under stress, and (c) geometric variables used in the analysis.

The axial height of the strand is h and θ is the tilting angle from the vertical, which is assumed to be small. $h\sin\theta$ is the lever arm on which the force acts to produce the torque. The tilting causes the material of the ring to stretch above its axial center and to compress below it. This produces stresses in the ring, which in turn produce a torque that opposes the torque calculated earlier.

To calculate this opposing torque, let y measure the distance above the axial center of the strand as shown in Figure 10.6b. The increase in radius at distance y above the centerline is $y\tan\theta$. Therefore, the strain at position y is

$$\varepsilon = \frac{y\tan\theta}{R} \tag{10.21}$$

where ε is an azimuthal strain. This produces an azimuthal tensile stress (hoop stress) given by

$$\sigma = E\varepsilon = \frac{Ey\tan\theta}{R} \tag{10.22}$$

where E is Young's modulus for the conductor material. This hoop stress results in an inward force, F_r, on the section of strand given by (Figure 10.6c)

$$F_r = 2\sigma(t\Delta y)\sin\left(\frac{\Delta\varphi}{2}\right) \cong \sigma(t\Delta y)\Delta\varphi = \sigma(t\Delta y)\frac{\Delta\ell}{R} \tag{10.23}$$

In (10.23), $t\Delta y$ is the area at height y over which the stress σ acts and $\Delta\varphi$ the angle subtended by the azimuthal section of strand of length $\Delta\ell$. The small-angle approximation of the sin function, $\sin(\Delta\varphi/2) \approx \Delta\varphi/2$, and also $\Delta\varphi = \Delta\ell/R$ has been used in (10.23).

The inward force F_r produces a counter torque $\Delta\tau$ on the small section of height Δy that is given by

$$\Delta\tau = F_r y = \frac{\sigma t\Delta\ell y\Delta y}{R} \tag{10.24}$$

Using (10.22), this becomes

$$\Delta\tau = \frac{Et\Delta\ell\tan\theta y^2\Delta y}{R^2} \tag{10.25}$$

Letting Δy become infinitesimal and integrating from $y = 0$ to $h/2$, we get

$$\tau = \frac{Et\Delta\ell\tan\theta h^3}{24R^2} \tag{10.26}$$

Analyzing the portion of the strand below the centerline, the hoop stress is compressive and will result in an outward force on the strand. This will create a torque of the same magnitude and sense as that in (10.26), so to take the whole strand into account, we just need to multiply (10.26) by 2. In equilibrium, this resulting torque equals the applied torque given by (10.20). Equating these two expressions and using, for small θ, $\sin\theta \approx \tan\theta \approx \theta$, we get

$$P_c = \frac{E}{12}\left(\frac{h}{R}\right)^2 \tag{10.27}$$

When the conductor strand has squared ends, there is an additional resistance to tilting as a result of the ends digging into the key spacers or paper. This results in an additional resisting torque of magnitude [Ste62], [Wat66]:

$$C\frac{f_{ks}t^3 \sin\theta \Delta\ell}{6} \tag{10.28}$$

where
    C is a constant depending on the spacer material
    f_{ks} is the fraction of the coil's average circumference occupied by key spacers

If there are no key spacers, then set $f_{ks} = 1$. This should be added to $2 \times$ Equation 10.26 and their sum equated to Equation 10.20, resulting in a tilting pressure of

$$P_c = \frac{E}{12}\left(\frac{h}{R}\right)^2 + \frac{C}{6}\left(\frac{f_{ks}t^2}{h}\right) \tag{10.29}$$

If the strand has rounded corners of radius R_{corn}, then t in this formula is reduced by $2R_{corn}$, so only its flat portion is considered. The resulting critical axial pressure is therefore

$$P_c = \frac{E}{12}\left(\frac{h}{R}\right)^2 + \frac{C}{6}\left[\frac{f_{ks}(t - 2R_{corn})^2}{h}\right] \tag{10.30}$$

In this equation, E is the initial Young's modulus for copper and is essentially independent of the hardness. It has a value of 1.10×10^5 MPa. C has been taken to equal 2.44×10^3 MPa/mm. The second term in (10.30) is called the bedding term.

This equation is often modified to take into account other factors based on experimental data such as the proof strength of copper and whether the cable is made of magnet wire or transposed cable (CTC). A possible modified version is

$$P_c = C_1 C_2 \left\{ \frac{E}{8\pi}\left(\frac{h}{R}\right)^2 + \frac{C_3}{6}\left[\frac{f_{ks}(t - 2R_{corn})^2}{h}\right] \right\} \tag{10.31}$$

where C_1 depends on the proof strength of the copper and varies from 1.0 for regular copper to 1.4 for very hard copper. For magnet wire, $C_2 = 1$ for $f_{ks} = 1$ and $C_2 = 1.4$ for $f_{ks} < 1$. For CTC, $C_2 = 1.2$ for $f_{ks} = 1$, and $C_2 = 2.0$ for $f_{ks} < 1$. $C_3 = 270$ MPa/mm for magnet wire and 132 MPa/mm for CTC.

To compare this with the applied maximum axial compressive force, multiply (10.30) or (10.31) by the radial surface area of the strands in one horizontal layer, A_{layer}. This is

$$A_{layer} = 2\pi R t N_d N_h N_{st,radial} \tag{10.32}$$

where
    N_d is the number of turns in a disk or section
    N_h is the number of cables in parallel radially/turn
    $N_{st,radial}$ is the number of strands in the radial direction/cable

$N_{st,radial}$ = 1 for magnet wire and, as shown in Figure 10.4b, $(N_{st} - 1)/2$ for CTC. Thus, the critical axial force is

$$F_{cr} = P_c A_{layer} \qquad (10.33)$$

This applies to unbonded cable. For bonded cable, we take $F_{cr} = \infty$ since it is assumed in this case that tilting is very unlikely. We compare (10.33) with the maximum applied axial compressive force, F_c, by taking the ratio F_{cr}/F_c. This ratio must be >1 for a viable design, that is, F_c must be <F_{cr}.

We should note that the expression for A_{layer} does not take into account the fractional area coverage of the key spacers since the force is applied to the key spacer area. However, except for the bedding term, the calculation involved an averaging process over all the strands in the layer. The bedding term already includes the fractional key spacer area coverage.

10.3.4 Stress in the Tie Bars

The tie bars or flitch plates are used to join the upper and lower clamping structures that keep the coils under compression. These are generally long rectangular bars of steel, which are placed along the sides of the core legs. They are under mild tension during normal transformer operation. During short circuit, the tensile stresses can increase considerably. Also, when the transformer is lifted, the tie bars support the entire weight of the core and coils.

The short circuit stress in the tie bars is due to the total end thrust produced by all the coils on a single leg. This is the sum of the total upward or downward forces acting on the coils. Because of the 3-phase current, the maximum force will not appear on all legs simultaneously. For a single line to ground fault, one of the legs will have the highest currents and forces. This leg is therefore singled out for the stress calculation. For a 3-phase fault, the maximum force acts on all three legs but not simultaneously so that, even in this case, the force associated with only one leg may be used in the stress analysis. Although this force acts on a single leg, one can assume that all the tie bars participate in countering this force. Other assumptions or even tests could be made to determine how this force is distributed among the tie bars.

The total end thrust is the result of a static force calculation. Because of possible dynamic effects associated with the sudden application of a force to an elastic system, the end thrust could be considerably higher for a short period after the force application. To see what the dynamic force enhancement might be, we analyze an elastic bar subject to a suddenly applied force as illustrated in Figure 10.7. Let x be the change in length produced by the force, relative to the unstressed bar of length L. A bar under stress stores elastic energy, U, given by [Tim55]

$$U = \frac{EA}{2L} x^2 \qquad (10.34)$$

where
  E is Young's modulus for the bar material
  A is the cross-sectional area of the bar

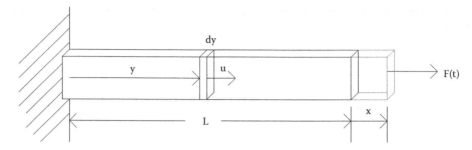

FIGURE 10.7
Elastic bar fixed at one end and subject to an applied force at the other end.

The applied force causes the bar material to move so that it acquires a kinetic energy. Since each portion of the bar moves with a different velocity—the bar is fixed at one end and moves with maximum velocity at the other end—it is necessary to integrate the kinetic energy of each segment along the bar to get the total. In Figure 10.7, a bar segment at distance y from the fixed end of thickness dy is isolated. The parameter u measures the displacement of this bar segment, and its velocity is therefore du/dt. But the strain ε is uniform along the bar so we have $\varepsilon = u/y = x/L$. Therefore, $u = xy/L$ and $du/dt = (y/L)$ dx/dt. The segment's kinetic energy is

$$d(KE) = \frac{1}{2}\rho A dy \left(\frac{y}{L}\frac{dx}{dt}\right)^2 = \frac{1}{2}\frac{\rho A}{L^2}\left(\frac{dx}{dt}\right)^2 y^2 dy \tag{10.35}$$

where ρ is the mass density of the bar material. Integrating over the bar, we get

$$KE = \frac{1}{2}\frac{\rho AL}{3}\left(\frac{dx}{dt}\right)^2 \tag{10.36}$$

This says that effectively 1/3 of the mass of the bar is moving with the end velocity dx/dt.

We use Lagrange's method to obtain the equation of motion of the bar. The Lagrangian is $L = KE - U$ and the equation of motion is

$$\frac{d}{dt}\left[\frac{\partial L}{\partial(dx/dt)}\right] - \frac{\partial L}{\partial x} = F(t) \tag{10.37}$$

where

$$L = \frac{1}{2}\frac{\rho AL}{3}\left(\frac{dx}{dt}\right)^2 - \frac{EA}{2L}x^2$$

Carrying out the differential operations in (10.37), this becomes

$$\frac{d^2x}{dt^2} + \frac{3E}{\rho L^2}x = \left(\frac{3}{\rho AL}\right)F(t) \tag{10.38}$$

The force applied to the tie bars during a fault is produced by the coils. Although the forces applied to the coils are proportional to the current squared, because of the coil's internal structure, the force transmitted to the tie bars may be modified. However, assuming the coils are well clamped, we expect that the force transmitted to the tie bars is also proportional to the current squared to a good approximation. The fault current has the approximate form [Wat66]

$$I = I_o \left(e^{-at} - \cos \omega t \right) u(t) \tag{10.39}$$

where
   a is a constant, which is a measure of the resistance in the circuit
   ω is the angular frequency
   u(t) is the unit step function, which is zero for times t < 0 and 1 for times t ≥ 0

The force has the form

$$F = F_o \left(\frac{1}{2} + e^{-2at} - 2e^{-at} \cos \omega t + \frac{1}{2} \cos 2\omega t \right) u(t) \tag{10.40}$$

which results from squaring (10.39) and using a trigonometric identity. This function is plotted in Figure 10.8a for a = 22.6 and $\omega = 2\pi(60)$.

To simplify matters and in a worst-case position with little damping, take a = 0 so (10.40) reduces to

$$F = F_o \left(\frac{3}{2} - 2\cos \omega t + \frac{1}{2} \cos 2\omega t \right) u(t) \tag{10.41}$$

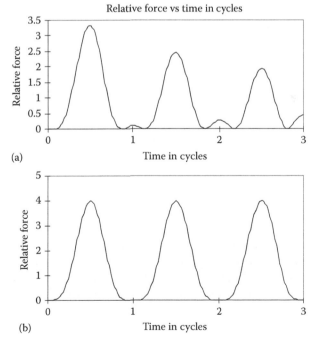

FIGURE 10.8
Graphs of some tie bar forces versus time in cycles. (a) Plot of Equation 10.40 and (b) plot of Equation 10.41.

This is plotted in Figure 10.8b. It achieves a maximum of $F_{max} = 4F_o$, whereas (10.40) reaches a maximum value of about $3.3F_o$ for the parameters used. If this maximum force were acting in a steady-state manner, the bar's displacement would be, according to Hooke's law,

$$x_{max} = \frac{L}{EA} 4F_o = \frac{L}{EA} F_{max} \tag{10.42}$$

This should be compared with the dynamical solution of (10.38) to get the enhancement factor.

To solve (10.38), take Laplace transforms, LaT, using the boundary conditions $x(t = 0^-) = 0$ and $dx/dt(t = 0^-) = 0$. We obtain

$$\left(s^2 + \frac{3E}{\rho L^2}\right) LaT(x) = \left(\frac{3}{\rho AL}\right) LaT(F) \tag{10.43}$$

The Laplace transform of (10.41) is

$$LaT(F) = F_o\left[\frac{3}{2}\left(\frac{1}{s}\right) - 2\left(\frac{s}{s^2 + \omega^2}\right) + \frac{1}{2}\left(\frac{s}{s^2 + 4\omega^2}\right)\right] \tag{10.44}$$

Substituting into (10.43), we obtain for the Laplace transform of x

$$LaT(x) = \left(\frac{3}{\rho AL}\right)\frac{F_o}{s^2 + b^2}\left[\frac{3}{2}\left(\frac{1}{s}\right) - 2\left(\frac{s}{s^2 + \omega^2}\right) + \frac{1}{2}\left(\frac{s}{s^2 + 4\omega^2}\right)\right]$$
$$\text{where } b^2 = \frac{3E}{\rho L^2} \tag{10.45}$$

Using some algebra to rewrite (10.45) and taking inverse transforms, we obtain

$$x(t) = \left(\frac{L}{EA}\right)F_{max}u(t)\left\{\frac{1}{4}\left[\frac{3}{2}(1 - \cos bt) + \frac{2}{1 - (\omega/b)^2}(\cos bt - \cos \omega t)\right.\right.$$
$$\left.\left. - \frac{1}{2}\left(\frac{1}{1 - 4(\omega/b)^2}\right)(\cos bt - \cos 2\omega t)\right]\right\} \tag{10.46}$$

where the quantity in curly brackets is the enhancement factor.

We need to compare the natural angular frequency b with the applied angular frequencies $\omega = 2\pi f = 377$ rad/s, assuming $f = 60$ Hz, and $2\omega = 754$ rad/s to see whether a resonance problem might occur. For steel bars, $E = 2.07 \times 10^5$ MPa, $\rho = 7837$ kg/m³. Using (10.45) for b, we obtain

$$b = \sqrt{\frac{3E}{\rho L^2}} = \frac{8902}{L \text{ (m)}} \text{ rad/s} \tag{10.47}$$

For a 2.54 m long tie bar, which is typical, this gives b = 3504 rad/s. Since this is much larger than ω or 2ω, we are far from resonance. Thus, $(\omega/b)^2$ can be ignored relative to 1 in (10.46) and it simplifies to

$$x(t) = \left(\frac{L}{EA}\right) F_{max} u(t) \frac{1}{4}\left(\frac{3}{2} - 2\cos\omega t + \frac{1}{2}\cos(2\omega t)\right) \tag{10.48}$$

Hence, x_{max} is the same as in the steady-state case and the enhancement factor is 1. This occurs at $t = \pi/\omega$. Thus, unless the applied or twice the applied frequency is close to the tie bar's natural frequency, there is no dynamic enhancement. By numerically checking over a large grid of times and electrical frequencies, a maximum enhancement of about 1.66 is produced if the time does not exceed one period ($\omega t < 2\pi$).

Another consideration with respect to tie bar strength is the matter of the lifting force on the tie bars when the core and coils must be moved during the production process. Since the tie bars must support the weight of the core and coils during lifting, it is necessary to check the stress in the tie bars produced by lifting. During lifting, the tie bars associated with the outer legs are mainly stressed since this is where the lifting hooks are positioned. Both the short circuit dynamic stresses and the lifting stresses must be below a maximum allowable stress in the tie bar material. We take this maximum allowable stress to be 620 MPa if low-carbon steel is used and 414 MPa if a stainless steel is used for the tie bar material.

10.3.5 Stress in the Pressure Ring

The pressure ring receives the total end thrust of the windings. It is often made of press-board of about 38–63.5 mm thickness. The ring covers the radial build of the windings with a little overhang. During a fault, it must support the full dynamic end thrust of the windings, which according to the last section could be as high as 1.7 times the maximum total end thrust per leg.

The end thrust or force is distributed over the end ring, producing an effective pressure of

$$P_{ring} = \frac{F_{ring}}{A_{ring}} \tag{10.49}$$

where we use ring to label the end force, F_{ring}, and ring area, A_{ring}. This area is given by

$$A_{ring} = \frac{\pi}{4}\left(D_{ring,out}^2 - D_{ring,in}^2\right) \tag{10.50}$$

in terms of the outer and inner ring diameters. The ring is supported on radial blocks with space between for the leads. This produces an unsupported span of a certain length, L_u. To a good approximation, the problem is similar to that discussed previously for the axial bending of a strand of wire. Thus, we can use formula (10.16) for the maximum stress in the end ring, with $L = L_u$, $t = (1/2)(D_{ring,out} - D_{ring,in})$ the radial build of the ring, $h = h_{ring}$ the ring's thickness, and $q = P_{ring}t$ the force/unit length along the unsupported span. We obtain

$$\sigma_{x,max} = \frac{P_{ring}}{2}\left(\frac{L_u}{h_{ring}}\right)^2 = \frac{F_{ring}}{2A_{ring}}\left(\frac{L_u}{h_{ring}}\right)^2 \tag{10.51}$$

Using a maximum bending permissible stress for pressboard rings of $\sigma_{\text{bend,max}} = 103$ MPa and substituting this value for $\sigma_{x,\text{max}}$ in (10.51), we find

$$F_{\text{ring,max}} = \sigma_{\text{bend,max}} \left(\frac{\pi}{2}\right)\left(D^2_{\text{ring,out}} - D^2_{\text{ring,in}}\right)\left(\frac{h_{\text{ring}}}{L_u}\right)^2 \tag{10.52}$$

This is the maximum end force the pressure ring can sustain. It must be greater than the applied maximum end force.

The maximum deflection can be obtained by appropriate substitution into (10.19).

10.3.6 Hoop Stress

The maximum radial pressure acting on the winding as obtained from the force calculation creates a hoop stress in the winding conductor. The hoop stress is tensile or compressive, depending on whether the pressure acts radially outward or inward, respectively. In Figure 10.9, we treat the winding as an ideal cylinder or ring subjected to a radially inward pressure, P_r. Let R_m be the mean radius of the cylinder and H its axial height. In part (b) of the figure, we show two compressive reaction forces F in the winding, sustaining the force applied to half the cylinder. The x-directed force produced by the pressure P_r cancels out by symmetry, and the net y-directed force acting downward is given by

$$F_y = P_r H R_m \int_0^\pi \sin\varphi\, d\varphi = 2P_r H R_m \tag{10.53}$$

This is balanced by a force of 2F acting upward, so equating these we get $F = P_r H R_m$. Dividing by the cross-sectional area A of the material sustaining the force, we get the compressive stress in the material

$$\sigma_{\text{hoop}} = \frac{F}{A} = \frac{P_r H R_m}{A} \tag{10.54}$$

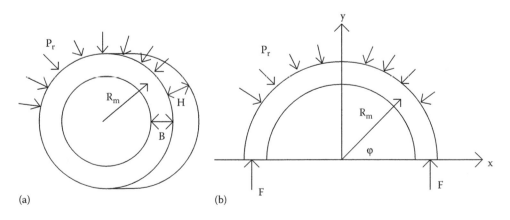

FIGURE 10.9
Geometry for determining the hoop stress in a cylinder acted on by a radially inward pressure. (a) Full cylinder geometry and (b) force diagram.

For A = HB, where B is the radial build of the cylinder, we get

$$\sigma_{hoop} = \frac{P_r R_m}{B} \tag{10.55}$$

This last formula assumes the cylinder is made of a homogeneous material. If the cylinder is made of conductors and insulating materials, the conductors primarily support the forces. In this case, A should equal the cross-sectional area of all the conductors in the winding, $A = A_t N_t$, where A_t is the cross-sectional area of the conductors in a turn and N_t is the total number of turns in the winding. If the winding is center fed, that is, consists of two parallel windings on the same leg, and N_t refers to the total turns/leg or twice the number of electrical turns, then A_t should be 1/2 the turn area = the area of the cable in either half of the winding. Substituting into (10.54) and using $D_m = 2R_m$, we obtain

$$\sigma_{hoop} = \frac{P_r H D_m}{2 A_t N_t} \tag{10.56}$$

for the hoop stress. This is compressive for P_r acting inward and tensile for P_r acting outward. In either case, this stress should not exceed the proof stress of the winding material.

When the radial pressure acts inward, the winding is apt to buckle before the proof stress is exceeded. This inward radial buckling is a complex process to analyze and will be discussed later.

10.3.7 Radial Bending Stress

Windings have inner radial supports called sticks made of pressboard or other material, which are spaced uniformly along their circumference and extend at least to the height of the winding. When an inward radial pressure acts on the winding, the sections of the winding between supports act like a curved beam subjected to a uniform loading. A similar situation occurs in the case of a rotating flywheel with radial spokes. In the flywheel case, the loading (centrifugal force) acts outward but otherwise the analysis is similar. The flywheel example is analyzed in Timoshenko [Tim55], which we follow here with minor changes.

We need to make use of Castigliano's theorem. This states that if the material of a system follows Hooke's law, that is, remains within the elastic limit, and if the displacements are small, then the partial derivative of the strain energy with respect to any force equals the displacement corresponding to the force. Here force and displacement have a generalized meaning, that is, they could refer to torques or moments and angular displacements as well as their usual meanings of force and length displacements. Also, the strain energy must be expressed as a quadratic function of the forces. For example, the strain energy, $U_{tensile}$, associated with tensile or compressive forces N in a beam of length L with elastic modulus E is

$$U_{tensile} = \int_0^L \frac{N^2}{2EA} dx \tag{10.57}$$

where N, A, and E can be functions of position along the beam, x. The strain energy, $U_{bending}$, associated with a bending moment M in a beam of length L is

$$U_{bending} = \int_0^L \frac{M^2}{2EI} dx \tag{10.58}$$

where M, I, and E can be functions of position, x. Here I is the area moment of inertia.

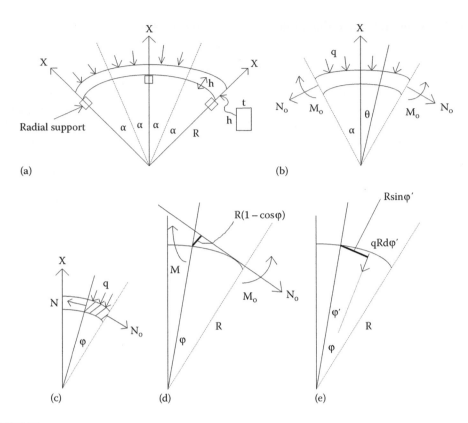

FIGURE 10.10
Geometry for determining the radial bending stresses.

In Figure 10.10a, we show a portion of the winding with the inner radial supports spaced an angle 2α apart. There is a normal force X acting radially outward at the supports, which counters the inward pressure, which has been converted to a force/unit length q acting on the coil section. The coil section is assumed to form a closed ring of radial build h, axial height t, and mean radius R.

In Figure 10.10b, we further isolate a portion of the ring, which extends between adjacent midsections between the supports. The reason for doing this is that there is no radially directed (shearing) force acting on these mid cross sections. This is because by symmetry, the distributed load between the midsections must balance the outward force at the included support. Thus, the only reactions at the midsections are an azimuthally directed force, N_o, and a couple, M_o, which need to be found.

Balancing the vertical forces, we have

$$X - 2N_o \sin\alpha - 2qR \int_0^\alpha \cos\theta d\theta = 0 \qquad (10.59)$$

Performing the integral and solving for N_o, we obtain

$$N_o = \frac{X}{2\sin\alpha} - qR \qquad (10.60)$$

At any cross section as shown in Figure 10.10c, measuring angles from the midsection position with the variable φ, we can obtain the normal force, N, from the static equilibrium requirement:

$$N = N_o \cos\varphi - qR \int_0^\varphi \sin\varphi' d\varphi' \tag{10.61}$$

Integrating and substituting for N_o from (10.60), we get

$$N = \frac{X\cos\varphi}{2\sin\alpha} - qR \tag{10.62}$$

Similarly, using Figure 10.10d and e, we can obtain the bending moment at the cross section of angle φ from the midsection by balancing the moments:

$$M = M_o - N_o R(1 - \cos\varphi) - qR^2 \int_0^\varphi \sin\varphi' d\varphi' \tag{10.63}$$

Performing the integration and substituting for N_o from (10.60), we obtain

$$M = M_o + \frac{XR(\cos\varphi - 1)}{2\sin\alpha} \tag{10.64}$$

Equations 10.62 and 10.64 express the normal force and bending moments as functions of position along the beam (arc in this case). These can be used in the energy expressions (10.57) and (10.58). Castigliano's theorem can then be used to solve for the unknowns. However, we are missing the strain energy associated with the supports. The radial supports consist of several different materials as illustrated in Figure 10.11. We assume they can be treated as a column of uniform cross-sectional area A_{stick}. In general, the column

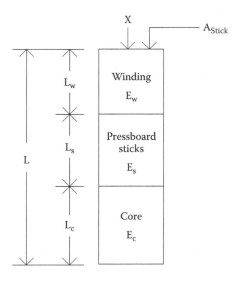

FIGURE 10.11
Radial support structure.

consists of winding material (copper), pressboard sticks, and core steel. However, for an innermost winding, the winding material is not present as part of the support column. For such a composite structure, we derive an equivalent Young's modulus, E_{eq}, by making use of the fact that the stress is the same throughout the column. Only the strain differs from material to material. We obtain

$$E_{eq} = \frac{L}{\left(L_w/E_w\right)+\left(L_s/E_s\right)+\left(L_c/E_c\right)} \tag{10.65}$$

where
  L_w is the length of the winding portion
  E_w its Young's modulus

Similarly, s refers to the stick and c to the core. $L = L_w + L_s + L_c$ is the total column length.
  The total strain energy for our system, retaining only the portion shown in Figure 10.10b, since the entire ring energy is simply a multiple of this, can be written as

$$U = 2\int_0^\alpha \frac{N^2 R}{2EA}\,d\varphi + 2\int_0^\alpha \frac{M^2 R}{2EI}\,d\varphi + \frac{X^2 L}{2E_{eq}A_{stick}} \tag{10.66}$$

where
  N and M are given by (10.62) and (10.64)
  A is the cross-sectional area of the ring, $A = th$
  I its bending moment, $I = th^3/12$
  the infinitesimal length along the bar, $Rd\varphi$, is used

The two unknowns are X and M_o. At the fixed end of the support column (center of the core), the displacement is zero; hence, by Castigliano's theorem, $\partial U/\partial X = 0$. Also, the bending moment at the midsection of the span between the supports produces no angular displacement by symmetry. Hence, by Castigliano's theorem, $\partial U/\partial M_o = 0$. Differentiating (10.66), we obtain

$$\frac{\partial U}{\partial X} = \frac{2R}{EA}\int_0^\alpha N\frac{\partial N}{\partial X}\,d\varphi + \frac{2R}{EI}\int_0^\alpha M\frac{\partial M}{\partial X}\,d\varphi + \frac{XL}{E_{eq}A_{stick}} = 0 \tag{10.67}$$

and

$$\frac{\partial U}{\partial M_o} = \frac{2R}{EA}\int_0^\alpha N\frac{\partial N}{\partial M_o}\,d\varphi + \frac{2R}{EI}\int_0^\alpha M\frac{\partial M}{\partial M_o}\,d\varphi = 0 \tag{10.68}$$

Substituting for N and M from (10.62) and (10.64), we obtain

$$\frac{2R}{EA}\int_0^\alpha \left(\frac{X\cos\varphi}{2\sin\alpha}-qR\right)\frac{\cos\varphi}{2\sin\alpha}\,d\varphi + \frac{2R}{EI}\int_0^\alpha \left[M_o + \frac{XR(\cos\varphi-1)}{2\sin\alpha}\right]\frac{R(\cos\varphi-1)}{2\sin\alpha}\,d\varphi$$

$$+ \frac{XL}{E_{eq}A_{stick}} = 0 \tag{10.69}$$

and

$$\frac{2R}{EI} \int_0^\alpha \left[M_o + \frac{XR(\cos\varphi - 1)}{2\sin\alpha} \right] d\varphi = 0 \tag{10.70}$$

Integrating these expressions, we obtain

$$M_o = \frac{XR}{2} \left(\frac{1}{\sin\alpha} - \frac{1}{\alpha} \right) \tag{10.71}$$

and

$$X = \frac{qR}{\dfrac{1}{4\sin^2\alpha} \left(\alpha + \dfrac{\sin 2\alpha}{2} \right) + \dfrac{AR^2}{I} \left[\dfrac{1}{4\sin^2\alpha} \left(\alpha + \dfrac{\sin 2\alpha}{2} \right) - \dfrac{1}{2\alpha} \right] + \dfrac{EA}{E_{eq}A_{stick}}} \tag{10.72}$$

where we have used the fact that L = R. Substituting into (10.62) and (10.64) and defining

$$f_1(\alpha) = \frac{1}{4\sin^2\alpha} \left(\alpha + \frac{\sin 2\alpha}{2} \right)$$

$$f_2(\alpha) = \frac{1}{4\sin^2\alpha} \left(\alpha + \frac{\sin 2\alpha}{2} \right) - \frac{1}{2\alpha} \tag{10.73}$$

we obtain for N and M

$$N = qR \left\{ \frac{\cos\varphi}{2\sin\alpha} \left[\frac{1}{f_1(\alpha) + \dfrac{AR^2}{I}f_2(\alpha) + \dfrac{EA}{E_{eq}A_{stick}}} \right] - 1 \right\} \tag{10.74}$$

$$M = \frac{qR^2}{2} \left(\frac{\cos\varphi}{\sin\alpha} - \frac{1}{\alpha} \right) \left[\frac{1}{f_1(\alpha) + \dfrac{AR^2}{I}f_2(\alpha) + \dfrac{EA}{E_{eq}A_{stick}}} \right] \tag{10.75}$$

We also have

$$\frac{AR^2}{I} = 12 \left(\frac{R}{h} \right)^2 \tag{10.76}$$

where h is the radial build of the ring.

N gives rise to a normal stress

$$\sigma_N = \frac{N}{A} = \frac{qR}{A}\left\{\frac{\cos\varphi}{2}\left(\frac{1}{\sin\alpha}\right)\left[\frac{1}{f_1(\alpha)+12\left(\frac{R}{h}\right)^2 f_2(\alpha)+\frac{EA}{E_{eq}A_{stick}}}-1\right]\right\} \quad (10.77)$$

and M gives rise to a bending stress, which varies over the cross section, achieving a maximum tensile or compressive value of (see formula (10.14))

$$\sigma_M = \frac{Mh}{2I} = \frac{qR}{A}\left(\frac{3R}{h}\right)\left(\cos\varphi - \frac{\sin\alpha}{\alpha}\right)\left(\frac{1}{\sin\alpha}\right)$$

$$\times\left[\frac{1}{f_1(\alpha)+12\left(\frac{R}{h}\right)^2 f_2(\alpha)+\frac{EA}{E_{eq}A_{stick}}}\right] \quad (10.78)$$

We have factored out the term qR/A since this can be shown to be the hoop stress in the ring. (In formula (10.54), PH corresponds to q and R_m to R in our development here.) Thus, σ_{hoop} will be substituted in the following formulas for qR/A, where σ_{hoop} is given by (10.56).

We now need to add σ_N and σ_M in such a way as to produce the worst-case stress in the ring. The quantity

$$\frac{1}{\sin\alpha\left[f_1(\alpha)+12\left(\frac{R}{h}\right)^2 f_2(\alpha)+\frac{EA}{E_{eq}A_{stick}}\right]} \quad (10.79)$$

occurs in both stress formulas. We have tabulated $f_1(\alpha)$, $f_2(\alpha)$, and sinα times these in Table 10.1 for a range of α values.

TABLE 10.1

Tabulated Values for $f_1(\alpha)$ and $f_2(\alpha)$ and sinα Times These Quantities

# Sticks/Circle	α (°)	$f_1(\alpha)$	$f_2(\alpha)$	sinα $f_1(\alpha)$	sinα $f_2(\alpha)$
2	90	0.3927	0.07439	0.3927	0.07439
4	45	0.6427	0.006079	0.4545	0.004299
6	30	0.9566	0.001682	0.4783	0.0008410
8	22.5	1.2739	0.0006931	0.4875	0.0002652
10	18	1.5919	0.0003511	0.4919	0.0001085
12	15	1.9101	0.0002020	0.4944	0.00005228
18	10	2.8648	0.00005942	0.4975	0.00001032
24	7.5	3.8197	0.00002500	0.4986	0.00000326
30	6	4.7747	0.00001279	0.4991	0.00000134

It can be seen that for any choice of the other parameters, Equation 10.79 is positive. In (10.78), the magnitude of σ_M for fixed α is determined by the term $(\cos\varphi - \sin\alpha/\alpha)$, where φ can range from 0 to α. This achieves a maximum in absolute value at $\varphi = \alpha$. The stress can have either sign depending on whether it is on the inner radial or outer radial surface of the ring. In (10.77), the magnitude of σ_N achieves a maximum at $\varphi = \alpha$ for virtually any choice of the other parameters. It has a negative sign consistent with the compressive nature of the applied force. Thus, σ_N and σ_M should be added with each having a negative sign at $\varphi = \alpha$ to get the maximum stress. We obtain

$$\sigma_{max} = \sigma_{hoop} \left\{ \frac{\cos\alpha}{2\sin\alpha} \left[\frac{1}{f_1(\alpha) + 12\left(\dfrac{R}{h}\right)^2 f_2(\alpha) + \dfrac{EA}{E_{eq}A_{stick}}} - 1 \right] \right.$$

$$\left. + \frac{3R}{h}\left(\cos\alpha - \frac{\sin\alpha}{\alpha}\right)\left(\frac{1}{\sin\alpha}\right)\left[\frac{1}{f_1(\alpha) + 12\left(\dfrac{R}{h}\right)^2 f_2(\alpha) + \dfrac{EA}{E_{eq}A_{stick}}} \right] \right\} \qquad (10.80)$$

This stress is negative although it is usually quoted as a positive number. It occurs at the support.

We have analyzed a ring subjected to a hoop stress having radial supports. A coil is usually not a monolithic structure but consists of a number of cables radially distributed. The cables could consist of a single strand of conductor as in the case of magnet wire or be multistranded. The latter could also be bonded. The average hoop stress in the winding will be nearly the same in all the cables since the paper insulation tends to equalize it. We will examine this in more detail in a later section. The radial thickness, h in the formulas, should refer to a single cable. If it is a magnet wire, then its radial thickness should be used. If a multistranded transposed cable, then something less than its radial thickness should be used since this is not a homogeneous material. If unbonded, we use twice the thickness of an individual strand as its effective radial build. If bonded, we use 80% of its actual radial thickness as its effective radial build.

10.4 Radial Buckling Strength

Buckling occurs when a sufficiently high force causes a structure to deform its shape to the point where it becomes destabilized and may collapse. The accompanying stresses may in fact be small and well below the proof stress of the material. An example is a slender column subjected to an axial compressive force. At a certain value of the force, a slight lateral bulge in the column could precipitate a collapse. Another example, which we will discuss here, is that of a thin ring subjected to a uniform compressive radial pressure. A slight deformation in the circular shape of the ring could cause a collapse if the radial pressure is high enough. This critical radial pressure produces a hoop stress (see Section 10.3.6). It is called the critical hoop stress, which could be well below the proof stress of the material.

In general, the smaller the ratio of the radial build to the radius of the ring, the smaller the critical hoop stress. Thus, buckling is essentially an instability problem and is analyzed by assuming a small distortion in the shape of the system under study and determining under what conditions this leads to collapse.

We wish to examine the possible buckling of a winding subjected to an inward radial pressure. We will treat an individual cable of the winding as a closed ring, as was done in the last section, since the cables are not bonded to each other. In free unsupported radial buckling, it is assumed that there are no inner supports. Thus, the sticks spaced around the inside circumference of a winding are assumed to be absent. It is argued that there is sufficient looseness in this type of support that the onset of buckling occurs as if these supports were absent and, once started, the buckling process continues toward collapse or permanent deformation. It could also be argued that even though buckling may begin in the absence of supports, before it progresses very far, the supports are engaged and from then on it becomes a different type of buckling, called forced or constrained buckling. The key to the last argument is that even though free buckling has begun, the stresses in the material are quite low, resulting in no permanent deformation, and the process is halted before collapse can occur.

10.4.1 Free Unsupported Buckling

Timoshenko [Tim55] analyzes free buckling of a circular ring in some detail and quotes the results for forced buckling. The lowest-order shape distortion away from a circle is taken to be an ellipse. This is shown in Figure 10.12 along with the parameters used to describe the system. R is the radius of the ring, h its radial thickness, and q the uniform inward force per unit length along the ring's outer surface. The analysis assumes a thin ring so that R can be taken to be a mean radius. It also assumes that the ring is uniform in the direction along the ring's axis.

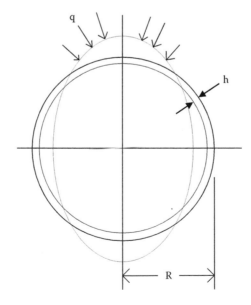

FIGURE 10.12
Buckling of a circular ring. The distorted elliptical shape, which is the lowest buckling mode, and the system parameters are shown.

For this lowest buckling mode, the critical load, that is, the critical force per unit length q_{crit}, which will precipitate buckling, is given by

$$q_{crit} = \frac{3EI}{R^3} \tag{10.81}$$

where
  E is Young's modulus of the hoop material
  I is the area moment of inertia, $I = th^3/12$, and t is the thickness of the ring in the direction of the ring's axis

We can convert q_{crit} to a critical pressure by dividing by t. Calling this pressure P_{crit}, it is given by

$$P_{crit} = \frac{3EI}{tR^3} \tag{10.82}$$

We can then use this in (10.55), substituting h for B, to obtain

$$\sigma_{crit} = \frac{3EI}{thR^2} = \frac{1}{4}E\left(\frac{h}{R}\right)^2 \tag{10.83}$$

This is the critical hoop stress. It depends geometrically only on the ratio of the radial build to the radius of the ring.

Since buckling occurs after the stress has built up in the ring to the critical value and increases incrementally beyond it, the appropriate modulus to use in (10.83) is the tangential modulus since this is associated with incremental changes. This argument for using the tangential modulus is based on an analogous argument for the buckling of slender columns together with supporting experimental evidence given in [Tim55]. We will assume it applies to thin rings as well.

The tangential modulus can be obtained graphically from the stress–strain curve for the material as illustrated in Figure 10.13a. However, for copper, which is the material of

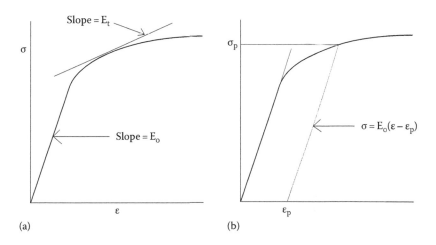

FIGURE 10.13
Stress–strain curve and derived quantities: (a) initial, E_o, and tangential, E_t, modulus and (b) proof stress illustrated.

interest here, the stress–strain curve can be parametrized for copper of different hardnesses according to [Tho79] by

$$\sigma = \frac{E_o \varepsilon}{\left[1 + k \left(\sigma / \sigma_o \right)^m \right]} \tag{10.84}$$

where $k = 3/7$, $m = 11.6$, and $E_o = 1.10 \times 10^5$ MPa. σ_o depends on the copper hardness. The tangential modulus obtained from this is

$$E_t = \frac{d\sigma}{d\varepsilon} = \frac{E_o}{\left[1 + \gamma \left(\sigma / \sigma_o \right)^m \right]} \tag{10.85}$$

where $\gamma = k(m + 1) = 5.4$. Substituting E_t from (10.85) for E in (10.83), we obtain a formula for self-consistently determining the critical stress:

$$\sigma_{crit}^{m+1} + \frac{\sigma_o^m}{\gamma} \sigma_{crit} - \frac{\sigma_o^m}{4\gamma} E_o \left(\frac{h}{R} \right)^2 = 0 \tag{10.86}$$

This can be solved by Newton–Raphson iteration.

The parameter σ_o is generally not provided by the wire or cable supplier. It could be obtained by fitting a supplied stress–strain curve. Alternatively and more simply, it can be obtained from the proof stress of the material, which is generally provided or specified. As one moves along a stress–strain curve and then removes the stress, the material does not move back toward zero stress along the same curve it followed when the stress increased but rather it follows a straight line parallel to the initial slope of the stress–strain curve as illustrated in Figure 10.13b. This leaves a residual strain in the material, labeled ε_p in the figure, corresponding to the stress σ_p, which was the highest stress achieved before it was removed. For $\varepsilon_p = 0.002$ (0.2%), σ_p is called the proof stress. (Some people use $\varepsilon_p = 0.001$ in this definition.)

Thus, the proof stress is determined from the intersection of the recoil line:

$$\sigma = E_o \left(\varepsilon - \varepsilon_p \right) \tag{10.87}$$

with the stress–strain curve given by (10.84). Solving (10.84) and (10.87) simultaneously, we find

$$\sigma_o = \left(\frac{k}{E_o \varepsilon_p} \right)^{1/m} \sigma_p^{(m+1)/m} \tag{10.88}$$

This permits us to find σ_o from a given proof stress σ_p corresponding to the appropriate ε_p. For a proof stress of 170 MPa at 0.2% strain ($\varepsilon_p = 0.002$), we obtain $\sigma_o = 154.5$ MPa for copper.

10.4.2 Constrained Buckling

When the axial supports (sticks) are engaged in the buckling process, we have forced or constrained buckling. Since there is some looseness in the supports because they do not rigidly clamp the winding cable, they can be regarded as a hinged type of support. Because this type of support allows a freer pivoting action, this type of buckling may also

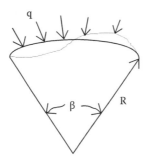

FIGURE 10.14
Buckling of a circular hinged arch.

be referred to as free supported buckling. In this case, the lowest-order constrained buckling mode is shown in Figure 10.14. The corresponding critical force/unit length, q_{crit}, is given by [Tim55]

$$q_{crit} = \frac{EI}{R^3}\left[\left(\frac{2\pi}{\beta}\right)^2 - 1\right] \tag{10.89}$$

where β is the angle between the outer supports. This force/unit length produces a critical hoop stress of

$$\sigma_{crit} = \frac{E_t}{12}\left(\frac{h}{R}\right)^2\left[\left(\frac{2\pi}{\beta}\right)^2 - 1\right] \tag{10.90}$$

where h is the radial build of the arch. This will exceed the free buckling critical stress, Equation 10.86, when $\beta < \pi$. $\beta = \pi$ corresponds to 2 end supports. Thus, for most cases where $\beta \ll \pi$, the constrained buckling stress will be much larger than the free buckling stress. Both buckling types depend on the radial build of the ring or arch. We will adopt the same procedure for determining the effective radial build of a cable, as was done for the radial bending stress determination at the end of Section 10.3.7.

Since a loose or hinged support can also be imagined as existing at the center point of the arch in Figure 10.14, β should be taken as twice the angle between supports (sticks). Letting n be the number of radial supports around the winding circumference so that $\beta = 4\pi/n$, Equation 10.90 can be written as

$$\sigma_{crit} = \frac{E_t}{12}\left(\frac{h}{R}\right)^2\left[\frac{n^2}{4} - 1\right] \tag{10.91}$$

Substituting the tangential modulus from (10.85) into this formula, we obtain a formula for self-consistently determining the critical stress:

$$\sigma_{crit}^{m+1} + \frac{\sigma_o^m}{\gamma}\sigma_{crit} - \frac{\sigma_o^m}{12\gamma}E_o\left(\frac{h}{R}\right)^2\left[\frac{n^2}{4} - 1\right] = 0 \tag{10.92}$$

This is similar to the procedure used to get (10.86). This equation can be solved by Newton–Raphson iteration.

Arched buckling with this value of β appears to provide a more realistic value of buckling strength in practice than totally free unsupported buckling. There is increasing experimental evidence for this [Sar00]. One such experiment will be described in the next section.

10.4.3 Experiment to Determine Buckling Strength

This section is largely based on a talk presented at a Doble Conference [Del01]. A small (5 MVA) 3-phase, 2-winding/phase, 3-leg core transformer was designed with the inner LV windings on each leg wound with different conductors. It had terminal line-to-line voltages of 13.2 kV for the HV and 3.6 kV for the LV and was Y–Y connected. The leakage impedance was 5.8% at 5 MVA. There were no taps so the windings were uniform except for crossovers. The outer HV winding was the same for all the legs. Details concerning the HV winding and the two inner windings, which failed by buckling, are given in Table 10.2. The LV windings, which failed by buckling, were on the outer 2 legs of the transformer, labeled LV1 and LV2.

The unit was tested at the Powertech Labs in Surrey, British Columbia, Canada. The phases (legs) were tested separately and sequentially at each power level, starting from 80% of the design voltage and increasing in steps of ~10% until failure occurred. The voltage was applied to the HV winding, with the LV winding shorted to ground before the unit was energized. The voltage application was timed to produce maximum offset in the current. A typical voltage–current readout is shown in Figure 10.15. The total duration of the short circuit event was ~0.25 s. Two shots/leg were performed before going to the next voltage level. A visual inspection of the windings was conducted after each shot via a Plexiglas window in the transformer tank. In addition, the impedance was calculated after each test from the ratio of voltage to current.

A frequency response analysis (FRA) measurement was performed on each leg after the two shot sequences or after a single shot if an impedance change occurred. The FRA measurement was performed by applying a sharp impulse to the HV

TABLE 10.2

Geometric and Other Parameters for the Windings

Winding	HV	LV1	LV2
Inner diameter	444.5	343	343
Radial build	35	25.4	25.4
Height	1041	1067	1067
# Disks	72	76	76
# Electrical turns/disk	4	1	1
Total turns	288	76	76
Wire type	Magnet	Magnet	Magnet
Copper strand thickness	3.56	3.68	2.67
Copper strand width	10.9	10.2	10.2
Paper thickness (2 sided)	0.762	0.305	0.305
# Strands/disk	8	6	8
# Radial supports (sticks)	12	12	12
Proof stress (MPa)	207	194	348

The dimensions are in mm.

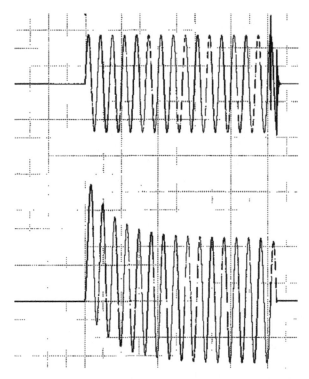

FIGURE 10.15
Short circuit HV applied phase voltage (top) and resulting current (bottom) for a typical test.

winding and measuring the current in the LV winding. The results were displayed as a transfer admittance versus frequency. The transferred admittance plot was then compared with that obtained initially before the tests began. Figure 10.16 shows such a comparison when the leg was considered to have passed the short circuit event, while Figure 10.17 shows a comparison when the unit failed. No further tests were

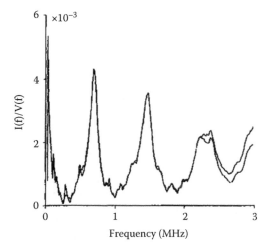

FIGURE 10.16
Comparison of the admittance versus frequency plot measured after a shot with the initial admittance plot for a passing case.

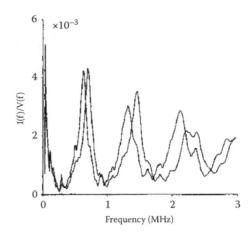

FIGURE 10.17
Comparison of the admittance versus frequency plot measured after a shot with the initial admittance plot for a failed case.

performed after a failure based on the FRA occurred. An impedance change of 2% and 13% also occurred in the two legs when an FRA failure was indicated. Prior to failure, the impedance changes were <1% and were possibly due to measurement uncertainties.

The middle leg failed in an axial mode. This failure, in addition to a large impedance and FRA change, was visually obvious since the outer winding of this leg collapsed. The failure mode of the two outer legs was only evident from the impedance and FRA changes, since the outer windings appeared normal after the failure. We shipped the unit back to our plant and disassembled it in order to find the cause of the failure. As Figures 10.18 and 10.19 show, the failure mode was radial buckling of the inner winding. It is also clear, from the position of the sticks relative to the winding distortion, that the sticks were actively involved in the failure. We can see in both figures that the winding bulges outward between two adjacent sticks and inward between adjacent sticks on opposite sides of the bulge.

FIGURE 10.18
View of the inner winding of LV1, which buckled over about the middle third of its axial height.

FIGURE 10.19
View of the inner winding of LV2, which buckled in two different areas that were both radially and axially displaced from each other. One area is facing the viewer and the other can be seen in silhouette on the upper right of the winding. Each buckling area covered about a third of the winding's axial height.

It can also be seen that all the turns participate in the buckling, that is, the entire radial build appears to buckle as a unit. The torn paper in the figures occurred while disassembling the windings.

Both inner windings failed in buckling at nearly the same current, which had a peak value at maximum offset in the HV winding of 13,300 amps. Using this current, together with the LV current required for amp-turn balance, in our computer program that calculates the magnetic flux, forces, and stresses, we obtain the maximum compressive hoop stresses in the LV windings given in Table 10.3. A calculation of how the stress is distributed across the radial build of the winding, taking the paper layers into account, produced a variation of ±2%, with the higher stress on the inner conductor. This calculation is similar to that in [Tho79].

In applying the buckling formulas, h was chosen to be the radial build of a bare copper strand and R was chosen as the average winding radius. The tangential modulus, E_t, was obtained by the method proposed here. It is given in the table as a ratio with respect to the initial modulus E_o, where $E_o = 1.10 \times 10^5$ MPa for copper.

As can be seen in Table 10.3, the free unsupported buckling stress is far below the actual hoop stress at failure. This buckling mode also has an elliptical shape, which is not observed.

TABLE 10.3

Measured and Calculated Buckling Hoop Stresses in the Inner Windings

Winding and Modulus Ratio	LV1	E_t/E_o	LV2	E_t/E_o
Max hoop stress at test failure	88.4		90.4	
Buckling stress				
Free unsupported	11.0	1.0	5.8	1.0
Constrained ($\beta = 360°/6$)	121.2	0.942	67.5	1.0
Constrained ($\beta = 360°/12$)	164.8	0.313	244.1	0.885

The stresses are in MPa.

The constrained buckling mode stress with β = 360°/12, the angle between the radial supports, is much too high compared with the actual failure stress. The buckling shape also does not agree with the observed shape. The constrained buckling stress with β = 360°/6, twice the angle between the radial supports, is fairly close to the actual failure hoop stress. In addition, the buckling shape for this mode agrees with observation.

10.5 Stress Distribution in a Composite Wire–Paper Winding Section

The hoop stress previously calculated for a winding section or disk (Section 10.3.6) was an average over the disk. In reality, for the innermost winding, the axial magnetic field varies from nearly zero on the inside of the winding to close to its maximum value at the outer radius of the winding. Since the current density is uniform, the force density also varies in the same fashion as the magnetic field. Thus, we might expect higher hoop stresses in the outermost turns as compared with the inner turns. However, because of the layered structure with paper insulation between turns, the stresses tend to be shared more equally by all the turns. This effect will be examined here in order to determine the extent of the stress nonuniformity so that, if necessary, corrective action can be taken.

We analyze an ideal ring geometry as shown in Figure 10.20. In the figure, r_{ci} denotes the inner radius of the ith conductor layer and r_{pi} the inner radius of the ith paper layer,

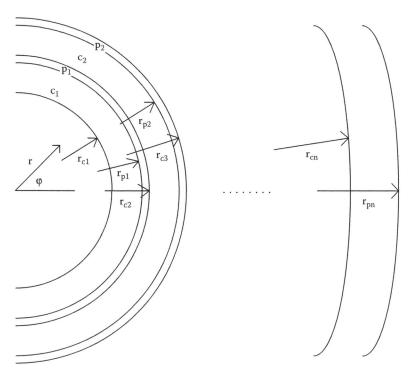

FIGURE 10.20
Conductor–paper layered ring winding section. ci refers to conductor i and pi to paper layer i.

where $i = 1, \ldots, n$ for the conductors and $i = 1, \ldots, n - 1$ for the paper layers. Because of the assumed close contact, the outer radius of the ith conductor layer equals the inner radius of the ith paper layer and the outer radius of the ith paper layer equals the inner radius of the $i + 1$th conductor layer. We do not need to include the innermost or outermost paper layers since they are essentially stress-free.

We assume that the radial force density varies linearly from the innermost to outermost conductor layers. Thus,

$$f_{ci} = \frac{f_o i}{n} \tag{10.93}$$

where f_o is the maximum force density at the outermost conductor. We wish to express this in terms of the average force density, f_{ave}. We have

$$f_{ave} = \frac{1}{n} \sum_{i=1}^{n} f_{ci} = \frac{f_o}{n^2} \sum_{i=1}^{n} i = \frac{f_o (n+1)}{2n} \tag{10.94}$$

Solving for f_o and substituting into (10.93), we get

$$f_{ci} = \frac{2f_{ave}}{n+1} i \tag{10.95}$$

It is even more convenient to relate f_{ci} to the average hoop stress in the winding, resulting from these radial forces. We related this stress to the radial pressure in (10.55). But the average force density is just the average radial pressure divided by the winding radial build. Therefore, we find, from (10.55),

$$\sigma_{ave,hoop} = \frac{P_r R_m}{B} = f_{ave} R_m \tag{10.96}$$

where R_m is the mean radius of the winding. Thus, Equation 10.95 can be written as

$$f_{ci} = \frac{2\sigma_{ave,hoop}}{R_m (n+1)} i \tag{10.97}$$

We assume that the winding section can be analyzed as a 2D stress distribution problem, that is, stress variations in the axial direction are assumed to be small. The governing equation for this type of problem in polar coordinates when only radial forces are acting and the geometry is cylindrically symmetric is [Tim70]

$$\frac{\partial \sigma_r}{\partial r} + \frac{\sigma_r - \sigma_\varphi}{r} + f_r = 0 \tag{10.98}$$

where
  σ_r is the radial stress
  σ_φ is the azimuthal stress
  f_r is the radial force density

The stresses are related to the strains for the 2D plane stress case by

$$\sigma_r = \frac{E}{1-v^2}\left(\varepsilon_r + v\varepsilon_\varphi\right), \quad \sigma_\varphi = \frac{E}{1-v^2}\left(\varepsilon_\varphi + v\varepsilon_r\right) \tag{10.99}$$

where
  E is Young's modulus
  v is Poisson's ratio ($v = 0.25$ for most materials)

The radial and azimuthal strains, ε_r and ε_φ, are related to the radial displacement u, in the cylindrically symmetric case, by

$$\varepsilon_r = \frac{du}{dr}, \quad \varepsilon_\varphi = \frac{u}{r} \tag{10.100}$$

Substituting (10.99) and (10.100) into (10.98), we obtain

$$r^2\frac{d^2u}{dr^2} + r\frac{du}{dr} - u + \frac{f_r\left(1-v^2\right)}{E}r^2 = 0 \tag{10.101}$$

with the general solution

$$u = Ar + \frac{B}{r} + Kr^2 \tag{10.102}$$

where A and B are constants to be determined by the boundary conditions and

$$K = -\frac{f_r\left(1-v^2\right)}{3E} \tag{10.103}$$

For our problem, f_r is negative (radially inward) so that K is positive. In the paper layers, K = 0 since there is no force density there. Using (10.102) for u in (10.100), we obtain for (10.99)

$$\sigma_r = \frac{E}{1-v^2}\left[A(1+v) - \frac{B}{r^2}(1-v) + Kr(2+v)\right]$$
$$\sigma_\varphi = \frac{E}{1-v^2}\left[A(1+v) + \frac{B}{r^2}(1-v) + Kr(1+2v)\right] \tag{10.104}$$

  The solution (10.102) through (10.104) applies to each layer of conductor or paper. We therefore need to introduce labels to distinguish the layers. Let A_{ci}, B_{ci} apply to conductor layer i and A_{pi}, B_{pi} apply to paper layer i. Let ci and pi also label the displacements, u, and the stresses σ_r, σ_φ for the corresponding layer. At the conductor–paper interface, the displacements must match:

$$u_{ci}\left(r_{pi}\right) = u_{pi}\left(r_{pi}\right), \quad i = 1,\ldots,n-1$$
$$u_{ci}\left(r_{ci}\right) = u_{p(i-1)}\left(r_{ci}\right), \quad i = 2,\ldots,n \tag{10.105}$$

There are 2(n − 1) such equations. (See Figure 10.20 for the labeling.) Also, at the interface, the radial stresses must match:

$$\sigma_{r,ci}(r_{pi}) = \sigma_{r,pi}(r_{pi}), \quad i = 1, \ldots, n-1$$
$$\sigma_{r,ci}(r_{ci}) = \sigma_{r,p(i-1)}(r_{ci}), \quad i = 2, \ldots, n \tag{10.106}$$

There are also 2(n − 1) such equations. We also have, at the innermost and outermost radii,

$$\sigma_{r,c1}(r_{c1}) = 0, \quad \sigma_{r,cn}(r_{pn}) = 0 \tag{10.107}$$

This provides two more equations. Thus, altogether, we have 4n − 2 equations. There are two unknowns, A_{ci} and B_{ci}, associated with each conductor layer for a total of 2n unknowns and two unknowns, A_{pi} and B_{pi}, associated with each paper layer for a total of 2(n − 1) unknowns since there are only n − 1 paper layers. Thus, there are altogether 4n − 2 unknowns to solve for and this matches the number of equations. In (10.104), we must use the appropriate material constants for the conductor or paper layer, that is, $E = E_c$ or E_p is the conductor's or paper's Young's modulus and $\nu = \nu_c$ or ν_p for the conductor's or paper's Poisson's ratio. In addition, K needs to be labeled according to the layer, that is, K_{ci}, and in the case of a paper layer $K_{pi} = 0$.

The resulting set of 4n − 2 equations in 4n − 2 unknowns is a linear system and can be solved by standard methods. Once the solution is obtained, Equation 10.104 can be used to find the stresses. The σ_φ can then be compared with the average hoop stress to see how much deviation there is from a uniform distribution. We show a sample calculation in Table 10.4. The average stresses are calculated for the conductor and paper and then the stresses in the layers are expressed as multipliers of this average stress. These multipliers are averages for the layers since the stresses vary across a layer. The input is the geometric data and the average hoop stress in the conductor, which is obtained from the radial pressure via (10.96). The calculated average hoop stress in the conductor does not quite agree with the input, probably because of numerical approximations in the averaging method. The hoop stress in the conductors varies by about 20% from the inner to the outer layers. The other stresses are small in comparison although they show considerable variation across the winding. The stresses are shown as positive, even though they are compressive and therefore negative.

10.6 Additional Mechanical Considerations

During a short circuit, the leads or busbars are subjected to an increased force due to the higher fault current they carry together with its interaction with the higher leakage flux from the main windings and from nearby leads. These forces will depend on the detailed positioning of the leads with respect to the main windings and with respect to each other. They will therefore vary considerably from design to design. The leads must be braced properly so that they do not deform or move much during a fault. The leakage flux in the vicinity of the lead can be obtained from a flux mapping calculation. The flux produced by neighboring leads can be determined from the Biot–Savart law. From these, the forces on the leads can be determined and the adequacy of the bracing checked. In general,

TABLE 10.4

Sample Stress Distribution in a Composite Conductor–Paper Disk

Input			
Wdg inner radius	381 mm	Conductor layer radial thick	6.35 mm
Wdg outer radius	457 mm	Paper layer radial thick	1.27 mm
# Radial turns	10	Average hoop stress	124.1 MPa
$E_c = 1.1 \times 10^5$ MPa		$E_p = 2.07 \times 10^2$ MPa	$\nu_c = \nu_p = 0.25$

Output	
Average hoop stress in conductor layers	127.6 MPa
Average hoop stress in paper layers	1.08 MPa
Average radial stress in conductor layers	3.03 MPa
Average radial stress in paper layers	3.38 MPa

Average Stress Multipliers

Layer	Conductor		Paper	
	Hoop	Radial	Hoop	Radial
1	0.90	0.28	0.60	0.51
2	0.91	0.78	0.89	0.89
3	0.93	1.13	1.10	1.15
4	0.96	1.36	1.23	1.30
5	0.99	1.47	1.27	1.35
6	1.03	1.47	1.24	1.30
7	1.06	1.35	1.12	1.14
8	1.08	1.12	0.92	0.87
9	1.09	0.76	0.63	0.50
10	1.08	0.28		

the bracing should contain sufficient margin based on past experience so that the above rather laborious analysis will only be necessary for unusual or novel designs.

We have neglected gravitational forces in the preceding sections except for the effect of the core and coil weight on the tie bar stress during lifting. Gravitational forces will affect the compressive force on the key spacers and on the downward end thrust, which acts on the bottom pressure ring. Another force that was neglected is the compressive force, which is initially placed on the coils by pretensioning the tie bars. This force adds to the compressive force on the key spacers and to the top and bottom thrust on the pressure ring as well as adding some initial tension to the tie bars. Since the compressive forces on the key spacers are involved in conductor tilting, this design criterion will also be affected. These additional forces are present during normal operation and will add to the fault forces when a fault occurs.

We have treated the axial and radial stress calculations independently, whereas in reality axial and radial forces are applied simultaneously, resulting in a biaxial stress condition. Results of such a combined analysis for a circular arched wire segment between supports have been reported and show good agreement with experiment [Ste72]. In this type of analysis, the worst-case stress condition is not necessarily associated with the largest axial or radial forces since these would not usually occur at the same position along the winding. It is rather due to some combination of the two, which would have to be examined at each position along the winding at which the forces are calculated.

As long as the materials remain linear, that is, obey Hooke's law, and the displacements are small, the axial and radial analyses can be performed separately. The resulting stresses can then be combined appropriately to get the overall stress state. Various criteria for failure can then be applied to this overall stress state. Our strategy of looking at the worst-case stresses produced by the axial and radial forces separately and applying a failure criterion to each is probably a good approximation to that obtained from a combined analysis, especially since the worst-case axial and radial forces typically occur at different positions along the winding. The radial forces are produced by axial flux, which is high in the middle of the winding, whereas the axial forces are produced by radial flux, which is high at the ends of the winding.

Dynamical effects have been studied by some authors, particularly the axial response of a winding to a suddenly applied short circuit current [Bos72], [Ste72]. They found that the level of prestress is important. When the prestress was low, ~10% of normal, the winding literally bounced against the upper support, resulting in a much higher than expected force. The enhancement factor over the expected nondynamical maximum force was about 4. However, when the prestress was normal or above, there was no enhancement over the expected maximum force. The prestress also affects the natural frequency of the winding to oscillations in the axial direction. Higher prestress tends to shift this frequency toward higher values, away from the frequencies in the short circuit forces. Hence, little dynamical enhancement is expected under these conditions. Thus, provided sufficient prestress is applied to clamp the windings in the axial direction, there should be little or no enhancement of the end thrust over the expected value based on the maximum fault currents. However, as the unit ages, the prestress could decrease. With modern precompressed pressboard key spacers, this effect should be small. Even so, we allow an enhancement factor of 1.7 in design.

An area of some uncertainty is how to treat transposed cable, with or without bonding, in the stress calculations. It is not exactly a solid homogeneous material, yet it is not simply a loose collection of individual strands. We believe we have taken a conservative approach in our calculations. However, this is one area, among others, where further experimental work would be useful.

11

Electric Field Calculations

11.1 Simple Geometries

It is often possible to obtain a good estimation of the electric field in a certain region of a transformer by idealizing the geometry to such an extent that the field can be calculated analytically. This has the advantage of exhibiting the field as a function of several parameters so that the effect of changing these and how this affects the field can be appreciated. Such insight is often worth the price of the slight inaccuracy that may exist in the numerical value of the field.

11.1.1 Planar Geometry

As a first example, we consider a layered insulation structure having a planar geometry as shown in Figure 11.1. This could represent the major insulation structure between two cylindrical windings having large radii compared with the gap separation. We are further approximating the layered structure as a smooth surface so that the resulting field calculation would be representative of the field away from any corner effects. We treat such a corner field in the next section.

We use one of Maxwell's equations in integral form to solve this:

$$\oint_S \mathbf{D} \cdot \mathbf{dS} = q \tag{11.1}$$

where
  $\mathbf{D}$ is the displacement vector
  $\mathbf{dS}$ is a vectorial surface area with an outward normal
  q is the charge enclosed by the closed surface S

Because of the assumed ideal planar geometry, the surface charge density on the electrode at potential V is uniform and is designated σ in the figure. An opposite surface charge of $-\sigma$ exists on the ground electrode. Note that both electrode potentials could be raised by an equal amount without changing the results. Only the potential difference matters. We will also assume that the materials have linear electrical characteristics so that

$$\mathbf{D} = \varepsilon \mathbf{E} \tag{11.2}$$

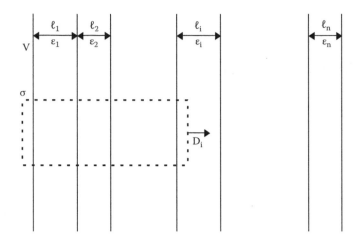

FIGURE 11.1
Geometry of a planar layered insulation structure.

holds within each material where ε is the permittivity, which can differ within the various layers as shown in the figure, and **E** is the electric field vector.

Because of the planar geometry, the **D** and **E** fields are directed perpendicular to the planes of the electrodes and layers are uniform along these planes. If we take our closed surface as the dotted rectangle shown in the figure, which has some depth into the figure, so that the two vertical and horizontal sides represent surfaces, then the only contribution to the integral in (11.1), which is nonzero, is the part over the right vertical surface. This is because the dot product of the displacement vector with the surface normal is zero on the top and bottom dotted surfaces and zero on the left vertical dotted surface because it is inside the metallic electrode. Therefore, the displacement vector has the uniform value labeled D_i in the figure on the right vertical surface of the dotted rectangle. Thus, for this closed surface, Equation 11.1 becomes

$$D_i S = q = \sigma S \quad \Rightarrow \quad D_i = \sigma \tag{11.3}$$

Using (11.2) applied to layer i, Equation 11.3 becomes

$$E_i = \frac{\sigma}{\varepsilon_i} \tag{11.4}$$

In terms of the potential V, we can write by definition

$$V = -\int \mathbf{E} \cdot \mathbf{d}\ell \tag{11.5}$$

where the line integral starts at the 0 potential electrode and ends on the V potential electrode. In terms of the E fields in the different materials and their thicknesses ℓ_i, Equation 11.5 becomes

$$V = E_1 \ell_1 + E_2 \ell_2 + \cdots = \sum_{j=1}^{n} E_j \ell_j \tag{11.6}$$

Using (11.4), this can be written as

$$V = \sigma \sum_{j=1}^{n} \frac{\ell_j}{\varepsilon_j} \qquad (11.7)$$

Solving for σ and substituting into (11.4), we get

$$E_i = \frac{V}{\varepsilon_i \left(\sum_{j=1}^{n} \ell_j / \varepsilon_j \right)} \qquad (11.8)$$

Letting ℓ be the total distance between the electrodes so that $\ell = \ell_1 + \ell_2 + \cdots + \ell_n$ and defining the fractional lengths, $f_i = \ell_i / \ell$, Equation 11.8 can be expressed as

$$E_i = \frac{(V/\ell)}{\varepsilon_i \left(\sum_{j=1}^{n} f_j / \varepsilon_j \right)} = \frac{E_o}{\varepsilon_i \left(\sum_{j=1}^{n} f_j / \varepsilon_j \right)} \qquad (11.9)$$

where E_o is the electric field that would exist between the electrodes if there was only one layer of uniform material between them. E_i / E_o is called an enhancement factor for the material of permittivity ε_i. It is given by

$$\frac{E_i}{E_o} = \frac{1}{\varepsilon_i \left(\sum_{j=1}^{n} f_j / \varepsilon_j \right)} \qquad (11.10)$$

It is the value of the field when there are multiple dielectrics between the conducting planes relative to the field when there is a single dielectric between them. Notice that only the ratios of permittivities are involved in the last few formulas so that only relative permittivities are required for calculation purposes.

Let's apply these results to an oil–pressboard insulation system. Even if there are many layers of pressboard used to subdivide the oil gap, only the total fractional thickness, f_{press}, matters in the calculation since they all have the same permittivity. Similarly, the subdivided oil gap's total fractional thickness, $f_{oil} = 1 - f_{press}$, is all that is needed to perform the calculation. For this situation, Equation 11.9 becomes

$$E_{oil} = \frac{E_o}{\varepsilon_{oil} \left(\dfrac{f_{oil}}{\varepsilon_{oil}} + \dfrac{f_{press}}{\varepsilon_{press}} \right)}, \quad E_{press} = \frac{E_o}{\varepsilon_{press} \left(\dfrac{f_{oil}}{\varepsilon_{oil}} + \dfrac{f_{press}}{\varepsilon_{press}} \right)} \qquad (11.11)$$

Since the relative permittivities of pressboard and oil are $\varepsilon_{press} \cong 4.4$ and $\varepsilon_{oil} \cong 2.2$, the electric field in the oil is about twice as high as the electric field in the pressboard for a given oil–pressboard combination. Thus, the oil's electric field is usually the most important to know for purposes of breakdown estimation. This is further reinforced by the fact that oil breaks down at a significantly lower field than pressboard. We plot E_{oil}/E_o versus f_{press} in Figure 11.2. We see from the figure that for a given oil gap, the lowest field results when

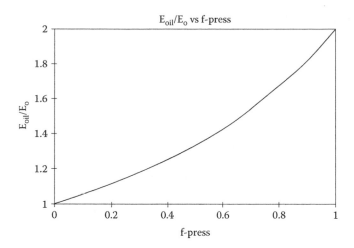

FIGURE 11.2
Relative electric field in the oil in a planar oil gap as a function of the fractional amount of pressboard. E_{oil}/E_o is the enhancement factor for the oil.

there is no pressboard. As more pressboard displaces the oil, the field in the oil increases, approaching a value of twice its all oil value when the gap is nearly filled with pressboard. When the gap is totally filled with pressboard, there is no enhancement factor. On the other hand, there is no oil in the gap in this case.

11.1.2 Cylindrical Geometry

A somewhat better estimate can be made of the electric field in the major insulation structure between two coils if we consider approximating the geometry as an ideal cylindrical geometry. This geometry is also useful for approximating the field around a long cable of circular cross section. We consider the general case of a multilayered concentric cylindrical insulation structure as shown in Figure 11.3. The innermost cylinder is at potential V and the outermost cylinder at zero potential, although it is only their potential difference that matters.

From symmetry, we see that the **D** and **E** fields are directed radially. We assume there is a surface charge per unit length λ along the inner cylinder (into the page in Figure 11.3). We apply (11.1) to the dashed line cylindrical surface in Figure 11.3, which is assumed to extend a distance L along the axis with disklike surfaces on either end. The only contribution to the integral is along the dotted cylindrical surface of the cylinder since the dot product is zero on end cap surfaces at either end. We find

$$2\pi r L D_i = \lambda L \quad \Rightarrow \quad D_i = \frac{\lambda}{2\pi r} \tag{11.12}$$

Assuming the linear relation (11.2), for layer i we get

$$E_i = \frac{\lambda}{2\pi \varepsilon_i r} \tag{11.13}$$

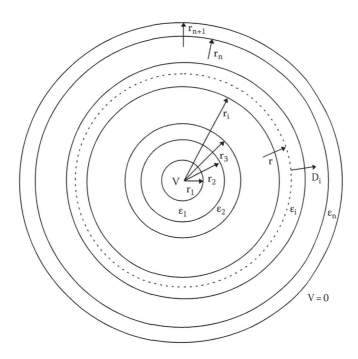

FIGURE 11.3
Ideal layered cylindrical insulation structure. This same drawing can be reinterpreted to refer to a spherical geometry.

Using the definition (11.5) and starting the line integral from the outer zero potential electrode and ending on the innermost cylinder, which is the surface of the conductor, we have, using (11.13),

$$V = -\frac{\lambda}{2\pi}\left[\frac{1}{\varepsilon_n}\int_{r_{n+1}}^{r_n}\frac{dr}{r} + \frac{1}{\varepsilon_{n-1}}\int_{r_n}^{r_{n-1}}\frac{dr}{r} + \cdots + \frac{1}{\varepsilon_1}\int_{r_2}^{r_1}\frac{dr}{r} \right]$$

$$= \frac{\lambda}{2\pi}\sum_{j=1}^{n}\frac{1}{\varepsilon_j}\ln\left(\frac{r_{j+1}}{r_j}\right) \tag{11.14}$$

Solving for λ and substituting into 11.13, we get

$$E_i = \frac{V}{\varepsilon_i r\left[\sum_{j=1}^{n}\frac{1}{\varepsilon_j}\ln\left(\frac{r_{j+1}}{r_j}\right)\right]} \tag{11.15}$$

For any given layer, the maximum field, $E_{i,max}$, occurs at its inner radius, so we have

$$E_{i,max} = \frac{V}{\varepsilon_i r_i\left[\sum_{j=1}^{n}\frac{1}{\varepsilon_j}\ln\left(\frac{r_{j+1}}{r_j}\right)\right]} \tag{11.16}$$

We see from the last two equations that, for a given layer, the electric field is inversely proportional to the permittivity of that layer. Thus, an oil layer at a given position will see about twice the electric field of a pressboard layer at the same position. Since the maximum field in a layer is also inversely proportional to the radius of the layer, we see that this field can be reduced by increasing the layer's radius. Thus, for a cable surrounded by solid insulation, such as paper, and immersed in oil, the critical field will probably occur in the oil at the outer surface of the paper. To reduce this, one could increase its radius by adding more paper or start with a larger radius cylindrical conductor to begin with.

If there were a single dielectric between the inner cylinder of radius r_1 and outer cylinder of radius r_{n+1}, then (11.15) would become

$$E_o(r) = \frac{V}{r \ln\left(\dfrac{r_{n+1}}{r_1}\right)} \tag{11.17}$$

Substituting V from (11.17) into (11.15), we get

$$E_i(r) = \frac{\ln\left(\dfrac{r_{n+1}}{r_1}\right)}{\varepsilon_i \left[\displaystyle\sum_{j=1}^{n} \frac{1}{\varepsilon_j} \ln\left(\dfrac{r_{j+1}}{r_j}\right)\right]} E_o(r) \tag{11.18}$$

The enhancement factor for the material in layer i with permittivity ε_i for this geometry is given by

$$\frac{E_i(r)}{E_o(r)} = \frac{\ln\left(\dfrac{r_{n+1}}{r_1}\right)}{\varepsilon_i \left[\displaystyle\sum_{j=1}^{n} \frac{1}{\varepsilon_j} \ln\left(\dfrac{r_{j+1}}{r_j}\right)\right]} \tag{11.19}$$

11.1.3 Spherical Geometry

Another geometry of some interest is the spherical geometry. The general case of a multi-layered spherical insulation structure is shown in Figure 11.3 by interpreting it as a cross section through a spherical system of insulators. In this case, the D and E fields are again directed radially by symmetry and the charge on the inner conductor at potential V is taken as q. The dashed circle in the figure now defines a spherical surface and (11.1) applied to this results in

$$E_i = \frac{q}{4\pi\varepsilon_i r^2} \tag{11.20}$$

From (11.5) and (11.20), we obtain

$$V = -\frac{q}{4\pi}\left[\frac{1}{\varepsilon_n}\int_{r_{n+1}}^{r_n}\frac{dr}{r^2} + \frac{1}{\varepsilon_{n-1}}\int_{r_n}^{r_{n-1}}\frac{dr}{r^2} + \cdots + \frac{1}{\varepsilon_1}\int_{r_2}^{r_1}\frac{dr}{r^2}\right]$$

$$= \frac{q}{4\pi}\sum_{j=1}^{n}\frac{1}{\varepsilon_j}\left(\frac{1}{r_j} - \frac{1}{r_{j+1}}\right) \tag{11.21}$$

Solving for q and substituting into (11.20), we obtain

$$E_i = \frac{V}{\varepsilon_i r^2 \left[\sum_{j=1}^{n} \frac{1}{\varepsilon_j r_j} \left(1 - \frac{r_j}{r_{j+1}} \right) \right]} \tag{11.22}$$

The maximum field in layer i, $E_{i,max}$, occurs at the inner radius and is given by

$$E_{i,max} = \frac{V}{\varepsilon_i r_i^2 \left[\sum_{j=1}^{n} \frac{1}{\varepsilon_j r_j} \left(1 - \frac{r_j}{r_{j+1}} \right) \right]} \tag{11.23}$$

This expression can be used to approximate the field near a sharp bend in a cable immersed in oil with the various radii defined appropriately. We see that as the radius of the layer increases, the field in it decreases. Thus, the field in the oil can be reduced by adding more insulation or decreasing the sharpness of the bend.

When there is a single dielectric between the inner and outer electrodes, Equation 11.22 becomes

$$E_o(r) = \frac{V}{r^2 \frac{1}{r_1} \left(1 - \frac{r_1}{r_{n+1}} \right)} \tag{11.24}$$

Substituting for V from (11.24) into (11.22), we obtain

$$E_i(r) = \frac{\frac{1}{r_1} \left(1 - \frac{r_1}{r_{n+1}} \right)}{\varepsilon_i \left[\sum_{j=1}^{n} \frac{1}{\varepsilon_j r_j} \left(1 - \frac{r_j}{r_{j+1}} \right) \right]} E_o(r) \tag{11.25}$$

Therefore, the enhancement factor for material in layer i for this geometry is given by

$$\frac{E_i(r)}{E_o(r)} = \frac{\frac{1}{r_1} \left(1 - \frac{r_1}{r_{n+1}} \right)}{\varepsilon_i \left[\sum_{j=1}^{n} \frac{1}{\varepsilon_j r_j} \left(1 - \frac{r_j}{r_{j+1}} \right) \right]} \tag{11.26}$$

11.1.4 Cylinder–Plane Geometry

The cylinder–plane geometry is a reasonable approximation to a commonly occurring configuration in transformers where a cable or lead runs parallel to the tank wall. The geometry and its parameterization are shown in Figure 11.4. The cylinder and plane extend infinitely far in the direction perpendicular to the page. The cylinder radius is R and the distance of its center from the plane is h. The cylinder is at a potential V_o while the plane is at 0 potential.

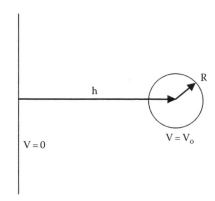

FIGURE 11.4
Cylinder–plane geometry and parameterization.

The potential and electric field distribution for this problem can be obtained by the method of images [Pug62], [Kul04]. Two equal and opposite line charges can be placed symmetrically about the zero potential plane as shown in Figure 11.5. Their charge per unit length is $\pm\lambda$ and their distance from the plane is s. The potential and electric field are desired at a point P shown in the figure, which is a distance r_1 from the negative charge and r_2 from the positive charge. The negative charge is behind the plane and the positive charge will be found to be located inside the charged cylinder. Therefore, these image charges do not occur within the solution space of interest, which is the space outside the cylinder and to the right of the zero potential plane. Therefore, we have from Maxwell's equations in this region

$$\nabla \cdot \mathbf{E} = 0, \quad \mathbf{E} = -\nabla V \quad \Rightarrow \quad \nabla^2 V = 0 \tag{11.27}$$

The last equation in (11.27) is Laplace's equation and is valid in the space of interest here. The solution of this equation, which satisfies the boundary conditions, is unique.

The electric field due to a line charge is

$$\mathbf{E} = \frac{\lambda}{2\pi\varepsilon_0 r} \mathbf{a}_r \tag{11.28}$$

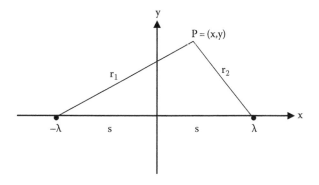

FIGURE 11.5
Image line charges for the cylinder–plane geometry. s is the positive distance of either charge from x = 0 along the x-axis.

where $\mathbf{a}_r$ is the outward unit vector from the charge. The potential relative to a distant point, say, at $r = C$ is

$$V = -\int_C^r \mathbf{E} \cdot \mathbf{dr} = -\frac{\lambda}{2\pi\varepsilon_o} \int_C^r \frac{dr'}{r'} = \frac{\lambda}{2\pi\varepsilon_o} \left[\ln(C) - \ln(r) \right] \tag{11.29}$$

Thus, the potential due to the equal and opposite line charges at distances r_1 and r_2 from the charges as shown in Figure 11.5 is given by

$$V = \frac{\lambda}{2\pi\varepsilon_o} \left[\ln(r_1) - \ln(r_2) \right] = \frac{\lambda}{2\pi\varepsilon_o} \ln\left(\frac{r_1}{r_2} \right) \tag{11.30}$$

Here, we are assuming that the distance C is the same from either line charge since it is assumed to be much larger than their separation. In any event, an overall constant may be added to a potential distribution without affecting any quantities of physical interest. The sign difference in (11.30) comes from the different signs of λ on the two line charges.

We see immediately from (11.30) that when $r_1 = r_2$, the potential equals 0. Thus, the boundary condition of zero potential on the plane between the 2 charged lines is satisfied. We need to satisfy the other boundary condition of potential V_o on the cylinder, which is on the right side of the zero potential plane. Let us rewrite the potential in (11.30) at the generic point P in Figure 11.5 as

$$V = A \ln\left(\frac{r_1}{r_2} \right) = \frac{A}{2} \ln\left(\frac{r_1}{r_2} \right)^2 = \frac{A}{2} \ln\left(\frac{r_1^2}{r_2^2} \right) = \frac{A}{2} \ln\left[\frac{(s+x)^2 + y^2}{(s-x)^2 + y^2} \right] \tag{11.31}$$

since the vector $\mathbf{r}_1 = (s + x, y)$ and $\mathbf{r}_2 = (s - x, y)$ in coordinate form. For V to be a constant on the cylinder, we need to have (r_1/r_2) equal a constant on the cylinder, say, m. Thus,

$$\left(\frac{r_1}{r_2} \right)^2 = \left[\frac{(s+x)^2 + y^2}{(s-x)^2 + y^2} \right] = m^2 \tag{11.32}$$

Working through the algebra, we obtain the equation

$$\left[x - \left(\frac{m^2 + 1}{m^2 - 1} \right) s \right]^2 + y^2 = \left(\frac{2ms}{m^2 - 1} \right)^2 \tag{11.33}$$

But this is just the equation of a circle centered at $x = ((m^2+1)/(m^2-1))s$, $y = 0$ with a radius of $2ms/(m^2-1)$. This is what we need to satisfy the boundary condition on the cylinder. For this to happen, as Figure 11.4 indicates, we set

$$h = \left(\frac{m^2 + 1}{m^2 - 1} \right) s, \quad R = \frac{2ms}{m^2 - 1} \tag{11.34}$$

Solving these last two equations for s and m, we obtain

$$s = h\sqrt{1-\left(\frac{R}{h}\right)^2}, \quad m = \frac{h}{R}\left[1+\sqrt{1-\left(\frac{R}{h}\right)^2}\right] = \frac{h+s}{R} \tag{11.35}$$

The first equation in (11.35) shows that $s < h$. We can also show that $s > h - R$. Thus, the positive line charge is inside the cylinder and out of the region where the potential distribution is required. The negative line charge is also outside this region.

For the value of m in (11.35), we require $V = V_o$. Using this in (11.31), we can get the value of A:

$$V_o = A\ln(m) = A\ln\left(\frac{h+s}{R}\right) \quad \Rightarrow \quad A = \frac{V_o}{\ln\left(\frac{h+s}{R}\right)} \tag{11.36}$$

where $s = h\sqrt{1-(R/h)^2}$.

Thus, the potential distribution is given by

$$V = \frac{V_o}{2\ln\left(\frac{h+s}{R}\right)}\ln\left[\frac{(x+s)^2+y^2}{(x-s)^2+y^2}\right] \tag{11.37}$$

The electric field is obtained from the potential according to

$$\mathbf{E} = -\nabla V = -\frac{\partial V}{\partial x}\mathbf{i} - \frac{\partial V}{\partial y}\mathbf{j} \tag{11.38}$$

where $\mathbf{i}$ and $\mathbf{j}$ are unit vectors in the x and y directions. Carrying out the differentiations, we find for the x and y components of $\mathbf{E}$

$$E_x = \frac{2sV_o}{\ln\left(\frac{h+s}{R}\right)}\left[\frac{(x^2-s^2)-y^2}{(x^2-s^2)^2+2y^2(x^2+s^2)+y^4}\right]$$
$$E_y = \frac{2sV_o}{\ln\left(\frac{h+s}{R}\right)}\left[\frac{2xy}{(x^2-s^2)^2+2y^2(x^2+s^2)+y^4}\right] \tag{11.39}$$

Using these, we can show that

$$\nabla^2 V = \frac{\partial^2 V}{\partial x^2}+\frac{\partial^2 V}{\partial y^2} = 0 \tag{11.40}$$

so that Laplace's equation is satisfied.

We see from (11.39) that when $y = 0$, $E_y = 0$ so there is only an x component of the field. The x component of $\mathbf{E}$ for $y = 0$ is given by

$$E_x\left(y = 0\right) = \frac{2sV_o}{\ln\left(\dfrac{h+s}{R}\right)}\left(\frac{1}{x^2 - s^2}\right) \qquad (11.41)$$

We notice that, along the x-axis from the zero potential plane to the cylinder, $x < s$, so that E_x is negative if V_o is positive. This is as expected since $\mathbf{E}$ should point away from the positive potential toward the zero potential plane.

As a check on these formulas, the cylinder–plane geometry was modeled with a finite element analysis program [Ansoft]. The geometry modeled was for $R = 10$ mm and $h = 100$ mm. We also chose $V_o = 1000$ V. The equipotential contours are shown in Figure 11.6, and the electric field vectors are shown in Figure 11.7.

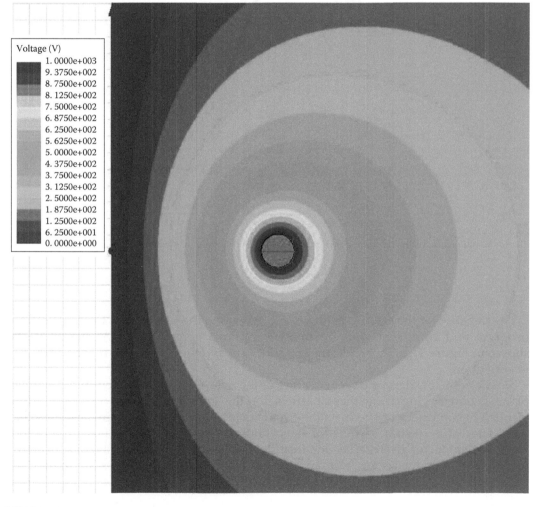

FIGURE 11.6
Cylinder–plane equipotential contours obtained with a finite element program.

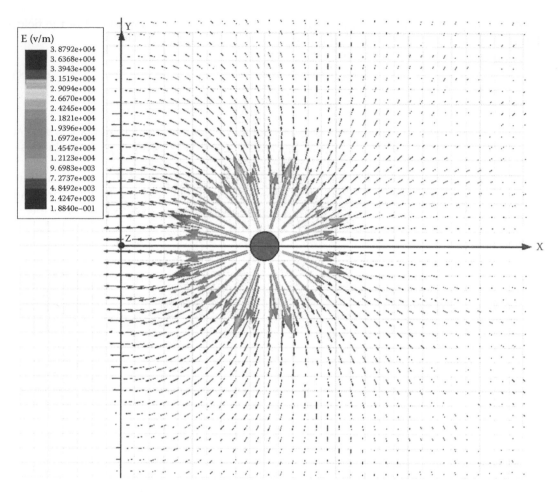

FIGURE 11.7
Electric field vectors for the cylinder–plane geometry modeled with finite elements.

The negative of the electric field was plotted along the x-axis from the finite element program and compared with the analytic formula given in (11.41). The results are shown in Figure 11.8. The fields are given in V/m and the distance in m. The agreement is very good.

No dielectric barriers or paper covering of the cylinder were included in the cylinder–plane geometry. Including these barriers for the other geometries considered earlier resulted in enhancement effects on the electric fields. One could approximate the enhancement effect due to such barriers for the cylinder–plane geometry by using the cylinder enhancement factor given earlier and treating the plane as a cylinder at a radial distance of h from the cylinder center. This approach could be refined, if necessary, based on a parametric study using a finite element program.

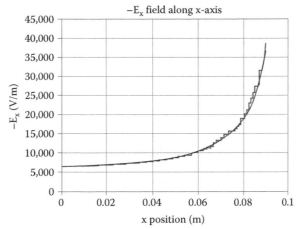

FIGURE 11.8

Negative of the x component of the electric field along the x-axis from the zero potential plane at x = 0 to the cylinder at x = 90 mm = 0.09 m. The finite element and analytic results are compared.

11.2 Electric Field Calculations Using Conformal Mapping

11.2.1 Mathematical Basis

In a region of space without charge, Maxwell's electrostatic equation is

$$\nabla \cdot \mathbf{D} = 0 \tag{11.42}$$

where $\mathbf{D}$ is the electric displacement. If the region has a uniform permeability ε, then $\mathbf{D} = \varepsilon \mathbf{E}$, where $\mathbf{E}$ is the electric field. Hence, Equation 11.42 becomes

$$\nabla \cdot \mathbf{E} = 0 \tag{11.43}$$

Introducing a potential function V, where

$$\mathbf{E} = -\nabla V \tag{11.44}$$

we obtain from (11.43)

$$\nabla^2 V = 0 \tag{11.45}$$

In two dimensions, using Cartesian coordinates, this last equation reads

$$\frac{\partial^2 V}{\partial x^2} + \frac{\partial^2 V}{\partial y^2} = 0 \tag{11.46}$$

The solution of (11.46), including boundary conditions, can then be used to determine the electric field via (11.44).

Boundary conditions are generally of two types, Dirichlet or Neumann. A Dirichlet boundary condition specifies the voltage along a boundary. This voltage is usually a constant as would be appropriate for a metallic surface. A Neumann boundary condition specifies the normal derivative along a boundary. The normal derivative is usually taken to be 0, which says that the equipotential surfaces (surfaces of constant V) or lines in 2D intersect the boundary at right angles. This type of boundary is often used to enforce a symmetry condition.

Functions satisfying Equation 11.46 are called harmonic functions. In the theory of functions of a complex variable, analytic functions play a special role. These are functions that are continuous and differentiable in some region of the complex plane. It turns out that the real and imaginary parts of analytic functions are harmonic functions. In addition, analytic mappings from one complex plane to another have properties that allow a solution of (11.46) in a relatively simple geometry in one complex plane to be transformed to a solution of this equation in a more complicated geometry in another complex plane. We briefly describe some of the important properties of these functions that are needed in the present application. See Reference [Chu60] for further details.

11.2.2 Conformal Mapping

Let $z = x + iy$ denote a complex variable where $i = \sqrt{-1}$ is the imaginary unit (i rather than j is used for the imaginary unit in this context). A function $f(z)$ can be written in terms of its real and imaginary parts as

$$f(z) = u(x, y) + iv(x, y) \qquad (11.47)$$

where u and v are real functions of two real variables. If f is analytic, it is differentiable at points z in its domain of definition. Since we are in the z-plane, which is 2D, the derivative can be taken in many directions about a given point and the value must be independent of direction. Taking this derivative in the x direction and then in the iy direction and equating the real and imaginary components of the results, we obtain the Cauchy–Riemann equations:

$$\frac{\partial u}{\partial x} = \frac{\partial v}{\partial y}, \quad \frac{\partial u}{\partial y} = -\frac{\partial v}{\partial x} \qquad (11.48)$$

Differentiating the first of these equations with respect to x and the second with respect to y, we have

$$\frac{\partial^2 u}{\partial x^2} = \frac{\partial^2 v}{\partial x \partial y}, \quad \frac{\partial^2 u}{\partial y^2} = -\frac{\partial^2 v}{\partial y \partial x} \qquad (11.49)$$

Since the mixed partial derivatives on the right-hand sides of these equations are equal for differentiable functions, when we add these equations, we obtain

$$\frac{\partial^2 u}{\partial x^2} + \frac{\partial^2 u}{\partial y^2} = 0 \qquad (11.50)$$

Thus, u is a harmonic function. By differentiating the first of Equations 11.48 with respect to y, the second with respect to x, and adding, we similarly find that v is a harmonic function.

The solution of the potential problem, Equation 11.46 with boundary conditions, is often needed in a rather complicated region geometrically. The idea behind using complex variable theory is to formulate the problem in a simpler geometric region where the solution is easy and then use an analytic function to map the easy solution onto the more complicated geometry of interest. The possibility of doing this derives from several additional properties of analytic functions.

First of all, an analytic function of an analytic function is also analytic. Since the real and imaginary components of analytic functions are harmonic, this says that analytic functions transformed by means of analytic transformations also have real and imaginary harmonic components. In terms of formulas, if f(z) is an analytic function of z and z = g(w) expresses z in terms of an analytic mapping from the w-plane, where w = u + iv, then f(g(w)) is an analytic function of w. The real and imaginary parts of f are transformed into harmonic functions of the new variables u and v.

Given an analytic mapping from the z- to w-plane, w = f(z), the inverse mapping z = F(w) is analytic at points where f'(z) = df/dz ≠ 0. Moreover, at such points,

$$F'(w) = \frac{1}{f'(z)} \quad \text{or} \quad \frac{dz}{dw} = \frac{1}{dw/dz} \tag{11.51}$$

This result will be useful in later applications to the electrostatic problem.

Perhaps the most important characteristic of analytic mappings in the present context is that they are conformal mappings. This means that if two curves intersect at an angle α in the z-plane, their images in the w-plane under an analytic mapping intersect at the same angle α at the transformed point. To see this, we use the fact that an analytic function can be expanded about a point z_0 using a Taylor's series expansion:

$$
\begin{aligned}
w = f(z) &= f(z_0) + f'(z_0)(z - z_0) + \cdots \\
&= w_0 + f'(z_0)(z - z_0) + \cdots
\end{aligned}
\tag{11.52}
$$

where we assume that $f'(z_0) \neq 0$. Writing $\Delta w = w - w_0$ and $\Delta z = z - z_0$, this last equation becomes for small Δz

$$\Delta w = f'(z_0)\Delta z \tag{11.53}$$

If Δz is an incremental distance along a curve in the z-plane, Δw is the corresponding incremental distance along the transformed curve in the w-plane (see Figure 11.9). Since we can write any complex number in polar form,

$$z = |z|e^{i\phi} \tag{11.54}$$

in terms of its magnitude $|z|$ and argument ϕ, Equation 11.53 can be expressed as

$$|\Delta w|e^{i\beta} = |f'(z_0)||\Delta z|e^{i(\psi_0 + \alpha)} \tag{11.55}$$

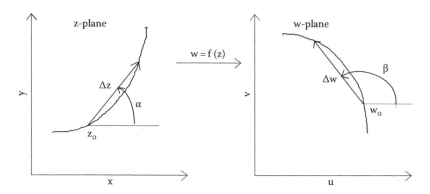

FIGURE 11.9
Analytic mapping of a curve from the z- to the w-plane.

where ψ_0 = argument $(f'(z_0))$ and α and β are shown in Figure 11.9. Thus, we see from (11.55) that

$$\beta = \psi_0 + \alpha \qquad (11.56)$$

As Δz and Δw approach zero, their directions approach that of the tangent to their respective curves. Thus, the angle that the transformed curve makes with the horizontal axis β is equal to the angle that the original curve made with its horizontal axis α rotated by the amount ψ_0. Since ψ_0 is characteristic of the derivative $f'(z_0)$, which is independent of the direction of the tangent vector of whatever curve passing through z_0 is mapped by f, this says that its tangent will be rotated by the same angle ψ_0. Thus, since any two intersecting curves are rotated by the same angle under the transformation, the angle between the curves will be preserved.

In terms of the electrostatic problem, this last mapping characteristic says that a set of equipotential (nonintersecting) curves in one geometry will remain nonintersecting in the transformed geometry obtained from the first by means of an analytic map. Similarly, the orthogonal relationship between the equipotentials and the electric field lines will be preserved in the new geometry.

Finally, we need to look at the boundary conditions. If H is a harmonic function, which is constant along some curve or boundary,

$$H(x, y) = C \qquad (11.57)$$

then, changing variables by means of z = f(w), this becomes, in the new variables,

$$H(x(u, v), y(u, v)) = C \qquad (11.58)$$

that is, a transformed curve along which H has the same constant value. Thus, Dirichlet boundary conditions are transformed into Dirichlet boundary conditions with the same boundary value along the transformed curve.

A Neumann boundary condition means that the normal derivative of H along the boundary vanishes. Since the normal derivative is the scalar product of the gradient and unit normal vector to the boundary surface, the vanishing of this derivative means that the gradient vector points along the boundary, that is, is tangential to it. However, the gradient vector is perpendicular to curves along which H is constant. Therefore, these curves of constant H are perpendicular to the boundary curve. Under a conformal mapping, this perpendicularity is preserved so that the transformed boundary curve is normal to the transformed curves of constant H. Therefore, the gradient of these transformed curves is parallel to the transformed boundary so that their normal derivative vanishes in the transformed geometry. Thus, Neumann boundary conditions are preserved under analytic transformations.

Since a solution of (11.46) that satisfies Dirichlet or Neumann boundary conditions is unique, under an analytic mapping, the transformed solution subject to the transformed boundary conditions will also be unique in the new geometry.

11.2.3 Schwarz–Christoffel Transformation

This transformation is an analytic mapping (except for a few isolated points) from the upper half plane to the interior of a closed polygon. The closed polygon can be degenerate in the sense that some of its vertices may be at infinity. This type of polygon includes the type of interest here.

Consider the general case as illustrated in Figure 11.10. Part of the x-axis from x_1 to $x_n = \infty$ is mapped onto the boundary of a closed polygon in the w-plane. We have also drawn unit tangent vectors **s** and **t** along corresponding boundary curves in the z- and w-planes. We showed earlier that the angle that the transformed curve makes with the horizontal axis at a point is given by the angle that the original curve makes with its horizontal plus the argument of the derivative of the mapping. In this case, the original curve is along the x-axis in the positive sense and so makes zero angle with this axis. Therefore, the transformed curve makes an angle with its axis given by $\arg(f'(z))$, where arg denotes argument of. Thus, if the mapping has a constant argument between two consecutive points along the x-axis, the transformed boundary curve will have a constant argument also and therefore be a straight line. However, as the w value moves along the boundary through a point w_i where the polygon transitions from one side to another, the argument of the tangent abruptly changes value. At these points, the mapping cannot be analytic (or conformal). However, there are only n such points for an n-sided polygon.

A mapping that has the above characteristics is given, in terms of its derivative, by

$$f'(z) = A(z - x_1)^{-k_1}(z - x_2)^{-k_2}\cdots(z - x_{n-1})^{-k_{n-1}} \tag{11.59}$$

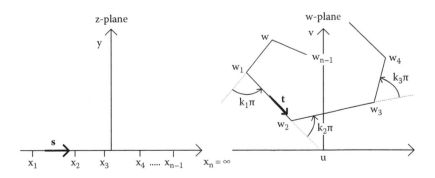

FIGURE 11.10
Schwarz–Christoffel mapping geometry.

where $x_1 < x_2 < \cdots < x_{n-1}$. Since

$$\left(z - x_i\right)^{-k_i} = \left[\left|z - x_i\right| e^{\arg(z - x_i)}\right]^{-k_i} = \left|z - x_i\right|^{-k_i} e^{-k_i \arg(z - x_i)} \tag{11.60}$$

and since the argument of a product of terms is the sum of their arguments,

$$\arg\left(f'(z)\right) = \arg A - k_1 \arg\left(z - x_1\right) - k_2 \arg\left(z - x_2\right) - \cdots - k_{n-1} \arg\left(z - x_{n-1}\right) \tag{11.61}$$

When $z = x < x_1$,

$$\arg\left(x - x_1\right) = \arg\left(x - x_2\right) = \cdots = \arg\left(x - x_{n-1}\right) = \pi \tag{11.62}$$

However, when $z = x$ moves to the right of x_1, $\arg(x - x_1) = 0$ but $\arg(x - x_i) = \pi$ for $i > 2$. Thus, the argument of $f(z)$ abruptly changes by $k_1\pi$ as $z = x$ moves to the right of x_1. This is shown in Figure 11.10. Similarly, when $z = x$ passes through x_2, $\arg(x - x_1) = \arg(x - x_2) = 0$ and all the rest equal π so that $\arg(f'(z))$ jumps by $k_2\pi$. These jumps are the exterior angles of the polygon traced in the w-plane. As such, they can be restricted to $-\pi \le k_i\pi \le \pi$ so that

$$-1 \le k_i \le 1 \tag{11.63}$$

Since the sum of the exterior angles of a closed polygon equals 2π, we have for the point at infinity

$$k_n\pi = 2\pi - \left(k_1 + k_2 + \cdots + k_{n-1}\right)\pi \tag{11.64}$$

so that

$$k_1 + k_2 + \cdots + k_n = 2 \tag{11.65}$$

Note that the point x_n could be a finite point, in which case it must be included in Equation 11.59. However, the transformation is simplified if it is at infinity.

To obtain the Schwarz–Christoffel transformation, we must integrate (11.59) to get

$$w = f(z) = A \int^z \left(\zeta - x_1\right)^{-k_1} \left(\zeta - x_2\right)^{-k_2} \cdots \left(\zeta - x_{n-1}\right)^{-k_{n-1}} d\zeta + B \tag{11.66}$$

The complex constants A and B and the x_i values can be chosen to achieve the desired map. There is some arbitrariness in the choice of these values, and this freedom should be used to simplify the calculations. With this brief background, we now proceed to determine the mapping of interest here. Reference [Chu60] or other standard books on complex variables contain further details and proofs.

11.2.4 Conformal Map for the Electrostatic Field Problem

Figure 11.11 shows the w-plane geometry of interest for the electrostatic field problem and the corresponding z-plane boundary points. The disk-to-disk separation is 2g and the disk-to-ground plane gap is h. The x, y and u, v coordinate systems are indicated in the figure,

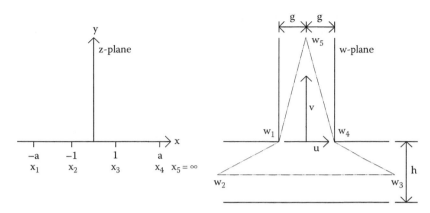

FIGURE 11.11
Schwarz–Christoffel transformation for the electrostatic problem.

and their respective origins are where the axes intersect. Note that some of the image points, w_2, w_3, and w_5, are at infinity, albeit in different directions in the complex plane. These are shown at finite positions in the figure for representational purposes. In assigning values to x_1, x_2, x_3, and x_4, we have taken advantage of the symmetry in the w-plane geometry. In fact, the coordinate systems were chosen to exploit this symmetry. Note that, in Figure 11.10, the positive direction for the exterior angles was chosen so that if we are moving along a side toward the next vertex, the angle increases if we make a left turn and decreases if we turn toward the right.

In Figure 11.11, as we move from w_5, which is at ∞, through the vertex w_1, we turn toward the right by 90°; hence, $k_1 = -1/2$. When a point is at ∞ like w_5 in the figure, it should be considered to be at the vertex of an extremely tall and narrow triangle so that motion along either side is possible. Moving through w_2, also at ∞, which we approach from w_1, we see that we make a 180° turn to the left in going from one side of the long narrow triangle to the other side, so $k_2 = 1$. Passing through w_3, we again make a 180° left turn so $k_3 = 1$. We turn right by 90° in going back from w_3 through w_4, so $k_4 = -1/2$. From w_4 through w_5, the angular change is 180° to the left so $k_5 = 1$. Thus, $k_1 + k_2 + k_3 + k_4 + k_5 = -1/2 + 1 + 1 - 1/2 + 1 = 2$ as required for a closed polygon.

The derivative of the transformation is, from (11.59),

$$f'(z) = A\frac{(z+a)^{1/2}(z-a)^{1/2}}{(z+1)(z-1)} = A\frac{(z^2-a^2)^{1/2}}{(z^2-1)} \tag{11.67}$$

where $a > 1$. The integral can be more easily evaluated by writing

$$\frac{1}{z^2-1} = \frac{1}{2}\left(\frac{1}{z-1} - \frac{1}{z+1}\right) \tag{11.68}$$

Substituting into (11.67), we get

$$f'(z) = \frac{A}{2}\left[\frac{(z^2-a^2)^{1/2}}{z-1} - \frac{(z^2-a^2)^{1/2}}{z+1}\right] \tag{11.69}$$

Using tables of integrals and the definitions of the complex version of the standard functions (see [Dwi61]), we can integrate this to obtain

$$w = f(z) = A\left\{ \ln\left(2\sqrt{z^2 - a^2} + 2z\right) - \frac{\sqrt{a^2 - 1}}{2}\left[\sin^{-1}\left(\frac{z - a^2}{a(z-1)}\right) + \sin^{-1}\left(\frac{z + a^2}{a(z+1)}\right) \right] \right\} + B \quad (11.70)$$

where ln is the natural logarithm. This solution can be verified by taking its derivative.

In order to fix the constants, we must match the image points w_1, w_2, etc., to their corresponding x-axis points. Thus, we must get $w_1 = -g$ when $z = x_1 = -a$. Substituting into (11.70), we find

$$-g = A\left\{ \ln(-2a) - \frac{\sqrt{a^2 - 1}}{2}\left[\sin^{-1}(1) + \sin^{-1}(-1) \right] \right\} + B \quad (11.71)$$

Since

$$\ln z = \ln|z| + i\arg(z)$$

where we are using the principal value of the logarithm, Equation 11.71 becomes

$$-g = A\left[\ln(2a) + i\pi \right] + B \quad (11.72)$$

since $\arg(-2a) = \pi$.

Similarly, we must have $w_4 = g$ corresponding to $x_4 = a$. Substituting into (11.70), we find

$$g = A\left\{ \ln(2a) - \frac{\sqrt{a^2 - 1}}{2}\left[\sin^{-1}(-1) + \sin^{-1}(1) \right] \right\} + B$$

$$= A\ln(2a) + B \quad (11.73)$$

Subtracting (11.73) from (11.72), we obtain

$$A = i\frac{2g}{\pi} \quad (11.74)$$

Adding (11.72) and (11.73) and using (11.74), we get

$$B = g - i\frac{2g}{\pi}\ln(2a) \quad (11.75)$$

Substituting A and B into (11.70), we obtain

$$w = f(z) = i\frac{2g}{\pi}\left\{ \ln\left(2\sqrt{z^2 - a^2} + 2z\right) - \frac{\sqrt{a^2 - 1}}{2}\left[\sin^{-1}\left(\frac{z - a^2}{a(z-1)}\right) + \sin^{-1}\left(\frac{z + a^2}{a(z+1)}\right) \right] \right\}$$

$$+ g - i\frac{2g}{\pi}\ln(2a) \quad (11.76)$$

The log terms can be combined, resulting in

$$w = f(z) = i\frac{2g}{\pi}\left\{ \ln\left(\frac{\sqrt{z^2-a^2}+z}{a}\right) - \frac{\sqrt{a^2-1}}{2}\left[\sin^{-1}\left(\frac{z-a^2}{a(z-1)}\right) + \sin^{-1}\left(\frac{z+a^2}{a(z+1)}\right)\right]\right\} + g \quad (11.77)$$

Further manipulation of the log term leads to

$$\ln\left(\frac{\sqrt{z^2-a^2}+z}{a}\right) = \ln\left(\sqrt{\left(\frac{z}{a}\right)^2 - 1} + \left(\frac{z}{a}\right)\right) = \ln\left(i\sqrt{1-\left(\frac{z}{a}\right)^2} + \left(\frac{z}{a}\right)\right)$$

$$= \ln\left[i\left(\sqrt{1-\left(\frac{z}{a}\right)^2} - i\left(\frac{z}{a}\right)\right)\right] = i\frac{\pi}{2} + \ln\left[\sqrt{1-\left(\frac{z}{a}\right)^2} - i\left(\frac{z}{a}\right)\right] \quad (11.78)$$

where we have used $\ln(i) = \ln(1) + i\,\arg(i) = i\,\pi/2$. Using the identity from [Dwi61],

$$\sin^{-1}C = i\ln\left(\pm\sqrt{1-C^2} - iC\right) \quad (11.79)$$

where C is a complex number, Equation 11.78 becomes

$$\ln\left(\frac{\sqrt{z^2-a^2}+z}{a}\right) = i\frac{\pi}{2} - i\sin^{-1}\left(\frac{z}{a}\right) \quad (11.80)$$

Substituting into (11.77), we get

$$w = \frac{2g}{\pi}\sin^{-1}\left(\frac{z}{a}\right) - i\frac{g\sqrt{a^2-1}}{\pi}\left[\sin^{-1}\left(\frac{z-a^2}{a(z-1)}\right) + \sin^{-1}\left(\frac{z+a^2}{a(z+1)}\right)\right] \quad (11.81)$$

To determine the constant a, we must use the correspondence between another set of points, such as x_2, w_2 or x_3, w_3. Using the x_3, w_3 pair, we note that as x approaches $x_3 = 1$ from below, w_3 must approach $\infty - ih$. In (11.81), therefore, let $z = x \to 1$ from below. Here we use another expression for the complex inverse sine function [Dwi61]:

$$\sin^{-1}(x \pm iy) = \sin^{-1}\left(\frac{2x}{p+q}\right) \pm i\cosh^{-1}\left(\frac{p+q}{2}\right)$$

$$\text{where } p = \sqrt{(1+x)^2 + y^2}, \; q = \sqrt{(1-x)^2 + y^2} \text{ and } y \geq 0 \quad (11.82)$$

where p and q are assumed to be positive. In (11.82), we have taken the principal value of the inverse sine function, corresponding to $n = 0$ in [Dwi61]. The values of p and q for $y = 0$, which is the case here, are as follows:

If $y = 0$ and $x > 1$, then $p = 1 + x$, $q = -(1 - x)$, and $p + q = 2x$.
If $y = 0$ and $x < -1$, then $p = -(1 + x)$, $q = 1 - x$, and $p + q = -2x$.
If $y = 0$ and $|x| < 1$, then $p = 1 + x$, $q = 1 - x$, and $p + q = 2$.

This latter case corresponds to the real sine inverse function. Note that the sign of the cosh^{-1} term in (11.82) is the sign of the imaginary part of the complex argument of the sin^{-1} on the left-hand side of that equation.

Substituting $z \to 1$ into (11.81), we get

$$\infty - ih = \frac{2g}{\pi} \sin^{-1}\left(\frac{1}{a}\right) - i \frac{g\sqrt{a^2-1}}{\pi}\left[\sin^{-1}(\infty) + \sin^{-1}\left(\frac{1+a^2}{2a}\right)\right] \qquad (11.83)$$

where the infinities must be interpreted in a limiting sense. These infinities are real infinities and can be grouped with the real part of this last equation. The imaginary part can be handled separately. The imaginary part is what is used to find the value of a. Keeping this in mind and using (11.82), we find

$$\sin^{-1}(\infty) = \sin^{-1}(1) + i\cosh^{-1}(\infty) = \frac{\pi}{2} + i\cosh^{-1}(\infty) \qquad (11.84)$$

For the other inverse sine term, since

$$\frac{1+a^2}{2a} = \frac{2a+(a-1)^2}{2a} = 1 + \frac{(a-1)^2}{2a} > 1 \qquad (11.85)$$

we have

$$\sin^{-1}\left(\frac{1+a^2}{2a}\right) = \sin^{-1}(1) + i\cosh^{-1}\left(\frac{1+a^2}{2a}\right) = \frac{\pi}{2} + i\cosh^{-1}\left(\frac{1+a^2}{2a}\right) \qquad (11.86)$$

Since a > 1, the first inverse sine term in (11.83) is an ordinary real inverse sine function. Adding the two sin^{-1} terms in the square bracket in (11.83), as given by (11.84) and (11.86), and substituting back into (11.83) and neglecting finite real terms compared with ∞, we get

$$\infty - ih = \infty - ig\sqrt{a^2-1} \qquad (11.87)$$

Therefore, equating the imaginary parts of both sides of this equation, we get $\sqrt{a^2-1} = h/g$ and solving for a

$$a = \sqrt{1 + \left(\frac{h}{g}\right)^2} \qquad (11.88)$$

Thus, Equation 11.81 becomes

$$w = \frac{g}{\pi}\left\{2\sin^{-1}\left(\frac{z}{a}\right) - i\frac{h}{g}\left[\sin^{-1}\left(\frac{z-a^2}{a(z-1)}\right) + \sin^{-1}\left(\frac{z+a^2}{a(z+1)}\right)\right]\right\} \qquad (11.89)$$

with a given by (11.88). Other points of correspondence can be checked for consistency by similar procedures. In this derivation, the infinities contribute to the real part of (11.81), whereas only the imaginary part is needed to obtain the value of a.

11.2.4.1 Electric Potential and Field Values

In the w-plane, which is the plane of interest for the electrostatic problem, the boundary values of the potential are shown in Figure 11.12. Two conductors are at potentials V_1 and V_2 relative to a plane at zero potential. This can be done without loss of generality since if the plane were not at zero potential, its value could be subtracted from all the potential values without altering the values of the electric field. The corresponding boundary values in the z-plane are also shown in Figure 11.12. The idea is to solve the problem in the simpler z-plane geometry and then use the conformal map given by (11.89) to transfer the solution to the w-plane. A method of doing this is shown in Figure 11.13. The angles θ_1 and θ_2 are

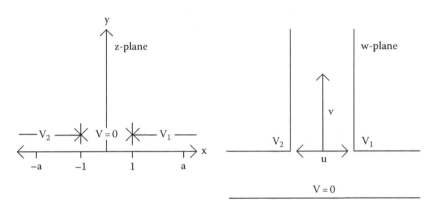

FIGURE 11.12
Correspondence between potential boundary values for the z- and w-planes.

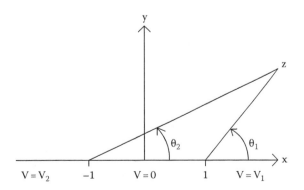

FIGURE 11.13
The potential function in the z-plane.

angles between the vectors from points 1 and −1 to z. In terms of these angles, a potential that satisfies the boundary conditions in the z-plane is given by

$$V = V_1 - \frac{V_1}{\pi}\theta_1 + \frac{V_2}{\pi}\theta_2 \tag{11.90}$$

When $z = x > 1$, θ_1 and θ_2 are zero, so $V = V_1$. When $-1 < z = x < 1$, $\theta_1 = \pi$ and $\theta_2 = 0$, so $V = 0$. When $z = x < -1$, $\theta_1 = \theta_2 = \pi$, so $V = V_2$. Thus, the boundary conditions are satisfied.

We must now show that V is a harmonic function. Note that

$$\theta_1 = \arg(z-1), \quad \theta_2 = \arg(z+1) \tag{11.91}$$

Note also that

$$\ln(z \pm 1) = \ln|z \pm 1| + i\arg(z \pm 1) \tag{11.92}$$

Since the log function is analytic, its real and imaginary components are harmonic functions, as was shown previously for analytic functions. Thus, θ_1 and θ_2 are harmonic and so is V since sums of harmonic functions are also harmonic.

In terms of x and y where $z = x + iy$, Equation 11.90 can be written as

$$V = V_1 - \frac{V_1}{\pi}\tan^{-1}\left(\frac{y}{x-1}\right) + \frac{V_2}{\pi}\tan^{-1}\left(\frac{y}{x+1}\right) \tag{11.93}$$

Since the transformation (11.89) is analytic except at a few points, the inverse transformation is defined and analytic at all points where $w' = dw/dz \neq 0$. Thus, V can be considered a function of u, v through the dependence of x and y on these variables via the inverse transformation. This will enable us to obtain expressions for the electric field in the w-plane. We have

$$E_u = -\frac{\partial V}{\partial u}, \quad E_v = -\frac{\partial V}{\partial v} \tag{11.94}$$

Referring to (11.93), note that

$$\frac{\partial}{\partial u}\tan^{-1}\left(\frac{y}{x-c}\right) = \frac{(x-c)\dfrac{\partial y}{\partial u} - y\dfrac{\partial x}{\partial u}}{(x-c)^2 + y^2} \tag{11.95}$$

Substituting $c = \pm 1$, corresponding to the two terms in (11.93), and applying a similar formula for the v derivative, we get for the electric field components

$$E_u = \frac{V_1}{\pi}\left[\frac{(x-1)\dfrac{\partial y}{\partial u} - y\dfrac{\partial x}{\partial u}}{(x-1)^2 + y^2}\right] - \frac{V_2}{\pi}\left[\frac{(x+1)\dfrac{\partial y}{\partial u} - y\dfrac{\partial x}{\partial u}}{(x+1)^2 + y^2}\right]$$

$$E_v = \frac{V_1}{\pi}\left[\frac{(x-1)\dfrac{\partial y}{\partial v} - y\dfrac{\partial x}{\partial v}}{(x-1)^2 + y^2}\right] - \frac{V_2}{\pi}\left[\frac{(x+1)\dfrac{\partial y}{\partial v} - y\dfrac{\partial x}{\partial v}}{(x+1)^2 + y^2}\right] \tag{11.96}$$

The derivatives in these formulas can be determined by means of (11.51), (11.67), and (11.74). Thus,

$$\frac{dz}{dw} = \frac{1}{dw/dz} = -i\frac{\pi}{2g}\frac{(z^2-1)}{(z^2-a^2)^{1/2}} \tag{11.97}$$

Because the square root of a complex number with argument θ changes sign when θ increases to $\theta + 2\pi$, there is an ambiguity in the sign of (11.97). Thus, it is necessary to make a branch cut in the complex plane, keeping the + sign for values on one side of the cut and − on the other. This cut is made along a ray from the origin, typically along $\theta = 0$. Expressing the square root term in the denominator of (11.97) in polar form, we get

$$\left(z^2-a^2\right)^{1/2} = \left(x^2-y^2-a^2+2ixy\right)^{1/2}$$
$$= \left(\sqrt{\left(x^2-y^2-a^2\right)^2+4x^2y^2}\right)^{1/2} e^{\frac{1}{2}\tan^{-1}\left(\frac{2xy}{x^2-y^2-a^2}\right)} \tag{11.98}$$

We see that the argument is 0 along the line $x = 0$. It also changes sign when x crosses this line from positive to negative or vice versa. Thus, we choose the $x = 0$ line as a branch cut. In our geometry, this is a line from the origin along the positive y-axis. We take the positive sign for $x \geq 0$ and the negative sign for $x < 0$.

Expressing (11.97) in terms of x and y, converting the numerator and denominator to the polar form of a complex number, combining the exponentials, and separating into real and imaginary parts, we get

$$\frac{dz}{dw} = \frac{\pi}{2g}\left\{\frac{\left(x^2-y^2-1\right)^2+4x^2y^2}{\left[\left(x^2-y^2-a^2\right)^2+4x^2y^2\right]^{1/2}}\right\}^{1/2}$$
$$\times\left\{\sin\left[\tan^{-1}\left(\frac{2xy}{x^2-y^2-1}\right)-\frac{1}{2}\tan^{-1}\left(\frac{2xy}{x^2-y^2-a^2}\right)\right]\right.$$
$$\left.-i\cos\left[\tan^{-1}\left(\frac{2xy}{x^2-y^2-1}\right)-\frac{1}{2}\tan^{-1}\left(\frac{2xy}{x^2-y^2-a^2}\right)\right]\right\} \tag{11.99}$$

Now, using the uniqueness of the derivative when taken in the u or iv directions, which led to the Cauchy–Riemann equations mentioned earlier, we can write

$$\frac{dz}{dw} = \frac{\partial x}{\partial u}+i\frac{\partial y}{\partial u} = \frac{\partial y}{\partial v}-i\frac{\partial x}{\partial v} \tag{11.100}$$

This, together with (11.99), can be used to extract the appropriate derivatives for use in (11.96). Thus,

$$\frac{\partial x}{\partial u} = \frac{\partial y}{\partial v} = \frac{\pi}{2g} \left\{ \frac{\left(x^2 - y^2 - 1\right)^2 + 4x^2 y^2}{\left[\left(x^2 - y^2 - a^2\right)^2 + 4x^2 y^2\right]^{1/2}} \right\}^{1/2}$$

$$\times \sin\left[\tan^{-1}\left(\frac{2xy}{x^2 - y^2 - 1}\right) - \frac{1}{2}\tan^{-1}\left(\frac{2xy}{x^2 - y^2 - a^2}\right)\right]$$

$$\frac{\partial y}{\partial u} = -\frac{\partial x}{\partial v} = -\frac{\pi}{2g} \left\{ \frac{\left(x^2 - y^2 - 1\right)^2 + 4x^2 y^2}{\left[\left(x^2 - y^2 - a^2\right)^2 + 4x^2 y^2\right]^{1/2}} \right\}^{1/2}$$

$$\times \cos\left[\tan^{-1}\left(\frac{2xy}{x^2 - y^2 - 1}\right) - \frac{1}{2}\tan^{-1}\left(\frac{2xy}{x^2 - y^2 - a^2}\right)\right]$$

(11.101)

These apply when $x \geq 0$, and their negatives must be used when $x < 0$.

Thus, the electric field in the w-plane can be expressed completely in terms of x and y via (11.96) and (11.101). However, we must invert Equation 11.89 to do this. The first step is to express (11.89) in terms of x and y and in terms of its real and imaginary parts. Thus, we write the first $\sin^{-1}$ term, using (11.82), as

$$\sin^{-1}\left(\frac{z}{a}\right) = \sin^{-1}\left(\frac{x}{a} + i\frac{y}{a}\right) = \sin^{-1}\left[\frac{2(x/a)}{p+q}\right] + i\cosh^{-1}\left(\frac{p+q}{2}\right)$$

(11.102)

where

$$p = \sqrt{\left(1 + \frac{x}{a}\right)^2 + \left(\frac{y}{a}\right)^2}, \quad q = \sqrt{\left(1 - \frac{x}{a}\right)^2 + \left(\frac{y}{a}\right)^2}$$

The second $\sin^{-1}$ term in (11.89) can be written similarly as

$$\sin^{-1}\left[\frac{z - a^2}{a(z-1)}\right] = \sin^{-1}\left\{\frac{\left(x - a^2\right) + iy}{a\left[(x-1) + iy\right]}\right\}$$

$$= \sin^{-1}\left\{\frac{\left(x - a^2\right)(x-1) + y^2 + iy\left(a^2 - 1\right)}{a\left[(x-1)^2 + y^2\right]}\right\}$$

$$= \sin^{-1}\left\{\frac{2\left[\left(x - a^2\right)(x-1) + y^2\right]}{a\left[(x-1)^2 + y^2\right](p_1 + q_1)}\right\} + i\cosh^{-1}\left(\frac{p_1 + q_1}{2}\right)$$

(11.103)

$$\text{where } \left.\begin{matrix} p_1 \\ q_1 \end{matrix}\right\} = \sqrt{\left\{1 \pm \frac{\left[\left(x - a^2\right)(x-1) + y^2\right]}{a\left[(x-1)^2 + y^2\right]}\right\}^2 + \left\{\frac{y\left(a^2 - 1\right)}{a\left[(x-1)^2 + y^2\right]}\right\}^2}$$

with the upper sign referring to p_1 and the lower sign to q_1. Treating the third $\sin^{-1}$ term in (11.89) similarly, we get

$$\sin^{-1}\left[\frac{z+a^2}{a(z+1)}\right] = \sin^{-1}\left\{\frac{(x+a^2)+iy}{a[(x+1)+iy]}\right\}$$

$$= \sin^{-1}\left\{\frac{(x+a^2)(x+1)+y^2-iy(a^2-1)}{a[(x+1)^2+y^2]}\right\}$$

$$= \sin^{-1}\left\{\frac{2[(x+a^2)(x+1)+y^2]}{a[(x+1)^2+y^2](p_2+q_2)}\right\} - i\cosh^{-1}\left(\frac{p_2+q_2}{2}\right) \qquad (11.104)$$

$$\text{where} \quad \left.\begin{array}{c}p_2\\q_2\end{array}\right\} = \sqrt{\left\{1\pm\frac{[(x+a^2)(x+1)+y^2]}{a[(x+1)^2+y^2]}\right\}^2 + \left\{\frac{y(a^2-1)}{a[(x+1)^2+y^2]}\right\}^2}$$

Again, the upper sign refers to p_2 and the lower sign to q_2.

Using (11.102) through (11.104) in (11.89), we obtain

$$u = \frac{g}{\pi}\left\{2\sin^{-1}\left[\frac{2(x/a)}{p+q}\right] + \frac{h}{g}\left[\cosh^{-1}\left(\frac{p_1+q_1}{2}\right) - \cosh^{-1}\left(\frac{p_2+q_2}{2}\right)\right]\right\}$$

$$v = \frac{g}{\pi}\left\{2\cosh^{-1}\left(\frac{p+q}{2}\right) - \frac{h}{g}\left[\sin^{-1}\left(\frac{2[(x-a^2)(x-1)+y^2]}{a[(x-1)^2+y^2](p_1+q_1)}\right)\right.\right.$$

$$\left.\left. +\sin^{-1}\left(\frac{2[(x+a^2)(x+1)+y^2]}{a[(x+1)^2+y^2](p_2+q_2)}\right)\right]\right\} \qquad (11.105)$$

where

$$a = \sqrt{1+\left(\frac{h}{g}\right)^2}$$

$$\left.\begin{array}{c}p\\q\end{array}\right\} = \sqrt{\left(1\pm\frac{x}{a}\right)^2 + \left(\frac{y}{a}\right)^2}$$

$$\left.\begin{array}{c}p_1\\q_1\end{array}\right\} = \sqrt{\left\{1\pm\frac{[(x-a^2)(x-1)+y^2]}{a[(x-1)^2+y^2]}\right\}^2 + \left\{\frac{y(a^2-1)}{a[(x-1)^2+y^2]}\right\}^2}$$

$$\left.\begin{array}{c}p_2\\q_2\end{array}\right\} = \sqrt{\left\{1\pm\frac{[(x+a^2)(x+1)+y^2]}{a[(x+1)^2+y^2]}\right\}^2 + \left\{\frac{y(a^2-1)}{a[(x-1)^2+y^2]}\right\}^2}$$

The upper and lower signs refer to the p's and q's, respectively.

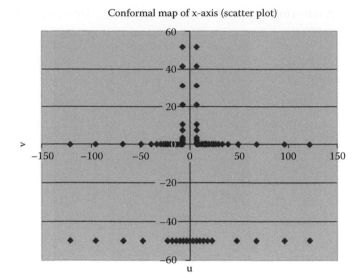

FIGURE 11.14
Map of (11.105) for points along the x-axis in the z-plane to points in the w-plane.

At this point, it is interesting to map (11.105) to check it visually to see how it behaves, particularly at the critical points of x = ±1 and x = ±a, y = 0. This is shown in Figure 11.14 for a number of points along the x-axis and for g = 7 mm and h = 50 mm.

Let us trace a point on the x-axis as it moves from +1,000,000 to −1,000,000. At x = +1,000,000, the point in the w-plane is only 51.4 mm above the u-axis on the right side boundary of the gap between the two disks. As x decreases and approaches +a from the right, the point moves down the right boundary parallel to the v-axis and reaches the right corner at x = a. As x passes through a, the point in the w-plane then moves right along the positive u-axis increasing toward +∞ as x approaches +1. As x continues to decrease and passes through +1, it abruptly switches from v = 0 to v = −h = −50 mm in this case. That is, it jumps from the top of the horizontal gap to the bottom of the gap. The point then proceeds along the bottom of the horizontal gap toward −∞ as x approaches −1 from the right. As x passes through −1, the w-plane point abruptly switches to the top of the horizontal gap where v = 0 and u = −∞. As x continues toward −a, the mapped point heads toward the left corner point at u = −g and v = 0. Then as x heads toward −∞, the point moves up the left side of the vertical gap between the two disks, reaching v = 51.4 mm at x = −1,000,000.

To see how this map behaves for y > 0, we plot the same x-axis points but with y = 0.001 in Figure 11.15 and with y = 0.1 in Figure 11.16.

Given u and v, Equation 11.105 can be inverted by means of a Newton–Raphson procedure. Using this method, the problem reduces to one of solving the following equations:

$$f_1(x,y) = u(x,y) - u_o = 0$$
$$f_2(x,y) = v(x,y) - v_o = 0$$

(11.106)

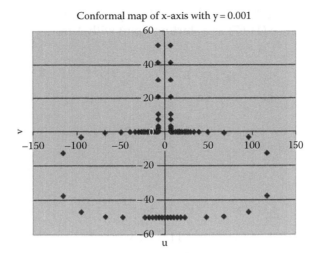

FIGURE 11.15
Map of (11.105) for points along the x-axis with y = 0.001 to points in the w-plane.

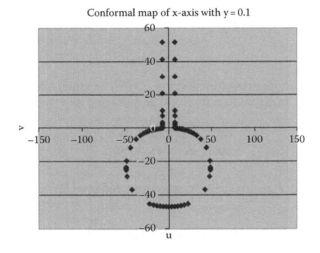

FIGURE 11.16
Map of (11.105) for points along the x-axis with y = 0.1 to points in the w-plane.

where u_0 and v_0 are the desired coordinates at which to evaluate the field in the w-plane. Since $\partial f_1/\partial x = \partial u/\partial x$, etc., the Newton–Raphson equations for the increments Δx, Δy are

$$\begin{pmatrix} \Delta x \\ \Delta y \end{pmatrix} = \frac{-1}{\left[\left(\dfrac{\partial u}{\partial x}\right)^2 + \left(\dfrac{\partial u}{\partial y}\right)^2\right]} \begin{pmatrix} \dfrac{\partial u}{\partial x} & -\dfrac{\partial u}{\partial y} \\ \dfrac{\partial u}{\partial y} & \dfrac{\partial u}{\partial x} \end{pmatrix} \begin{pmatrix} f_1(x, y) \\ f_2(x, y) \end{pmatrix} \tag{11.107}$$

where we have used the Cauchy–Riemann equations (11.48). At each iteration, we let $x_{new} = x_{old} + \Delta x$ and $y_{new} = y_{old} + \Delta y$ and stop when Δx and Δy are sufficiently small.

The derivatives in (11.107) can be obtained by a procedure similar to that used to derive (11.101). Basically, we use (11.67) as a starting point with A from (11.74):

$$f'(z) = \frac{dw}{dz} = i\frac{2g}{\pi}\frac{\left(z^2 - a^2\right)^{1/2}}{\left(z^2 - 1\right)} = i\frac{2g}{\pi}\left\{\frac{\left[\left(x + iy\right)^2 - a^2\right]^{1/2}}{\left(x + iy\right)^2 - 1}\right\}$$

$$= i\frac{2g}{\pi}\left\{\frac{\left[\left(x^2 - y^2 - a^2\right) + i2xy\right]^{1/2}}{\left(x^2 - y^2 - 1\right) + i2xy}\right\}$$

$$= \frac{2g}{\pi}\left\{\frac{\left[\left(x^2 - y^2 - a^2\right)^2 + 4x^2y^2\right]^{1/2}}{\left(x^2 - y^2 - 1\right)^2 + 4x^2y^2}\right\}^{1/2}$$

$$\times \left\{-\sin\left[\frac{1}{2}\tan^{-1}\left(\frac{2xy}{x^2 - y^2 - a^2}\right) - \tan^{-1}\left(\frac{2xy}{x^2 - y^2 - 1}\right)\right]\right.$$

$$\left. + i\cos\left[\frac{1}{2}\tan^{-1}\left(\frac{2xy}{x^2 - y^2 - a^2}\right) - \tan^{-1}\left(\frac{2xy}{x^2 - y^2 - 1}\right)\right]\right\} \qquad (11.108)$$

Again, because of the square root function in (11.108), we must make a branch cut along the line $x = 0$ and take the positive sign for $x \geq 0$ and the negative sign for $x < 0$. Since

$$\frac{dw}{dz} = \frac{\partial u}{\partial x} + i\frac{\partial v}{\partial x} = \frac{\partial v}{\partial y} - i\frac{\partial u}{\partial y} \qquad (11.109)$$

the partial derivatives for use in (11.107) can be extracted from (11.108), using (11.109). We find

$$\frac{\partial u}{\partial x} = -\frac{2g}{\pi}\left\{\frac{\left[\left(x^2 - y^2 - a^2\right)^2 + 4x^2y^2\right]^{1/2}}{\left(x^2 - y^2 - 1\right)^2 + 4x^2y^2}\right\}^{1/2}$$

$$\times \sin\left[\frac{1}{2}\tan^{-1}\left(\frac{2xy}{x^2 - y^2 - a^2}\right) - \tan^{-1}\left(\frac{2xy}{x^2 - y^2 - 1}\right)\right]$$

$$\frac{\partial u}{\partial y} = -\frac{2g}{\pi}\left\{\frac{\left[\left(x^2 - y^2 - a^2\right)^2 + 4x^2y^2\right]^{1/2}}{\left(x^2 - y^2 - 1\right)^2 + 4x^2y^2}\right\}^{1/2} \qquad (11.110)$$

$$\times \cos\left[\frac{1}{2}\tan^{-1}\left(\frac{2xy}{x^2 - y^2 - a^2}\right) - \tan^{-1}\left(\frac{2xy}{x^2 - y^2 - 1}\right)\right]$$

Again, these are used when $x \geq 0$ and their negatives when $x < 0$.

11.2.4.2 Calculations and Comparison with a Finite Element Solution

A computer program was written to implement this procedure for calculating the electric field. According to the formulas given earlier, the field is infinite at the corners of the conductors. However, because real corners are not perfectly sharp, the field remains finite in practice. In transformer applications, the conductors are usually covered with an insulating layer of paper and the remaining space is filled with transformer oil. Because the paper has a much higher breakdown stress than the oil, the field in the oil is usually critical for design purposes. The highest oil fields will occur at least a paper's thickness away from the corner of the highest potential conductor, which we assume is at potential V_1. We have accordingly calculated the field at the three points shown in Figure 11.17, which are a paper's thickness, d, away from the corner of the V_1 potential conductor.

It should be noted that the presence of the paper, which has a different dielectric constant than oil, will modify the electric field. In general, it will increase it in the oil and decrease it in the paper. We will attempt to estimate this oil enhancement factor later. Here, we wish to compare the analytic results as given by the formulas earlier with results from a finite element calculation without the presence of paper but with the fields calculated at the points shown in Figure 11.17, a paper's distance away from the corner. The magnitudes of the fields, E, are compared, where

$$E = \sqrt{E_u^2 + E_v^2} \qquad (11.111)$$

The results are shown in Table 11.1 for several different potential combinations and conductor separations. The agreement is very good, especially with the sharp corner finite element results. Also shown for comparison are finite element results for a 0.5 mm radius on the corners. These are also reasonably close to the analytic results and show that the sharpness or smoothness of the corners is washed out at distances greater than or equal to a paper's distance away. Although not shown in the table, the field deep in the gap between the V_1 and V_2 conductors was also calculated with the analytic formulas and produced the expected result, $E = (V_2 - V_1)/2g$.

The agreement between the conformal mapping calculation and the finite element calculation, with or without rounded corners, as shown in Table 11.1 is very good.

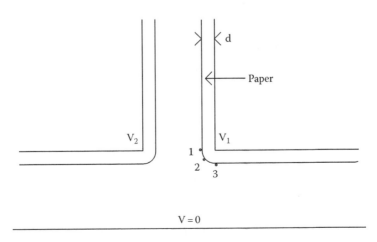

FIGURE 11.17
Points near the corner of the highest potential conductor where the electric field is calculated.

TABLE 11.1

Comparison of Analytic and Finite Element Electric Field
Magnitudes at Points 1, 2, and 3 of Figure 11.17

	Lengths (mm)	Potentials (kV)	Fields (kV/mm)
	Conformal Mapping	Finite Element (Sharp Corner)	Finite Element (Rounded Corner)[a]
(a) $V_1 = 825$, $V_2 = 725$, $g = 6.6$, $h = 63.5$, $d = 0.635$			
E_1	27.7	27.7	27.6
E_2	27.9	28.2	28.6
E_3	28.3	28.5	28.4
(b) $V_1 = 850$, $V_2 = 750$, $g = 5.08$, $h = 63.5$, $d = 0.508$			
E_1	32.7	32.7	32.4
E_2	32.4, 28.7[b]	29.0[b]	32.9
E_3	32.6	32.6	32.6
(c) $V_1 = 800$, $V_2 = 800$, $g = 5.08$, $h = 63.5$, $d = 0.508$			
E_1	11.3	11.4	12.3
E_2	13.3, 12.2[b]	12.4[b]	13.5
E_3	15.0	15.2	14.6
(d) $V_1 = 100$, $V_2 = -50$, $g = 12.7$, $h = 63.5$, $d = 0.508$			
E_1	17.8	17.6	17.2
E_2	16.9, 15.0[b]	15.1[b]	16.9
E_3	16.1	16.0	16.0

Enhancement factors are shown in parentheses.
[a] The rounded corner conductor radius = 0.5.
[b] $d = 0.71$.

The conformal mapping technique, however, does not apply to the paper–oil situation so other approximate approaches must be employed to account for oil enhancement. The equipotential contour plot for a finite element calculation with paper present is shown in Figure 11.18. This is for the conditions given in Table 11.1a with sharp corners. Figure 11.19 is a plot of the vector field for the same case.

11.2.4.3 Estimating Enhancement Factors

For a paper–oil layering in a planar geometry as shown in Figure 11.20a, all the paper can be lumped into one layer for calculational purposes. Letting 1 refer to the paper layer and 2 to the oil layer, the enhancement factor for the oil, η = the ratio of the field in the oil with and without a paper layer, is

$$\eta = \frac{1}{\varepsilon_2 \left(\dfrac{f_1}{\varepsilon_1} + \dfrac{f_2}{\varepsilon_2} \right)} \tag{11.112}$$

where $f_1 = \ell_1/\ell$, $f_2 = \ell_2/\ell$ with $\ell = \ell_1 + \ell_2$ are the fractional lengths of materials 1 and 2, respectively. It is assumed that when the paper is absent, it is replaced by oil, keeping the total distance ℓ between the metal surfaces the same.

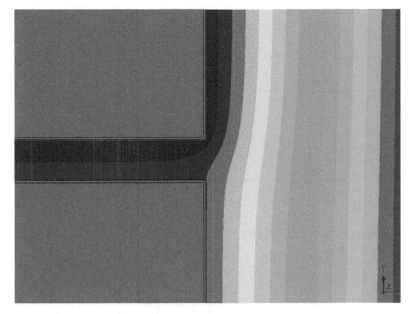

FIGURE 11.18
Equipotential contour plot in the oil from a finite element calculation showing the geometry near the corner for the conditions given in Table 11.1a with sharp corners and paper cover. The potential varies from high (red) near the upper disk to low (blue) near the zero potential plane.

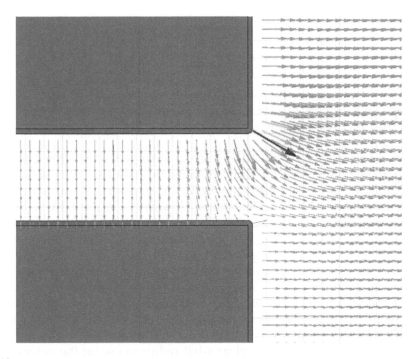

FIGURE 11.19
Electric field vector plot from a finite element calculation showing an enlarged region near the conductor's corners for the same conditions as Figure 11.18.

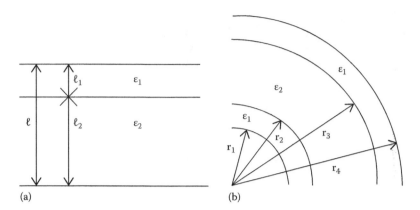

FIGURE 11.20
Paper–oil configurations for (a) planar and (b) cylindrical geometries resulting in enhanced fields in the oil compared with the all oil case.

In the cylindrical case, shown in Figure 11.20b, the position of the paper layer or layers is important in calculating the enhancement factors. We will assume that one paper layer is next to the inner conductor and one next to the outer conductor. When additional layers are present, such as pressboard barriers, their position must be known and they can be included in the calculation. The oil enhancement factor for the situation shown in Figure 11.20b is

$$\eta = \frac{\ln\left(\dfrac{r_4}{r_1}\right)}{\varepsilon_2\left[\dfrac{1}{\varepsilon_1}\ln\left(\dfrac{r_2}{r_1}\right) + \dfrac{1}{\varepsilon_2}\ln\left(\dfrac{r_3}{r_2}\right) + \dfrac{1}{\varepsilon_1}\ln\left(\dfrac{r_4}{r_3}\right)\right]} \tag{11.113}$$

The cylindrical enhancement factor is larger than the planar enhancement factor for the same paper and oil layer thicknesses. It reduces to the planar case when r_1 becomes large.

The enhancement factors given earlier refer to ideal geometries. For the geometry of interest shown in Figure 11.17, the enhancement factors reduce to the planar case for field points away from the corner and deep into the $V_1 - V_2$ gap or the $V_1 - 0$ or $V_2 - 0$ gaps. This is borne out by the finite element calculations. However, near the corner, the geometry is closer to the cylindrical case.

For a sharp corner, corresponding to $r_1 = 0$, Equation 11.113 shows that $\eta \rightarrow \varepsilon_1/\varepsilon_2 = 1.82$ for the oil–paper case. This is an upper limit on the enhancement factor. It is too high as shown by a comparison of the finite element fields in the oil, calculated with a paper cover, with the conformal mapping fields a paper distance away from the corner. The enhancement factor varied from 1.03 to 1.2 for different cases. For our purposes, the rounded corner enhancement factors are the most relevant ones. Table 11.1 shows that these are nearly the same for the 3 corner points. Treating the rounded corner case as a cylindrical geometry, we can take $r_1 = 0.5$ mm. It is less clear what radius to use for the outer conductor. In Table 11.2, we calculate the planar and cylindrical oil enhancement factors for the two

TABLE 11.2

Calculated Oil Enhancement Factors for the Cases in Table 11.1

	Planar		Cylindrical		Weighted Cylindrical	Table 11.1 Ave Rounded
Case	$V_1 - V_2$ Gap	$V_1 - 0$ Gap	$V_1 - V_2$ Gap	$V_1 - 0$ Gap		
(a)	1.095	1.004	1.182	1.081	1.137	1.133
(b)	1.099	1.004	1.174	1.069	1.125	1.140
(c)	1.099	1.004	1.174	1.069	1.125	1.096
(d)	1.037	1.004	1.113	1.069	1.094	1.137

different gaps, where the outer conductor radius is taken to be the inner conductor radius plus the gap length that is 2g or h for the two gaps (see Figure 11.11). Thus, referring to Figure 11.20b,

$$r_4 \left(V_1 - V_2 \ \text{gap} \right) = r_4' = r_1 + g$$
$$r_4 \left(V_1 - 0 \ \text{gap} \right) = r_4'' = r_1 + h \qquad (11.114)$$

Note that the $V_1 - 0$ gap has only one layer of paper. We also show a weighted cylindrical enhancement factor, which uses a weighted average of the two radii given earlier. The weighting is taken to be inversely proportional to the radius so that the V_2 conductor, which is closest to the corner, is given the highest weighting. This results in an effective outer conductor radius of

$$r_4 = \frac{2r_4' r_4''}{r_4' + r_4''} \qquad (11.115)$$

These weighted enhancements come closest to the corner enhancements determined by the finite element calculations in most cases. We therefore prefer a cylindrical enhancement factor with this weighted outer conductor radius for design calculations if it is above the $V_1 - V_2$ gap planar enhancement factor. Otherwise, the $V_1 - V_2$ gap planar enhancement factor is preferred.

Summarizing these results, the conformal mapping calculations of the electric field near the corner of one of a pair of conductors at different potentials relative to a neighboring ground plane agree well with calculations obtained by means of a finite element program. Since the electric field is infinite at the corner of a perfectly sharp conductor, the calculations were compared a small distance from the corner, taken to be the thickness of a paper layer in an actual transformer winding. The finite element calculations were made with sharp and rounded corners, and the resulting fields were nearly the same a paper's thickness away from the corner. Since the oil breakdown fields are the most critical in transformer design and these occur beyond the paper's thickness, this result shows that a conformal mapping calculation is appropriate for determining these fields.

The only problem with the conformal mapping approach is that it does not take into account the different dielectric constants of oil and paper. We have proposed a method to take these approximately into account by means of an oil enhancement factor. By comparing with a finite element calculation, a formula was developed to obtain this enhancement factor.

11.3 Finite Element Electric Field Calculations

Finite element methods permit the calculation of electric potentials and fields for complicated geometries. Modern commercial finite element codes generally provide a set of drawing tools, which allow the user to input the geometry in as much detail as desired. More sophisticated versions even allow parametric input so that changes in one or more geometric parameters such as the distance between electrodes can be easily accomplished without redoing the entire geometry. Both 2D and 3D versions are available although the input to the 3D versions is, of course, more complicated. For many problems, a 2D geometry can be an adequate approximation to the real configuration. 2D versions usually allow an axisymmetric geometry by inputting a cross section of it. In this sense, it is really solving a 3D problem that happens to have cylindrical symmetry. The x–y 2D geometry is really modeling an infinitely long object having the specified 2D cross section.

The basic geometry, which the user inputs, is then subdivided into a triangular mesh. Smaller triangles are used in regions where the potential is expected to change most rapidly. Larger triangles can adequately describe more slowly varying potential regions. Some programs automatically perform the triangular meshing and, through an iterative process, even refine the mesh in critical regions until the desired solution accuracy is achieved. When linear triangles are used, the potential is solved for only at the triangle nodes and a linear interpolation scheme is used to approximate it inside the triangle. For higher-order triangles, additional nodes are added per triangle and higher-order polynomial approximations are used to find the potential inside the triangles. Some programs use only second-order triangles since these provide sufficient accuracy for reasonable computer memory and execution times.

Some art is required even for the geometric input. Very often, complete detail is unnecessary to a determination of the fields in critical regions. Thus, the user must know when it is reasonable to ignore certain geometric details that are irrelevant to the problem. This not only saves on the labor involved in inputting the geometry but it can also considerably reduce required computer memory and solution times.

Finite element programs require that the user input sources and the appropriate boundary conditions for the problem at hand. In the case of electric potential calculations, the sources are electric charges and the boundary conditions include specifying the voltage at one or more electrode surfaces. These are often the surfaces of metallic objects and thus have a constant potential throughout. It is therefore unnecessary to model their interiors. Typically, the program will allow the user to declare such metallic objects nonexistent or perfect conductors so that the solution is not solved for over their interiors. Their surface, however, is still included as an equipotential surface. Sometimes the equipotential surface is a boundary surface so it already has no interior. A metallic object can also be allowed to float so that its potential is part of the problem solution.

On external boundaries where no specification is made, the assumption is that these have natural or Neumann boundary conditions. This means that the normal derivative of the potential vanishes along them. This implies that the potential lines (in 2D) or surfaces (in 3D) enter the boundary at right angles. These types of boundary are usually used to express some symmetry condition. For instance, a long conducting cylinder centered inside a rectangular grounded box can be modeled by means of a circle inside the box. However, by symmetry, only a quarter of the geometry centered on the circle need be modeled and natural boundary conditions imposed on the new boundaries created by isolating this region. This is depicted in Figure 11.21.

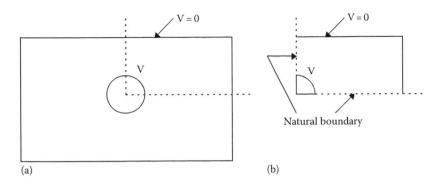

FIGURE 11.21
Using symmetry to simplify the finite element problem: (a) full geometry and (b) 1/4 geometry.

The method works here because the potential lines are concentric ovoids about the conducting cylinder and therefore enter the dotted line boundaries in the 1/4 geometry at right angles. Although the problem depicted in Figure 11.21 is relatively simple to solve in the full geometry so that the use of symmetry does not save much in input or solution times, this technique can save much effort when more complicated geometries are involved.

Another boundary condition, which some programs allow, is the balloon boundary. This type of boundary is essentially specifying that the boundary doesn't exist and the solution continues beyond it as if there were empty space out far in that direction. There is a practical necessity for this type of boundary condition since the finite element technique requires that the entire solution space of interest be subdivided into triangles or elements. When this solution space extends far enough that the geometric region of interest would be dwarfed relative to the whole space, it is convenient to specify such balloon boundaries rather than model vast regions of empty space. In practice, the program actually models a much larger region of space but with much coarser elements far away from the geometry of interest. This process is transparent to the user.

We now give some examples of solving electrostatic problems with a finite element program. We use Ansoft's Maxwell® 2D software [Ansoft]. The first is a varistor stack assembly shown in Figure 11.22. This is an axisymmetric geometry. The varistors themselves are modeled as two continuous cylinders separated by a metallic region in the center, which is floating. The ends consist of shaped metal electrodes to help reduce the end fields. There are pressboard disks that mechanically hold the assembly together inside a pressboard cylinder. There is another pressboard cylinder outside followed by the ground cylinder that defines the outermost boundary. The top and bottom boundaries are balloon boundaries. Although the geometry of the varistor stack itself is really cylindrical, the ground may not, in fact, be a concentric cylinder as modeled. For instance, it may be a tank wall. However, it is far enough away that a cylindrical approximation is reasonable.

The varistor cylinders, the pressboard elements, and the oil were given appropriate permittivities. The metallic end caps and center region were declared perfect conductors. Different voltages were specified on the two end caps and the outer boundary cylinder was given a zero voltage. The center metallic conductor was floating. The equipotential lines obtained by solving this problem are shown. Since the electric field is the gradient of the potential, where the lines are closely spaced, the field is highest. The field itself can be

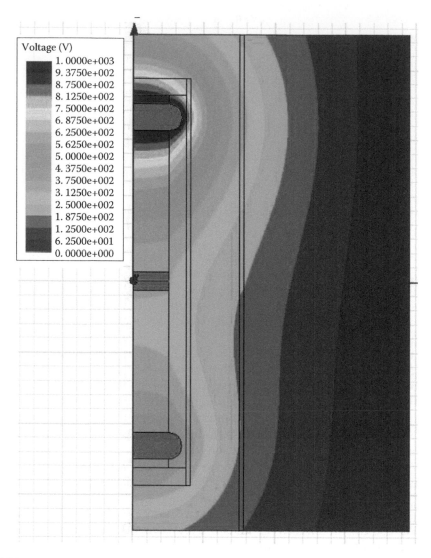

FIGURE 11.22
Axisymmetric model of a varistor stack with equipotential contours. The top metal cap was at 1000 V and the bottom cap at 500 V.

calculated and displayed vectorially as shown in Figure 11.23. The field can also be obtained numerically at specified points, along specified lines, or within specified regions.

A second example is of a region near the tops of the LV and HV windings with a static ring on top of the HV winding. These rings are used to moderate the electric fields near the tops of windings by providing a surface of relatively large curvature compared with the sharper corners, which occur on windings disks. The static rings are at the same potential as the top disk. These are most effective when close to the top disk. In the example shown in Figure 11.24, it is a few millimeters from the top disk. Since only the fields near the top disks of the two windings and static ring are of interest, the rest of the windings were modeled as solid blocks at the same potential as the top disks. Also shown are pressboard

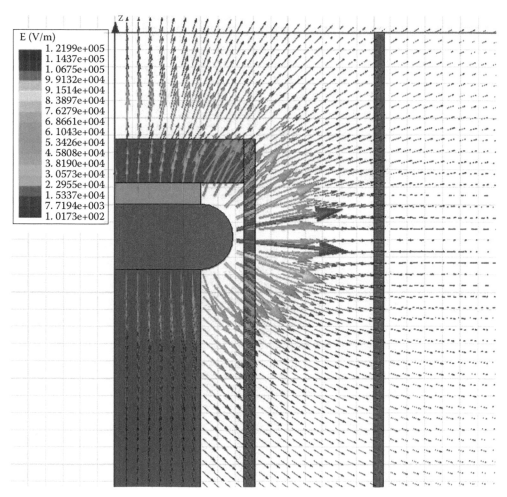

FIGURE 11.23
Electric field vectors near the top metal cap.

collars and cylinders, which are used to subdivide the large oil gaps to lower the electrical breakdown probabilities. A cylindrically symmetric model was used, but only a blowup of the region near the top of the two windings is shown in the figure.

Notice that the static ring has diminished the corner field at the top of the HV winding on the right but that the corner field on the LV winding on the left is fairly high. It could probably also benefit by having its own static ring or thicker paper cover.

A third example is that of a bushing shield that surrounds a lead entering a bushing, which is attached to the transformer tank top. The shield and the tank portion of the lead are immersed in the tank oil in the model shown in Figure 11.25. Only the lower part of the model is studied for electrical purposes. Also the model is axisymmetric about the center-line at the left of the figure. Both the top and left boundary lines have natural or Neumann boundary conditions. Details of any structure inside the shield are also omitted since they are not relevant to the purpose of the study, which is to determine the electric stresses in the oil at the outer surface of the shield. This is a very practical problem that needs to be addressed, particularly when the lead is at a high voltage.

FIGURE 11.24
Electric field magnitude contours in the oil near the tops of an LV winding at ground potential on the left and an HV winding at 450 kV on the right topped by a static ring at the same potential. The model is axisymmetric with centerline along the center of the core leg. The key is in kV/m.

The electric stress contour plot in the oil surrounding a bushing shield is shown in the figure. The shield is at 1050 kV. The metal part of the shield is 330 mm from the vertical tank wall at the right, which is a 0 potential boundary line. The bottom boundary line is also at 0 potential. The distance from the metal part of the shield to the first pressboard barrier is 19 mm. The stress is in kV/m, although the plot shows V/m. Therefore, the maximum stress is 17.4 kV/mm and occurs along the rounded part of the shield. Avoiding sharp corners by rounding is an important device for reducing electric field strength at the surface of a bare metal part at high voltage.

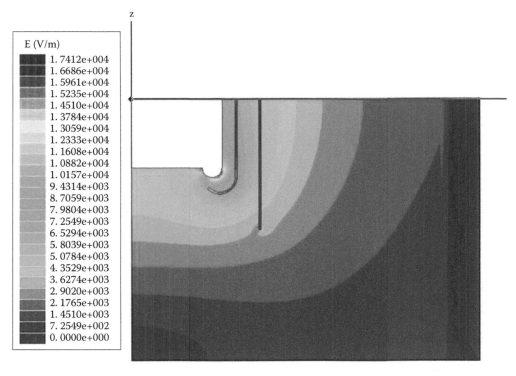

FIGURE 11.25
Simplified model of a bushing shield for studying electric stress in the surrounding oil. The model is axisymmetric with centerline along the center of the shield. The key is in kV/m, not V/m as shown.

12

Capacitance Calculations

12.1 Introduction

Under impulse conditions, very fast voltage pulses are applied to a transformer. These contain high-frequency components, eliciting capacitative effects that are absent at normal operating frequencies. Thus, in order to simulate the behavior of a transformer under impulse conditions, capacitances must be determined for use in circuit or traveling wave models [Mik78], [Rud40].

Usually, the highest electrical stresses occur at the high-voltage end of the winding so that modifications are sometimes made to the first few disks to meet voltage breakdown limits. These modifications commonly take the form of the addition of one or more static rings so that it is important to determine their effect on disk capacitances. Other methods such as the use of wound-in-shields and interleaving are effective in increasing the disk capacitance, which is desirable for better voltage distribution. Wound-in-shields are discussed in this chapter, but interleaving is treated only for the particular variety used in multi-start tap windings.

We employ an energy method to determine the capacitance or, in general, the capacitance matrix. This method is a generalization of the method used by Stein to determine the disk capacitance of a disk embedded in a winding of similar disks [Ste64]. It utilizes a continuum model of a disk so that disks having many turns are contemplated. We also compare capacitances determined in this manner with capacitances determined using a more conventional approach. The conventional method also works for helical windings, that is, windings having one turn per disk, so that it is useful in its own right.

For completeness, we also calculate winding-to-winding, winding-to-core, and winding-to-tank capacitances, which are all qualitatively similar. These are based on a simpler model based on an infinitely long coil with no end effects. They are really capacitances per unit length.

12.2 Distributive Capacitance along a Winding or Disk

We try to be as general as possible so that we may apply the results to a variety of situations. Thus, we consider a disk or coil section having a series capacitance per unit length, c_s, and shunt capacitances per unit length, c_a and c_b, to the neighboring objects on either

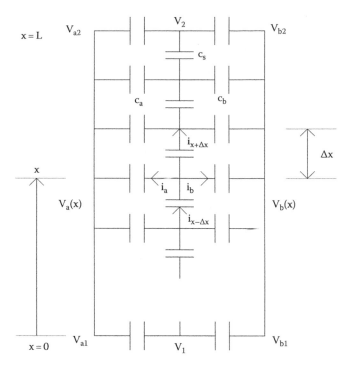

FIGURE 12.1
Approximately continuous capacitance distribution of a coil section.

side of the disk or coil section as shown in Figure 12.1. These neighboring objects are assumed, for generality, to have linear voltage distributions given by

$$V_a(x) = V_{a1} + \frac{x}{L}(V_{a2} - V_{a1})$$
$$V_b(x) = V_{b1} + \frac{x}{L}(V_{b2} - V_{b1})$$

(12.1)

where
  L is the length of the disk or coil section
  x measures distances from the high-voltage end at V_1 to the low-voltage end at V_2

Setting $V_{a1} = V_{a2} = V_a$ and $V_{b1} = V_{b2} = V_b$ gives these neighboring objects a constant voltage.
  Applying current conservation to the node at x in Figure 12.1 and assuming the current directions shown, we have

$$i_{x-\Delta x} - i_{x+\Delta x} = i_a + i_b$$

(12.2)

Using the current–voltage relationship for a capacitor, $i = C dV/dt$, and letting V(x) denote voltages along the vertical centerline of Figure 12.1, we can rewrite (12.2) as

$$c_s \Delta x \frac{d}{dt}\left[V(x - \Delta x) - V(x)\right] - c_s \Delta x \frac{d}{dt}\left[V(x) - V(x + \Delta x)\right] = c_a \Delta x \frac{d}{dt}\left[V(x) - V_a(x)\right]$$
$$+ c_b \Delta x \frac{d}{dt}\left[V(x) - V_b(x)\right]$$

(12.3)

Rearranging, we find

$$\frac{d}{dt}\left\{c_s\left[V(x+\Delta x)-2V(x)+V(x-\Delta x)\right]-(c_a+c_b)V(x)+c_aV_a(x)+c_bV_b(x)\right\}=0 \quad (12.4)$$

Thus, the quantity in curly brackets is a constant in time and may be set to 0 for all practical purposes. (For an applied pulse, all the voltages are zero after a very long time so the constant must be zero.) Dividing the constant curly bracket term in (12.4) by $c_s(\Delta x)^2$, we obtain

$$\frac{V(x+\Delta x)-2V(x)+V(x-\Delta x)}{(\Delta x)^2}-\frac{(c_a+c_b)}{c_s(\Delta x)^2}V(x)+\frac{\left[c_aV_a(x)+c_bV_b(x)\right]}{c_s(\Delta x)^2}=0 \quad (12.5)$$

The first term in (12.5) is the finite difference approximation to d^2V/dx^2. The combination

$$\frac{(c_a+c_b)}{c_s(\Delta x)^2} \quad (12.6)$$

can be expressed in terms of the total series capacitance C_s and total shunt capacitances C_a and C_b using

$$C_s=\frac{c_s\Delta x}{N}, \quad C_a=c_a\Delta xN, \quad C_b=c_b\Delta xN \quad (12.7)$$

where N is the number of subdivisions of the total length L into units of size Δx, $N=L/\Delta x$. Using this value for N in (12.7), we get

$$C_s=\frac{c_s(\Delta x)^2}{L}, \quad C_a=c_aL, \quad C_b=c_bL \quad (12.8)$$

Substituting into (12.6), we obtain

$$\frac{(c_a+c_b)}{c_s(\Delta x)^2}=\frac{C_a+C_b}{C_sL^2}=\frac{\alpha^2}{L^2} \quad (12.9)$$

where

$$\alpha=\sqrt{\frac{C_a+C_b}{C_s}} \quad (12.10)$$

Similarly,

$$\frac{c_a}{c_s(\Delta x)^2}=\frac{C_a}{C_sL^2}, \quad \frac{c_b}{c_s(\Delta x)^2}=\frac{C_b}{C_sL^2} \quad (12.11)$$

Substituting into (12.5) and taking the limit as $\Delta x \to 0$ result in the differential equation

$$\frac{d^2 V}{dx^2} - \left(\frac{\alpha}{L}\right)^2 V = -\frac{1}{C_s L^2}\left[C_a V_a(x) + C_b V_b(x)\right] \tag{12.12}$$

The solution to the homogeneous part of (12.12) is

$$V = A e^{(\alpha/L)x} + B e^{-(\alpha/L)x} \tag{12.13}$$

where A and B are constants to be determined by the boundary conditions. For the inhomogeneous part of (12.12), we try

$$V = F + Gx \tag{12.14}$$

where F and G are determined by substituting into (12.12) and using (12.1). Doing this, we get

$$-\left(\frac{\alpha}{L}\right)^2 (F + Gx) = -\frac{1}{C_s L^2}\left\{C_a V_{a1} + C_b V_{b1} + \frac{x}{L}\left[C_a(V_{a2} - V_{a1}) + C_b(V_{b2} - V_{b1})\right]\right\} \tag{12.15}$$

Collecting terms, we find

$$F = \frac{1}{\alpha^2 C_s}(C_a V_{a1} + C_b V_{b1})$$
$$G = \frac{1}{\alpha^2 C_s L}\left[C_a(V_{a2} - V_{a1}) + C_b(V_{b2} - V_{b1})\right] \tag{12.16}$$

Thus, the general solution to (12.12) is

$$V(x) = A e^{(\alpha/L)x} + B e^{-(\alpha/L)x} + \frac{1}{\alpha^2 C_s}(C_a V_{a1} + C_b V_{b1}) + \frac{1}{\alpha^2 C_s}\left(\frac{x}{L}\right)\left[C_a(V_{a2} - V_{a1}) + C_b(V_{b2} - V_{b1})\right] \tag{12.17}$$

Using the boundary conditions $V = V_1$ at $x = 0$ and $V = V_2$ at $x = L$, we can solve (12.17) for A and B. Performing the algebra, the solution can be cast in the form

$$V(x) = (\gamma_a V_{a1} + \gamma_b V_{b1}) + \left(\frac{x}{L}\right)\left[\gamma_a(V_{a2} - V_{a1}) + \gamma_b(V_{b2} - V_{b1})\right]$$
$$+ \frac{1}{\sinh \alpha}\left\{\left[V_2 - (\gamma_a V_{a2} + \gamma_b V_{b2})\right]\sinh\left(\alpha \frac{x}{L}\right)\right.$$
$$\left. + \left[V_1 - (\gamma_a V_{a1} + \gamma_b V_{b1})\right]\sinh\left[\alpha\left(1 - \frac{x}{L}\right)\right]\right\} \tag{12.18}$$

where

$$\gamma_a = \frac{C_a}{C_a + C_b}, \quad \gamma_b = \frac{C_b}{C_a + C_b} \tag{12.19}$$

so that $\gamma_a + \gamma_b = 1$. The derivative of this expression is also needed for evaluating the energy and for obtaining turn–turn voltages. It is

$$
\begin{aligned}
\frac{dV}{dx} = &\frac{1}{L}\left[\gamma_a\left(V_{a2} - V_{a1}\right) + \gamma_b\left(V_{b2} - V_{b1}\right)\right] \\
&+ \frac{\alpha}{L\sinh\alpha}\left\{\left[V_2 - \left(\gamma_a V_{a2} + \gamma_b V_{b2}\right)\right]\cosh\left(\alpha\frac{x}{L}\right)\right. \\
&\left. - \left[V_1 - \left(\gamma_a V_{a1} + \gamma_b V_{b1}\right)\right]\cosh\left[\alpha\left(1 - \frac{x}{L}\right)\right]\right\}
\end{aligned}
\tag{12.20}
$$

To determine the capacitance, the stored electrostatic energy must be evaluated. This energy is given by $C = (1/2)C(\Delta V)^2$, where C is the capacitance and ΔV the voltage across the capacitor. Reverting to the original discrete notation, the energy in the series capacitance is

$$E_{series} = \frac{1}{2}\sum_1^N c_s\Delta x(\Delta V)^2 \tag{12.21}$$

Using (12.8), this becomes

$$E_{series} = \frac{C_s L}{2}\sum_1^N \frac{1}{\Delta x}(\Delta V)^2 \tag{12.22}$$

Substituting $\Delta V = (dV/dx)\Delta x$ into (12.22) and letting $\Delta x \to 0$, we obtain

$$E_{series} = \frac{C_s L}{2}\int_0^L \left(\frac{dV}{dx}\right)^2 dx \tag{12.23}$$

The energy in the shunt capacitances can be found similarly:

$$E_{shunt} = \frac{1}{2}\sum_1^N c_a\Delta x\left[V(x) - V_a(x)\right]^2 + \frac{1}{2}\sum_1^N c_b\Delta x\left[V(x) - V_b(x)\right]^2 \tag{12.24}$$

Again, using (12.8) and taking the limit as $\Delta x \to 0$, we get

$$E_{shunt} = \frac{C_a}{2L}\int_0^L \left[V(x) - V_a(x)\right]^2 dx + \frac{C_b}{2L}\int_0^L \left[V(x) - V_b(x)\right]^2 dx \tag{12.25}$$

Combining (12.23) and (12.25), the total energy is

$$E = \frac{C_s L}{2} \int_0^L \left(\frac{dV}{dx} \right)^2 dx + \frac{C_a}{2L} \int_0^L \left[V(x) - V_a(x) \right]^2 dx + \frac{C_b}{2L} \int_0^L \left[V(x) - V_b(x) \right]^2 dx \quad (12.26)$$

Substituting (12.18) and (12.20) into (12.26) and performing the integrations, we obtain

$$E = \frac{C_s}{2} \left\{ \left[(V_1 - \beta_1)^2 + (V_2 - \beta_2)^2 \right] \frac{\alpha}{\tanh \alpha} - 2(V_1 - \beta_1)(V_2 - \beta_2) \frac{\alpha}{\sinh \alpha} - \eta^2 - 2\eta(V_1 - V_2) \right.$$
$$+ \alpha^2 \gamma_a \left[(\beta_1 - V_{a1})^2 - (\beta_1 - V_{a1}) \left[\eta - (V_{a2} - V_{a1}) \right] - \frac{1}{3} \left[\eta - (V_{a2} - V_{a1}) \right]^2 \right]$$
$$\left. + \alpha^2 \gamma_b \left[(\beta_1 - V_{b1})^2 + (\beta_1 - V_{b1}) \left[\eta - (V_{b2} - V_{b1}) \right] - \frac{1}{3} \left[\eta - (V_{b2} - V_{b1}) \right]^2 \right] \right\} \quad (12.27)$$

where

$$\beta_1 = \gamma_a V_{a1} + \gamma_b V_{b1}, \quad \beta_2 = \gamma_a V_{a2} + \gamma_b V_{b2}, \quad \eta = \beta_2 - \beta_1 \quad (12.28)$$

For most of the applications of interest here, the side objects on which the shunt capacitances terminate are at a constant potential. Thus, as mentioned previously, to achieve this we should set

$$V_{a1} = V_{a2} = V_a, \quad V_{b1} = V_{b2} = V_b \quad (12.29)$$

From this and (12.28), we see that

$$\beta_1 = \beta_2 = \beta = \gamma_a V_a + \gamma_b V_b, \quad \eta = 0 \quad (12.30)$$

For this case, Equation 12.18 becomes

$$V(x) = \frac{1}{\sinh \alpha} \left\{ (V_2 - \beta) \sinh \left(\alpha \frac{x}{L} \right) + (V_1 - \beta) \sinh \left[\alpha \left(1 - \frac{x}{L} \right) \right] \right\} + \beta \quad (12.31)$$

Also (12.20) becomes

$$\frac{dV}{dx} = \frac{\alpha}{L \sinh \alpha} \left\{ (V_2 - \beta) \cosh \left(\alpha \frac{x}{L} \right) - (V_1 - \beta) \cosh \left[\alpha \left(1 - \frac{x}{L} \right) \right] \right\} \quad (12.32)$$

The energy expression in (12.27) reduces to

$$E = \frac{C_s}{2} \left\{ \left[(V_1 - \beta)^2 + (V_2 - \beta)^2 \right] \frac{\alpha}{\tanh \alpha} - 2(V_1 - \beta)(V_2 - \beta) \frac{\alpha}{\sinh \alpha} \right.$$
$$\left. + \alpha^2 \left[\gamma_a (\beta - V_a)^2 - \gamma_b (\beta - V_b)^2 \right] \right\} \quad (12.33)$$

12.3 Stein's Disk Capacitance Formula

As an example, we can apply the results obtained to the case considered by Stein consisting of a disk embedded in a coil of similar disks [Ste64]. The situation is shown in Figure 12.2. In this application, a single disk is a substitute for the coil in Section 12.2. Thus, V is the voltage drop across an individual disk, so we can take $V_1 = V$ and $V_2 = 0$. This applies to the middle disk in the figure. Starting from the right side of the bottom disk at potential $-V$ and proceeding toward the left in the direction of voltage increase, we see that the voltage on the left side of the bottom disk is zero. This is the same as the voltage on the left side of the middle disk since they are connected at that point. Proceeding along the middle disk, the voltage reaches V on the right side of that disk. But this is connected to the outside of the top disk at that point, so its voltage is also V. Proceeding along the top disk toward the left, we reach a voltage of 2 V at its left side.

Here, we are assuming that neighboring disks have the same voltage drop. Thus, we can imagine equipotential planes between the disks with the values shown by dotted lines in the figure for the middle disk, namely, $V_a = V$ and $V_b = 0$. As we move along the middle disk, these values are the average of the potential on the middle disk and the potential on the neighboring disk and can be taken as representing the potential value midway between them. This middle value remains constant as we move along one of the dotted lines, namely, V or 0.

The capacitances to the midplane are twice the disk–disk capacitance, $C_a = C_b = 2C_{dd}$, since they are equal and in series. Since the two capacitances are equal, this means that $\gamma_a = \gamma_b = 1/2$ and $\beta = V/2$. For these values, including $V_1 = V$ and $V_2 = 0$, the energy from (12.33) becomes

$$E = \frac{C_s V^2}{2}\left[\frac{1}{2}\frac{\alpha}{\tanh\alpha} + \frac{1}{2}\frac{\alpha}{\sinh\alpha} + \frac{\alpha^2}{4}\right] \tag{12.34}$$

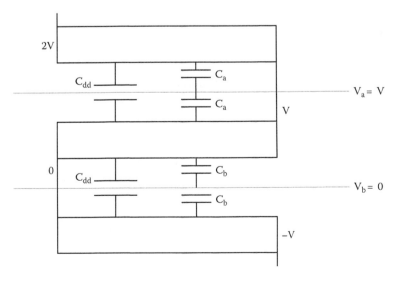

FIGURE 12.2
Disk embedded in a coil of similar disks with V, the voltage drop along a disk.

The effective disk capacitance is found from $E = 1/2\, CV^2$ and is

$$C_{Stein} = C_s \left[\frac{1}{2} \frac{\alpha}{\tanh \alpha} + \frac{1}{2} \frac{\alpha}{\sinh \alpha} + \frac{\alpha^2}{4} \right] \tag{12.35}$$

where C_s is the series capacitance of the disk and

$$\alpha = \sqrt{\frac{C_a + C_b}{C_s}} = \sqrt{\frac{4C_{dd}}{C_s}} \tag{12.36}$$

where C_{dd} is the disk–disk capacitance.

Let us compare this result with the more conventional approach. This assumes that the voltage drop along the disk is linear so that

$$V(x) = V\left(1 - \frac{x}{L}\right) \tag{12.37}$$

Substituting the derivative of (12.37) into (12.23), we get the energy in the series turns:

$$E_{series} = \frac{1}{2} C_s V^2 \tag{12.38}$$

We must consider the shunt energy with respect to the equipotential midplanes as before since we want the energy associated with a single disk. This is, using (12.25),

$$E_{shunt} = \frac{1}{2}\left(\frac{2C_{dd}}{L}\right) \int_0^L \left\{ \left[V\left(1 - \frac{x}{L}\right) - V \right]^2 + \left[V\left(1 - \frac{x}{L}\right) - 0 \right]^2 \right\} dx \tag{12.39}$$

Evaluating this expression, we find

$$E_{shunt} = \frac{1}{2}\left(\frac{4}{3} C_{dd}\right) V^2 \tag{12.40}$$

Combining (12.38) and (12.40) and extracting the effective disk capacitance, we obtain

$$C_{conv} = C_s + \frac{4}{3} C_{dd} \tag{12.41}$$

This expression can be applied to a helical winding (one turn/disk) by taking $C_s = 0$.

In order to compare (12.35) with (12.41), let us normalize by dividing both by C_s. Thus,

$$\frac{C_{Stein}}{C_s} = \frac{1}{2} \frac{\alpha}{\tanh \alpha} + \frac{1}{2} \frac{\alpha}{\sinh \alpha} + \frac{\alpha^2}{4} \tag{12.42}$$

and

$$\frac{C_{conv}}{C_s} = 1 + \frac{4}{3} \frac{C_{dd}}{C_s} = 1 + \frac{\alpha^2}{3} \tag{12.43}$$

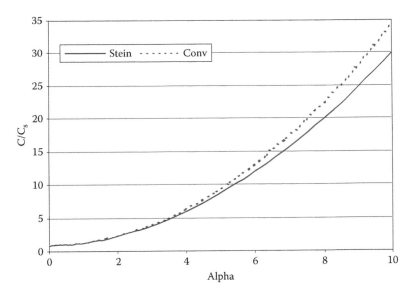

FIGURE 12.3
Comparison of Stein's and conventional capacitance formulas.

using (12.36) for α. Thus, the right-hand side of (12.42) is a function of α, as is the right-hand side of (12.43). For small α, it can be shown that (12.42) approaches (12.43). For larger α, the comparison is shown graphically in Figure 12.3. The difference becomes noticeable at values of $\alpha > 5$. At $\alpha = 10$, the conventional capacitance is about 15% larger than Stein's.

The voltage distribution along the disk, using Stein's method, can be obtained from (12.31); substituting $V_1 = V$, $V_2 = 0$, and $\beta = V/2$,

$$V(x) = \frac{V}{2}\left\{1 + \frac{1}{\sinh\alpha}\left[\sinh\left[\alpha\left(1-\frac{x}{L}\right)\right] - \sinh\left(\alpha\frac{x}{L}\right)\right]\right\} \tag{12.44}$$

This is plotted in Figure 12.4 in normalized form, that is, as $V(x)/V$. As can be seen, the voltage becomes increasingly less uniform as α increases. The $\alpha = 1$ case closely approximates the linear distribution given in (12.37) used for the conventional capacitance calculation.

The voltage gradient is obtained from (12.32) with the appropriate substitutions and is

$$\frac{dV}{dx} = -\frac{\alpha V}{2L\sinh\alpha}\left\{\cosh\left[\alpha\left(1-\frac{x}{L}\right)\right] + \cosh\left(\alpha\frac{x}{L}\right)\right\} \tag{12.45}$$

This is always negative as Figure 12.4 indicates. Its largest value, in absolute terms, occurs at either end of the disk:

$$\left.\frac{dV}{dx}\right|_{max} = \frac{\alpha V}{2L\sinh\alpha}(1+\cosh\alpha) = \frac{\alpha V}{2L\tanh(\alpha/2)} \tag{12.46}$$

As $\alpha \to 0$, this approaches the uniform value of V/L. This voltage gradient is equal to the stress (electric field magnitude) in the turn–turn insulation, which must be able to handle

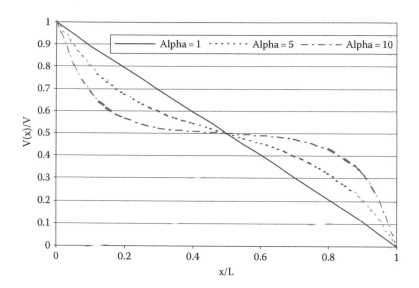

FIGURE 12.4
Normalized voltage along the disk for various values of alpha in Stein's example.

it without breakdown. From the figure, it can be seen that the absolute value of the voltage gradient has its smallest value at the center of the disk.

12.3.1 Determining Practical Values for the Series and Shunt Capacitances, C_s and C_{dd}

In this and later applications, the series C_s and shunt or disk–disk C_{dd} capacitances must be determined for use in practical design applications. As shown in Figure 12.5, a typical disk consists of N_t turns, usually rectangular in cross-section, with paper thickness τ_p between turns. τ_p is the two-sided paper thickness of a turn. The disks are separated by means of key spacers of thickness τ_{ks} and width w_{ks} spaced around the circumference.

The turn–turn capacitance, C_{tt}, is given approximately by

$$C_{tt} = \varepsilon_o \varepsilon_p 2\pi R_{ave} \frac{\left(h + 2\tau_p\right)}{\tau_p} \tag{12.47}$$

where
   ε_o is the permittivity of vacuum = 8.8542×10^{-12} F/m in the SI system
   ε_p is the relative permittivity of paper ($\cong 3.5$ for oil-soaked paper)
   R_{ave} is the average radius of the disk
   h is the bare copper or conductor height

The addition of $2\tau_p$ to h is designed to take fringing effects into account. There are $N_t - 1$ turn–turn capacitances in series, which results in a total series capacitance of $C_{tt}/(N_t - 1)$.

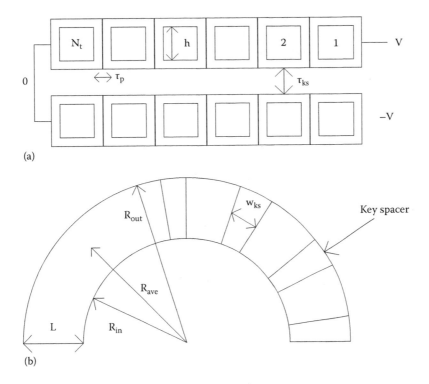

FIGURE 12.5
Geometry of a practical disk coil: (a) side view and (b) top view.

However, they do not see the full disk voltage drop but only the fraction $(N_t - 1)/N_t$. Thus, the capacitive energy is

$$E = \frac{1}{2}\left(\frac{C_{tt}}{N_t-1}\right)\left(\frac{N_t-1}{N_t}\right)^2 V^2 \tag{12.48}$$

so that, extracting the full series capacitance, we get

$$C_s = C_{tt}\frac{(N_t-1)}{N_t^2} \tag{12.49}$$

This makes sense because for $N_t = 1$, we get $C_s = 0$.

The disk–disk capacitance can be considered to be two capacitances in parallel, namely, the capacitance of the portion containing the key spacers and the capacitance of the remainder containing an oil or air thickness instead of key spacers. Let f_{ks} be the key spacer fraction

$$f_{ks} = \frac{N_{ks}w_{ks}}{2\pi R_{ave}} \tag{12.50}$$

where N_{ks} is the number of key spacers spaced around the circumference and w_{ks} their width. Typically, $f_{ks} \approx 1/3$. The key spacer fraction of the disk–disk space is filled with two dielectrics, paper, and a pressboard, the latter being the usual key spacer material. For a planar capacitor containing two dielectric layers in series of permittivity ε_1 and ε_2, it follows from electrostatic theory that the capacitance is

$$C = \frac{A}{\left(\left(\ell_1/\varepsilon_1\right)+\left(\ell_2/\varepsilon_2\right)\right)} \tag{12.51}$$

where
  A is the area
  ℓ_1 and ℓ_2 are the thicknesses of the layers

Applying this to the disk–disk capacitance, we obtain

$$C_{dd} = \varepsilon_0 \pi \left(R_{out}^2 - R_{in}^2\right)\left[\frac{f_{ks}}{\left(\left(\tau_p/\varepsilon_p\right)+\left(\tau_{ks}/\varepsilon_{ks}\right)\right)} + \frac{\left(1-f_{ks}\right)}{\left(\left(\tau_p/\varepsilon_p\right)+\left(\tau_{ks}/\varepsilon_{oil}\right)\right)}\right] \tag{12.52}$$

where
  ε_{ks} is the relative permittivity of the key spacer material (=4.5 for an oil-soaked pressboard)
  ε_{oil} is the oil-relative permittivity (=2.2 for transformer oil)
  R_{in} and R_{out} are the inner and outer radii of the disk, respectively

A finite element simulation to test the Stein capacitance formula (12.35) was conducted. A 16-turn disk was modeled as shown in Figure 12.6. Static rings are used to define the equipotential surfaces assumed in Stein's method. The upper static ring is at potential V and the bottom at V = 0 V. The rightmost turn is at potential V and the leftmost turn at V = 0.

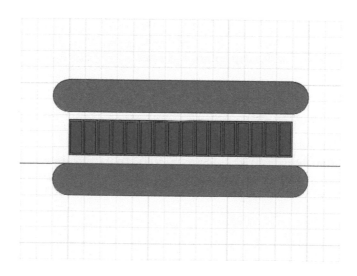

FIGURE 12.6
Disk coil of 16 turns with static rings on either side supplying equipotential surfaces at the voltages assumed in Stein's method.

The model shown is an enlargement of the region near the disk. The full model is axisymmetric and the distance from the leftmost side of the disk to the centerline of the core leg is 300 mm. The radial build of the disk is 70 mm and the disk height, including paper, is 11 mm.

The bare conductor dimensions of a single turn are 10 mm high by 3.375 mm wide. The 2-sided paper thickness is 1 mm. The distance from the paper to the nearest equipotential is 2.5 mm so that the disk–disk spacing is 5 mm. Since the model is axisymmetric, the disk–disk material was chosen to be all oil (no key spacers).

From these dimensions, we can calculate C_{tt} given in (12.47). The parameters in the formula are in SI units, so we have $R_{ave} = 0.335$ m, $h = 0.01$ m, $\tau_p = 0.001$ m, and $\varepsilon_p = 3.5$. ε_o was given previously. We obtain $C_{tt} = 7.8275 \times 10^{-10}$ F. Since there are $N_t = 16$ turns, we get $C_s = 4.5864 \times 10^{-11}$ F from (12.49).

We can obtain C_{dd} from (12.52) using $R_{in} = 0.3$ m, $R_{out} = 0.37$ m, $f_{ks} = 0$, $\tau_p = 0.001$ m, $\varepsilon_p = 3.5$, $\tau_{ks} = 0.005$ m, and $\varepsilon_{oil} = 2.2$. We get $C_{dd} = 5.0991 \times 10^{-10}$ F. Using (12.36), we see that $\alpha = 6.6687$. Inserting this value of α and C_s into Stein's capacitance formula (12.35), we get $C_{Stein} = 6.6322 \times 10^{-10}$ F.

For the finite element calculation, we chose $V = 100$ V. We distributed the voltage along the turns according to (12.44) using $\alpha = 6.6687$. This can be visualized with reference to Figure 12.4. The equipotential contour plot is shown in Figure 12.7. We chose balloon boundary conditions for the outer boundaries. These, along with the centerline of the axisymmetric geometry, are not shown.

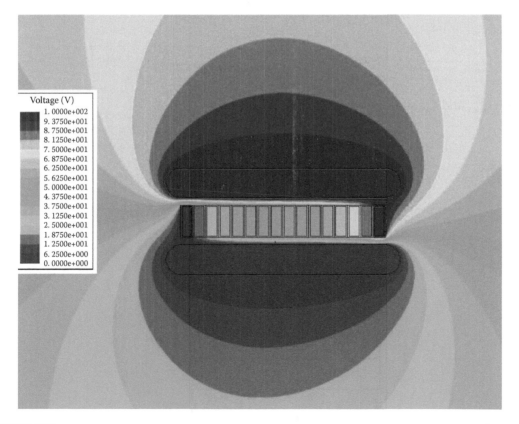

FIGURE 12.7
Equipotential plot near the disk as a test of Stein's capacitance formula.

From the solution, the total energy was calculated as 4.1864×10^{-6} J. The capacitance can be extracted from the formula Energy $= 1/2C(\Delta V)^2$. We get $C_{\text{Finite Element}} = 8.373 \times 10^{-10}$ F. This is 26% higher than the value obtained using Stein's formula. We notice from Figure 12.3 that, for this value of α, the conventional capacitance formula is about 10% higher than Stein's formula, so it would apparently give better agreement with the finite element result. However, the conventional capacitance formula assumes a linear voltage distribution along the winding. If we use a linear voltage distribution along the winding in the finite element calculation, we find that the capacitance is about 26% higher than the conventional capacitance formula so that there is no apparent advantage in using the conventional versus Stein's formula, based on the finite element results. Perhaps some of the discrepancy can be attributed to the fact the finite element model shows some fringing flux whereas Stein's model assumes no fringing flux.

12.4 General Disk Capacitance Formula

More generally, if the disk–disk spacings on either side of the main disk are unequal so that $C_a \neq C_b$ as shown in Figure 12.8, then we have, from (12.33),

$$C_{\text{general}} = C_s\left[\left(\gamma_a^2 + \gamma_b^2\right)\frac{\alpha}{\tanh\alpha} + 2\gamma_a\gamma_b\frac{\alpha}{\sinh\alpha} + \gamma_a\gamma_b\alpha^2\right] \tag{12.53}$$

where

$$\gamma_a = \frac{C_{dd1}}{C_{dd1} + C_{dd2}}, \quad \gamma_b = \frac{C_{dd2}}{C_{dd1} + C_{dd2}}, \quad \alpha = \sqrt{\frac{2\left(C_{dd1} + C_{dd2}\right)}{C_s}} \tag{12.54}$$

with C_{dd1} and C_{dd2} the unequal disk–disk capacitances.

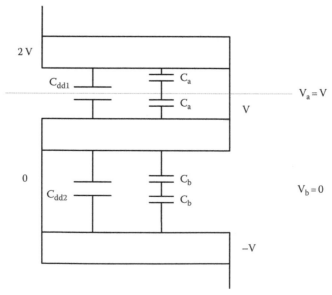

FIGURE 12.8
A more general case of a disk embedded in a coil of similar disks.

12.5 Coil Grounded at One End with Grounded Cylinders on Either Side

An early impulse model for a coil assumed that the coil consisted of a uniformly distributed chain of series capacitances connected to ground cylinders on either side by shunt capacitances, that is, the same model shown in Figure 12.1 but with $V_{a1} = V_{a2} = V_{b1} = V_{b2} = 0$ [Blu51]. The V_1 terminal was impulsed with a voltage V and the V_2 terminal was grounded. From (12.30), we have $\beta = 0$ and (12.31) becomes

$$V(x) = V \frac{\sinh\left[\alpha\left(1-(x/L)\right)\right]}{\sinh\alpha} \tag{12.55}$$

with $\alpha = \sqrt{C_g/C_s}$. Here, C_g is the total ground capacitance (both sides) and C_s the series capacitance of the coil. This is shown in normalized form in Figure 12.9 for several values of α.

This voltage distribution is expected to apply immediately after the application of the impulse voltage, before inductive effects come into play. Later, oscillations can cause voltage swings above the values shown in the figure.

The voltage gradient is given by

$$\frac{dV}{dx} = -\frac{\alpha V}{L\sinh\alpha}\cosh\left[\alpha\left(1-\frac{x}{L}\right)\right] \tag{12.56}$$

The maximum gradient occurs at the line end ($x = 0$) and is

$$\left|\frac{dV}{dx}\right|_{max} = \frac{\alpha V}{L\tanh\alpha} \tag{12.57}$$

Thus, the maximum disk–disk voltage immediately after impulse is approximately (12.57) multiplied by the disk–disk spacing.

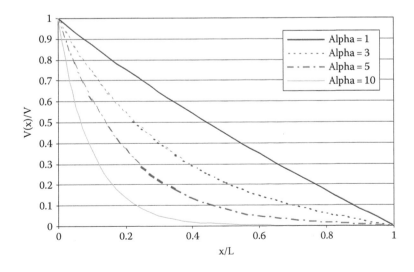

FIGURE 12.9
Graph of normalized voltage along a coil for several values of α.

The coil's total capacitance to ground is, from (12.33),

$$C_{coil} = C_s \frac{\alpha}{\tanh \alpha} \tag{12.58}$$

In this as well as previous formulas in this section, the series capacitance is due to N_d disks in series and if the disk capacitances are obtained by Stein's formula, we have

$$C_s = \frac{C_{Stein}}{N_d} \tag{12.59}$$

For an inner coil, the surfaces of the neighboring coils are usually taken to be the ground cylinders. For the innermost coil, the core determines the ground on one side, while for an outermost coil, the tank is the ground on one side. In general, the distance to ground is filled with various dielectric materials, including oil or air. One such structure is shown in Figure 12.10. There are usually multiple pressboard layers, but for convenience these are grouped into a single layer.

The sticks provide spacing for cooling oil or air to flow. This composite structure is similar to that analyzed previously for the disk–disk capacitance. The ground spacing is usually small relative to the coil radius for power transformers so that an approximately planar geometry may be assumed. We obtain for the ground capacitance on one side of the coil, C_{g1}

$$C_{g1} = \varepsilon_o 2\pi R_{gap} H \left[\frac{f_s}{\left(\left(\tau_{press} / \varepsilon_{press} \right) + \left(\tau_s / \varepsilon_s \right) \right)} + \frac{\left(1 - f_s \right)}{\left(\left(\tau_{press} / \varepsilon_{press} \right) + \left(\tau_s / \varepsilon_{oil} \right) \right)} \right] \tag{12.60}$$

where f_s is the fraction of the space occupied by sticks (vertical spacers),

$$f_s = \frac{N_s w_s}{2\pi R_{gap}} \tag{12.61}$$

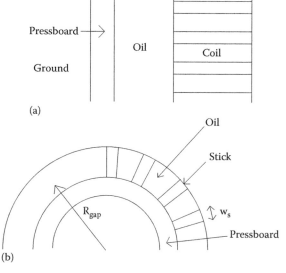

FIGURE 12.10
Ground capacitance geometry: (a) side view and (b) top view.

where
  R_{gap} is the mean gap radius
  w_s is the stick width
  N_s is the number of sticks around the circumference
  H is the coil height
  τ_{press} is the pressboard thickness
  ε_{press} is the pressboard relative permittivity
  τ_s and ε_s are the corresponding quantities for the sticks

The ground capacitance of both gaps C_{g1} and C_{g2} should be added to obtain the total ground capacitance, C_g.

12.6 Static Ring on One Side of a Disk

If a static ring is present on one side of a disk and connected to the terminal voltage as shown in Figure 12.11, then we have a situation similar to that considered in the general disk capacitance section. The only difference is that C_a is the disk–static ring capacitance since the static ring is an equipotential surface. Thus, (12.53) applies with

$$\gamma_a = \frac{C_a}{C_a + 2C_{dd}}, \quad \gamma_b = \frac{2C_{dd}}{C_a + 2C_{dd}}, \quad \alpha = \sqrt{\frac{C_a + 2C_{dd}}{C_s}} \tag{12.62}$$

where
  C_{dd} is the disk–disk capacitance to the lower disk
  C_s is the single disk series capacitance

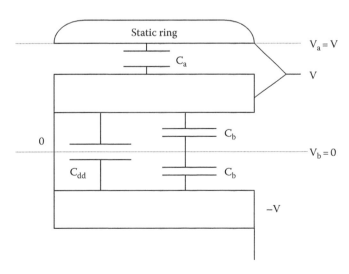

FIGURE 12.11
Static ring on one side of a disk at the end of a coil.

This case would be identical to Stein's if the static ring were spaced at 1/2 the normal disk–disk spacing or whatever is required to achieve $C_a = 2C_{dd} = C_b$. Then $\gamma_a = \gamma_b = 1/2$, $\alpha = \sqrt{4C_{dd}/C_s}$ and (12.53) would reduce to (12.35). Thus, the end disk would have the same capacitance as any other disk.

12.7 Terminal Disk without a Static Ring

In case the end disk does not have an adjacent static ring, we assume that the shunt capacitance on the end side is essentially 0. Then we have the situation shown in Figure 12.12a. We have $C_a = 0$ so that $\gamma_a = 0$ and $\gamma_b = 1$, and from (12.53),

$$C_{end} = C_s \frac{\alpha}{\tanh \alpha} \qquad (12.63)$$

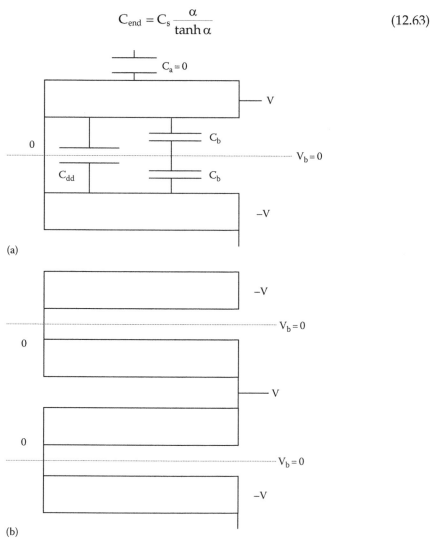

(a)

(b)

FIGURE 12.12
Terminal disk without a static ring: (a) top fed and (b) center fed.

with

$$\alpha = \sqrt{\frac{2C_{dd}}{C_s}}$$ (12.64)

This situation also applies to a center-fed winding without a static ring. In this case, as Figure 12.12b shows, there is no capacitative energy between the two center disks so that effectively $C_a = 0$. This result would also follow if both center disks were considered as a unit and the energy divided equally between them.

12.8 Capacitance Matrix

Before proceeding to other cases of interest, we need to introduce the capacitance matrix. For a system of conductors having voltages V_i and total electrostatic energy E from the general theory of linear capacitors [Smy68], we have

$$\frac{\partial E}{\partial V_1} = Q_1 = C_{11}V_1 + C_{12}V_2 + \cdots$$

$$\frac{\partial E}{\partial V_2} = Q_2 = C_{21}V_1 + C_{22}V_2 + \cdots$$ (12.65)

$$\vdots$$

or, in matrix form,

$$\begin{pmatrix} \dfrac{\partial E}{\partial V_1} \\[2mm] \dfrac{\partial E}{\partial V_2} \\[2mm] \vdots \\[2mm] \dfrac{\partial E}{\partial V_n} \end{pmatrix} = \begin{pmatrix} C_{11} & C_{12} & \cdots & C_{1n} \\ C_{21} & C_{22} & \cdots & C_{2n} \\ \vdots & \vdots & \ddots & \vdots \\ C_{n1} & C_{n2} & \cdots & C_{nn} \end{pmatrix} \begin{pmatrix} V_1 \\ V_2 \\ \vdots \\ V_n \end{pmatrix}$$

The C's can be grouped into a capacitance matrix that is symmetric, $C_{ij} = C_{ji}$. The diagonal terms are positive, while the off-diagonal terms are negative. This follows because if V_k is a positive voltage while all other voltages are 0, that is, the other conductors are grounded, then the charge on conductor k must be positive, $Q_k = C_{kk}V_k > 0 \Rightarrow C_{kk} > 0$. By charge conservation, since $V_k > 0$ and all other V's = 0,

$$\sum_i Q_i = \sum_{i,j} C_{ij}V_j = \sum_i C_{ik}V_k = 0$$ (12.66)

where
  Q_i is the charge on conductor i
  C_{ii} is the self-capacitance of conductor i
  C_{ij} is the mutual capacitance between conductors i and j

From the last equality in (12.66), we have

$$C_{kk} = -\sum_{i \neq k} C_{ik} \qquad (12.67)$$

that is, the negative of the sum of the off-diagonal terms in column k of the capacitance matrix equals the diagonal term. Since this matrix is symmetric, this also applies to the off-diagonal terms in the rows.

Let us apply this to the general energy expression (12.27). Consider the V_1 voltage node:

$$\frac{\partial E}{\partial V_1} = Q_1 = \frac{C_s}{2}\left[2(V_1 - \beta_1)\frac{\alpha}{\tanh \alpha} - 2(V_2 - \beta_2)\frac{\alpha}{\sinh \alpha} - 2(\beta_2 - \beta_1)\right]$$

$$= C_s\left[V_1\frac{\alpha}{\tanh \alpha} - V_2\frac{\alpha}{\sinh \alpha} - \gamma_a\left(\frac{\alpha}{\tanh \alpha} - 1\right)V_{a1} - \gamma_a\left(1 - \frac{\alpha}{\sinh \alpha}\right)V_{a2}\right.$$

$$\left. - \gamma_b\left(\frac{\alpha}{\tanh \alpha} - 1\right)V_{b1} - \gamma_b\left(1 - \frac{\alpha}{\sinh \alpha}\right)V_{b2}\right] \qquad (12.68)$$

Using the labeling scheme 1, 2, 3, 4, 5, 6 ↔ 1, 2, a1, a2, b1, b2, the off-diagonal mutual capacitances are, from (12.68),

$$C_{1,2} = -C_s\frac{\alpha}{\sinh \alpha}$$

$$C_{1,a1} = -C_s\gamma_a\left(\frac{\alpha}{\tanh \alpha} - 1\right), \quad C_{1,a2} = -C_s\gamma_a\left(1 - \frac{\alpha}{\sinh \alpha}\right) \qquad (12.69)$$

$$C_{1,b1} = -C_s\gamma_b\left(\frac{\alpha}{\tanh \alpha} - 1\right), \quad C_{1,b2} = -C_s\gamma_b\left(1 - \frac{\alpha}{\sinh \alpha}\right)$$

These are all negative and the negative of their sum is C_{11}, which is

$$C_{11} = C_s\frac{\alpha}{\tanh \alpha} \qquad (12.70)$$

The capacitance diagram corresponding to this situation is shown in Figure 12.13, which

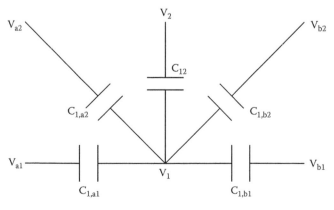

FIGURE 12.13
Lumped capacitance model of a general lattice capacitor network. The C's are taken to be positive.

shows only the capacitances attached to voltage node V_1. The other mutual capacitances can be filled in by a similar procedure. If the side voltages are constant so that $V_{a1} = V_{a2} = V_a$ and $V_{b1} = V_{b2} = V_b$, then there is only one mutual capacitance connecting 1 to a and it is given by

$$C_{1,a} = C_{1,a1} + C_{1,a2} = -C_s\gamma_a\left(\frac{\alpha}{\tanh\alpha} - \frac{\alpha}{\sinh\alpha}\right) \tag{12.71}$$

and similarly for $C_{1,b}$. For small α, this approaches

$$C_{1,a} = -C_s\gamma_a\frac{\alpha^2}{2} = -\frac{C_a}{2} \tag{12.72}$$

so that, in the case of small α, on the capacitance diagram, $1/2$ the shunt capacitance is attached to the V_1 node. If we carried through the analysis, we would find that $1/2$ of the shunt capacitance would also be attached to the V_2 node, producing a π capacitance diagram.

12.9 Two End Static Rings

When two static rings are positioned at the end of a coil, they are situated as shown in Figure 12.14. Both are attached to the terminal voltage V_1. This situation also applies to a center-fed coil with three static rings since adjacent pairs are configured similarly

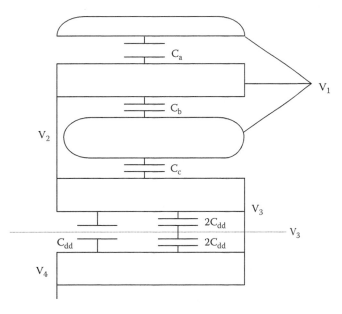

FIGURE 12.14
Two static rings at the end of a coil.

with respect to the top or bottom coil. It is necessary to analyze more than one disk at a time since their electrostatic energies are coupled via the static rings. We are allowing for the possibility of different spacings between the static rings and adjacent disks and between disks by letting the disk–static ring and disk–disk capacitances be different.

The energy associated with the first or top disk is found, using (12.33) with $V_a = V_b = V_1$ so that $\beta = V_1$:

$$E_{First\,disk} = \frac{C_s}{2}\left(V_2 - V_1\right)^2 \frac{\alpha}{\tanh \alpha} \tag{12.73}$$

where $\alpha = \sqrt{\left(C_a + C_b\right)/C_s}$, C_a and C_b being the disk–static ring capacitances. The energy of the second disk, assuming the same series capacitance C_s, is obtained from (12.33) with the substitutions $V_a = V_1$, $V_b = V_3$ so that $\beta = \gamma_a V_1 + \gamma_b V_3$. Notice that for this disk, the high-voltage end is at voltage V_2 and the low-voltage end at voltage V_3. Therefore, in (12.33), we must associate $V_1 \leftrightarrow V_2$ and $V_2 \leftrightarrow V_3$. We find

$$
\begin{aligned}
E_{Second\,disk} = \frac{C_s}{2}\Bigg\{ & \left[\left(V_2 - \gamma_a V_1 - \gamma_b V_3\right)^2 + \left(V_3 - \gamma_a V_1 - \gamma_b V_3\right)^2\right]\frac{\alpha_1}{\tanh \alpha_1} \\
& - 2\left(V_2 - \gamma_a V_1 - \gamma_b V_3\right)\left(V_3 - \gamma_a V_1 - \gamma_b V_3\right)\frac{\alpha_1}{\sinh \alpha_1} \\
& + \alpha_1^2\left[\gamma_a\left(V_1 - \gamma_a V_1 - \gamma_b V_3\right)^2 + \gamma_b\left(V_3 - \gamma_a V_1 - \gamma_b V_3\right)^2\right]\Bigg\}
\end{aligned}
\tag{12.74}
$$

where

$$\alpha_1 = \sqrt{\left(C_c + 2C_{dd}\right)/C_s}$$
$$\gamma_a = C_c/\left(C_c + 2C_{dd}\right)$$
$$\gamma_b = 2C_{dd}/\left(C_c + 2C_{dd}\right)$$

Using $\gamma_a + \gamma_b = 1$, this can be simplified as

$$
\begin{aligned}
E_{Second\,disk} = \frac{C_s}{2}\Bigg\{ & \left[\left(V_2 - \gamma_a V_1 - \gamma_b V_3\right)^2 + \gamma_a^2\left(V_3 - V_1\right)^2\right]\frac{\alpha_1}{\tanh \alpha_1} \\
& + 2\gamma_a\left(V_2 - \gamma_a V_1 - \gamma_b V_3\right)\left(V_1 - V_3\right)\frac{\alpha_1}{\sinh \alpha_1} + \alpha_1^2\gamma_a\gamma_b\left(V_1 - V_3\right)^2\Bigg\}
\end{aligned}
\tag{12.75}
$$

Thus, the total energy in the first two disks with static rings is

$$
\begin{aligned}
E_{Both\,disk} = \frac{C_s}{2}\Bigg\{ & \left(V_2 - V_1\right)^2\frac{\alpha}{\tanh \alpha} + \left[\left(V_2 - \gamma_a V_1 - \gamma_b V_3\right)^2 + \gamma_a^2\left(V_3 - V_1\right)^2\right]\frac{\alpha_1}{\tanh \alpha_1} \\
& + 2\gamma_a\left(V_2 - \gamma_a V_1 - \gamma_b V_3\right)\left(V_1 - V_3\right)\frac{\alpha_1}{\sinh \alpha_1} + \alpha_1^2\gamma_a\gamma_b\left(V_1 - V_3\right)^2\Bigg\}
\end{aligned}
\tag{12.76}
$$

The lumped capacitance network associated with this configuration can be obtained by the procedure described in the previous section. Thus,

$$
\begin{aligned}
\frac{\partial E_{\text{Both disk}}}{\partial V_1} &= C_s\left\{(V_1 - V_2)\frac{\alpha}{\tanh\alpha} + \left[-\gamma_a(V_2 - \gamma_a V_1 - \gamma_b V_3) + \gamma_a^2(V_1 - V_3)\right]\frac{\alpha_1}{\tanh\alpha_1}\right. \\
&\quad \left. + \gamma_a\left[(V_2 - \gamma_a V_1 - \gamma_b V_3) - \gamma_a(V_1 - V_3)\right]\frac{\alpha_1}{\sinh\alpha_1} + \alpha_1^2\gamma_a\gamma_b(V_1 - V_3)\right\} \\
&= C_s\left\{\left[\frac{\alpha}{\tanh\alpha} + 2\gamma_a^2\left(\frac{\alpha_1}{\tanh\alpha_1} - \frac{\alpha_1}{\sinh\alpha_1}\right) + \alpha_1^2\gamma_a\gamma_b\right]V_1\right. \\
&\quad - \left[\frac{\alpha}{\tanh\alpha} + \gamma_a\left(\frac{\alpha_1}{\tanh\alpha_1} - \frac{\alpha_1}{\sinh\alpha_1}\right)\right]V_2 \\
&\quad \left. - \left[\gamma_a(2\gamma_a - 1)\left(\frac{\alpha_1}{\tanh\alpha_1} - \frac{\alpha_1}{\sinh\alpha_1}\right) + \alpha_1^2\gamma_a\gamma_b\right]V_3\right\}
\end{aligned} \tag{12.77}
$$

Reading off the mutual capacitances from (12.77), we find

$$
\begin{aligned}
C_{12} &= -C_s\left[\frac{\alpha}{\tanh\alpha} + \gamma_a\left(\frac{\alpha_1}{\tanh\alpha_1} - \frac{\alpha_1}{\sinh\alpha_1}\right)\right] \\
C_{13} &= -C_s\left[\gamma_a(2\gamma_a - 1)\left(\frac{\alpha_1}{\tanh\alpha_1} - \frac{\alpha_1}{\sinh\alpha_1}\right) + \alpha_1^2\gamma_a\gamma_b\right]
\end{aligned} \tag{12.78}
$$

These are both negative and the negative of their sum is the self-capacitance C_{11}.

We also need to find C_{23}. This is obtained by differentiating (12.76) with respect to V_2:

$$
\begin{aligned}
\frac{\partial E_{\text{Both disk}}}{\partial V_2} &= C_s\left\{-(V_1 - V_2)\frac{\alpha}{\tanh\alpha} + (V_2 - \gamma_a V_1 - \gamma_b V_3)\frac{\alpha_1}{\tanh\alpha_1} + \gamma_a(V_1 - V_3)\frac{\alpha_1}{\sinh\alpha_1}\right\} \\
&= C_s\left\{-\left[\frac{\alpha}{\tanh\alpha} + \gamma_a\left(\frac{\alpha_1}{\tanh\alpha_1} - \frac{\alpha_1}{\sinh\alpha_1}\right)\right]V_1 + \left[\frac{\alpha}{\tanh\alpha} + \frac{\alpha_1}{\tanh\alpha_1}\right]V_2\right. \\
&\quad \left. - \left[\gamma_b\frac{\alpha_1}{\tanh\alpha_1} + \gamma_a\frac{\alpha_1}{\sinh\alpha_1}\right]V_3\right\}
\end{aligned} \tag{12.79}
$$

Extracting the mutual capacitances, we find

$$
\begin{aligned}
C_{21} &= -C_s\left[\frac{\alpha}{\tanh\alpha} + \gamma_a\left(\frac{\alpha_1}{\tanh\alpha_1} - \frac{\alpha_1}{\sinh\alpha_1}\right)\right] \\
C_{23} &= -C_s\left[\gamma_b\frac{\alpha_1}{\tanh\alpha_1} + \gamma_a\frac{\alpha_1}{\sinh\alpha_1}\right]
\end{aligned} \tag{12.80}
$$

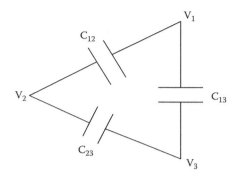

FIGURE 12.15
Lumped capacitance diagram for the 2-static ring configuration. The C's are taken to be positive.

These are negative and the negative of their sum is C_{22}. Moreover, we see that $C_{21} = C_{12}$ as expected since the capacitance matrix is symmetric. Differentiating (12.76) with respect to V_3 would give us no new information.

Thus, the lumped capacitance diagram for this configuration is shown in Figure 12.15. We assume the capacitances shown are positive (the negative of the mutual capacitances). Hence, the total capacitance between the V_1 and V_3 terminals is given by

$$C_{\text{Both disks}} = C_{13} + \frac{C_{12}C_{23}}{C_{12} + C_{23}} \tag{12.81}$$

In model impulse–voltage calculations, one could treat the first two disks as a unit having the capacitance given by (12.81). Then the voltages across each disk could be obtained from the overall voltage difference $(V_1 - V_3)$ via

$$V_1 - V_2 = \frac{C_{23}}{C_{12} + C_{23}}(V_1 - V_3)$$

$$V_2 - V_3 = \frac{C_{12}}{C_{12} + C_{23}}(V_1 - V_3) \tag{12.82}$$

12.10 Static Ring between the First Two Disks

Sometimes, a static ring is placed between the first two disks. This is usually only considered for center-fed windings so that there are two symmetrically spaced static rings, one for each of the two stacked coils. This case is very similar to the previous case. The only difference is that the energy in the first disk is given by

$$E_{\text{First disk}} = \frac{C_s}{2}(V_1 - V_2)^2 \frac{\alpha}{\tanh \alpha} \tag{12.83}$$

with $\alpha = \sqrt{C_b/C_s}$ since $C_a = 0$. Thus, the formulas of the last section apply with this value of α.

12.11 Winding Disk Capacitances with Wound-in-Shields*

In order to improve the voltage distribution along a transformer coil, that is, to reduce the maximum disk–disk voltage gradient, it is necessary to make the distribution constant, α, as small as possible, where $\alpha = \sqrt{C_g/C_s}$ with C_g, the total ground capacitance, and C_s, the series capacitance of the coil. One way of accomplishing this is to increase C_s. Common methods for increasing the series capacitance include interleaving [Nuy78] and the use of wound-in-shields [For69]. Both of these techniques rely on inductive effects. Geometric methods for increasing C_s, such as decreasing turn–turn or disk–disk clearances, are generally ruled out by voltage withstand or cooling considerations.

Interleaving can produce large increases in C_s, which may be necessary in very-high-voltage applications. However, its implementation can be very labor-intensive and in practice tends to be limited to magnet wire applications. Wound-in-shields tend to produce more modest increases in C_s compared with interleaving. Moreover, the installation of the shields requires less labor and is suitable for use with transposed cable. In addition, a suitable positioning of the shields can be used to provide a tapered capacitance profile to match the voltage stress profile of the winding.

In this section, we present a simple analytic formula for calculating the disk capacitance with a variable number of wound-in-shield turns. Since this formula rests on certain assumptions, a detailed circuit model is developed to test these assumptions. Finally, experiments are carried out to check the formula under a wide variety of conditions.

12.11.1 Analytic Formula

Figure 12.16 shows the geometry of a pair of disks containing wound-in-shields. Also shown is the method of labeling turns of the coil and shield. Since the shield spans two disks, it is necessary to calculate the capacitance of the pair. Each disk has N turns and n wound-in-shield turns, where $n \leq N - 1$. The voltage across the pair of disks is V and we assume the rightmost turn of the top disk, i = 1, is at voltage V, and the rightmost turn of the bottom disk, j = 1, is at 0 V, so that the coil is wound in a positive sense from the

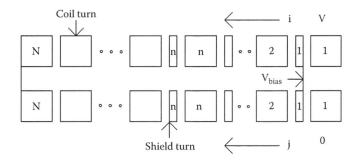

FIGURE 12.16
Disk pair with wound-in-shields with labeling and other parameters indicated. (Reprinted with permission from Del Vecchio, R.M. et al., *IEEE Trans. Power Deliv.*, 13(2), April 1998, 503–509. © 1998 IEEE.)

---

* This section follows closely our published paper [Del98], reprinted with permission from *IEEE Trans. Power Deliv.*, 13(2), April 1998, 503–509. © 1998 IEEE.

outer to inner turn on the bottom disk and from the inner to outer turn on the top disk. The shield turns are placed between the coil turns and are wound in the same sense as the coil. However, their cross-over is at the outermost turn rather than the innermost one, as is the case for the coil. This means that the positive voltage sense for the shield is from turn j = 1 to j = n on the bottom disk and then from the leftmost turn, i = n, to turn i = 1 on the top disk.

We assume the voltage rise per turn is ΔV where

$$\Delta V = \frac{V}{2N} \tag{12.84}$$

We assume that the same volts per turn applies to the shield as well. For definiteness, we assume the voltage at the shield cross-over point is V_{bias} as shown in Figure 12.16. Taking the voltage at the midpoint of a turn, we have for the top disk coil turns

$$V_c(i) = V - (i - 0.5)\Delta V \tag{12.85}$$

where $V_c(i)$ is the coil voltage for turn i, i = 1, …, N. Also for the top disk wound-in-shield turns,

$$V_w(i) = V_{bias} - (i - 0.5)\Delta V \tag{12.86}$$

where $V_w(i)$ is the shield voltage for shield turn i, i = 1, …, n.

For the bottom disk, we have

$$V_c(j) = (j - 0.5)\Delta V \tag{12.87}$$

for j = 1, …, N and

$$V_w(j) = V_{bias} + (j - 0.5)\Delta V \tag{12.88}$$

for j = 1, …, n.

Letting c_w be the capacitance between a coil turn and an adjacent shield turn, the energy stored in the capacitance between shield turn i and its adjacent coil turns is

$$\frac{1}{2}c_w\left\{\left[V_c(i) - V_w(i)\right]^2 + \left[V_c(i+1) - V_w(i)\right]^2\right\} \tag{12.89}$$

The two terms reflect the fact that there are two adjacent coil turns for every shield turn. Using the previous expressions for the V's, Equation 12.89 becomes

$$\frac{1}{2}c_w\left[\left(V - V_{bias}\right)^2 + \left(V - V_{bias} - \Delta V\right)^2\right] \tag{12.90}$$

This does not depend on i and so is the same for all n shield turns on the top disk. There are n such turns in the energy. For the bottom disk, Equation 12.89 applies with j replacing i, so the capacitative energy between shield turn j and its surrounding coil turns is

$$\frac{1}{2}c_w\left\{\left[V - V_{bias} - 2N\Delta V\right]^2 + \left[V - V_{bias} - (2N-1)\Delta V\right]^2\right\} \tag{12.91}$$

This again does not depend on j and so is the same for all n shield turns on the bottom disk. Thus, there are n such terms in the energy.

To the said energies, we must add the capacitative energy of the turns without wound-in-shields between them. Letting c_t be the turn–turn capacitance between regular winding turns, this energy is simply

$$\frac{1}{2}c_t(\Delta V)^2 \tag{12.92}$$

There are $2(N - n - 1)$ such terms in the energy. In addition, there is energy stored in the disk–disk capacitance, c_d. Since we are assuming the voltage varies linearly along the disks, this capacitative energy is given by

$$\frac{1}{2}\left(\frac{c_d}{3}\right)V^2 \tag{12.93}$$

We are ignoring capacitative coupling to other disk pairs at this stage since our experimental setup consisted of an isolated disk pair. However, if this disk pair were embedded in a larger coil of similar disks, then (12.93) would need to be doubled before adding to the energy.

The total capacitative energy of the disk pair is found by adding the previously mentioned contributions. To simplify the formula, we define

$$\beta = \frac{V - V_{bias}}{\Delta V} \tag{12.94}$$

In terms of this parameter and the expression for ΔV given in (12.84), the total capacitative energy, E_{tot}, is given by

$$
\begin{aligned}
E_{tot} &= \frac{1}{2}\left\{ c_w n(\Delta V)^2\left[\beta^2 + (\beta - 1)^2 + \left(\frac{V_{bias}}{\Delta V}\right)^2 + \left(\frac{\Delta V - V_{bias}}{\Delta V}\right)^2 \right] \right\} \\
&\quad + \frac{1}{2}c_t(\Delta V)^2 \, 2(N - n - 1) + \frac{1}{2}\left(\frac{c_d}{3}\right)V^2 \\
&= \frac{1}{2}V^2\left\{ c_w\left(\frac{n}{4N^2}\right)\left[\beta^2 + (\beta - 1)^2 + (\beta - 2N)^2 + (\beta + 1 - 2N)^2 \right] + c_t\left(\frac{N - n - 1}{2N^2}\right) + \frac{c_d}{3} \right\} \tag{12.95}
\end{aligned}
$$

Extracting the equivalent or total capacitance from (12.95), we get

$$C_{tot} = c_w\left(\frac{n}{4N^2}\right)\left[\beta^2 + (\beta - 1)^2 + (\beta - 2N)^2 + (\beta + 1 - 2N)^2 \right] + c_t\left(\frac{N - n - 1}{2N^2}\right) + \frac{c_d}{3} \tag{12.96}$$

At this point, V_{bias} and, hence, β are unspecified. This will depend on whether the shield is floating or whether it is attached at some point to a coil voltage. If the shield is floating, we expect $V_{bias} = V/2$. If the shield is attached at the cross-over to the high-voltage terminal, then $V_{bias} = V$. If the leftmost or end shield turn on the top disk (i = n) is attached to the high-voltage terminal, then $V_{bias} = V + (n - 0.5)\Delta V$. In terms of β,

$$
\begin{array}{lll}
\text{Shield floating} & \beta = N & \\
\text{Shield attached at cross-over to V} & \beta = 0 & (12.97) \\
\text{Shield attached at top left end to V} & \beta = -(n-0.5) &
\end{array}
$$

Other situations can be considered as well.

For the turn–turn capacitance, we used the expression (in SI units)

$$
c_t = \frac{2\pi\varepsilon_o\varepsilon_p R_{ave}\left(h + 2\tau_p\right)}{\tau_p} \tag{12.98}
$$

where
  R_{ave} is the average radius of the coil
  h is the bare copper height of a coil turn in the axial direction
  τ_p is the 2-sided paper thickness of a coil turn
  $\varepsilon_o = 8.8542 \times 10^{-12}$ F/m
  ε_p is the relative permittivity of paper

The addition of $2\tau_p$ to h is designed to take fringing effects into account. Also, the use of R_{ave} is an approximation, which is reasonably accurate for coils with radial builds small compared with their radii. For the capacitance c_w, the same expression was used but with τ_p replaced by $0.5(\tau_p + \tau_w)$ where τ_w is the 2-sided paper thickness of a shield turn. Since there are key spacers separating the disks and gaps between them, the disk–disk capacitance is given by

$$
c_d = \pi\varepsilon_o\left(R_o^2 - R_i^2\right)\left[\frac{f_{ks}}{\left(\left(\tau_p/\varepsilon_p\right)+\left(\tau_{ks}/\varepsilon_{ks}\right)\right)} + \frac{\left(1-f_{ks}\right)}{\left(\left(\tau_p/\varepsilon_p\right)+\left(\tau_{ks}/\varepsilon_{air}\right)\right)}\right] \tag{12.99}
$$

where
  R_i and R_o are the inner and outer radii of the disk, respectively
  f_{ks} is the fraction of the disk–disk space occupied by key spacers
  τ_{ks} is the key spacer thickness
  ε_{ks} is the key spacer relative permittivity
  ε_{air} is the relative permittivity of air (=1) since the coils were tested in air

12.11.2 Circuit Model

Since quite a few assumptions went into deriving the capacitance formula in the last section, we decided to model the disk pair by means of a circuit model including capacitive, inductive, and resistive effects as sketched in Figure 12.17. We include all mutual couplings

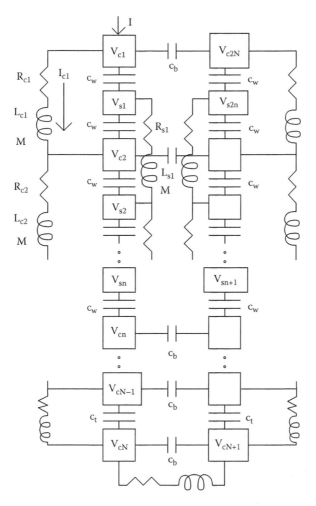

FIGURE 12.17
Circuit model for a disk pair with wound-in-shields. The labeling scheme and circuit parameters are indicated.

between the turns of the coil and wound-in-shield. The circuit is assumed to be excited by a current source, which is nearly a step function. The circuit equations are

$$C\frac{d\mathbf{V}}{dt} = A\mathbf{I}$$

$$M\frac{d\mathbf{I}}{dt} = B\mathbf{V} - R\mathbf{I}$$

(12.100)

where
  C is a capacitance matrix
  M an inductance matrix
  R a diagonal resistance matrix
  A and B are matrices of ±1's and 0's

The voltage and current vectors, $\mathbf{V}$ and $\mathbf{I}$, include the coil and shield turn voltages and currents. These equations are solved by means of a Runge–Kutta solver.

Since the experimental setup was in air, we used air core inductance and mutual inductance expressions. The mutual inductance between two thin wire coaxial loops of radii, r_1 and r_2, spaced a distance d apart is given by [Smy68] in MKS units:

$$M = \frac{2\mu_o}{k} \sqrt{r_1 r_2} \left[\left(1 - \frac{k^2}{2} \right) K(k) - E(k) \right]$$

with   (12.101)

$$k^2 = \frac{4r_1 r_2}{\left[\left(r_1 + r_2 \right)^2 + d^2 \right]}$$

K(k) and E(k) are complete elliptic integrals of the first and second kinds, respectively, and $\mu_o = 4\pi \times 10^{-7}$ H/m. For rectangular cross-section coils, Lyle's method in conjunction with (12.101) could be used for a more accurate determination of the mutual inductance [Gro73]. However, for the turn–turn mutual inductances in our experimental coils, treating the turns as thin circular loops was nearly as accurate as Lyle's method. The self-inductance of a single-turn circular coil of square cross-section with an average radius of a and square side length of c is given by [Gro73] in MKS units:

$$L = \mu_o a \left\{ \frac{1}{2} \left[1 + \frac{1}{6} \left(\frac{c}{2a} \right)^2 \right] \ln \left[\frac{8}{\left(c/2a \right)^2} \right] + 0.2041 \left(\frac{c}{2a} \right)^2 - 0.84834 \right\}$$ (12.102)

This applies for $c/2a \leq 0.2$. When the cross-section is not square, it can be subdivided into a number of squares and (12.102) together with (12.101) can be applied to compute the self-inductance more accurately. In our experimental coil, the turn dimensions were such that the simple formula with c taken as the square root of the turn area agreed well with the more accurate calculation.

The turn–turn and turn–shield capacitances were the same as given in the last section. The capacitance c_b in Figure 12.17 was taken as c_d/N, where c_d is the disk-to-disk capacitance. We did not include the capacitance between shield turns on neighboring disks in our final calculations. Their inclusion had little effect on the total capacitance.

The resistances used in the circuit model were based on the wire dimensions but were multiplied by a factor to account for high-frequency losses. This factor may be estimated by examining the frequency dependence of the two main contributors to the coil loss, namely, the Joule, or I^2R loss, and the eddy current loss due to stray flux carried by the conductor strands. Based on a formula in [Smy68] for cylindrical conductors, the Joule loss is nearly independent of frequency at low frequencies, which includes 60 Hz for our conductor dimensions. At high frequencies, the loss divided by the dc or 60 Hz loss is given by

$$\frac{W_{Joule(f)}}{W_{Joule} \left(f = 0 \right)} = \frac{r_{cond}}{2} \sqrt{\pi \mu_o \sigma f}$$ (12.103)

where
    r_{cond} is the radius of the conductor or in our case an effective radius based on the wire dimensions
    σ is the wire's conductivity
    f is the frequency in Hz

Based on our cable dimensions and for a typical frequency encountered in our calculations and experiment of ~0.15 MHz, we estimate that $W_{Joule}(f)/W_{Joule}(f = 0) \sim 15.9$. The eddy current frequency dependence due to stray flux is given in [Lam66]. At low frequencies, this loss varies as f^2, whereas at high frequencies, it varies as $f^{0.5}$. Taking ratios of the high-to low-frequency eddy current loss, we obtain

$$\frac{W_{Eddy(f)}}{W_{Eddy}\left(60 \text{ Hz}\right)} = \frac{6\sqrt{f}}{b^3\left(\pi\mu_o\sigma\right)^{1.5}\left(60\right)^2} \tag{12.104}$$

where b is the thickness of the lamination or strand in a direction perpendicular to the stray magnetic field. In our case, the cable was made of rectangular strands with dimensions 1.4 mm × 4.06 mm. For the small dimension, Equation 12.104 gives a ratio of 68,086 while for the large dimension, we get a ratio of 2,766. A detailed field mapping is necessary to obtain the eddy current loss contribution at low frequencies but typically this is about 10% of the dc loss, depending on the cable construction. If we assume that the small- and large-dimension eddy losses are equal at low frequencies, that is, each is 5% of the dc loss, then we find that the ratio of high- to low-frequency total loss based on the said ratios is given by $W_{tot}(f)/W_{tot}(60 \text{ Hz}) = 3557$. This is only a crude estimate. Our data show that this ratio is about 1500 for our cable. Since the capacitance obtained from the simulation is nearly independent of the resistance used, we did not try to match the model's resistance with the experimental values. A sample output is shown in Figure 12.18.

Two methods were used to extract the equivalent total capacitance from the circuit model. Since the voltages of each coil and shield turn were determined at each time step, the capacitive energy was simply summed and the total capacitance determined via

$$E_{tot} = \frac{1}{2}C_{tot}V^2 \tag{12.105}$$

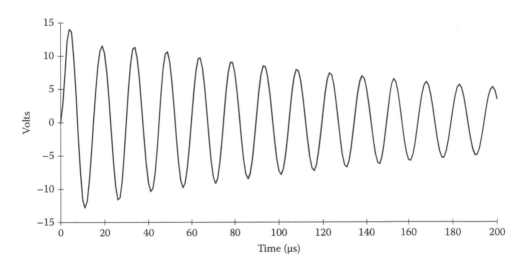

FIGURE 12.18
Voltage across the coil pair versus time from the detailed circuit model calculations. (Reprinted with permission from Del Vecchio, R.M. et al., *IEEE Trans. Power Deliv.*, 13(2), April 1998, 503–509. © 1998 IEEE.)

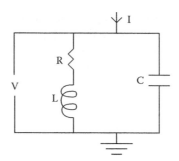

FIGURE 12.19
Simplified circuit model for capacitance determination. (Reprinted with permission from Del Vecchio, R.M. et al., *IEEE Trans. Power Deliv.*, 13(2), April 1998, 503–509. © 1998 IEEE.)

where V is the voltage of turn i = 1 at the particular time step. At the end of the total time duration of about 200 time steps, the average and median total capacitances were calculated. These two values generally agreed fairly closely. The other method consisted of extracting the total capacitance from a simplified equivalent circuit as shown in Figure 12.19. This latter method was also used to obtain the total capacitance experimentally.

Using Laplace transforms, the circuit of Figure 12.19 can be solved analytically, assuming a step function current input of magnitude I. We obtain

$$V(t) = IR\left\{1 - e^{-(R/2L)t}\left[\cos(\omega_o t) + \frac{1}{2}\left(\frac{R}{2L}\right)\left(\frac{1}{\omega_o} - \frac{\omega_o}{(R/2L)^2}\right)\sin(\omega_o t)\right]\right\} \quad (12.106)$$

for $t \geq 0$. Here,

$$\omega_o = \sqrt{\frac{1}{LC} - \left(\frac{R}{2L}\right)^2} \quad (12.107)$$

In the limiting case as $R \to 0$, Equation 12.106 becomes

$$V(t) = I\sqrt{\frac{L}{C}}\sin(\omega_o t) \quad (12.108)$$

In all the cases examined computationally or experimentally, the term $(R/2L)^2$ was much smaller than $1/LC$ so that it can be ignored in the expression for ω_o. Thus by measuring the oscillation frequency, we determine the combination LC. To obtain C, an additional capacitance, C_1, was placed in parallel with the coil and a new oscillation frequency, ω_1, determined. Since this added capacitance does not change L, we have, taking ratios and squaring (12.107),

$$\frac{C + C_1}{C} = \left(\frac{\omega_o}{\omega_1}\right)^2 \quad (12.109)$$

Since ω_o and ω_1 can be measured or determined from the output of the circuit model and C_1 is known, C can be obtained. We found good agreement between the two methods of determining C.

12.11.3 Experimental Methods

A coil containing two disk sections was made of transposed cable. There were 10 turns per disk. The cable turns were 11.7 mm radial build by 9.19 mm axial height, including a 0.76 mm (2-sided) paper cover. The inner radius was 249 mm. The outer radius depended on the number of shield turns but was approximately 392 mm. The wound-in-shield turns consisted of 3.55 mm radial build by 9.27 cm axial height magnet wire, including a 0.51 mm (2-sided) paper cover. The disks were separated by means of 18 key spacers equally spaced around the circumference. The key spacers were 44.5 mm wide by 4.19 mm thick.

The coil was excited by means of a current source, which produced a near step function current. A sample of the current input and coil voltage output is shown in Figure 12.20. This voltage vs time plot was Fourier analyzed to extract the resonant frequency. Frequencies were measured with and without an external capacitor of 10 nF across the coil in order to obtain the coil capacitance by the ratio method described previously.

Shield turns/disk of n = 3, 5, 7, 9 were tested, as well as the no shield case, n = 0. In addition, for each n value, tests were performed with the shield floating, attached to the high-voltage terminal at the cross-over and attached to the high-voltage terminal at the end shield turn on the top disk. Other methods of attachment were also made for the n = 9 case to test for expected symmetries.

Because the disk pair was not contained in a tightly wound coil, there tended to be some looseness in the winding. This could be determined by squeezing the disk turns tightly together and noting how much the radial build decreased. From this, we could determine how much looseness there was in the insulation and correct for it. This amounted to about a 5% correction.

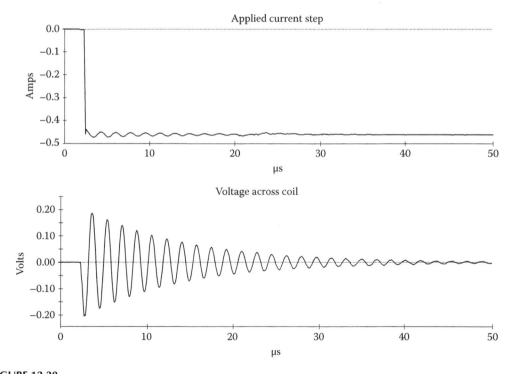

FIGURE 12.20
Experimental output from one of the test runs on the coil showing the input current and voltage across the coil vs time. (Reprinted with permission from Del Vecchio, R.M. et al., *IEEE Trans. Power Deliv.*, 13(2), April 1998, 503–509. © 1998 IEEE.)

12.11.4 Results

There are some uncertainties associated with the values of the relative permittivities to use in the capacitance calculations, particularly that of paper in air. For the pressboard key spacers, [Mos87] gives a method for determining its permittivity in terms of the board density, the permittivity of the fibers, and the permittivity of the substance filling the voids. For our key spacers in air, we obtain $\varepsilon_{ks} = 4.0$ using this method. (For pressboard in oil, the value is $\varepsilon_{ks} = 4.5$.) For paper, [Cla62] presents a graph of the dielectric constant vs density in air and oil. Unfortunately, paper wrapping on cable in not as homogeneous a substance as pressboard. Its effective density would depend on how the wraps overlap and how loosely or tautly it is wound. In addition, for transposed cable, the paper–copper interface is not a clean rectilinear one. This is because of the extra unpaired strand on one side of the strand bundle. We found that a value of $\varepsilon_p = 1.5$ was needed to get good overall agreement with the test data. This would correspond to an effective paper density of 0.5 g/cm^3 according to the graph in [Cla62]. At that density, paper in oil would have $\varepsilon_p = 3.0$ according to the same graph. For tightly wound paper in oil, values of $\varepsilon_p = 3.5$–4.0 are typically used. Thus, a paper permittivity of 1.5 for paper in air is not unreasonable.

Table 12.1 shows the test results along with the calculated values of the capacitance for different numbers of shield turns and different shield biasings. The overall agreement is good. Certainly, the trends are well reproduced. Even for the floating shield case, which has the lowest capacitive enhancement, there is a capacitance increase of a factor of 5 for 3 shield turns/disk and a factor of about 13 for 9 shield turns/disk over the unshielded case. By biasing the shield in different ways, even greater increases are achieved.

In Table 12.2, the coil turn and shield turn voltages as calculated by the circuit model at time t = 30 μs are shown for the n = 9 case with the shields floating. As can be seen, the coil

TABLE 12.1

Capacitance of Coil with Two Disks of 10 Turns/Disk and a Variable Number of Wound-in-Shield Turns/Disk

Shield Turns n	Calc Method	C_{tot} (nF)		
		Floating	Attached to V at Crossover	Attached to V at Top End
0	Measured	0.24		
	Analytic	0.264		
	Circuit model	0.287		
3	Measured	1.20	2.33	3.21
	Analytic	1.31	2.44	3.08
	Circuit model	1.41	2.66	3.31
5	Measured	1.78	3.53	6.07
	Analytic	2.03	3.94	6.04
	Circuit model	2.21	4.34	6.73
7	Measured	2.80	5.73	11.8
	Analytic	2.75	5.45	10.1
	Circuit model	2.99	6.01	11.4
9	Measured	3.17	6.82	16.1
	Analytic	3.49	6.99	15.5
	Circuit model	3.83	7.61	17.9

Source: Reprinted with permission from Del Vecchio, R.M. et al., *IEEE Trans. Power Deliv.*, 13(2), April 1998, 503–509. © 1998 IEEE.

TABLE 12.2

Turn Voltages at t = 30 μs for a Coil of Two Disks with 10 Turns/Disk and 9 Shield Turns/Disk with the Shield Floating

Coil Turns			Adjacent Shield Turn			Shld-Adj Coil
#	Volts	ΔV	#	Volts	ΔV	$|\Delta V|$
1	24.27		1	11.43		12.84, 11.49
2	22.92	1.35	2	10.12	1.31	12.80, 11.39
3	21.51	1.45	3	8.73	1.39	12.78, 11.35
4	20.08	1.43	4	7.30	1.43	12.78, 11.35
5	18.65	1.43	5	5.87	1.43	12.78, 11.46
6	17.24	1.41	6	4.45	1.42	12.79, 11.44
7	15.89	1.35	7	3.09	1.36	12.80, 11.53
8	14.62	1.27	8	1.79	1.30	12.83, 11.63
9	13.42	1.20	9	0.52	1.27	12.90, 11.77
10	12.29	1.13				
11	11.49	0.80				
12	10.50	0.99	10	23.43		12.93, 11.94
13	9.45	1.05	11	22.27	1.16	12.82, 11.77
14	8.28	1.17	12	21.08	1.19	12.80, 11.63
15	6.98	1.30	13	19.78	1.30	12.80, 11.50
16	5.64	1.34	14	18.39	1.39	12.75, 11.41
17	4.25	1.39	15	16.99	1.40	12.74, 11.35
18	2.83	1.42	16	15.56	1.43	12.73, 11.31
19	1.40	1.43	17	14.14	1.42	12.74, 11.31
20	0	1.40	18	12.75	1.39	12.75, 11.35

$V_{bias} = (V_{shield\ turn\ 1} + V_{shield\ turn\ 18})/2 = 12.09 \approx V/2 = 12.14$

Source: Reprinted with permission from Del Vecchio, R.M. et al., *IEEE Trans. Power Deliv.*, 13(2), April 1998, 503–509. © 1998 IEEE.

volts/turn is about the same as the shield volts/turn, verifying the assumption made in the analytic model. In addition, $V_{bias} \approx V/2$, as was also assumed in the simple model. Also, the voltage differences between the coil and shield turns fall into two groups of either ~ 12.8 or 11.5 V in this case. This also corresponds to the simple model prediction of either $N\Delta V = V/2$ or $(N - 1)\Delta V = V/2 - \Delta V$ volts. Thus, in the case where the shields are floating, the maximum shield turn–coil turn voltage is $V/2$. Based on the simple model, this holds regardless of the number of shield turns.

In Table 12.3, we show the corresponding output from the circuit model at t = 150 μs for the case where the shield is attached at the cross-over to the high-voltage terminal. (It was actually attached to shield turn i = 1 in the model.) We see again that the coil volts/turn ≈ shield volts/turn. According to the simple model prediction ($\beta = 0$), the voltage differences between the shield and adjacent turn on the top disk are 0 and ΔV as is also nearly the case for the circuit model. Along the bottom disk, the simple model gives voltage differences of $2N\Delta V = V$ and $(2N - 1)\Delta V = V - \Delta V$ and this is also nearly the case for the circuit model. Thus, the maximum coil-turn-to-shield-turn voltage is V. According to the simple model, this holds regardless of the number of shield turns.

The case where the end turn of the shield on the top disk is attached to the high-voltage terminal could be analyzed similarly, although this configuration is harder to achieve in practice. This configuration also has a higher coil-turn-to-shield-turn-voltage difference

TABLE 12.3

Turn Voltages at t =150 μs for a Coil of Two Disks with 10 Turns/Disk and 9 Shield Turns/Disk with the Shield Attached at the Crossover to the High-Voltage Terminal

#	Coil Turns Volts	ΔV	#	Adjacent Shield Turn Volts	ΔV	Shld-Adj Coil |ΔV|
1	5.13		1	5.13		0, 0.29
2	4.84	0.29	2	4.87	0.26	0.03, 0.33
3	4.54	0.30	3	4.58	0.29	0.04, 0.35
4	4.23	0.31	4	4.28	0.30	0.05, 0.35
5	3.93	0.30	5	3.99	0.29	0.06, 0.36
6	3.63	0.30	6	3.69	0.30	0.06, 0.34
7	3.35	0.28	7	3.41	0.28	0.06, 0.33
8	3.08	0.27	8	3.14	0.27	0.06, 0.30
9	2.84	0.24	9	2.90	0.24	0.06, 0.28
10	2.62	0.22				
11	2.42	0.20				
12	2.24	0.18	10	7.52		5.28, 5.10
13	2.02	0.22	11	7.30	0.22	5.28, 5.06
14	1.77	0.25	12	7.05	0.25	5.28, 5.03
15	1.50	0.27	13	6.79	0.26	5.29, 5.02
16	1.22	0.28	14	6.51	0.28	5.29, 5.01
17	0.92	0.30	15	6.23	0.28	5.31, 5.01
18	0.62	0.30	16	5.94	0.29	5.32, 5.02
19	0.31	0.31	17	5.66	0.28	5.35, 5.04
20	0	0.31	18	5.38	0.28	5.38, 5.07

Source: Reprinted with permission from Del Vecchio, R.M. et al., *IEEE Trans. Power Deliv.*, 13(2), April 1998, 503–509. © 1998 IEEE.

than the other methods of shield attachment. As expected, however, the price to pay for the higher capacitances is higher coil-turn-to-shield-turn voltage differences.

Several symmetric situations were noted in the experimental data. These symmetries were only checked for the n = 9 case but should apply to all n values according to the simple model. We found that attaching the high-voltage terminal to the end shield turn on the top disk (i = n) produced the same capacitance as attaching the low-voltage terminal, at 0 V in this case, to the end turn on the bottom disk (j = n). This symmetry can be seen in the simple formula (12.96) by using the appropriate values for β. Attaching the top-end shield turn to the low-voltage terminal produced the same capacitance as attaching the bottom-end shield turn to the high-voltage terminal. This can also be shown by means of the simple formula. Also the same capacitance was produced whether the shield cross-over was attached to the high- or low-voltage terminal. This also follows from the simple model.

For consistency, the inductance of our coil was extracted from the experimental data as well as from the circuit model output. We found L(data) = 280 μH and L(circuit model) = 320 μH. Using an algorithm in Grover [Gro73], we obtained L(calc) = 340 μH. These are all within reasonable agreement. The decay constant R/2L appearing in (12.106) could be extracted from the data by analyzing the envelope of the damped sinusoid (Figure 12.20) and we found that it could be eliminated in the formula for ω_0 (12.107) in comparison with the 1/LC term. We also observed a significant resistance change when the frequency was changed by adding the external capacitance.

12.12 Multi-Start Winding Capacitance

Multi-start windings are a simple example of an interleaved type of winding. They are commonly used as tap windings. In these windings, adjacent turns have voltage differences, which can differ from the usual turn–turn voltage drop along typical disk or helical windings. Although multi-start windings are helical windings, they can be thought of as a collection of superposed series-connected helical windings as shown in Figure 12.21.

These windings essentially start over again and again, hence the name. Each start represents a constituent winding having a certain number of turns called turns per start. The number of starts is the same as the number of constituent windings. By connecting taps between the starts, the turns per start become the number of tap turns. Their advantage as tap windings compared with the standard type is that they allow a more balanced force distribution regardless of the tap setting and do not require the thinning of adjacent windings.

Although shown as side-by-side windings in the figure for explanatory purposes, the windings are superposed into one helical type of winding. The bottom-to-top connections are made external to this winding. The constituent windings are meshed in such a way that the voltage difference between adjacent turns in the composite winding is kept to one or two times the voltage drop along a constituent winding, which is really the best that can be done. Letting 1, 2, 3, … label turns from the different constituent windings, acceptable meshing schemes are shown in Figure 12.22 for various numbers of starts, N_s.

The turns are arranged along the winding according to Figure 12.22. The voltage differences between adjacent turns, measured in units of the voltage drop between starts, are also indicated in the figure. A pattern can be seen in the organization: Start with turn 1 from start coil 1. Put turn 1 from start coil 2 at the end of the group. Put turn 1 from start coil 3 below turn 1. Put turn 1 from start coil 4 above turn 2. For $N = 4$ starts repeat this pattern using turn 2 from the different coils until you run out of turns per start. Otherwise, continue for $N = 5$ starts: Put turn 1 from start coil 5 below turn 1 from start coil 3 then continue to repeat this pattern until you run out of turns per start. For $N = 6$ starts, put turn 1 from start coil 6 above turn 4 and repeat until you run out of turns per start, and so on, until you run out of start coils.

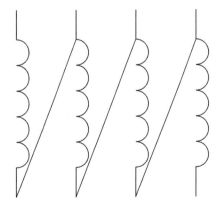

FIGURE 12.21
Schematic illustration of a multi-start winding.

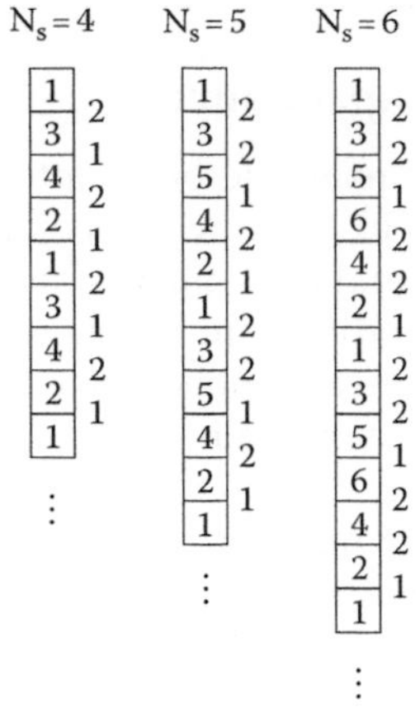

FIGURE 12.22
Winding schemes for multi-start windings. The block numbers indicate the coil, which is associated with the turn. The numbers beside the blocks are the voltage differences between turns in units of the voltage drop between starts.

The capacitance is obtained by summing up the capacitative energy associated with the winding configuration. Letting c_t be the turn-to-turn capacitance and ΔV_t the voltage difference between turns, the energy associated with a pair of adjacent turns is given by Energy(turn–turn) $= 1/2c_t(\Delta V_t)^2$. Letting the voltage drop between starts be ΔV_s, the total energy is obtained by summing all the turn–turn energies. As can be seen from Figure 12.22, each group of turns has two 1's except for the last group that has only 1. If there are n turns/ start, there are n such groups. Hence, the energy for the 1's is $2n - 1$ times the energy associated with voltage ΔV_s across capacitance c_t. The remaining turn–turn voltages in the group have a voltage of $2 \times \Delta V_s$ across them. There are $N_s - 2$ such turns in the group. Since there are n such groups, the energy associated with these is $n \times (N_s - 2)$ times the energy associated with a voltage of $2\Delta V_s$ across capacitance c_t. Combining these energies, we get a total energy of

$$E_{total} = (2n-1)\frac{1}{2}c_t(\Delta V_s)^2 + n(N_s - 2)\frac{1}{2}c_t(2\Delta V_s)^2$$
$$= \frac{1}{2}c_t(\Delta V_s)^2(4nN_s - 6n - 1) \qquad (12.110)$$

Since $\Delta V_s = V/N_s$, where V is the total voltage across the entire winding, by substituting this into (12.110), we can extract a total capacitance for the winding, C_{ms}:

$$C_{ms} = c_t\frac{(4nN_s - 6n - 1)}{N_s^2} \qquad (12.111)$$

c_t will depend on the insulation structure of the winding, that is, whether the turns are touching, paper to paper, in the manner of a layer winding or whether there is an oil gap with key spacers separating them. In any event, the capacitance is much higher than that across a comparable helical winding where the turn–turn voltages are much smaller.

13

Voltage Breakdown Theory and Practice

13.1 Introduction

A transformer's insulation system must be designed to withstand not only the a.c. operating voltages, with some allowance for an overvoltage of ~15%, but also the much higher voltages produced by lightning strikes or switching operations. These latter voltages can be limited by protective devices, such as lightning or surge arrestors, but these devices are usually set to protect at levels well above the normal a.c. operating voltage. Fortunately, the transformer's insulation can withstand higher voltages for the shorter periods of time characteristic of lightning or switching disturbances. Thus, insulation designed to be adequate at the operating voltage can also be sufficient for the short-duration higher voltages that may be encountered.

Insulation design is generally an iterative process. A particular winding type is chosen, such as disk, helix, or layer, for each of the transformer's windings. They must have the right number of turns to produce the desired voltage and must satisfy thermal, mechanical, and impedance requirements. The voltage distribution is then calculated throughout the windings, using a suitable electrical model together with the appropriate input such as a lightning impulse excitation. Voltage differences and/or electric fields are then calculated to determine if they are high enough to cause a breakdown, according to some breakdown criterion, across the assumed insulation structure. More elaborate path integrals are sometimes used to determine breakdown. If the breakdown criterion is exceeded at some location, the insulation is redesigned and the process repeated until the breakdown criteria are met. Insulation redesign can consist of adding more paper insulation to the wire or cable, increasing the size of the oil or air ducts or resorting to interleaving the winding conductors or adding wound-in-shields or other types of shields.

Although voltages and electric fields can be calculated to almost any desired accuracy, assuming the material properties are well known, the same cannot be said for breakdown fields in solids or liquids. The theory of breakdown in gases is reasonably well established, but the solid or liquid theory of breakdown is somewhat rudimentary. An oil breakdown model is presented here, which attempts to address this problem. Nevertheless, design rules have evolved based on experience. With suitable margins, these rules generally produce successful designs. Success is usually judged by whether a transformer passes a series of dielectric tests using standard impulse waveforms or a.c. power frequency voltages for specified time periods without breakdown or excessive corona. These tests have been developed over the years in an effort to simulate a typical lightning or switching waveshape.

13.2 Principles of Voltage Breakdown

We briefly discuss some of the proposed mechanisms of voltage breakdown in solids, liquids, and gases with primary emphasis on transformer oil. This is because in oil-filled transformers, due to the higher dielectric constant of the solid insulation, the highest electric stress tends to occur in the oil. In addition, the breakdown stress of the oil is generally much lower than that of the solid insulation. The same situation occurs in dry-type transformers, however the breakdown mechanism in the gas is much better understood.

In gases, breakdown is thought to occur by electron avalanche, also called the Townsend mechanism [Kuf84]. In this process, the electric field imparts sufficient energy to the electrons between collisions with the atoms or molecules of the gas that they release or ionize additional electrons upon subsequent collisions. These additional electrons, in turn, acquire sufficient energy between collisions to release more electrons in an avalanche process. The process depends in detail on the collision cross-sections for the specific gas. These collisions can lead to elastic scattering, ionization, and absorption, and the collision cross-sections are highly energy-dependent. They have been measured for a variety of gases. In principle, breakdown can be calculated from known collision cross-sections, together with corrections, due to the influence of other factors, such as the presence of positive ions, photo-excitation, etc. In practice, the theory has served to illuminate the parametric dependence of the breakdown process and is even in reasonable quantitative agreement for specific gases.

One of the major results of the theory of gaseous breakdown is that the breakdown voltage across a uniform gap depends on the product of pressure and gap thickness or, more generally, on the product of gas density and gap thickness. This relationship is called the Paschen curve. A sketch of such a curve is shown in Figure 13.1. A fairly common feature of such curves is the existence of a minimum. Thus for a given gap distance, as the pressure is lowered, assuming we are to the right of the minimum, the breakdown voltage will drop and rise again as the pressure is lowered past the minimum. Care must be taken that the gap voltage is sufficiently far from the minimum to avoid breakdown. For a given gas pressure or more accurately, density, the breakdown field depends only on the gap thickness.

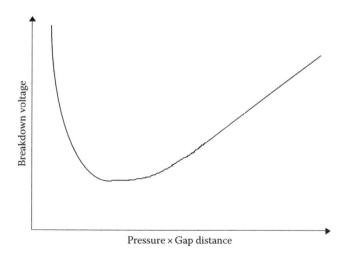

FIGURE 13.1
Schematic Paschen curve.

During the avalanche process, because the negatively charged electrons move much more rapidly than the positively charged ions and because they are pulled toward opposite electrodes, a charge separation occurs in the gas. When the excess charge is large enough, as can occur in a well-developed avalanche for large gap distances or high values of pressure × gap thickness, the electric field produced by the excess charge approaches the applied field. When this occurs, the Townsend mechanism of breakdown gives way to a streamer type of breakdown. In this process, secondary breakdown paths or plasma channels form at the front of the avalanche, leading to a more rapid breakdown than can be accounted for by the Townsend mechanism alone. Theoretical calculations, based on idealized charge configurations, can account approximately for this type of breakdown.

In solids and liquids where the distance between electron collisions is much shorter than in gases so that the electrons have a harder time acquiring enough energy to produce an avalanche, the Townsend mechanism is not considered to be operative except possibly for extremely pure liquids. A streamer type of mechanism is considered to be much more likely, but the theory is not as developed. Moreover, especially in liquids, there are usually many types of impurities whose presence even in small concentrations can lower the breakdown stress considerably. This is well established experimentally where further and further purifications lead to higher breakdown stress to the point where the so-called intrinsic breakdown stress of a pure liquid can be reached but has rarely been measured.

In solids and liquids, the breakdown stress does not appear to be strictly a function of the gap thickness but rather appears to depend on the area of the electrodes or the volume of the material under stress. This would argue against a strictly Townsend mechanism of breakdown according to which, with the nearly constant density of most solids and liquids, the breakdown should depend on the gap distance only. It should be noted, however, that the experimental evidence is fragmentary and sometimes contradictory.

An electrode area or volume dependence of breakdown is usually explained by means of a weak link theory. According to this theory, there is some weak spot, imperfection, or mechanism based on the presence of imperfections that causes the failure. Thus as the size of the specimen grows, weaker spots or more and greater imperfections are uncovered, resulting in failure at lower electric stress. Some support for this type of failure mechanism comes from studying the statistics of breakdown. There is much experimental evidence to show that breakdown probabilities follow an extreme value distribution, in particular, the Weibull distribution [Gum58]. This type of distribution is consistent with a weak link mechanism. In fact, Weibull invented it to account for failure statistics in fracture mechanics, which can be associated with material flaws. In general form, the distribution function giving the probability of failure for a voltage $\leq V$, $P(V)$, is

$$P(V) = \begin{cases} 1 - e^{-\left(\frac{V-V_o}{a}\right)^k}, & V \geq V_o \\ 0, & V < V_o \end{cases} \tag{13.1}$$

where V_o, a, and k are parameters >0. The probability density function, which is the derivative of the distribution function, when multiplied by ΔV gives the probability of failure in a small interval ΔV about V. It is

$$p(V) = \frac{dP(V)}{dV} = \frac{k}{a}\left(\frac{V-V_o}{a}\right)^{k-1} e^{-\left(\frac{V-V_o}{a}\right)^k} \tag{13.2}$$

The density function is, in general, asymmetric about the mode or most probable value and this asymmetry is usually taken as evidence that one is dealing with an extreme value distribution in contrast to a Gaussian density function, which is symmetric about the mode or mean.

According to advocates of an electrode area dependence of the breakdown stress, the weak link can be a protrusion on the electrode surface where the field will be enhanced or it can be an area of greater electron emissivity on the surface. Advocates of the volume dependence of breakdown emphasize impurities in the material, which increase with volume [Wil53]. According to Kok [Kok61], impurities in liquids such as transformer oil tend to have a higher dielectric constant than the oil, particularly if they have absorbed some water. Thus, they are attracted to regions of higher electric field by the presence of gradients in the field. They will tend to acquire an induced dipole moment so that other dipoles will attach to them in a chain-like fashion. Such a chain can lead to a relatively high conductivity link between the electrodes, leading to breakdown. The probability of forming such a chain increases with the amount of impurity present and hence with the volume of liquid.

Other mechanisms, which can account for some of the breakdown data, such as the formation of bubbles in the breakdown process, have been summarized in [Gall75]. We know from numerous experiments that the breakdown electric stress in transformer oil is lowered by the presence of moisture, by particles such as cellulose fibers shed by paper or pressboard insulation, and by the presence of dissolved gas. It was even shown in one experiment that the breakdown stress between two electrodes depended on whether the electrodes were horizontal or vertical. Presumably, the dissolved gas and its tendency to form bubbles when the electric stress was applied was influencing the results, since in one orientation the bubbles would be trapped, while in the other, they could float away. In other experiments, causing the oil to flow between the electrodes increased the breakdown stress compared with stationary oil. The said influences make it difficult to compare breakdown results from different investigators. However, for a given investigator, using a standardized liquid or solid preparation and testing procedure, observed trends in the breakdown voltage or stress with other variables are probably valid.

Breakdown studies not concerned with time as a variable are generally done under impulse or a.c. power frequency conditions. In the latter case, the time duration is usually 1 min. In impulse studies, the waveform is a unidirectional pulse having a rise time of $t_{rise} = 1\text{–}1.5$ μs and a fall time to 50% of the peak value of $t_{fall} = 40\text{–}50$ μs as sketched in Figure 13.2. In most such studies, the breakdown occurs on the tail of the pulse, but the breakdown voltage level is taken as the peak voltage. However, in front-of-wave breakdown studies, breakdown occurs on the rising part of the pulse and the breakdown voltage is taken as the voltage reached when breakdown occurs. Although the variability in times for the rise and fall of the pulse is not considered too significant when breakdown occurs on the tail of the pulse, front-of-wave breakdown voltages are generally higher than those occurring on the tail. The polarity of the impulse can also differ between studies, although the standard is negative polarity since that is the polarity of the usual lightning strike. There is also a fair degree of variability in the experimental conditions for a.c. power frequency breakdown studies. The frequency used can be 50 or 60 Hz, depending on the power frequency in the country where the study was performed. Holding times at voltage can be between 1 and 3 min or the breakdown occurs on a rising voltage ramp where the volts/second rise can differ from study to study. These differences can lead to differences in the breakdown levels reported, but they should not amount to more than a few percent. The ratio between the full wave impulse breakdown voltage and the a.c. rms breakdown

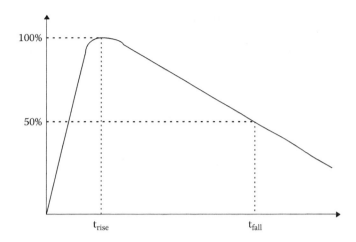

FIGURE 13.2
Impulse waveshape.

voltage is called the impulse ratio and is found to be in the range of 2–3. This ratio applies to a combination of solid (pressboard) insulation and transformer oil, although it is not too different for either to be considered separately.

There is some controversy concerning the breakdown mechanism in the different time regimes for transformer oil. Endicott and Weber [End57] found the same asymmetric probability density distributions for impulse and a.c. breakdown voltages, indicating that extreme value statistics are operative. They also found that both types of breakdown depended on the electrode area. On the other hand, Bell [Bel77] found that the impulse (front-of-wave) breakdown voltages had a symmetrical Gaussian probability density, whereas the a.c. breakdowns had an asymmetric probability density. They found, nevertheless, that the impulse breakdown levels depended on the stressed oil volume. According to this finding, the volume effect under impulse is not linked to extreme value statistics. One would have difficulty imagining a chain of dipoles aligning in the short times available during a front-of-wave impulse test. Palmer and Sharpley [Pal69] found that both impulse and a.c. breakdown voltages depended on the volume of stressed oil. However, they reported that both impulse and a.c. breakdown statistics followed a Gaussian distribution. It would appear from these and similar studies that the breakdown mechanism in transformer oil is not understood enough to conclude that different mechanisms are operative in the different time regimes.

One of the consequences of the theoretical uncertainty in transformer oil breakdown and to a certain extent in, breakdown in solids also is that there is no accepted way to parameterize the data. Thus, some authors present graphs of breakdown voltage or stress vs electrode area; others show breakdown vs stressed oil volume, and others breakdown vs gap thickness. An attractive compromise appears in the work of Danikas [Dan90]. In this work and others [Bel77], breakdown is studied as a function of both electrode area and gap thickness. This also allows for the possibility of a volume effect should the dependence be a function of area × gap thickness. In fact, in [Dan90] the area effect appears to saturate at large areas, that is, the breakdown level is unchanged as the area increases beyond a given value. Thus, breakdown becomes purely a function of gap spacing at large enough areas. Higake et al. [Hig75] found that the breakdown electric stress becomes constant for large gap distances as well as for large volumes. This would suggest caution in extrapolating experimental results either in the direction of larger or smaller parameter values from those covered by the experiment.

At this point, we present some of the breakdown data and parameter dependencies reported in the literature. When graphs are presented by the authors, we have converted the best fit into an equation. Also, voltage values are converted to electric field values when the voltage is applied across a uniform gap. For consistency, we use kV/mm units for breakdown stress, mm for gap thickness, and mm^2 or mm^3 for areas or volumes in the formulas. This will allow us to compare different results not only with respect to parameter dependencies but also with respect to numerical values. Generally, the breakdown voltages are those for which the probability of breakdown is 50%. Thus, some margin below these levels would be needed in actual design.

13.2.1 Breakdown in Solid Insulation

We begin with breakdown in solid insulation, namely, oil saturated paper and pressboard. Samples are prepared by drying and vacuum impregnation and tested under oil. For paper at 25°C, Blume et al. [Blu51] report breakdown voltage stress $E_{b,ac}$ versus thickness d for a.c. 60 Hz voltages:

$$E_{b,ac}\left(\frac{kV_{rms}}{mm}\right) = \frac{17.1}{d^{0.33}}\left(0.85 + \frac{0.15}{t^{1/4}}\right) \tag{13.3}$$

with d in mm and t, the duration of the voltage application, in minutes. Palmer and Sharpley [Pal69] report the impulse breakdown in paper, $E_{b,imp}$, versus thickness d in mm at 90°C as

$$E_{b,imp}\left(\frac{kV_{peak}}{mm}\right) = \frac{79.43}{d^{0.275}} \tag{13.4}$$

In going from 90°C to 20°C, the impulse breakdown stress increases by about 10% according to [Pal69]. Clark [Cla62] reports for Kraft paper at room temperature and a.c. test conditions

$$E_{b,ac}\left(\frac{kV_{rms}}{mm}\right) = \frac{32.8}{d^{0.33}} \tag{13.5}$$

The impulse ratio for paper, corrected for temperature and based on (13.4) and (13.5) is ~2.7.

Results from different investigators are difficult to compare because the thickness buildup is achieved by stacking thin layers of paper. The stacking processes could differ. Some could use lapping with different amounts of overlap as well as different thicknesses of the individual layers. Other possible differences could include the shape and size of the electrodes. Nevertheless, the exponent of the thickness dependence is nearly the same in different studies. Reference [Cla62] reports an area effect, but it is difficult to quantify.

For pressboard in oil at 25°C, Reference [Blu51] reports

$$E_{b,ac}\left(\frac{kV_{rms}}{mm}\right) = \frac{25.7}{d^{0.33}}\left(\frac{1.75}{f^{0.137}}\right)\left(0.5 + \frac{0.5}{t^{1/4}}\right) \tag{13.6}$$

with the frequency f in Hz and time duration t in minutes. The frequency dependence was only tested in the range of 25–420 Hz but is expected to hold for even higher frequencies.

For pressboard at room temperature, using 25 mm sphere electrodes, Moser [Mos79] reports both the a.c. and impulse breakdown level as

$$E_{b,ac}\left(\frac{kV_{rms}}{mm}\right) = \frac{33.1}{d^{0.32}}$$

$$E_{b,imp}\left(\frac{kV_{peak}}{mm}\right) = \frac{94.6}{d^{0.22}}$$

(13.7)

Thus, the impulse ratio for pressboard obtained from (13.7) is ~3.0.

Reference [Pal69] reports for pressboard in oil at 90°C the a.c. and impulse results as

$$E_{b,ac}\left(\frac{kV_{rms}}{mm}\right) = \frac{27.5}{d^{0.26}}$$

$$E_{b,imp}\left(\frac{kV_{peak}}{mm}\right) = \frac{91.2}{d^{0.26}}$$

(13.8)

The impulse ratio for pressboard based on (13.8) is ~3.3. The trend in the data between [Mos79] and [Pal69] is in the right direction since pressboard breakdown strength decreases with increasing temperature.

Based on the previously mentioned data, there does not appear to be much difference between the breakdown strength of oil-soaked paper or pressboard insulation whether under a.c. or impulse test conditions. Even the thickness dependencies are similar. Although Cygan and Laghari [Cyg87] found an area and thickness dependence on the dielectric strength of polypropylene films, it is not known how applicable this is to paper or pressboard insulation.

13.2.2 Breakdown in Transformer Oil

For transformer oil at 90°C, Reference [Pal69] reports breakdown electric fields that depend on volume Λ in mm³ according to

$$E_{b,ac}\left(\frac{kV_{rms}}{mm}\right) = 34.9 - 1.74 \ln\Lambda$$

$$E_{b,imp}\left(\frac{kV_{peak}}{mm}\right) = 82.5 - 3.69 \ln\Lambda$$

(13.9)

This would imply an impulse ratio for oil of ~2.5. In contrast to paper insulation, the breakdown strength of transformer oil increases slightly with temperature in the range of −5°C to 100°C [Blu51]. Nelson [Nel89] summarizes earlier work on the volume effect for breakdown in oil by means of the formula

$$E_{b,ac}\left(\frac{kV_{rms}}{mm}\right) = \frac{46.1}{\Lambda^{0.137}}$$

(13.10)

with Λ in mm³. We should note that the logarithmic dependence on volume given in (13.9) cannot be valid for very large volumes since the breakdown stress would eventually become negative. In (13.10), the breakdown stress becomes unrealistically close to zero as volume increases.

References [End57], [Web56] present breakdown strength in oil versus electrode area, A, in the form

$$\left(E_1 - E_2\right)_{b,ac}\left(\frac{kV_{rms}}{mm}\right) = 1.74\ln\left(\frac{A_2}{A_1}\right)$$
$$\left(E_1 - E_2\right)_{b,imp}\left(\frac{kV_{peak}}{mm}\right) = 4.92\ln\left(\frac{A_2}{A_1}\right) \tag{13.11}$$

In (13.11), E_1 refers to the breakdown voltage for area A_1, etc., for E_2 and A_2. The gap spacing in both of these studies was 1.9 mm. Although the impulse conditions in this study were front of wave, the voltage ramp was kept slow enough in an attempt to approximate full wave conditions. Thus, we can reasonably obtain an impulse ratio from (13.11) of ~2.8.

Reference [Mos79] gives the a.c. (50 Hz, 1 min) partial discharge inception electric stress, $E_{pd,ac}$, for oil as a function of the gap thickness only. While not strictly the breakdown strength, partial discharge maintained over a long enough time can lead to breakdown and is therefore undesirable. Separate curves are given for gas-saturated or degassed oil and for insulated and noninsulated electrodes. Fitting the curves, we obtain

$$E_{pd,ac}\left(\frac{kV_{rms}}{mm}\right) = \frac{14.2}{d^{0.36}} \quad \text{(gas-saturated oil, non insulated electrodes)}$$
$$E_{pd,ac}\left(\frac{kV_{rms}}{mm}\right) = \frac{17.8}{d^{0.36}} \quad \text{(degased oil, non insulated electrodes)}$$
$$E_{pd,ac}\left(\frac{kV_{rms}}{mm}\right) = \frac{19.0}{d^{0.38}} \quad \text{(gas-saturated oil, insulated electrodes)} \tag{13.12}$$
$$E_{pd,ac}\left(\frac{kV_{rms}}{mm}\right) = \frac{21.2}{d^{0.36}} \quad \text{(degassed oil, insulated electrodes)}$$

with d in mm.

Giao Trinh et al. [Tri82] analyzed transformer oil for dielectric strength dependence on both area and volume. They conclude that both area and volume effects can be present with the area effect becoming more important for ultraclean oil and the volume effect for oil having a higher particle content. Although the data show much scatter, they present curves for oils of different purities. For their technical-grade transformer oil, the middle grade of the three analyzed, the following formulas are approximate fits to their curves:

$$E_{b,ac}\left(\frac{kV_{rms}}{mm}\right) = 5\left(1 + \frac{10}{A^{0.2}}\right)$$
$$E_{b,imp}\left(\frac{kV_{peak}}{mm}\right) = 10\left(1 + \frac{23.7}{A^{0.25}}\right) \tag{13.13}$$

$$E_{b,ac}\left(\frac{kV_{rms}}{mm}\right) = 5\left(1 + \frac{7.9}{\Lambda^{0.14}}\right)$$
$$E_{b,imp}\left(\frac{kV_{peak}}{mm}\right) = 15\left(1 + \frac{9.7}{\Lambda^{0.18}}\right) \tag{13.14}$$

with A in mm^2 and Λ in mm^3. They seem to be saying that it is immaterial, whether one describes breakdown in terms of an area or volume effect. However, the consequences of these two approaches are quite different. For areas and volumes in the range of ~10^3–10^7 mm^2 or mm^3, the impulse ratio implied by the said formulas is in the range of ~2–3. In these formulas, as is also evident in the work of [Dan90], [Hig75], the dielectric strength approaches a constant value as the area or volume becomes very large. This is a reasonable expectation. On the other hand, dielectric strengths approach infinity as areas, volumes, and gap distances approach zero in all the said formulas. This is unlikely, although the formulas seem to hold for quite small values of these quantities.

It can be seen, by putting in typical values for d, A, and Λ as found in transformers in the said formulas, that the dielectric strength of paper or pressboard is approximately twice that of oil. However, the electric stress that occurs in the oil is typically greater than that which occurs in the paper or pressboard. For this reason, breakdown generally occurs in the oil gaps first. However, once the oil gaps break down, the solid insulation will see a higher stress so that it could, in turn, break down. Even if the solid insulation can withstand the higher stress, the destructive effects of the oil breakdown such as arcing or corona could eventually puncture the solid insulation. Thus, it would seem inappropriate to design a solid-oil insulation system so that the solid by itself could withstand the full voltage applied across the gap as has sometimes been the practice in the past, unless the insulation is all solid. In fact, excess solid insulation, because of its higher dielectric constant than oil, increases the stress in the oil above that of a design more sparing of the solid insulation. This is an example of the enhancement factor discussed in an earlier chapter.

Because of the gap and volume dependence of oil breakdown strength, it is common practice to subdivide large oil gaps, as occurs for example between the transformer windings, by means of one or more thin pressboard cylinders. Thus, a gap having a large distance or volume is reduced to several smaller gaps, each having higher breakdown strength. Hence, current practice favors a gap or volume effect over a pure area effect since gap subdivision would not be of benefit for an area effect. A combination of gap–area or gap–volume effect would also be consistent with current practice.

The breakdown data referred to previously apply to uniform gaps or gaps as reasonably uniform as practical. This situation is rarely achieved in design so the question arises as to how to apply these results in practice. In volume-dependent breakdown, it is suggested that only the oil volume encompassing electric field values between the maximum and 90% of the maximum be used [Pal69], [Wil53]. Thus for concentric cylinder electrodes, for example, only the volume between the inner cylinder and a cylinder at some fraction of the radial distance to the outer cylinder would be used in the volume-dependent breakdown formulas. For more complicated geometries, numerical methods such as finite elements could be used to determine this effective volume.

For gap-distance-dependent breakdown, the suggestion is to subdivide a possible breakdown path into equal-length subdivisions and to calculate the average electric field over each subdivision. The maximum of these average fields is then compared with the breakdown value corresponding to a gap length equal to the subdivision length. If it exceeds the breakdown value, breakdown along the entire path length is assumed to occur. This procedure is repeated for coarser and coarser subdivisions until a single subdivision consisting of the entire path is reached. Other possible breakdown paths are then chosen and the procedure is repeated [Nel89], [Franc].

Another type of breakdown, which can occur in insulation structures consisting of solids and liquids or solids and gases, is creep breakdown. This occurs along a solid surface in contact with a liquid or gas. These potential breakdown surfaces are nearly unavoidable in insulation design. For example, the oil gaps present in the region between windings are kept uniform by means of sticks placed around the circumference. The surfaces of these sticks bridge the gap, providing a possible surface breakdown path. Since breakdown along such surfaces generally occurs at a lower stress than breakdown in the oil or air through the gap itself, surface breakdown is often design limiting. Reference [Pal69] parameterizes the surface creep breakdown stress along pressboard surfaces in oil at power frequency, $E_{cb,ac}$, in terms of the creep area, A_c, in mm^2 according to

$$E_{cb,ac}\left(\frac{kV_{rms}}{mm}\right) = 16.0 - 1.09 \ln A_c \qquad (13.15)$$

On the other hand, Reference [Mos79] describes creep breakdown along pressboard surfaces in oil in terms of the creep distance along the surface, d_c, in mm according to

$$E_{cb,ac}\left(\frac{kV_{rms}}{mm}\right) = \frac{16.6}{d_c^{0.46}} \qquad (13.16)$$

For nonuniform field situations, the same procedure of path subdivision and comparison with the creep breakdown strength calculated by (13.16) is followed as described earlier for gap breakdown.

13.3 Geometric Dependence of Transformer Oil Breakdown

One of the difficulties apparent in the previous discussion is that there are different theories with different mechanisms to explain breakdown in transformer oil or other liquids. These theories do not provide detailed numerical predictions, as is the case for breakdown in gases where the mechanism of breakdown is better understood on a fundamental level [Ree73]. At best, they point to a parameter dependence of the breakdown process, such as a dependence on the gap length, electrode area, or stressed volume. Experimental data are then used to fit a formula with the desired parameter dependence. This raises a problem. Does one choose the parameter dependence to fit the different physical configurations encountered in practice? This appears to be ad hoc. The method of gap subdivision described earlier comes closest to providing a general procedure to apply to all situations. The problem with this approach is that it does not appear to have any rational physical basis.

Often, the same breakdown data can be fit to a gap, area, volume, or other parameter dependence. For example, in a series of papers detailing the breakdown process in transformer oil [Top02], [Les02], [Top02a], the authors find a dependence of the breakdown or corona inception field on the electrode area over a wide range of areas. However, their own data were also fit to a breakdown field dependence on the radius of curvature of the electrode in the point or rod-plane geometry used.

While statistically, area or volume effects seem to play some role in the breakdown process, they do not provide a breakdown mechanism that can account for all the data. Indeed, there does not seem to be any concrete breakdown mechanism associated with the area effect. In practice, designers of high-voltage core-form transformers subdivide the vertical gaps between the windings with a series of thin pressboard cylinder barriers, suggesting their belief in a gap dependence of the breakdown voltage. (This technique would also be effective for a volume-dependent breakdown voltage but not for an area-dependent one.) In this design process, reliance is placed on gap-dependent partial discharge inception curves such as those provided by [Mos79]. On the other hand, there are geometries within a transformer, which are not covered by a simple gap-dependent breakdown curve, for example, the stress at the corner of a disk, which is part of a disk or helical winding. This also applies to leads or cylindrical cables at high voltage next to the grounded transformer tank.

In this section, we develop a phenomenological theory of breakdown in oil that can be applied to all geometries. By this, we mean that we assume an underlying theory of breakdown but that the parameters of the theory are obtained from experiment. Basically, the theory derives its key parameters from planar gap breakdown but then can be used to predict breakdown in other geometries. We will mainly consider the extension to cylindrical and spherical geometries since these can be derived analytically. However, some discussion of the more general approach with examples will also be given.

We assume a Townsend type of breakdown mechanism for the oil but do not attempt to link the parameters of the theory to any fundamental electronic or molecular processes. As mentioned earlier, the parameters of the theory will be chosen to fit the gap-dependent planar breakdown data, and then these same parameters will be applied to the cylindrical and spherical geometries. Data for these geometries can then provide a test of the theory. We use the term *breakdown* in this section loosely to refer to actual breakdown or to corona or partial discharge inception voltage.

13.3.1 Theory*

We will focus on the electrons produced in the avalanche process and assume that they move with a velocity $\mathbf{v}$ proportional to the electric field $\mathbf{E}$:

$$\mathbf{v} = \mu\mathbf{E} \qquad (13.17)$$

where μ is the mobility assumed constant. (Here as elsewhere, when we use the term *electrons*, we mean any charged complex or entity.) Actually, the electrons move opposite to the electric field so the mobility can be considered negative or, in later discussions, the applied voltage can be considered negative. We assume that all the electrons, even the newly released ones, achieve this equilibrium velocity instantaneously or in extremely short times. Letting n be the number of electrons per unit volume, the continuity or conservation equation for electrons can be written [Dai73] as

$$\frac{\partial n}{\partial t} + \nabla \cdot (n\mathbf{v}) = \gamma \qquad (13.18)$$

where γ is the number of electrons produced per unit volume per unit time. Ignoring diffusion effects, which should be extremely slow compared with the breakdown process, and assuming steady-state conditions, Equation 13.18 becomes

$$\mathbf{v} \cdot \nabla n + n\nabla \cdot \mathbf{v} = \gamma \qquad (13.19)$$

---

* This section follows closely our published paper [Del04], *IEEE Trans. Power Deliv.*, 19(2), April 2004, 652–656. © 2004 IEEE.

We will assume that γ depends on the electric field and has the form

$$\gamma = \beta n v g(E) \tag{13.20}$$

where
  β is a constant having the units of inverse length
  $g(E)$ is a unitless field-dependent quantity
  n and v have their usual meanings

Note that $\beta g(E)$ is the number of electrons produced or released by a single electron in an electric field E per unit distance traveled. This depends on the magnitude of the electron velocity via (13.17). nv is the flux of electrons, that is, the number of electrons per unit area per unit time. Hence, the product γ is the number of secondary electrons released per unit volume per unit time.

13.3.2 Planar Geometry

Applying (13.19) to a planar geometry as shown in Figure 13.3, where the E-field is uniform in the x direction so that $\nabla \cdot \mathbf{v} = 0$ by virtue of (13.17) and all quantities only depend on x, we have

$$v \frac{dn}{dx} = \beta n v g(E) \tag{13.21}$$

Canceling out v and rearranging, we get

$$\frac{dn}{n} = \beta g(E) dx \tag{13.22}$$

Note that we are ignoring space charge effects since, in general, $\nabla \cdot \mathbf{E} = \rho/\varepsilon$, where ρ is the charge density and ε the permittivity. This effectively assumes that the electric field produced by any charge accumulation is small compared with the applied field and thus does

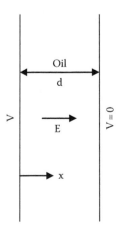

FIGURE 13.3
Planar geometry. (From Del Vecchio, R.M., *IEEE Trans. Power Deliv.*, 19(2), April 2004, 652–656. © 2004 IEEE.)

not significantly distort it. While this is probably a good approximation for the incipient breakdown conditions we are considering here, it would probably not apply if the details of the breakdown process itself were being modeled. Equation 13.22 can be integrated to get

$$\frac{n(d)}{n_0} = e^{\beta \int_0^d g(E)dx} \tag{13.23}$$

where n_0 is the number density of electrons at $x = 0$ and the integral is over the thickness of the gap, d. Since E is constant, this becomes

$$\frac{n(d)}{n_0} = e^{\beta g(E)d} \tag{13.24}$$

We expect breakdown to occur when $n(d)/n_0$ is greater than some large number. However, this is equivalent to stating that

$$\beta g(E)d > N \tag{13.25}$$

where $N \sim 10$ since e^{10} is a large number. The exact value is not significant since we will soon eliminate it from the equations. Since β is a constant, we can write more succinctly

$$g(E)d > \frac{N}{\beta} = \kappa \tag{13.26}$$

where κ is some characteristic constant having units of length. We can think of κ as the number of electrons, which must be released per unit electron times the distance over which this release must occur in order to cause breakdown. Thus, an energetic electron, which releases many secondary electrons, need only traverse a short distance to cause breakdown. On the other hand, a less energetic electron, which releases few secondary electrons, must traverse a longer distance to cause breakdown.

The planar gap breakdown data for a.c. voltages and degassed oil with noninsulated electrodes are given in [Mos79]. It has been fitted to the formula given previously in (13.12) and repeated here:

$$E_{ac,planar}\left(\frac{kV_{rms}}{mm}\right) = \frac{17.8}{d^{0.36}} \tag{13.27}$$

where the rms breakdown field is in kV/mm and the gap distance d is in mm. Comparing (13.26) and (13.27), we see that g(E) must be of the form

$$g(E) = \left(\frac{E}{E_o}\right)^m \tag{13.28}$$

where E_o is a constant chosen to make g(E) unitless. Using (13.28) in (13.26), we get for the breakdown field

$$E = E_o \left(\frac{\kappa}{d}\right)^{1/m} \tag{13.29}$$

We see that we should take $1/m = 0.36$ or $m = 2.778$ in order to reproduce the exponential dependence in (13.27). If we take the arbitrary constant $E_o = 100$ kV/mm, we find that

$$E = E_o \left(\frac{\kappa}{d} \right)^{0.36} = \frac{17.8}{d^{0.36}} \quad \Rightarrow \quad \kappa^{0.36} = \frac{17.8}{100} \quad \Rightarrow \quad \kappa = 0.0083$$

Thus, $\kappa = 0.0083$ mm is needed in order to fit (13.27).

For impulse waveforms, the impulse breakdown voltage for a given gap thickness has been shown by numerous experiments as cited earlier to be a constant multiple of the a.c. rms breakdown voltage, which on average is about 2.8. This is called the impulse ratio. Using this impulse ratio, the planar impulse breakdown is given in (13.30), with d in mm

$$E_{Imp,planar} \left(\frac{kV_{peak}}{mm} \right) = \frac{50}{d^{0.36}} \tag{13.30}$$

In order to fit this equation, m remains the same but now $\kappa = 0.146$ mm. This would imply that a greater electron avalanche is required to produce breakdown under the short-duration impulse conditions as compared with the much longer a.c. breakdown conditions for a given distance of electron travel. This has some intuitive appeal.

It is expected that β and, hence, κ will depend on oil quality or oil type. Also, m may depend on the type or quality of the oil. It is certainly well-known that impurities in the oil affect the breakdown value.

13.3.3 Cylindrical Geometry

We now apply these same principles to a cylindrical geometry as shown in Figure 13.4. The electric field in the gap between the inner cylinder of radius a and the outer cylinder of radius b is given by

$$E = \frac{V}{r \ln(b/a)} a_r \tag{13.31}$$

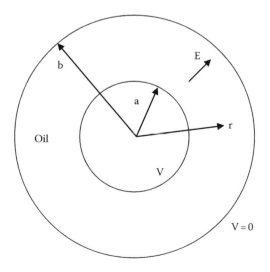

FIGURE 13.4
Cylindrical or spherical geometry. (From Del Vecchio, R.M., *IEEE Trans. Power Deliv.*, 19(2), April 2004, 652–656. © 2004 IEEE.)

where

    a_r is a unit radial vector
    r is the radial coordinate
    V is the applied voltage

Since **v** is proportional to **E** by (13.17), **v** is also a radial r-dependent vector. In fact, because of the cylindrical symmetry, all quantities depend only on r. This greatly simplifies (13.19) when expressed in cylindrical coordinates. Using (13.17) and (13.31) and ignoring space charge effects, we have $\nabla \cdot \mathbf{v} = 0$ in cylindrical coordinates since

$$\nabla \cdot \mathbf{v} = \nabla \cdot \mu \mathbf{E} = \frac{1}{r}\frac{\partial}{\partial r}\left(r\, \frac{\mu V}{r \ln(b/a)} \right) a_r = 0$$

Expressing the remaining terms in (13.19) in cylindrical coordinates, we have simply

$$\frac{dn}{dr} = \beta n g(E) \tag{13.32}$$

This is identical in form to (13.22) except that g(E) now depends on r through E. Thus, the breakdown condition becomes

$$\int_a^b \left(\frac{E(r)}{E_o} \right)^m dr > \frac{N}{\beta} = \kappa \tag{13.33}$$

where m has been given numerically. Getting E(r) from (13.31), we can perform the integral in (13.33) to get as a breakdown condition

$$\left(\frac{E_a}{E_o} \right)^m \left(\frac{a}{m-1} \right)\left[1 - \left(\frac{a}{b}\right)^{m-1} \right] = \kappa \tag{13.34}$$

where

$$E_a = \frac{V}{a \ln(b/a)} \tag{13.35}$$

is the electric field at the inner cylinder as well as the maximum field in the oil gap between the cylinders. Solving for E_a, we find

$$E_a = E_o \left\{ \frac{(m-1)\kappa}{a\left[1 - \left(\dfrac{a}{b}\right)^{m-1} \right]} \right\}^{1/m} \tag{13.36}$$

For the same choice of E_o, κ, and m used in the planar case, we get

$$E_{a,ac,cylindrical}\left(\frac{kV_{rms}}{mm} \right) = \frac{21.9}{\left\{ a\left[1 - \left(\dfrac{a}{b}\right)^{1.778} \right] \right\}^{0.36}} \tag{13.37}$$

and for impulse conditions, assuming an impulse ratio of 2.8,

$$E_{a,Imp,cylindrical}\left(\frac{kV_{peak}}{mm}\right) = \frac{61.3}{\left\{a\left[1-\left(\frac{a}{b}\right)^{1.778}\right]\right\}^{0.36}}$$

(13.38)

a and b are in mm in the preceding formulas.

13.3.4 Spherical Geometry

The spherical geometry is also depicted in Figure 13.4 where a and b are now the radii of the inner and outer spheres, respectively. The electric field in the spherical gap is given by

$$E = \frac{V}{r^2 \frac{1}{a}\left(1-\frac{a}{b}\right)} \mathbf{a_r}$$

(13.39)

where the symbols have their usual meanings. Because of the spherical symmetry, all quantities depend only on r. Also, in view of (13.17) and (13.39) and ignoring space charge effects, $\nabla \cdot \mathbf{v} = 0$ in spherical coordinates since

$$\nabla \cdot \mathbf{v} = \nabla \cdot \mu \mathbf{E} = \frac{1}{r^2}\frac{\partial}{\partial r}\left(r^2 \frac{\mu V}{r^2 \frac{1}{a}\left(1-\frac{a}{b}\right)}\right)\mathbf{a_r} = 0$$

Therefore, Equation 13.19 becomes, in these coordinates,

$$\frac{dn}{dr} = \beta ng(E)$$

(13.40)

Thus, (13.40) is of the same form as (13.22) except that E is a function of r. Thus, the breakdown condition becomes identical to (13.33), except that the E(r) dependence is different. Substituting from (13.39) and performing the integral, we get

$$\left(\frac{E_a}{E_o}\right)^m \left(\frac{a}{2m-1}\right)\left[1-\left(\frac{a}{b}\right)^{2m-1}\right] = \kappa$$

(13.41)

where

$$E_a = \frac{V}{a\left(1-\frac{a}{b}\right)}$$

(13.42)

is the E-field at the inner sphere and the maximum field in the oil gap. Solving for E_a,

$$E_a = E_o \left\{ \frac{(2m-1)\kappa}{a\left[1-\left(\dfrac{a}{b}\right)^{2m-1}\right]} \right\}^{1/m} \tag{13.43}$$

Choosing E_0, κ, and m from the planar fit, we have

$$E_{a,ac,spherical}\left(\frac{kV_{rms}}{mm}\right) = \frac{30.8}{\left\{a\left[1-\left(\dfrac{a}{b}\right)^{4.556}\right]\right\}^{0.36}} \tag{13.44}$$

For impulse waveforms, we have

$$E_{a,Imp,spherical}\left(\frac{kV}{mm}\right) = \frac{86.2}{\left\{a\left[1-\left(\dfrac{a}{b}\right)^{4.556}\right]\right\}^{0.36}} \tag{13.45}$$

with a and b in mm.

13.3.5 Comparison with Experiment

Note that for an isolated cylinder or sphere, where $b \gg a$, we have from (13.36) and (13.43)

$$E_{a,cylinder} = E_o\left[\frac{(m-1)\kappa}{a}\right]^{1/m} \tag{13.46}$$

$$E_{a,sphere} = E_o\left[\frac{(2m-1)\kappa}{a}\right]^{1/m} \tag{13.47}$$

Taking the ratio

$$\frac{E_{a,sphere}}{E_{a,cylinder}} = \left(\frac{2m-1}{m-1}\right)^{1/m} \tag{13.48}$$

This ratio should be constant, independent of the radius of the cylinder or sphere and independent of the waveform, provided they are the same for the two cases. For m = 2.778, this ratio is 1.4.

Reference [Wil53], citing data for oil obtained by Peek in 1915 [Pee29], shows this ratio to be nearly constant for different radii. However, the ratio found is 1.6. This could reflect a different type of oil used then since m could depend on the oil type or quality. Or it could reflect the fact that the necessarily finite length cylinders used in the experiments were not a sufficiently close approximation to the ideal infinite-length ones analyzed here.

In the case of an isolated sphere (b ≫ a) under impulse conditions, we have from (13.45)

$$E_{a,Imp,sphere}\left(\frac{kV}{mm}\right) = \frac{86.2}{a^{0.36}} \tag{13.49}$$

Breakdown data for impulse conditions obtained by [Les02] for point and rod electrodes reasonably far from the ground plane were fit to the expressions

$$E_{positive\ polarity}\left(\frac{kV}{mm}\right) = \frac{93}{a^{0.35}}$$

$$E_{negative\ polarity}\left(\frac{kV}{mm}\right) = \frac{78}{a^{0.35}} \tag{13.50}$$

where a is the radius of curvature of the tip in mm. Although the geometry is not quite that of an isolated sphere, it is reasonably close so that the favorable comparison with (13.49) is meaningful.

13.3.6 Generalization

We should note that the breakdown formulas for the cylindrical and spherical geometries reduce to the planar case when a and b become large. This is expected since locally the cylindrical and spherical gaps start looking more and more like planar gaps. However, for small a, the breakdown formulas depart significantly from the planar case. In fact, for b large compared with a, as (13.49) indicates, the formulas depend only on the radius of curvature of the inner electrode. The gap or distance between a and b plays no role. As a decreases, the breakdown strength increases. Apart from the confirmation of this effect provided by [Les02], transformer designers know that the breakdown stress at the corners of the individual disks in a disk winding, where small radii of curvature are present, can be much higher than the allowable breakdown stress in the gaps between windings.

Although expressions for breakdown have been found for idealized geometries, in principle, Equation 13.19 could be solved for any geometry. This requires first solving for the electric field. Then **v** and g(E) become known so that n can be found throughout the volume where **E** exists. By examining the boundary values of n, one could then see if it has been multiplied sufficiently from its starting electrode value to cause breakdown. n would need to be set at some value, say 1, on the starting electrode.

While the described procedure requires a considerable calculational effort, it may be possible to justify a simpler procedure for a general geometry. For example, having solved for the E-field, one might perform the integral

$$\int \left(\frac{E(s)}{E_o}\right)^m ds \tag{13.51}$$

along possible breakdown paths and then compare it with κ where E_o = 100 kV/mm, κ = 0.0083 mm, and m = 2.778 are reasonable choices for standard transformer oil and a.c. voltages. Assuming an impulse ratio of 2.8, one needs to use κ = 0.146 mm for impulse conditions. In performing the integral in (13.51), one should always remain tangential to the E-field or perpendicular to the equipotential surfaces since the electrons are assumed to move in this fashion. More specifically, the breakdown condition would become

$$\int_P \left(\frac{E(s)}{E_o} \right)^m ds \geq \kappa \tag{13.52}$$

where the integral is taken along a most likely breakdown path, P. This would then generalize (13.33), which was used for planar, cylindrical, and spherical geometries, although in these three cases, the most likely breakdown path is clear. Note that the integral in (13.52) is in mm, that is, the same unit as κ. Also, if there is insulation present and E(s) is calculated without including the enhancement effect of this insulation, then an enhancement factor should multiply E(s) in (13.52). If this can be considered constant along the path, then it can be factored out of the integral as an overall multiplier. Thus, if the constant enhancement factor is η, then η^m should multiply the integral.

$$\int_P \left(\frac{\eta E(s)}{E_o} \right)^m ds = \eta^m \int_P \left(\frac{E(s)}{E_o} \right)^m ds \geq \kappa \quad \Rightarrow \quad \int_P \left(\frac{E(s)}{E_o} \right)^m ds \geq \frac{\kappa}{\eta^m} \tag{13.53}$$

This shows that the presence of an enhancement factor lowers the threshold for breakdown.

On the other hand, if one wants to determine the voltage that produces breakdown for a given situation, it is necessary to assume that the E(s)'s in (13.53) are proportional to the voltage V that is producing the E-fields. This is the case for the three geometries considered in the equation. Thus to ensure that equality holds in (13.53) for the given E(s)'s used in the equation, we need to multiply E(s) by a factor X to be determined. This leads to the breakdown condition

$$\int_P \left(\frac{XE(s)}{E_o} \right)^m ds = \frac{\kappa}{\eta^m} \quad \Rightarrow \quad \int_P \left(\frac{E(s)}{E_o} \right)^m ds = \frac{\kappa}{(X\eta)^m} \quad \Rightarrow$$

$$X\eta = \left(\frac{\kappa}{\int_P \left(\frac{E(s)}{E_o} \right)^m ds} \right)^{1/m} \quad \Rightarrow \quad X = \frac{1}{\eta} \left(\frac{\kappa}{\int_P \left(\frac{E(s)}{E_o} \right)^m ds} \right)^{1/m} \tag{13.54}$$

Breakdown equality was assumed in the special geometries considered thus far, although these did not include an enhancement factor. This equality would apply to new or non-ideal geometries when the breakdown voltage needs to be determined.

13.3.6.1 Breakdown for the Cylinder-Plane Geometry

Another case where the breakdown path is clear is for the cylinder-plane geometry. Here, the most likely breakdown path is from the point on the cylinder nearest the ground plane along the shortest straight line to the ground plane. The electric field along this line is given

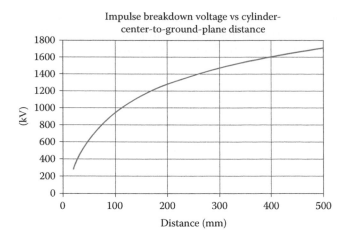

FIGURE 13.5
Breakdown kV with a 20% margin versus the distance from the cylinder center to the ground plane for a cylinder of radius = 10 mm and with a paper cover of 5 mm.

by Equation 11.41. As a numerical example, let the radius of the cylinder be 10 mm with a 5 mm thick paper covering. The distance from the cylinder center to the ground plane is allowed to vary. The breakdown path is therefore along a straight line from the ground plane to the paper cover. In this case, we use an enhancement factor for a cylinder geometry given in Equation 11.19. The enhancement factor varied with the distance to the ground plane but was about 1.05. The gap is filled with oil and the only solid insulation is the paper cover. The breakdown voltage level in kV was calculated for impulse breakdown using (13.54) and includes a 20% margin. It is plotted versus the distance from the cylinder center to the ground plane in Figure 13.5. To get the a.c. breakdown levels with a 20% margin, divide the kV values on the figure by 2.8.

13.3.6.2 Breakdown for the Disk–Disk-to-Ground Plane Geometry

In Chapter 11, we calculated the electric field near the corner of the disk–disk-to-ground plane geometry using conformal mapping. The field was calculated a paper distance away from the corner and included an enhancement factor to account for the paper and other insulation. We can use this solution method here to calculate the field along a possible breakdown path starting at the outer surface of the corner paper.

In this method, the initial E-field vector defines a direction away from the paper surface. We set a sufficiently small step size and find the next point a step size away from the surface in the direction of the initial E-field. At this new point, we calculate a new E-field and use its direction to find the next point a step size away from the second field. Using the direction determined by the second E-field, we get a third point another step size away from the second point and calculate the third E-field here. This process is continued until we encounter a surface such as the ground plane or the paper surface of the second disk. An example of such a path is shown in Figure 13.6. In this example, the disk–disk separation was 10 mm and the 2-sided paper cover was 1 mm so that the gap from disk metal to disk metal was 11 mm. The gap from the disk metal to the ground plane was 100.5 mm, including the 1-sided paper cover. The 2 disks were at 100 and 80 kV. The path started on the outer corner paper of the right disk at 100 kV and ended on the ground plane.

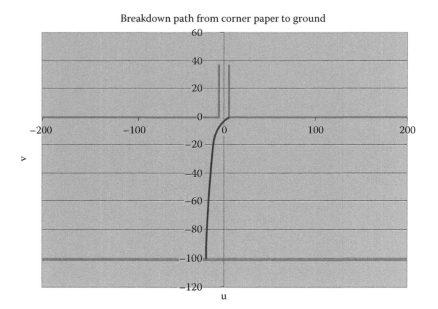

FIGURE 13.6
Breakdown path from the outer paper cover of the disk–disk-to-ground plane geometry to the ground plane. Disk–disk gap = 10 mm, disk–ground plane = 100 mm, 2-sided paper thickness = 1 mm. Disk voltages were 100 and 80 kV.

The maximum corner E-field was selected as the starting point from the three positions shown in Chapter 11.

The integral in (13.51) was performed along this path for the impulse value of $\kappa = 0.146$. Using the enhancement factor described in Chapter 11, which was 1.09 for this case, the calculated breakdown field was 30.0 kV/mm. The maximum field occurs at the outer paper surface at the disk corner.

As another example, the disk–disk gap was set to 21 mm, including a 1 mm 2-sided paper cover, and the disk-to-ground plane was set to 150.5 mm including a 1-sided paper cover. The disk voltages were 100 and 50 kV. In this case, the breakdown path was from one disk to the other as shown in Figure 13.7. Here, the enhancement factor was 1.08 and the breakdown field was 34.0 kV/mm for impulse conditions.

Another example is with a gap spacing of 4.5 mm, including a 0.5 mm thick 2-sided paper cover, and a distance to the ground plane of 50.25 mm, including a 0.25 mm thick 1-sided paper cover. The disk voltages were 100 and 80 kV. The breakdown path is shown in Figure 13.8. In this case, the enhancement factor was 1.07 and the breakdown E-field was 40.0 kV/mm for impulse conditions.

It should be noted that the calculated E-fields for the larger disk–ground plane gaps are probably unrealistically high. The E-field along the breakdown path for this last case is plotted in Figure 13.9. This E-field does not include the enhancement factor or the multiplier required to reach the breakdown level. This is the E-field tangential to the path, which is also the E-field magnitude since it is directed along the path.

It should be noted that the calculated breakdown fields for the larger disk–ground plane gaps are probably unrealistically low. This is because the two disks are assumed to extend indefinitely in the horizontal directions, maintaining their voltage levels. This means that higher fields are present over the longer lengths of the breakdown paths than would normally occur in practice. This is not a concern when the breakdown path is between the

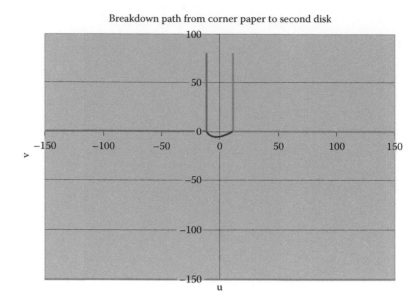

FIGURE 13.7
Breakdown path from corner outer paper cover of one disk to the other disk. Disk–disk gap = 20 mm, disk–ground plane = 150 mm, 2-sided paper thickness = 1 mm. Disk voltages were 100 and 50 kV.

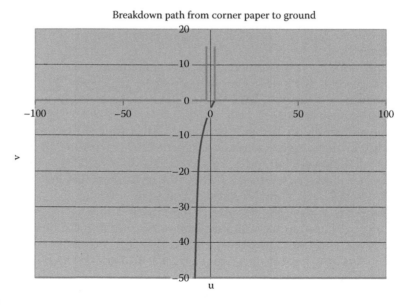

FIGURE 13.8
Breakdown path from corner outer paper cover to the ground plane. Disk–disk gap = 4 mm, disk–ground plane = 50 mm, 2-sided paper thickness = 0.5 mm. Disk voltages were 100 and 80 kV.

two disks since, although the gap between the disks extends indefinitely in the vertical direction, the breakdown path length is relatively short and is confined to the region near the disk corner where the field calculation is more realistic. In practice, there is often a limit placed on the corner E-field, which is in the range of 30–35 kV/mm. This is consistent with the previously described results.

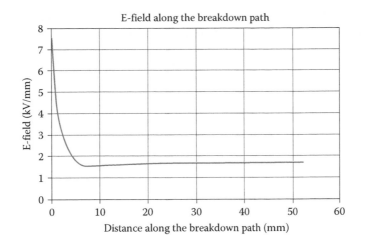

FIGURE 13.9
E-field along the breakdown path for the case shown in Figure 13.8.

When pressboard or other barriers are present, the breakdown path should only be taken up to the barrier because the breakdown avalanche would stop at this point. This is consistent with the practice of subdividing gaps to increase the breakdown electric field levels.

13.3.7 Discussion

We have shown how it is possible to start from the planar geometry breakdown data, assuming a simple Townsend breakdown mechanism, to arrive at a prediction of the breakdown data in other geometries, which agrees quite well with the experimental data in these geometries. Although we referred to electrons participating in the breakdown mechanism, this could clearly refer to any charged complex or entity and almost certainly not to bare electrons. Furthermore, although we refer to breakdown, this could also be interpreted as corona or partial discharge inception voltage, depending on what the planar breakdown data refer to.

This entire exercise highlights the unity of the breakdown process, regardless of the geometry in which it occurs. It is clearly unphysical to invoke area or volume effects to explain breakdown in curved geometries but not in planar geometries. Most of the data support a gap effect for planar geometries, which is characteristic of a Townsend-type mechanism. In addition, the common and effective practice of inserting pressboard barriers in large oil gaps to increase the breakdown voltage is a tacit acknowledgement of a gap effect.

This work has emphasized the need for more oil breakdown data in as close to ideal as possible cylindrical and spherical geometries, where mathematical methods may be applied without much approximation. The startling conclusion given in (13.48) that the breakdown voltage ratio for spherical and cylindrical electrodes of the same radius subject to the same voltage waveform is a constant, independent of the radius of the electrode, provided they are sufficiently isolated, and should be rechecked experimentally. Although ideal spherical geometries can be reasonably approximated, an ideal cylinder (infinitely long) is not easily accessible. Perhaps an equation similar to (13.48) can be derived for finitely long cylinders, provided breakdown occurs near the center. This should approach the infinite case as the cylinder length/radius increases and would at least provide a guide as to when the cylinder is long enough.

There are essentially two parameters in the methodology discussed here, m and κ. (E_o is really arbitrary.) The parameter m appears to be rather insensitive to experimental breakdown conditions, while κ certainly depends on the breakdown waveform, whether one is considering full or partial breakdown, and also on oil quality. Whether these parameters can be obtained in a more fundamental way from the properties of oil remains to be seen. At present, one may expect to parameterize m and κ in terms of waveform duration, the type of breakdown, and oil impurity levels based on experimental data.

13.4 Insulation Coordination

Insulation coordination concerns matching the insulation design to the protective devices used to limit the voltages applied to the terminals of a transformer by lightning strikes or switching surges and possibly other potentially hazardous events. Since the insulation must withstand the normal operating voltages that are present continuously as well as lightning or switching events that are of short duration, how breakdown depends on the time duration of the applied voltage is of major importance in insulation coordination. We have seen previously that the impulse ratio is ~2–3 for oil-filled transformer insulation. This means that the breakdown strength of a short-duration impulse voltage lasting ~5–50 μs is much higher than the breakdown strength of a long-duration (~1 min) a.c. voltage. It appears that, in general, breakdown voltages or stresses decrease with the time duration of the voltage application for transformer insulation. The exact form of this breakdown vs time dependence is still somewhat uncertain, probably because of variations in material purity and/or experimental methods among different investigators.

References [Blu51], [Cla62] show a time-dependent behavior of relative strength of oil or pressboard that is schematically illustrated in Figure 13.10. Three regions are evident on the curve, labeled A, B, and C. For short durations, <10 μs, there is a rapid fall-off in the strength with increasing time. This is followed by a flat portion, B, which extends to about 1,000 μs for oil and to about 20,000 μs for pressboard. In region A, there is a more gradual

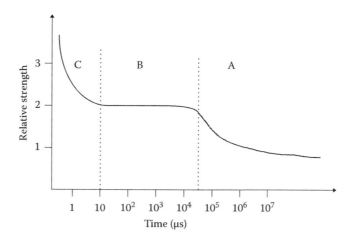

FIGURE 13.10
Oil or pressboard breakdown relative strength vs time—schematic.

fall-off approaching an asymptotic value at long times. The combination of an oil press-board gap also produces a curve similar to Figure 13.10.

Using statistical arguments, that is, equating the probability of breakdown for times less than t for a fixed voltage across an oil gap under corona-free conditions, which is $1 - \exp(-t/t_o)$ where t_o is the average time to breakdown, to Equation 13.1 with V_o taken to be 0, Reference [Kau68] obtains the breakdown voltage vs time relationship for an oil gap in the form

$$\frac{t_2}{t_1} = \left(\frac{V_1}{V_2}\right)^k \tag{13.55}$$

Experimentally, they find that for times from a few seconds to a few weeks, k is between 15 and 30. This relationship cannot, however, hold for very long times since it predicts breakdown at zero voltage at infinite time. To correct this, one could use the same statistical argument to show that

$$\frac{t_2}{t_1} = \left(\frac{V_1 - V_o}{V_2 - V_o}\right)^k \tag{13.56}$$

where V_o is the infinite time breakdown voltage. This can be rearranged to the form of

$$V_2 = V_o + \frac{(V_1 - V_o)t_1^{1/k}}{t_2^{1/k}} \tag{13.57}$$

Picking a V_1, t_1 pair, this can be written for a general V, t pair:

$$V = V_o\left(1 + \frac{K}{t^{1/k}}\right) \tag{13.58}$$

Dividing by the gap thickness, this can be written in terms of the breakdown stress, E_b, as

$$E_b = E_{b,o}\left(1 + \frac{K}{t^{1/k}}\right) \tag{13.59}$$

where $E_{b,o}$ is the infinite time breakdown stress. This last equation is very similar to (13.6) in its time dependence. As shown in Figure 13.10, the time-dependent behavior of transformer oil breakdown is more complicated over a time span from µs to years than the formulas discussed would suggest.

At short times in the µs region, the breakdown voltage or stress vs time characteristic changes rapidly with time. Impulse waveshapes as shown in Figure 13.2 are somewhat complicated functions of time. In order to extract a single voltage–time duration pair from this waveshape for comparison with the breakdown curve, the common practice is to take the peak voltage and the time during which the voltage exceeds 90% of its peak value. Thus, a standard impulse wave, which rises to its peak value in 1.2 µs and decays to 50% of its peak value in 50 µs, spends about 10 µs above its 90% voltage level. A switching surge test waveshape is similar to that of Figure 13.2, but the rise time to peak value is ~100 µs and the fall time to the 50% level is ~500 µs. The time spent above the 90% of peak voltage level is ~200 μs. Therefore, we would expect breakdown to occur on a switching surge test at a lower voltage than for a full wave impulse test.

Two other types of impulse test are the chopped wave test and front-of-wave test, although the latter is sometimes considered unnecessary in view of modern methods of protection. A standard chopped wave, as shown in Figure 13.11, rises to its peak in ~1.2 μs and abruptly falls to zero with a slight undershoot at the chop time of ~3 μs. It is above its 90% of peak voltage for ~3 μs. The front of wave is chopped on the rising part of the wave and is above 90% of its maximum value for ~0.5 μs. The generally accepted breakdown levels corresponding to these different times, normalized to the full wave breakdown level, are given in Table 13.1.

We have included in Table 13.1 two other points, the 1 min a.c. test point and the essentially infinite time nominal voltage point. For the 1 min test point, we have used an impulse ratio of 2.8. Thus, the normalized breakdown peak voltage level is $1/2.8 \times \sqrt{2} = 0.5$ and the duration above 90% of the peak voltage is 0.287×60 s $= 17.2$ s. We have taken the nominal system voltage to be half the 1 min test voltage. Other values for these could have been chosen with equal justification. For instance, an impulse ratio of 2.4 is commonly assumed in setting test values and the nominal voltage can be a factor of 2.5–3.0 below the 1 min test level. Note that the nominal system voltage used here is from terminal to ground.

We can produce a reasonable fit to these tabular values with an equation of the form (13.58). Letting $V_{b,rel}$ be the breakdown levels relative to the full wave level, we obtain

$$V_{b,rel} = 0.25\left(1 + \frac{3.8}{t^{1/12}}\right) \tag{13.60}$$

with t in μs. Table 13.1 is based on practical experience in testing transformers. The numbers can only be regarded as approximate, and different choices are often made within a reasonable range about the values shown. This would influence the fit given in (13.60) or possibly

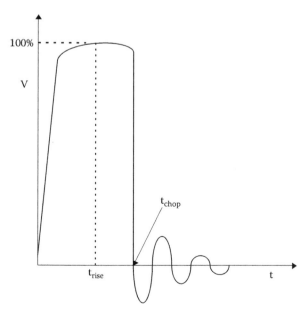

FIGURE 13.11
Chopped wave impulse waveshape.

TABLE 13.1

Breakdown Voltage Durations and Normalized
to the Full Wave Impulse Level

Type	Duration (μs)	Breakdown Level
Front of wave	0.5	1.3
Chopped wave	3	1.1
Full wave	10	1.0
Switching surge	200	0.83
1 min a.c.	17.2×10^6	0.5 peak
Nominal a.c.	∞	0.25 peak

require a different parameterization. This parameterization may have no fundamental significance since it is not based on any carefully controlled experiment. However, its form, which agrees with the more fundamentally based equation (13.58), is noteworthy.

In practical applications, a table like Table 13.1 provides a way of linking the various test voltages to the full wave impulse voltage that is also called the basic impulse level (BIL). Since the required impulse test levels are usually linked to the lightning or surge arrestor protection level, whereas the a.c. test levels or nominal voltage level are not, they can be specified independently. For example, a 500 kV_{rms} line-to-line transformer (= 288 kV_{rms} line to ground = 408.2 kV_{peak} line to ground) would correspond to a full wave impulse level of 1633 kV according to Table 13.1. However, the user may have sufficient protection that only a 1300 kV impulse test is required. Nevertheless, the insulation would have to be designed to withstand a full wave 1633 kV impulse test, even though not performed, since that level of protection is required to guarantee satisfactory operation at the nominal 500 kV_{rms} line-to-line voltage. As another example, one may have breakdown stress vs distance, area, or volume curves for 1 min a.c. test conditions. In order to compare these to the stress levels generated in a simulated impulse test calculation, one needs to know the relative breakdown level factors given in Table 13.1.

13.5 Continuum Model of Winding Used to Obtain the Impulse Voltage Distribution

For very short times after the application of a voltage to a winding terminal, the voltage distribution along the winding is governed primarily by capacitive coupling. This is because the winding inductance limits the flow of current initially.

13.5.1 Uniform Capacitance Model

In the uniform capacitance model, a winding is represented by the capacitive ladder diagram given in Figure 12.1. The discussion here is somewhat simpler since the voltages on the neighboring windings, core, or tank are assumed to be at zero potential. In this case, the differential equation derived in Equation 12.12 becomes

$$\frac{d^2V}{dx^2} - \left(\frac{\alpha}{L}\right)V = 0 \tag{13.61}$$

where $\alpha = \sqrt{C_g/C_s}$ with C_g, the total ground capacitance, and C_s, the series capacitance. L is the winding height and x is measured from the high-voltage end.

If a voltage V_o is suddenly applied at the line end and the other end of the winding is grounded, the voltage distribution along the winding, as given by (12.55), is

$$V(x) = V_o \frac{\sinh\left[\alpha\left(1-\dfrac{x}{L}\right)\right]}{\sinh\alpha} \tag{13.62}$$

When the nonimpulsed end of the winding is floating when the voltage V_o is applied to the line end, the voltage distribution as given by the solution of (13.61) is

$$V(x) = V_o \frac{\cosh\left[\alpha\left(1-\dfrac{x}{L}\right)\right]}{\cosh\alpha} \tag{13.63}$$

Equation 13.62 is replotted in Figure 13.12 for several values of α. This is done here for comparison with the solution of (13.63), which is plotted in Figure 13.13. As can be seen, the slope of the curve for (13.62) becomes very steep at the line end as α increases. This implies that the disk–disk voltage ΔV_d increases with increasing α since this is given approximately as

$$\Delta V_d = \left|\frac{dV}{dx}\right| w = V_o \alpha \left(\frac{w}{L}\right) \frac{\cosh\left[\alpha\left(1-\dfrac{x}{L}\right)\right]}{\sinh\alpha} \tag{13.64}$$

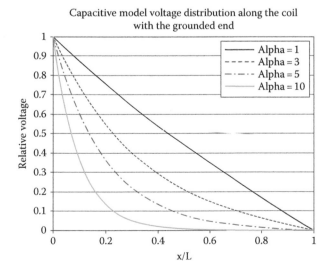

FIGURE 13.12
Initial voltage distribution along a winding grounded at one end for various values of α.

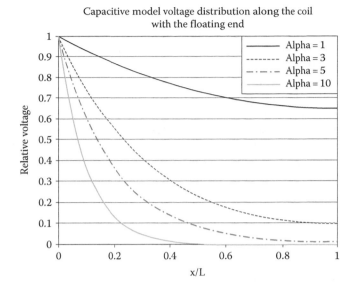

Capacitive model voltage distribution along the coil with the floating end

FIGURE 13.13
Initial voltage distribution along a winding floating at one end for various values of α.

where w is the disk–disk spacing, assumed small compared with L. This has its maximum value at the line end where $x = 0$,

$$\Delta V_{d,line} = V_o \left(\frac{w}{L} \right) \frac{\alpha}{\tanh \alpha} \tag{13.65}$$

For the ungrounded or floating-end winding, Equation 13.63 is plotted in Figure 13.13. The maximum stress again occurs at the line end and, using (13.63), is

$$\Delta V_{d,line} = V_o \left(\frac{w}{L} \right) \alpha \tanh \alpha \tag{13.66}$$

which is somewhat less than (13.65), since $|\tanh \alpha| \leq 1$. When α is large, say >3, $\tanh \alpha \approx 1$, so (13.65) and (13.66) are essentially equal. If α is small, say <0.1, $\Delta V_{d,line} = V_o(w/L)$ for the grounded-end case, while $\Delta V_{d,line} \approx 0$ for the floating-end case.

Since the disk–disk stress increases with α and by implication, the turn–turn stress also, this suggests that to reduce α we need to increase C_s or decrease C_g or both. This has led to winding schemes such as interleaving, which can dramatically increase the series capacitance. Interleaving schemes can be quite complicated and need not be applied to the whole winding. In addition, the scheme may vary within a given winding. A multistart winding is an example of a simple interleaving scheme for increasing the series capacitance of a tap winding. Another method of increasing the series capacitance is by means of wound-in-shields, which can be applied to a section of the winding near the line end where they are most needed. Decreasing C_g is not as promising since this would involve, at least in a simple approach, increasing the winding-to-winding or winding-to-tank distance for an outer winding or winding to core distance for and inner winding. Such changes could conflict with impedance or cooling requirements.

13.5.2 Traveling Wave Theory

An improvement over the static capacitance model is a model that includes the winding inductance in an approximate way. This is the traveling wave theory as developed by Rudenberg [Rud68], [Rud40]. This is also a continuum uniform winding model with the circuit parameters defined on a per-unit-length basis. A new variable, the inductance per unit length is introduced to take inductive effects into account. This can only be regarded as approximate since it ignores the mutual inductance between the different sections of the winding. However, this approximation can be justified to some extent mathematically and by comparison with experiment.

Without going into all the details, which can be obtained from the references given earlier, the disk–disk voltages along the winding are given by

$$\Delta V_d = V_o \alpha \left(\frac{w}{L} \right) \left(\frac{1}{2} e^{-\alpha(x/L)} + \frac{1}{\pi} \right) \tag{13.67}$$

with x, the distance along the winding, measured from the line end. This has its maximum at the line end where $x = 0$, which is given by

$$\Delta V_{d,\text{line}} = 0.818 V_o \alpha \left(\frac{w}{L} \right) \tag{13.68}$$

For $\alpha > 3$, which is somewhat typical, $\tanh \alpha \approx 1$, so that, comparing (13.68) with its counterparts in the purely capacitive cases, this disk–disk voltage drop is about 18% lower.

Norris [Nor48] has proposed an improved version of (13.67), which has the form

$$\Delta V_d = V_o \alpha \left(\frac{w}{L} \right) \left(0.6 e^{-\alpha(x/L)} + 0.4 K_g \right) \tag{13.69}$$

where $K_g \leq 1$ can be found as a solution of the equation [Rud68], [Rud40]

$$\frac{\alpha}{\pi L} x = \begin{cases} \dfrac{1}{K_g^2} \left(1 + \sqrt{1 - K_g^2} \right), & x > \dfrac{\pi L}{\alpha} \\[2mm] 1, & x \leq \dfrac{\pi L}{\alpha} \end{cases} \tag{13.70}$$

This reduces to a quartic equation, which can be solved analytically for K_g as a function of x. At $x = 0$ and for $\alpha > 3$, Equation 13.69 is equal to the disk–disk voltage drop in the capacitive case, as can be seen by comparison with (13.65). Norris has also suggested corrections for incoming waves having finite rise and fall times and gives procedures for handling nonuniform windings.

14

High-Voltage Impulse Analysis and Testing

14.1 Introduction

Lightning strikes and switching surges are probably the most serious threats to the electrical system of a transformer. Although protective gear is normally used to limit the severity of these threats, this protection typically limits the maximum voltage of the impulse to the basic impulse level (BIL) of a transformer. The BIL is therefore the design level and is often below the maximum lightning strike voltage. However, because it is a pulse of short duration, many high-frequency components of the voltage are present in the pulse. This in turn excites capacitive and inductive elements present in the transformer. For this reason, a circuit model of the transformer, which includes these elements along with resistive elements, is desirable.

14.2 Lumped Parameter Model for Transient Voltage Distribution

In this section, we develop a circuit model of the transformer that can be used to analyze impulse and other transient electrical shocks to a transformer. This model includes capacitive, inductive, and resistive elements. The inductive elements include mutual inductances between elements in the same winding and between different windings and the effects of the iron core. The coils can be subdivided into as fine or coarse a manner as is consistent with the desired accuracy. Moreover, the subdivisions can be unequal so that accuracy in certain parts of the coil can be increased relative to that in other parts. The approach taken is similar to that of Miki et al. [Mik78] except that it includes winding resistance and the effects of the iron core. Also, the differential equations describing the circuit are organized in such a way that circuit symmetries can be exploited and other circuit elements such as nonlinear varistors may be included.

14.2.1 Circuit Description

A transformer is approximated as a collection of lumped circuit elements as shown in Figure 14.1. Although not shown, mutual inductances between all the inductors are assumed to be present. The subdivisions may correspond to distinctly different sections of the coil, having different insulation thicknesses, for example, or may simply be present for increased accuracy. The core and tank are assumed to be at ground potential. The presence of the tank only affects the capacitance to ground of the outer coil but not the inductance

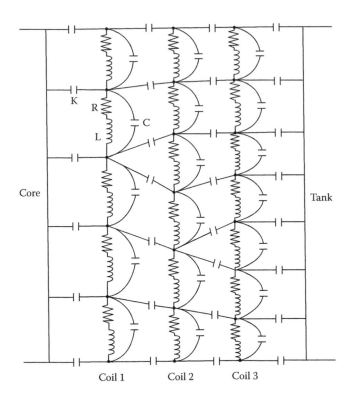

FIGURE 14.1
Circuit model of a transformer. The number of coils and subdivisions within a coil are arbitrary.

calculation. Other elements such as grounding resistors, capacitors, inductors, or nonlinear elements may be added. Terminals or nodes may be interconnected, shorted to ground, or connected to ground via a resistor, reactor, capacitor, or varistor.

In order to analyze the circuit of Figure 14.1, we isolate a representative portion as shown in Figure 14.2. We adopt a node numbering scheme starting from the bottom of the innermost coil and proceeding upward. Then continue from the bottom of the next coil, etc. Similarly, a section or subdivision numbering scheme is adopted, starting from the bottom of the innermost coil, proceeding upward, etc. These numbering schemes are related. In the figure, we have labeled the nodes along a winding with i, (p's, q's, and s's for nodes on adjacent windings) and the winding sections with j. Our circuit unknowns are the nodal voltages V_i and section currents I_j directed upward as shown.

Considering the voltage drop from node i − 1 to node i, we can write

$$L_j \frac{dI_j}{dt} + \sum_{k \neq j} M_{jk} \frac{dI_k}{dt} + R_j I_j = V_{i-1} - V_i \tag{14.1}$$

where M_{jk} is the mutual inductance between sections j and k. Defining a section current vector **I**, a nodal voltage vector **V**, an inductance matrix M, where $M_{jj} = L_j$, and a diagonal resistance matrix R, we can compress this formula as

$$M \frac{d\mathbf{I}}{dt} = B\mathbf{V} - R\mathbf{I} \tag{14.2}$$

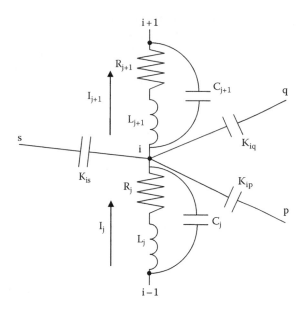

FIGURE 14.2
Representative portion of Figure 14.1.

where B is a rectangular matrix whose rows correspond to sections and columns to nodes such that $B_{ji-1} = 1$ and $B_{ji} = -1$, where nodes i − 1 and i bracket section j. The inductance matrix M is a symmetric positive definite matrix. Equation 14.2 holds independently of how the terminals are interconnected or what additional circuit elements are present.

Applying Kirchhoff's current law to node i, we obtain

$$\left(C_j + C_{j+1} + \sum_p K_{ip} \right) \frac{dV_i}{dt} - C_j \frac{dV_{i-1}}{dt} - C_{j+1} \frac{dV_{i+1}}{dt} - \sum_p K_{ip} \frac{dV_p}{dt} = I_j - I_{j+1} \tag{14.3}$$

where the p-sum is over all nodes connected to node i via shunt capacitances K_{ip}. C_j is set to zero when node i is at the bottom of the coil and $C_{j+1} = 0$ when node i is at the top of the coil, although a small value for capacitive coupling to the yokes could be used. $dV_p/dt = 0$ when node p is at ground potential such as when connected to the core or tank. Defining a capacitance matrix C, Equation 14.3 can be rewritten as

$$C \frac{dV}{dt} = AI \tag{14.4}$$

where A is a rectangular matrix whose rows correspond to nodes and columns to sections and where $A_{ij} = 1$ and $A_{ij+1} = -1$, with sections j and j + 1 on either side of node i. One of these terms is 0 when i is at the end of a coil.

When a nodal voltage V_i is specified, for example, at the impulsed terminal, as $V_i = V_s$, Equation 14.3 is replaced by

$$\frac{dV_i}{dt} = \frac{dV_s}{dt} \tag{14.5}$$

If $V_s = 0$ as for a grounded terminal, then (14.5) becomes $dV_i/dt = 0$. When node i is shorted to ground via a resistor R_s, a term V_i/R_s is added to the left-hand side of (14.3). If node i is grounded by means of a capacitor C_s, a term $C_s\, dV_i/dt$ is added to the left-hand side of (14.3). For a shorting inductor L_s uncoupled from all the other inductors, a term $\int V_i\, dt/L_s$ is added to the left-hand side of (14.3). Other situations such as a resistor or varistor joining two nodes can be accommodated.

When several nodes are joined together, say, nodes i, r, s, ..., their Kirchhoff's current law equations are simply added. The resulting equation replaces the node i equation. Then the node r, s, ... equations are replaced by

$$\frac{dV_r}{dt} - \frac{dV_i}{dt} = 0, \quad \frac{dV_s}{dt} - \frac{dV_i}{dt} = 0, \quad \ldots \tag{14.6}$$

The net result of all these circuit modifications is that the capacitance matrix in (14.4) may be altered and the right-hand side may acquire additional terms, for example, terms involving V for a resistor, $\int V\, dt$ for an inductor, and the impulsed voltage V_s. Thus, Equation 14.4 is replaced by

$$C'\frac{d\mathbf{V}}{dt} = A\mathbf{I} + \mathbf{f}(V, V_s) \tag{14.7}$$

where
  C' is the new capacitance matrix
  f depends on the added elements

This procedure lends itself to computer implementation.

Equations 14.2 and 14.7 can now be solved simultaneously by means of, for example, a Runge–Kutta algorithm starting from a given initial state. Linear equations must be solved at each time step to determine $d\mathbf{I}/dt$ and $d\mathbf{V}/dt$. Since M is a symmetric positive definite matrix, the Cholesky algorithm may be used to solve (14.2) while (14.7) may be solved by Gaussian elimination. LL^T or LU factorization is first performed on the respective matrices and the factors are used subsequently to solve the linear equations at each time step, saving much computation time.

14.2.2 Mutual and Self-Inductance Calculations

The transformer core and coil geometry including the iron yokes is assumed to have cylindrical symmetry. The iron is assumed to be infinitely permeable. A typical coil section is rectangular in cross section and carries a uniform current density azimuthally directed. The yokes extend outward to infinity and we ignore the tank walls. This geometry corresponds to that used in the Rabins' inductance and mutual inductance calculations of Section 9.4. There formulas are given for the mutual inductances between coil sections on the same or different coils and for the self-inductances of coil sections. These are used in the circuit model discussed here.

14.2.3 Capacitance Calculations

The series capacitances must take into account the type of coil, whether helix, disk, multi-start, or other. Series and shunt capacitances incorporate details of the paper and pressboard insulation, the placement of key spacers and sticks, and the oil duct geometry.

Except when wound-in-shields are present, the formula developed by Stein as discussed in Chapter 12 is used for the series disk capacitances. This formula was found to produce the best results in the work reported by Miki et al. [Mik78]. Simple standard formulas are used for helical and multi-start windings as well as for shunt capacitances as discussed in Chapter 12. Basically, a shunt capacitance per unit length is calculated, assuming the coils are infinitely long, and this is multiplied by the length of the section to get the section to section or section to ground shunt capacitances. The nearest-neighbor sections on adjacent windings are used in this calculation.

The series capacitance for a section containing several disks is obtained by adding the single disk capacitances in series, that is, dividing the single disk capacitance by the number of disks in the section. Since the subdivisions need not contain an integral number of disks, fractional disks are allowed in this calculation. Other subdivision schemes could be adopted that restrict the sections to contain an integral number of disks. When wound-in-shields are present, only an integral number of disk pairs are allowed in the subdivision. The finer the subdivisions or the greater their number, the greater is the expected accuracy. However, too fine a subdivision, for example, less than the height of a disk, could be counterproductive. Fine subdivisions can also lead to convergence problems for the Fourier sums in the mutual inductance calculation.

14.2.4 Impulse Voltage Calculations and Experimental Comparisons

The standard impulse voltage waveform can be mathematically represented as

$$V_s(t) = A\left(e^{-\alpha t} - e^{-\beta t}\right) \tag{14.8}$$

where α, β, and A are adjusted so that V_s rises to its maximum value in 1.2 μs and decays to half its value in 50 μs. It is desirable to have an analytic formula, which is smoothly differentiable such as (14.8) since the derivative is used in the solution process. Other waveforms such as a chopped wave may also be used. Procedures may be developed for extracting the parameters used in (14.8) from the desired rise and fall times and peak voltage. In our experimental recurrent surge oscillograph (RSO) tests discussed later, the actual rise and fall times were 3 and 44 μs, respectively, and the parameters in (14.8) were adjusted accordingly.

We tested a 45 MVA autotransformer having four windings as shown in Figure 14.3. The high-voltage (HV or series) winding consists of two coils in parallel with two ends joined at the center where the impulse is applied. The other two ends are joined at the autopoint with the low voltage (LV or common) and X-line tap windings. The HV and LV windings are disk windings. The tap winding is a multi-start winding and is grounded for the impulse test. The tertiary voltage (TV) winding is a helical winding and is grounded at both ends for the impulse test. These windings represent one phase of a 3-phase transformer, but the phases are essentially isolated from each other for the impulse test.

Both the HV and TV windings contain tapped sections along their lengths. The tap turns are out on the HV winding and in on the TV winding during the impulse test. In the calculations, a direct short is placed across the tapped out sections. In the vicinity of these tap sections, the turn density is lowered in the LV winding (thinning) for better short circuit strength. In the calculations, these thinned areas are specified as one or more separate subdivisions with their own properties, which differ from those of the subdivisions in the rest of the winding.

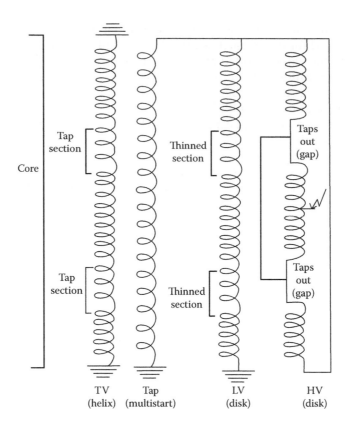

FIGURE 14.3
Schematic diagram of impulsed autotransformer.

Although nodal voltages and section currents are calculated throughout the transformer as a function of time, voltage differences can easily be obtained. In fact, disk–disk voltages are calculated throughout all the windings, and the maximum occurring over the time duration of the pulse is printed out for each winding. Similarly, the maximum voltage difference occurring between adjacent windings for the duration of the pulse is also printed out. Secondary quantities such as maximum electric fields in the oil and corner electric fields on the disks can likewise be obtained.

Experimentally, only voltages along the HV coil and at the tap positions on the tertiary winding were easily accessible in the tests conducted. In an RSO test, a repetitive series of pulses having the shape described by (14.8) are applied to, in our case, the HV terminal. This repetitive input results in persistent oscilloscope displays of the output voltages for easy recording. The peak applied voltage is usually low, typically several hundred volts. This test simulates an impulse test done at much higher voltages to the extent that the system is linear. This is likely to be a good assumption provided the core does not saturate and provided nonlinear circuit elements such as varistors do not come into play. In our RSO test, the transformer was outside the tank. Thus, the dielectric constant of air was used in the capacitance formulas. In addition, the tank distance, which affects the ground capacitance of the outer coil, was taken as very large.

For the tests conducted here, the computer outputs of interest were the voltages to ground at the experimentally measured points. We present a sufficient number of these in the following figures to indicate the level of agreement between calculation and experiment. We chose to normalize the input to 100 at the peak of the impulse waveform. The units can therefore be interpreted as a percent of the BIL.

It quickly became apparent in comparing the simulations with experiment that the section resistances that were obtained from the wire geometry and the d.c. or power frequency resistivity were not adequate to account for the damping observed in the output waveforms. In fact, we found it necessary to increase this resistivity by a factor of about 3000 to account for the damping. Such a factor is not unreasonable in view of the fact that the eddy currents induced by the stray flux associated with the impulse waveform contain high-frequency components that induce much higher losses than those occurring at power frequency. A factor of this magnitude was estimated for the experiment involving wound-in-shields in Chapter 12. A better approach, which we subsequently adopted, is to Fourier analyze the current waveforms and calculate the effective resistivity using the formula given in Chapter 12. This involves running the calculation at least twice, once to obtain the waveforms using an assumed resistivity, Fourier analyzing the waveforms to obtain a better estimate of the effective resistivity, and then rerunning the calculation with the recalculated resistivity.

Each of the two halves of the HV coil had 52 disks having 12 radial turns each. The impulse was applied to the center of the leg where the two halves are joined. Output voltages are recorded relative to this center point. Figure 14.4 shows the experimental and calculated voltage waveform four disks below the impulsed terminal. The impulse voltage is also shown for comparison. Figure 14.5 shows the voltage 12 disks below the impulsed terminal. There is good agreement in the major oscillations but some differences in the lower-amplitude higher-frequency oscillations. Figure 14.6 shows the experimental and theoretical voltages at the tap position, which is about 30 disks below the impulsed terminal. Figure 14.7 shows the voltage to ground at the upper tap position on the TV winding.

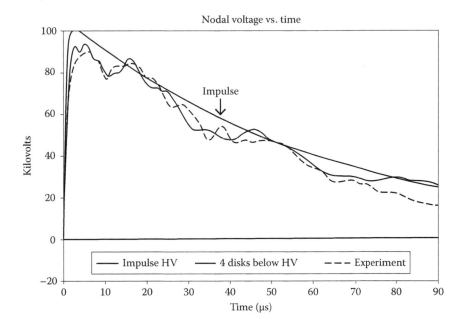

FIGURE 14.4
Experimental and calculated voltage to ground four disks below the HV impulsed terminal. The impulse voltage is also shown.

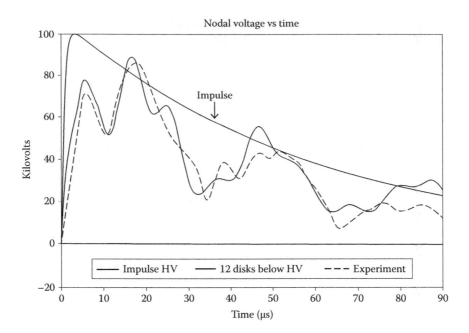

FIGURE 14.5
Experimental and calculated voltage to ground 12 disks below the HV impulsed terminal.

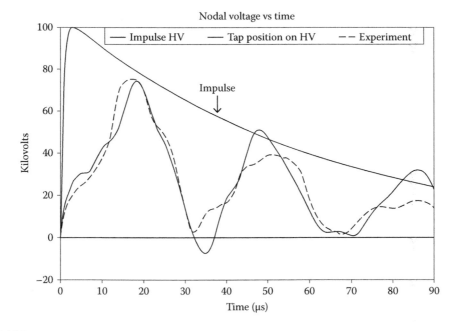

FIGURE 14.6
Experimental and calculated voltages to ground at the tap position on the HV winding, about 30 disks below the impulsed terminal.

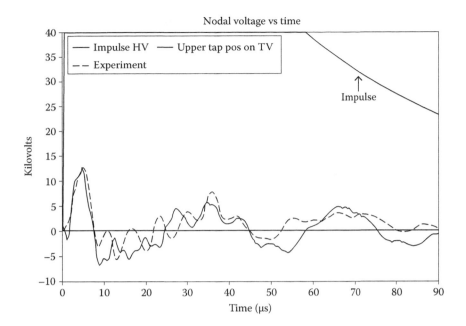

FIGURE 14.7
Experimental and calculated voltages to ground at the center of the upper tap position on the TV winding.

This is a transferred voltage via capacitive and mutual inductive effects. There is good agreement between theory and experiment in overall magnitude and major oscillations, but the higher-frequency ripple is not as well predicted. Figure 14.8 shows another way to present the calculations. This shows the voltage as a function of relative coil position along the top HV winding at various instants of time. Note the flat portion of the curves at a relative position of ~0.75 that corresponds to the tap section with the taps out. Note also that the high capacitance of the multi-start tap winding effectively grounds the autopoint. Maximum disk–disk voltages could be obtained by examining such curves although this is more easily done by programming.

14.2.5 Sensitivity Studies

We studied the sensitivity of the calculations to two of the inputs to which some uncertainty is attached, namely, the effective resistivity of the copper and the number of subdivisions along the coils. The resistivity was increased from its d.c. value to 3000 times its d.c. value and the waveform studied at several locations along the HV winding. The effect of increased resistivity was to increase the damping of the waveform at longer times but had little effect on the waveform at short times.

Generally, each coil has a minimum number of subdivisions dictated by the number of physically different coil sections. In our case, this was about 6–8 per coil. The multi-start coil, however, has only one section. Because of its construction, it cannot be meaningfully subdivided into axial sections as can the other coils. In fact, the capacitances to neighboring coils must be coupled equally to both the top and bottom nodes because its voltage is not a unique function of position. We increased the number of subdivisions in helical and disk

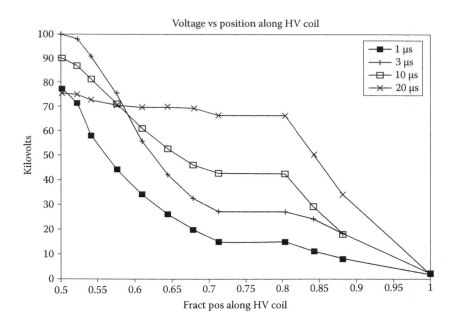

FIGURE 14.8
Calculated voltage profiles at various instants of time along the upper HV coil. Relative position 0.5 is the impulsed terminal and 1 the top of the winding.

coils from the minimum value of 7 up to 22. Once the number of subdivisions is above 12/coil, the results were fairly insensitive to any further refinement. Thus, one does not have to be overly concerned about the number of subdivisions used, provided they are above some reasonable minimum, which in this case is ~12/coil.

14.3 Setting the Impulse Test Generator to Achieve Close-to-Ideal Waveshapes*

An important issue, which arises when performing an impulse test, is how to select the impulse test generator settings in order to produce the ideal or nearly ideal waveshape. It is sometimes difficult to approximate the ideal waveform, particularly the full wave one, and a trial and error approach is often taken. Here, we describe an approach that couples a circuit model of the impulse generator with a transformer circuit model and show how, from this combination, we can arrive at close to ideal settings for the generator, provided such settings are possible. An alternate approach to determining the optimum impulse generator settings, which makes use of a genetic algorithm, is described in [Sam15].

In order to simulate the effect of a lightning strike on the power lines that are connected to a transformer, standards organizations have arrived at a standard voltage waveform to be applied to the various transformer terminals undergoing an impulse

---

* This section is largely based on [Del02], *IEEE Trans. Power Deliv.*, 17(1), January 2002, 142–148. © 2002 IEEE.

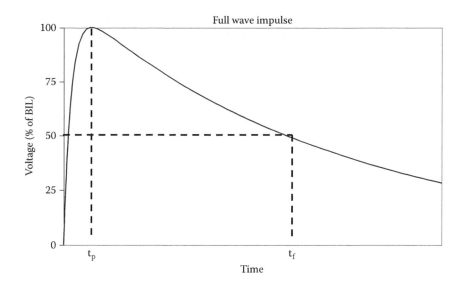

FIGURE 14.9
Standard full wave impulse waveform. (From Del Vecchio, R.M. et al., *IEEE Trans. Power Deliv.*, 17(1), January 2002, 142–148. © 2002 IEEE.)

test. Although these waveforms differ in peak value or BIL level, they have similar time characteristics. For a full wave, which is the only type we consider here, they have the general appearance shown in Figure 14.9. The key parameters are the rise time to reach the peak value, t_p, and the fall time to fall to 50% of the peak value, t_f. A common parameterization for this waveform is given by (14.8), where α, β, and A can be determined from t_p, t_f, and the BIL level. More specifically, the standards often specify $t_p = 1.2$ μs and $t_f = 50$ μs, that is, a 1.2 × 50 waveform.

Impulse generators for large power transformers are generally of the Marx type [Kuf84], that is, they employ a number of capacitor stages, which are charged in parallel at a lower voltage and discharged in series at a much higher voltage. We will only be concerned with the discharge configuration and with its equivalent circuit. Generally, each stage has a front resistor in series with the capacitor and a tail resistor in parallel with the capacitor. There are also charging resistors in the circuit, which are in parallel with the entire combination, although these are large and could be ignored. A circuit model of an individual stage is illustrated in Figure 14.10. In addition, there is some inductance associated with

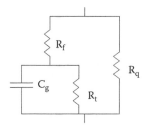

FIGURE 14.10
Circuit model of a generator stage. R_f is the front resistor, R_t the tail resistor, R_q the charging resistor, and C_g the stage capacitance. (From Del Vecchio, R.M. et al., *IEEE Trans. Power Deliv.*, 17(1), January 2002, 142–148. © 2002 IEEE.)

each stage, but it is unclear where this occurs. We will simply lump it into one inductor in the overall circuit model. We should note that these stages are not always series connected but that some of them may be connected in parallel in order to obtain a more desirable overall capacitance value.

By suitably choosing the front and tail resistances and the series–parallel combination of stages, it is generally possible to approximate the waveform of Figure 14.9. However, this usually requires a trial and error process or having experience with previous impulse tests on similar transformers. There are even cases where it does not seem possible to achieve the desired waveform by varying the generator parameters. This is usually a matter of the tail falling too quickly so that the 50% of peak voltage value is reached at times <50 μs. In such cases, it is sometimes permissible to add a grounding resistor to one of the non-impulsed windings to help boost the tail. In these cases, it would be helpful to know in advance that the proper waveform cannot be achieved by changing the generator param-eters alone before undue effort is spent on this process.

Here, we outline a method for obtaining a set of generator parameters, if they exist, which will produce an impulse waveform closely approximating the ideal one. This method requires a suitable circuit model of the impulse generator as well as one for the transformer. Such a combined model should come reasonably close to predicting the actual impulse waveform achieved when the generator parameters in the model are the same as those used for the test. Given such a combination, one is already ahead of the game because one can now perform trial and error parameter variations on the computer in an attempt to achieve the desired waveform rather than on the test floor. However, this tedious pro-cess can be circumvented by relatively little additional effort, as will be discussed later. First, we detail the generator circuit model and then briefly discuss the transformer circuit model before treating the combination.

14.3.1 Impulse Generator Circuit Model

Although we will discuss the Marx-type generator here, the method could be applied to any generator with the following provisos. The generator should be capable of producing the desired waveform, at least with a simple load, say, a capacitive load. Its overall circuit model description should also be available or reasonably possible to develop. To this end, we assume that each stage of the Marx generator is identical and that each series unit is made up of either one stage or the same number of stages in parallel. Thus, if we have n series units and each consists of m stages in parallel, there are altogether mn stages in the circuit.

Since the waveforms are not sinusoidal, we will use Laplace transforms to analyze the circuit. Thus, we replace circuit elements by their Laplace transforms so that resistances remain the same but a capacitance C becomes $1/sC$ and an inductance L becomes sL in the circuit analysis, where s is the Laplace variable. Thus, the impedance of a single stage shown in Figure 14.10, Z, is given by

$$Z = \frac{Z_1 R_q}{Z_1 + R_q} \tag{14.9}$$

where

$$Z_1 = R_f + \frac{R_t}{1 + sR_t C_g} \tag{14.10}$$

The impedance of the n series units, each consisting of m parallel stages, Z_{gen}, is given by

$$Z_{gen} = \left(\frac{n}{m}\right)Z = \frac{\left(\frac{n}{m}\right)Z_1 \left(\frac{n}{m}\right)R_q}{\left(\frac{n}{m}\right)Z_1 + \left(\frac{n}{m}\right)R_q} \tag{14.11}$$

Thus, we can define an equivalent impedance, $Z_{1,eq} = (n/m)Z_1$, and an equivalent charging resistor, $R_Q = (n/m)R_q$, so that

$$Z_{gen} = \frac{Z_{1,eq}R_Q}{Z_{1,eq} + R_Q} \tag{14.12}$$

We also have

$$Z_{1,eq} = \left(\frac{n}{m}\right)Z_1 = \left(\frac{n}{m}\right)R_f + \frac{\left(\frac{n}{m}\right)R_t}{1 + s\left(\frac{n}{m}\right)R_t\left(\frac{m}{n}\right)C_g} \tag{14.13}$$

We can therefore define an equivalent front resistor, $R_F = (n/m)R_f$, an equivalent tail resistor, $R_T = (n/m)R_t$, and an equivalent generator capacitance, $C_G = (m/n)C_g$, so that (14.13) becomes

$$Z_{1,eq} = R_F + \frac{R_T}{1 + sR_TC_G} \tag{14.14}$$

Thus, we see that (14.12) and (14.14) have the same form as (14.9) and (14.10) but with different values for the circuit parameters. They therefore have the same circuit diagram. We redraw this in Figure 14.11a for the overall generator.

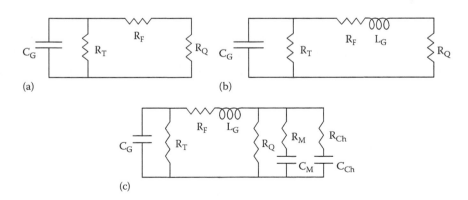

(a) (b)

(c)

FIGURE 14.11
Circuit models of the impulse generator. (a) Equivalent generator circuit built from the individual stages shown in Figure 14.10, (b) generator circuit with inductance, and (c) generator circuit with the measuring equipment represented by R_M, C_M, and the inactive chopping gap represented by R_{Ch}, C_{Ch}, added. (From Del Vecchio, R.M. et al., *IEEE Trans. Power Deliv.*, 17(1), January 2002, 142–148. © 2002 IEEE.)

We mentioned previously that the generator has some inductance. We include this as one lumped element as shown in Figure 14.11b. From the manufacturer's data, this inductance appears to have a contribution from each stage, adding a term proportional to (n/m) to the overall inductance, as well as a constant contribution. During the full wave test, there is measuring equipment at the generator output as well as an inactive chopping gap circuit. The latter, nevertheless, contributes something to the generator load. Both of these contributions can be adequately represented in the circuit model shown in Figure 14.11c.

Since the transformer circuit model is solved by means of coupled differential equations, we will develop a similar set of equations for the generator so that the circuits can be solved simultaneously. In addition, we can use these equations without the transformer attached to check the generator output at no load against test measurements and thus verify the accuracy of the generator model. We redraw the generator circuit, indicating the voltages and currents that must be solved for, in Figure 14.12. In this figure, I_L is the load current that enters the transformer and V is the impulse voltage. The initial condition is that the generator capacitor be initially charged to some voltage, V_o, that is, $V_G = V_o$ at t = 0.

Using Kirchhoff's laws, the circuit equations can be reduced to

$$\frac{dV_G}{dt} = -\frac{V_G}{R_T C_G} - \frac{I_F}{C_G} \tag{14.15}$$

$$\frac{dI_F}{dt} = \frac{(V_G - V - I_F R_F)}{L_G} \tag{14.16}$$

$$\frac{dV_M}{dt} = \frac{(V - V_M)}{R_M C_M} \tag{14.17}$$

$$\frac{dV_{Ch}}{dt} = \frac{(V - V_{Ch})}{R_{Ch} C_{Ch}} \tag{14.18}$$

$$I_L = I_F - \frac{V}{R_Q} - \frac{(V - V_M)}{R_M} - \frac{(V - V_{Ch})}{R_{Ch}} \tag{14.19}$$

The last equation must be combined with an equation for V from the transformer circuit when the transformer is attached. When the transformer is not attached, setting $I_L = 0$ in (14.19) allows one to solve for V in terms of the other unknowns. When V, obtained

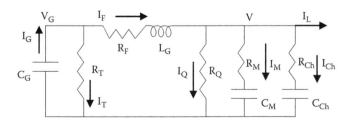

FIGURE 14.12
Generator circuit with unknown voltages and currents labeled. (From Del Vecchio, R.M. et al., *IEEE Trans. Power Deliv.*, 17(1), January 2002, 142–148. © 2002 IEEE.)

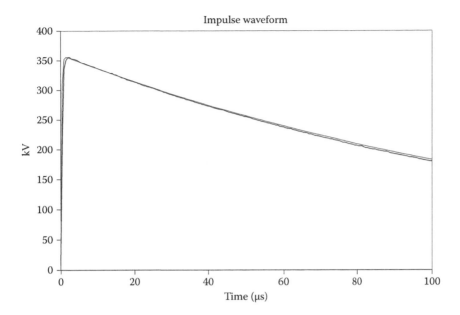

FIGURE 14.13
Comparison of calculated and measured impulse generator output without a transformer load. The two curves practically coincide. (From Del Vecchio, R.M. et al., *IEEE Trans. Power Deliv.*, 17(1), January 2002, 142–148. © 2002 IEEE.)

from (14.19) with $I_L = 0$, is substituted into Equations 14.15 through 14.18, we obtain a coupled set of four differential equations in four unknowns that can be solved, starting from the initial condition. Once solved, we can use (14.19) with $I_L = 0$ to find V, the output voltage without transformer. Figure 14.13 shows a comparison of the calculated and measured generator voltage output without transformer. Similar good agreement was obtained with a variety of generator settings.

14.3.2 Transformer Circuit Model

The transformer circuit model was discussed in Chapter 13. This model is very detailed. It includes the winding section's self-inductances and the mutual inductances between all the winding sections, the section resistances, section series capacitances, and section capacitances to the neighboring windings and/or core and tank.

As discussed in Chapter 13, the capacitance equations can be combined into a matrix equation for the nodal voltages V_i, where i ranges over all the nodes:

$$C\frac{dV}{dt} = AI \tag{14.20}$$

Here
  C is a capacitance matrix
  A is a matrix whose entries are ±1 or 0
  V is a vector of nodal voltages
  I is a vector of branch currents

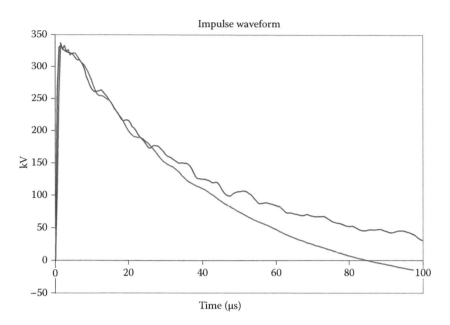

FIGURE 14.14
Comparison of a calculated with measured impulse waveform with transformer attached and without optimization. The smoother (faster falloff) curve is the measured waveform. (From Del Vecchio, R.M. et al., *IEEE Trans. Power Deliv.*, 17(1), January 2002, 142–148. © 2002 IEEE.)

These can be solved, along with a set of differential equations for the currents obtained from the inductive–resistive branches.

In the normal solution method, if node k is the impulsed node, the kth equation in (14.20), corresponding to V_k, is replaced by

$$\frac{dV_k}{dt} = \frac{d}{dt} \left(\text{Ideal waveform} \right) \tag{14.21}$$

When the impulse generator is attached to node k, instead of (14.21), we add I_L from (14.19) to the right-hand side of (14.20) for the kth voltage node and change V to V_k in (14.15) through (14.19). We then solve all these equations simultaneously, starting with $V_G = V_o$ on the generator capacitor. Figure 14.14 shows a comparison of a calculated versus measured impulse waveform for a particular choice of generator parameters with the transformer attached. The calculated curve is obtained without any optimization. From this comparison, we see that the method is capable of reproducing the test waveform fairly well, although it is far from the ideal waveform.

14.3.3 Determining the Generator Settings for Approximating the Ideal Waveform

With the transformer attached, we redraw the circuit of Figure 14.12 to include the transformer, which we represent as a lumped impedance Z_{Tr} in Figure 14.15. We do not need to know Z_{Tr} explicitly. For purposes of calculation, the impulsed transformer node is connected to node 1 in the figure so that the transformer circuit is in parallel with all the elements between 1 and 1'. We now lump all the impedances in the dashed box in Figure 14.15

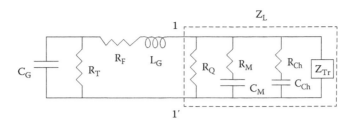

FIGURE 14.15
Circuit model of the generator with transformer attached, represented by the load Z_{Tr}. (From Del Vecchio, R.M. et al., *IEEE Trans. Power Deliv.*, 17(1), January 2002, 142–148. © 2002 IEEE.)

together into a load impedance Z_L. This impedance remains fixed as we consider variations in the generator parameters R_F, R_T, and C_G. Note that we are including the charging resistor as well as the measuring circuit and inactive chopping circuit in Z_L. These remain fixed for the given transformer and BIL level. At this point, we include the generator inductance in the circuit to the left of Z_L; however, its effect on the impulse waveform is small provided R_F is large enough.

With Z_L as a load, the remaining circuit can be analyzed by means of Thevenin's theorem as stated, for example, in [Des69]. For linear network elements, as is the case here, we can replace the network to the left of the terminals labeled 1 and 1′ in Figure 14.15 by an equivalent network consisting of a voltage source and an impedance in series with it. The voltage source is the open circuit voltage at terminals 1, 1′ and the series impedance is obtained by setting all independent sources to zero and all initial conditions to zero. There are no sources in our circuit, but there is an initial condition on the capacitor. The open circuit voltage is that on a capacitor initially charged to a voltage V_o in parallel with a resistor R_T. Thus, the open circuit voltage, e_{oc}, is

$$e_{oc} = V_o e^{-t/R_T C_G} \tag{14.22}$$

Its Laplace transform is

$$E_{oc} = \frac{V_o R_T C_G}{1 + s R_T C_G} \tag{14.23}$$

The Laplace transform of the equivalent impedance is

$$Z_{eq} = R_F + s L_G + \frac{R_T}{1 + s R_T C_G} \tag{14.24}$$

Substituting the Thevenin equivalent circuit, the circuit of Figure 14.15 becomes that of Figure 14.16. We have indicated in Figure 14.16 the location of the impulse voltage V.

We see that the impulse voltage, V, in Figure 14.16 is given by

$$V = E_{oc} \left(\frac{Z_L}{Z_{eq} + Z_L} \right) \tag{14.25}$$

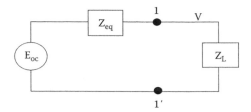

FIGURE 14.16
Thevenin equivalent circuit of Figure 14.15. (From Del Vecchio, R.M. et al., *IEEE Trans. Power Deliv.*, 17(1), January 2002, 142–148. © 2002 IEEE.)

Now consider two cases. First, the generator parameters are set to some starting values based on a reasonable estimate of what might produce the desired waveform and the resulting voltage V is recorded. Next, the ideal voltage V^I is produced by the unknown generator parameters, which we will label with a superscript I for ideal. Since E_{oc} and Z_{eq} depend on these unknown parameters, we also label them with a superscript I. However, Z_L remains the same since it does not depend on the variable generator parameters. Thus, we have

$$V^I = E_{oc}^I \left(\frac{Z_L}{Z_{eq}^I + Z_L} \right)$$

(14.26)

Solving (14.25) for Z_L and substituting into (14.26), we find

$$\frac{Z_{eq}^I}{\left(\dfrac{E_{oc}^I}{V^I} - 1 \right)} = \frac{Z_{eq}}{\left(\dfrac{E_{oc}}{V} - 1 \right)}$$

(14.27)

The ideal waveform is given in (14.8). Its Laplace transform, using superscript I to label the parameters, is

$$V^I = \frac{A^I \left(\beta^I - \alpha^I \right)}{\left(s + \alpha^I \right) \left(s + \beta^I \right)}$$

(14.28)

Z_{eq}^I is given by (14.24) with superscript I labeling the generator parameters. However, the presence of L_G makes it difficult to achieve the ideal waveform. This is pretty clear from [Kuf84], where it is shown that the ideal waveform is generated by the circuit to the left of the 1, 1′ terminals in Figure 14.15 without L_G and with a capacitive load. We cannot eliminate L_G, but it can be shown that its effect on the waveform will be reduced if

$$\frac{L_G}{R_F} \ll t_p$$

(14.29)

that is, if the inductive time constant is much less than the time to reach the peak voltage. We will assume this condition here and thus eliminate L_G from (14.24). The practical implications of this will be discussed later.

We also eliminate L_G in generating the waveform V for the initial simulation and we assume that it has the same form as (14.8) and therefore the same Laplace transform as given in (14.28) but without the I superscript labeling the parameters. Thus, Equations 14.15 through 14.19 need to be modified for the initial simulation by eliminating L_G from them. With these approximations, both sides of (14.27) will have the same form and differ only in the I superscripting labeling. Note that the waveform parameters, A, α, β, for the initial simulation can be determined by a variety of curve-fitting methods. After some algebra, Equation 14.27 becomes

$$\frac{A^I\left(\beta^I-\alpha^I\right)R_F^I}{V_o^I R_p^I C_G^I}\left\{\frac{1+sR_p^I C_G^I}{s^2+\left[\alpha^I+\beta^I-\dfrac{A^I\left(\beta^I-\alpha^I\right)}{V_o^I}\right]s+\alpha^I\beta^I-\dfrac{A^I\left(\beta^I-\alpha^I\right)}{V_o^I R_T^I C_G^I}}\right\}$$

$$=\frac{A\left(\beta-\alpha\right)R_F}{V_o R_p C_G}\left\{\frac{1+sR_p C_G}{s^2+\left[\alpha+\beta-\dfrac{A\left(\beta-\alpha\right)}{V_o}\right]s+\alpha\beta-\dfrac{A\left(\beta-\alpha\right)}{V_o R_T C_G}}\right\} \qquad (14.30)$$

where

$$R_p=\frac{R_F R_T}{R_F+R_T},\quad R_p^I=\frac{R_F^I R_T^I}{R_F^I+R_T^I} \qquad (14.31)$$

To achieve equality in (14.30), we require

$$\frac{A^I\left(\beta^I-\alpha^I\right)R_F^I}{V_o^I R_p^I C_G^I}=\frac{A\left(\beta-\alpha\right)R_F}{V_o R_p C_G} \qquad (14.32)$$

$$R_p^I C_G^I=R_p C_G \qquad (14.33)$$

$$\alpha^I+\beta^I-\frac{A^I\left(\beta^I-\alpha^I\right)}{V_o^I}=\alpha+\beta-\frac{A\left(\beta-\alpha\right)}{V_o} \qquad (14.34)$$

$$\alpha^I\beta^I-\frac{A^I\left(\beta^I-\alpha^I\right)}{V_o^I R_T^I C_G^I}=\alpha\beta-\frac{A\left(\beta-\alpha\right)}{V_o R_T C_G} \qquad (14.35)$$

Since the ideal waveform parameters are known, the only unknown parameters in these equations are R_F^I, R_T^I, C_G^I, and V_o^I. Thus, we have four equations in four unknowns that can be solved uniquely. Letting

$$x=1+\frac{V_o}{A\left(\beta-\alpha\right)}\left[\left(\alpha^I+\beta^I\right)-\left(\alpha+\beta\right)\right] \qquad (14.36)$$

$$y = 1 + \frac{V_p R_T C_G}{A(\beta - \alpha)}\left(\alpha^I \beta^I - \alpha\beta\right) \tag{14.37}$$

we find

$$R_F^I = \frac{R_F}{x} \tag{14.38}$$

$$C_G^I = \frac{C_G}{R_F^I\left[\dfrac{1}{R_p} - \left(\dfrac{y}{x}\right)\dfrac{1}{R_T}\right]} \tag{14.39}$$

$$R_T^I = \frac{R_T C_G}{C_G^I}\left(\frac{x}{y}\right) \tag{14.40}$$

$$V_o^I = V_o\left(\frac{A^I}{A}\right)\left(\frac{\beta^I - \alpha^I}{\beta - \alpha}\right)\frac{1}{x} \tag{14.41}$$

14.3.4 Practical Implementation

In order to get a reasonable set of starting parameters for the generator, we adopt the simpler model of the generator and load discussed in [Kuf84]. The circuit diagram is given in Figure 14.17. Starting with an initial generator capacitor voltage of V_o, the voltage across the load C_L is given by

$$V = \frac{V_o}{R_F C_L (\beta - \alpha)}\left(e^{-\alpha t} - e^{-\beta t}\right) \tag{14.42}$$

α and β can be expressed in terms of the circuit parameters, but since we are assuming an ideal waveform, α and β are known and we want to express the circuit parameters in terms of them. These are given by

$$R_F = \frac{1}{2C_L}\left(\frac{\alpha + \beta}{\alpha\beta}\right)\left[1 - \sqrt{1 - \frac{4\alpha\beta}{(\alpha + \beta)^2}\left(1 + \frac{C_L}{C_G}\right)}\right] \tag{14.43}$$

$$R_T = \frac{1}{2(C_G + C_L)}\left(\frac{\alpha + \beta}{\alpha\beta}\right)\left[1 + \sqrt{1 - \frac{4\alpha\beta}{(\alpha + \beta)^2}\left(1 + \frac{C_L}{C_G}\right)}\right] \tag{14.44}$$

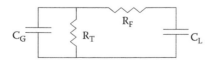

FIGURE 14.17
Simple generator–load model for obtaining starting generator parameters. (From Del Vecchio, R.M. et al., *IEEE Trans. Power Deliv.*, 17(1), January 2002, 142–148. © 2002 IEEE.)

From these equations, we see that we must assume some values for C_G and C_L. C_G can be selected based on BIL considerations or from experience. C_L can be calculated roughly from a simplified capacitive model of the transformer, but its selection is not critical.

Having chosen these initial parameters for the generator, the generator–transformer circuit model is solved to obtain the impulse voltage as a function of time. We do not include inductance in the generator model at this stage. The waveform is then analyzed to obtain A, α, β. To do this, we use a least squares curve-fitting procedure. The output waveform is only considered out to times where it remains positive. This is because (14.8) is inherently positive provided $\beta > \alpha$, which is always the case when the waveform begins with a positive slope. The new generator parameters are then calculated from (14.36) through (14.41).

The capacitance calculated from (14.39) is usually not obtainable exactly from the discrete possibilities arising from grouping the stages into series–parallel combinations. We therefore use the series–parallel combination that comes closest to achieving the desired capacitance. The front and tail resistances calculated from the procedure outlined earlier also cannot be realized exactly in practice. Their values must also be approximated, using the finite pool of available resistors.

We have found that for cases where the tail of the waveform droops significantly, R_T^I can become negative. This is due to y in Equation 14.37 becoming negative. In this case, we set y to a small value, say, $y = 0.1$, and recalculate R_T^I. The calculation is then repeated with the new parameters and the resulting waveform reanalyzed. If R_T^I is still negative, the process is repeated. After a few iterations, if the new R_T^I still remains negative, the process is stopped. A printout of the parameters before the last recalculation reveals that R_T^I has

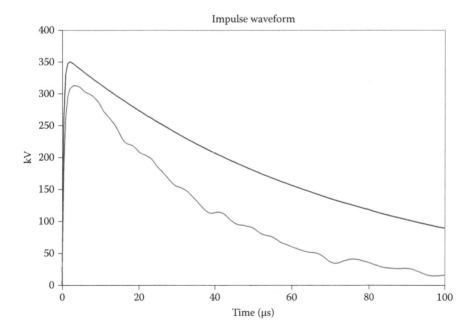

FIGURE 14.18
Waveform produced by the initial selection of generator parameters. The top curve is the ideal waveform. (From Del Vecchio, R.M. et al., *IEEE Trans. Power Deliv.*, 17(1), January 2002, 142–148. © 2002 IEEE.)

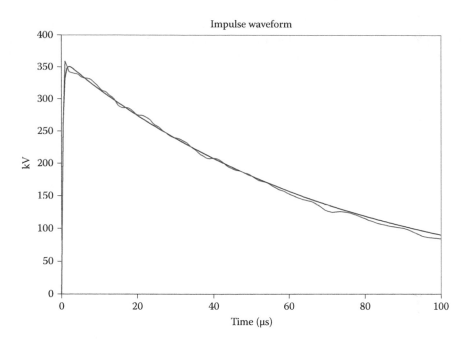

FIGURE 14.19
Waveform achieved with recalculated generator parameters. It nearly coincides with the ideal waveform. (From Del Vecchio, R.M. et al., *IEEE Trans. Power Deliv.*, 17(1), January 2002, 142–148. © 2002 IEEE.)

become extremely large and yet the waveform still has a drooping tail. This indicates that even if $R_T^I \rightarrow \infty$, an open circuit, the ideal waveform cannot be achieved. One is then left with the possibility of adding grounding resistors to one or more windings in order to obtain the desired waveform.

When the new generator parameters obtained from (14.36) through (14.41) are positive or after the last iteration in the case where the new R_T^I persists in being negative, the generator inductance is included in the circuit model. The transient calculation is then performed with these new parameters, including the inductance. Although the ideal waveform is usually well approximated when R_T^I does not become infinite, there is often a voltage spike near the peak of the waveform due to the generator inductance. Not only does this increase the peak voltage beyond the desired BIL level, but it tends to decrease the time to peak value t_p below the desired value. To counteract this, we have found that by artificially increasing the t_p value of the ideal waveform from, say, 1.2 to 2.0 μs, a larger value of R_F^I is obtained, thus reducing the L_G/R_F time constant, as discussed previously in the context of Equation 14.29. The inductance peak is thereby considerably reduced and the rise time increases to a value closer to the optimum.

Figure 14.18 shows an example of a calculated impulse waveform obtained using the initially estimated values for the generator parameters, $C_G = 500$ nF, $R_F = 177\ \Omega$, $R_T = 142\ \Omega$. The top curve is the ideal waveform. The generator waveform has the time parameters of $t_p = 3.15$ μs, $t_f = 29.5$ μs. Figure 14.19 shows the waveform resulting from the recalculated parameters of $C_G^I = 596$ nF, $R_F^I = 90\ \Omega$, and $R_T^I = 249\ \Omega$. As can be seen, it comes very close to the ideal curve. Its time parameters are $t_p = 1.10$ μs, $t_f = 50.5$ μs. We should note that the recalculated generator parameters depend to a limited extent on the initial choice for these parameters.

15

No-Load and Load Losses

15.1 Introduction

Transformer losses are broadly classified as no-load and load losses. No-load losses occur when the transformer is energized with its rated voltage at one set of terminals, but the other sets of terminals are open circuited so that no through or load current flows. In this case, full flux is present in the core and only the necessary exciting current flows in the windings. The losses are predominately core losses due to hysteresis and eddy currents produced by the time-varying flux in the core steel.

Load losses occur when the output is connected to a load so that current flows through the transformer from input to output terminals. When measuring load losses, the output terminals are shorted to ground and only a small impedance-related voltage is necessary to produce the desired full load current. In this case, the core losses are small because of the small core flux and do not significantly add to the measured losses. In operation, both types of losses are present and must be taken into account for cooling considerations.

Load losses are broadly classified as I^2R losses due to Joule heating produced by current flow in the coils and as stray losses due to the stray flux as it encounters metal objects such as tank walls, clamps or bracing structures, and the coils themselves. Because the coil conductors are often stranded and transposed, the I^2R losses are usually determined by the d.c. resistance of the windings. The stray losses depend on the conductivity, permeability, and shape of the metal object encountered. These losses are primarily due to induced eddy currents in these objects. Even though the object may be made of ferromagnetic material, such as the tank walls and clamps, their dimensions are such that hysteresis losses tend to be small relative to eddy current losses.

Although losses are usually a small fraction of the transformed power (<0.5% in large power transformers), they can produce localized heating that can compromise the operation of the transformer. Thus, it is important to understand how these losses arise and to calculate them as accurately as possible so that, if necessary, steps can be taken at the design stage to reduce them to a level that can be managed by the cooling system. Other incentives, such as the cost, which the customer attaches to the losses, can make it worthwhile to find ways of lowering the losses.

Modern methods of analysis, such as finite element or boundary element methods, have facilitated the calculation of stray flux losses in complex geometries. These methods are not yet routine in design because they require a fair amount of geometric input for each new geometry. However, some of these programs support the use of macros, which can quickly generate a standard geometry based on key parameters that can be changed for different designs. These methods can provide useful insights in cases where analytic methods are not available or are very crude. Occasionally, a parametric study using such methods can extend their usefulness beyond a specialized geometry. We will explore such methods, in particular the finite element method, when appropriate. However, we are largely concerned here with analytic methods, which can provide useful formulas covering wide parameter variations.

Typical power transformers have 3-phase cores so that the stray flux arises from the currents in all three phases. Analytic formulas are generally based on single-phase approximations. Although these formulas are based on physical principles, they generally require modifications obtained by comparison with test data. The necessity for such modifications can arise from the idealized geometries assumed in the calculation or from the ignored effects of 3-phase flux. 3D finite element methods are now available for modeling all three phases, including the effects of 3-phase flux. Even here, some approximations must be made to make the calculations tractable. We will explore this approach later and compare it with other methods.

15.2 No-Load or Core Losses

Cores in power transformers are generally made of stacks of electrical steel laminations. These are usually in the range of 0.23–0.46 mm in thickness and up to about 1 m wide or as wide as can be accommodated by the rolling mill. Modern electrical steels have a silicon content of about 3% that gives them a rather high resistivity, $\sim 50 \times 10^{-8}$ Ω m. Although higher silicon content can produce even higher resistivity, the brittleness increases with silicon content and this makes it difficult to roll them in the mill as well as to handle them after. Special alloying, rolling, and annealing cycles produce the highly oriented Goss texture (cube on edge) with superior magnetic properties such as high permeability along the rolling direction. Thus, it is necessary to consider the orientation of the laminations in relation to the flux direction when designing a core.

Although the thinness of the laminations and their high resistivity are desirable characteristics in reducing (classical) eddy current losses, the high degree of orientation (>95%) produces large magnetic domains parallel to the rolling direction as sketched in Figure 15.1.

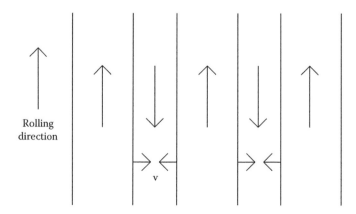

FIGURE 15.1
Idealized magnetic domain pattern in highly oriented electrical steel. The up and down arrows show the magnetization directions. The side-pointing small arrows show the direction of domain wall motion for increasing up magnetization. v is their velocity.

The lines between domains with magnetizations pointing up and down are called domain walls. These are narrow transition regions where the magnetization vector rotates through 180°. During an a.c. cycle, the up domains increase in size at the expense of the down domains during one part of the cycle and the opposite occurs during another part of the cycle. This requires the domain walls to move in the direction shown in the figure for increasing up magnetization. As the domain walls move, they generate eddy current losses. These losses were calculated by [Pry58] for the idealized situation shown in the figure. They found that these losses were significantly higher than the losses obtained from a classical eddy current calculation, which assumes a homogeneous mixture of many small domains. These nonclassical losses depend on the size of the domains in the zero magnetization state where there are equal-sized up and down domains. This is because the maximum distance the walls move and hence their velocity depend on the zero magnetization domain size. The larger this size, the greater the domain wall's velocity and the greater the loss.

In order to decrease the nonclassical eddy current losses, it is therefore necessary to reduce the domain size. This is accomplished in practice by laser or mechanical scribing. A laser or mechanical stylus is rastered across the domains (perpendicular to their magnetization direction) at a particular spacing. This introduces localized stress at the surface since the scribe lines are not very deep. The domain size is dependent on the stress distribution in the laminations. Localized stresses help to refine the domains. Thus, after scribing, the laminations are not annealed since this would relieve the stress. Figure 15.2 shows the domain pattern in an oriented electrical steel sample before and after laser scribing. The domain patterns are made visible by means of specialized optical techniques. One can clearly see the reduction in domain size as a result of laser scribing in this figure. The losses were reduced by ~12% as a result of laser scribing in this example.

Another type of loss in electrical steels is hysteresis loss. This results from the domain walls encountering obstructions during their motion. At an obstruction, which can be a crystal imperfection, an occlusion or impurity, or even a localized stress concentration,

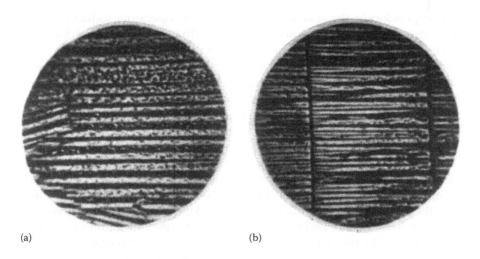

(a)　　　　　　　　　　　　　　　　　　　(b)

FIGURE 15.2
Effect of laser scribing on the domain wall spacing of oriented electrical steel: (a) before scribing and (b) after scribing. (Courtesy of Armco Inc.)

the domain wall is pinned temporarily. However, because of the magnetizing force driving its motion, it eventually breaks away from the pinning site. This process occurs very suddenly and the resulting high wall velocity generates localized eddy currents. These localized eddy current losses are thought to be the essence of what are called hysteresis losses. Thus, all losses in electrical steel are eddy current losses in nature. These hysteresis losses occur even at very low, essentially d.c., cycle rates. This is because, although the domain walls move very slowly until they encounter an obstacle, the breakaway process is still very sudden. Thus, in loss separation studies, the hysteresis losses can be measured independently by going to low cycle rates, whereas the total loss, including hysteresis, is measured at operational cycle rates. In high-quality electrical steel, the hysteresis and eddy current losses contribute about equally to the total loss.

A novel approach to eliminating hysteresis losses involved using the magnetic material at saturation. In this case, there are no domain walls and hence no wall pinning. Voltage is produced in coils surrounding the magnetic core by rotating the saturated flux vector so that the component linking the coil is sinusoidal [Kra88]. This could be accomplished by means of a toroidal core. This would not eliminate the classical eddy current losses and would require a material having a low excitation current to produce saturation in order to be practical.

The manufacturer or supplier of electrical steel generally provides the user with loss curves that show the total loss per kilogram or pound as a function of induction at the frequency of interest, usually 50 or 60 Hz. One of these curves is shown in Figure 15.3a. This curve is generally measured under ideal conditions, that is, low stress on the laminations, and uniform, unidirectional, and sinusoidal flux in the laminations, so that it represents the absolute minimum loss per kg or lb to be expected in service. Another useful curve, which the manufacturer can provide, is a curve of the exciting power per unit weight versus induction at the frequency of interest. A sample curve is shown in Figure 15.3b. Again, this is an idealized curve, but it can be useful in estimating the power and current needed to energize the transformer.

15.2.1 Building Factor

As mentioned previously, the specific core losses at the operating induction provided by the manufacturer are minimum expected losses so that multiplying by the core weight produces a total core loss, which is generally lower than what is measured in practice. The discrepancy is a result of the fact that stacked cores require joints where the induction must not only change direction but must bridge a gap between different laminations. Also stresses produced in the steel due to cutting and stacking operations increase the losses. There are other causes of this discrepancy such as burrs produced by cutting, but all of these can be lumped into a building factor, which multiplies the ideal core loss to produce the measured core loss. These building factors are generally in the range of 1.2–1.4 and are roughly constant for a given core building practice.

Many attempts have been made to understand these extra losses, particularly in the region of the core joints, and have led to improved joint designs such as a step-lapped joint where the joint is made gradually in a steplike manner. Studies have shown that near the joints where the flux changes direction by 90°, the induction vector rotates and this produces higher losses. Thus, another approach to implementing a building factor approach is to apply a multiplier to the ideal losses for the amount of steel in the joint region only. This joint multiplier would be higher than the average multiplier and could

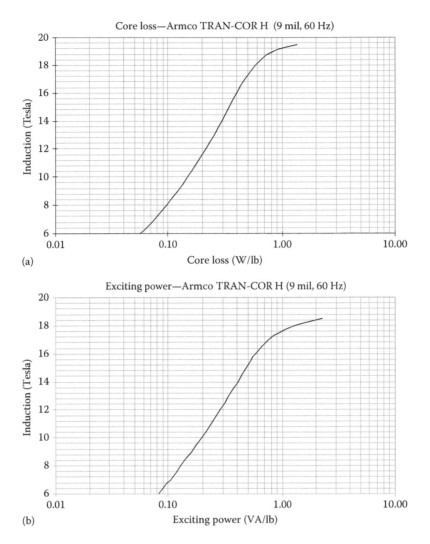

FIGURE 15.3
Graphs of core loss and exciting power based on a polynomial fit to data. (a) Manufacturer's curve of specific core loss versus induction. (b) Manufacturer's curve of specific exciting power versus induction. (Courtesy of Armco Inc., TRAN-COR H® is a registered trademark of Armco Inc.)

be as high as ~1.7. The advantage of this approach is that cores having different fractions of their overall weight in the joint regions should receive more accurate overall loss determinations.

15.2.2 Interlaminar Losses

The core laminations are coated with a glass-like insulating material. This is usually very thin, on the order of a few microns, to keep the space factor reasonably high (>96%). Like any other material, the coating is not a perfect insulator. Thus, eddy currents, driven by the bulk flux in the core, can flow perpendicularly through the stacked laminations, which

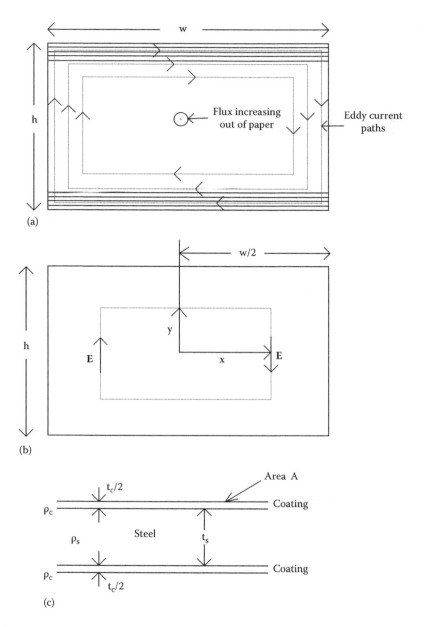

FIGURE 15.4
Interlaminar eddy currents produced by the bulk core flux. (a) Eddy current paths in laminated core, (b) geometry for loss calculation, and (c) geometry for surface resistivity calculation.

produces interlaminar losses. This is sketched in Figure 15.4 for a rectangular cross-sectional core. Of course, the eddy current paths are completed within the laminations where the resistance is much lower. The coating must be a good enough insulator to keep these losses low relative to the normal intralaminar losses. The insulative value of the coating is determined not only by the intrinsic resistivity of the coating material, which must be high, but also by its thickness. Although the thickness is generally not perfectly uniform, it should not vary so much that bald spots are produced. In high-quality electrical steels, two types of coatings are generally applied, a first glass-like coating and a second coat of special composition designed to apply a favorable stress to the steel. The two coatings make the occurrence of bald spots unlikely.

The insulating value of the coating is determined by measuring the resistance across a stack of laminations or ideally a single lamination as shown in Figure 15.4c. In terms of the parameters shown, the effective resistance across a lamination of area A is

$$R_{eff} = R_{coating} + R_{steel} = \frac{\rho_c t_c}{A} + \frac{\rho_s t_s}{A} = (\rho_c f_c + \rho_s f_s)\frac{t}{A} = \rho_{eff}\frac{t}{A} \tag{15.1}$$

where
  ρ_c is the coating resistivity and t_c its two-sided thickness
  ρ_s is the steel resistivity and t_s its thickness, $t = t_c + t_s$ is the combined thickness
  f_c and f_s are fractional thicknesses of the coating and steel, respectively

It is often convenient to express the result in terms of a surface resistivity, $\rho_{surf} = \rho_{eff}t$, where t is the thickness of one lamination. Its units are $\Omega\,m^2$ in the SI system.

We now estimate the interlaminar losses with the help of the geometry shown in Figure 15.4b. Assume a stack of core steel of width w and stack height h and infinitely long in the direction perpendicular to the page. A uniform sinusoidal flux with peak induction B_o and angular frequency ω flows through the stack perpendicular to the page. Using Faraday's law,

$$\oint \mathbf{E} \cdot d\ell = -\frac{d\Phi}{dt} = -\omega(4xy)B_o \tag{15.2}$$

applied to the rectangle of area 4xy shown dashed in the figure. By symmetry and Lenz's law, the electric field points as shown on the two vertical sides of the rectangle. The electric field is nearly zero along the horizontal sides since here the electric field is within the metallic laminations. Thus, from (15.2), we have

$$E(4y) = -\omega(4xy)B_o \quad \Rightarrow \quad E = -\omega B_o x \tag{15.3}$$

where E is the magnitude of the electric field, which points down on the right and up on the left sides of the rectangle. In these directions, we also have

$$E = \rho_{eff}J \tag{15.4}$$

where
  J is the current density
  ρ_{eff} is the effective resistivity perpendicular to the stack of laminations derived previously

The interlaminar losses are given by

$$Loss_{int} = \frac{1}{2}\int \rho_{eff}J^2 dV \tag{15.5}$$

where the volume integral is over the lamination stack, assumed uniform in the direction into the paper so that the losses calculated are losses per unit length. The factor of 1/2 comes from time averaging J, which is assumed to be expressed in terms of its peak value. Thus, from (15.3) through (15.5), we get

$$\text{Loss}_{int} = \frac{\omega^2 B_o^2 h}{\rho_{eff}} \int_0^{W/2} x^2 dx = \frac{\omega^2 B_o^2 h w^3}{24 \rho_{eff}} = \left(\frac{\pi^2}{6}\right) \frac{f^2 B_o^2 h w^3}{\rho_{eff}} \tag{15.6}$$

where we have used $\omega = 2\pi f$. The specific loss (loss per unit volume) is given by

$$P_{int} = \frac{\text{Loss}_{int}}{hw} = \left(\frac{\pi^2}{6}\right) \frac{f^2 B_o^2 h w^2}{\rho_{eff}} \tag{15.7}$$

since we assumed unit length in the other dimension. To find the loss per unit weight or mass, divide by the density of the core steel in the appropriate units. Equation 15.7 is in the SI system where w is in meters, B_o in Tesla, ρ_{eff} in Ω m, f in Hz, and P_{int} in W/m^3.

The interlaminar loss should be compared with the normal loss at the same peak induction. For typical values of the parameters, it is generally much smaller than the normal loss and can normally be ignored. As a numerical example, let f = 60 Hz, w = 0.75 m, B_o = 1.7 T, and ρ_{eff} = 20 Ω m. We get P_{int} = 481 W/m^3. The density of electrical steel is 7650 kg/m^3 so that P_{int} = 0.063 W/kg. This is a fairly small loss compared with the normal losses at 1.7 T of ~1.3 W/kg. However, a high enough interlaminar resistance must be maintained to achieve these low losses.

15.3 Load Losses

15.3.1 I²R Losses

I²R losses in the coil conductors are generally the dominant source of load losses. They are normally computed using the d.c. value of resistivity. However, in the case of wires with large cross-sectional areas carrying a.c. current, this normally requires that they be made of stranded and transposed conductors. To get a feeling for how a.c. current affects resistance, consider the resistance of an infinitely long cylinder of radius a, permeability μ, and d.c. conductivity $\sigma = 1/\rho$, where ρ is the d.c. resistivity. Let it carry current at an angular frequency $\omega = 2\pi f$. Then the ratio of a.c. to d.c. resistance is given by [Smy68]

$$\frac{R_{ac}}{R_{dc}} = \frac{x}{2} \left[\frac{\text{ber}\,x\,\text{bei}'\,x - \text{ber}'\,x\,\text{bei}\,x}{\left(\text{ber}'\,x\right)^2 + \left(\text{bei}'\,x\right)^2} \right], \quad x = a\sqrt{\omega\mu\sigma} \tag{15.8}$$

The ber and bei functions along with their derivatives ber' and bei' are given in [Dwi61]. This resistance ratio, which can also be regarded as the ratio of an effective a.c. to d.c. resistivity, is plotted in Figure 15.5.

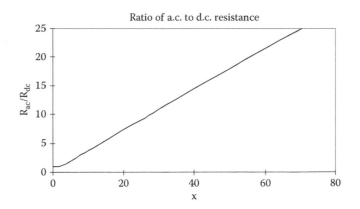

FIGURE 15.5
Plot of the a.c. to d.c. resistance ratio for an infinitely long cylinder.

As a numerical example, consider a copper cylinder with the following parameters: a = 2.52 cm, ρ = 2 × 10⁻⁸ Ω m, μ = 4π × 10⁻⁷ H/m, and f = 60 Hz resulting in x = 1.95. Then we find $R_{ac}/R_{dc} \approx 1.15$, that is, a 15% effect. Making the conductor out of strands, insulated from each other, is not sufficient to eliminate this effect. In addition, the strands must be transposed so that each strand occupies a given region of the cross-sectional area as often as any other strand. This is accomplished in modern transposed cables, which use typically 5–80 strands.

With this remedy or the use of small wire sizes, I²R losses can be calculated using the d.c. resistance formula

$$I^2R \ \text{Loss} = \frac{\rho \ell}{A} I^2 \tag{15.9}$$

where
  ρ is the resistivity at the temperature of interest
  ℓ is the length of the conductor, A its cross-sectional area = the sum of the areas of all wires in parallel
  I is the total current flowing into cross-sectional area A

The temperature-dependent resistivity obeys the formula

$$\rho(T) = \rho_o \left[1 + \alpha (T - T_o) \right] \tag{15.10}$$

over a wide range of temperatures, where ρ_o is the resistivity at T = T_o and α is the temperature coefficient of resistivity. For pure copper at T_o = 20°C, ρ_o = 1.72 × 10⁻⁸ Ω m and α = 0.0039. For pure aluminum at T_o = 20°C, ρ_o = 2.83 × 10⁻⁸ Ω m and α = 0.0039. Alloying these materials can change these resistivities considerably. The temperature dependence indicated in (15.10) is significant for copper and aluminum. For example, both copper and aluminum resistivities and hence I²R losses increase about 30% in going from 20°C to 100°C.

15.3.2 Stray Losses

These are losses caused by stray or leakage flux. Figure 15.6 shows the leakage flux pattern produced by the coil currents in the bottom half of a single phase or leg of a transformer, assuming cylindrical symmetry about the centerline of the core. This was generated with a 2D finite element program. The main components, core, coils, tank, and clamp, are shown. Shunts on the tank wall and clamp were given the material properties of transformer oil so they are not active. Figure 15.7 shows the same plot but with the tank and clamp shunts activated. These are made of the same laminated electrical steel as the core. The shunts divert the flux from getting into the tank or clamp walls so that the stray losses in Figure 15.7 are much less than those in Figure 15.6. Calculations show that the losses in the clamp were reduced by a factor of 8 with the shunts present relative to the losses without the shunts. For the tank wall, this reduction was about a factor of 40. The stray flux pattern depends

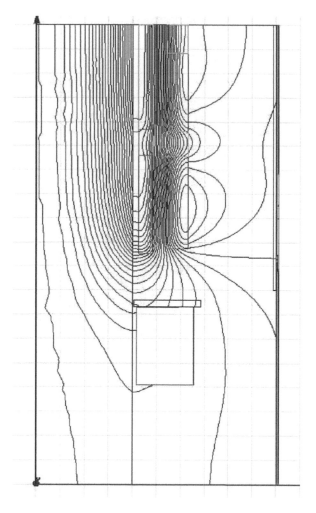

FIGURE 15.6
Stray flux in the lower half of a core leg with no shunts present. Although shown in the figure, they were given the magnetic properties of air (transformer oil).

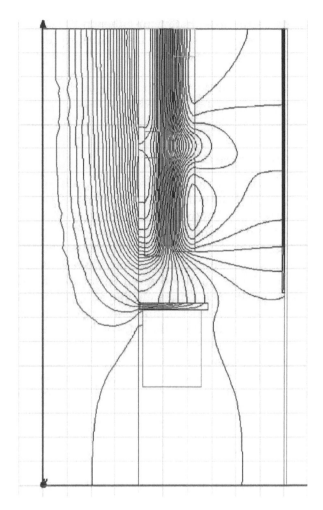

FIGURE 15.7
Stray flux in the lower half of a core leg with shunts present.

on the details of the winding sizes and spacings, the tank size, the clamp position, etc. The losses generated by this flux depend on whether shunts or shields are present as well as geometric and material parameters.

In addition to the coils' stray flux, there is also flux produced by the leads. This flux can generate losses, particularly if the leads are close to the tank wall or clamps.

As Figures 15.6 and 15.7 indicate, there is also stray flux within the coils themselves. This flux is less sensitive to the details of the tank and clamp position or whether shunts or shields are present. Therefore, other methods besides finite elements, such as Rabins' method, which uses a simplified geometry, can be used to accurately calculate this flux in or near the coils. The coil flux generates eddy currents in the wires or individual strands of cable conductors. The losses depend on the strand size as well as its orientation relative to the induction vector and the induction vector's magnitude. The localized losses are therefore different at different positions in the coil.

There are other types of stray loss that occur either in the case of an unusual design or when a manufacturing error occurs. In the latter category, extra losses are generated when a transposition is missed or misplaced in a coil made of two or more wires or cables in parallel.

We will examine many of these types of stray loss in this chapter, deriving analytic formulas or procedures for evaluating them where possible or relying on finite element studies or other numerical methods if necessary.

15.3.2.1 Eddy Current Losses in the Coils

In order to study the effect of stray flux on losses in the coils, we examine an individual wire or strand that could be part of a transposed cable. This is assumed to have a rectangular cross section. The magnetic field at the site of this strand segment will point in a certain direction relative to the strand's orientation. This vector can be decomposed into components parallel to each side of the rectangular cross section as shown in Figure 15.8a. (In a transformer with circular coils surrounding the core, there is little or no magnetic field directed azimuthally along the length of the wire.) We analyze the losses associated with each component of the magnetic field separately and add the results. This is accurate for linear magnetic materials such as copper or aluminum. We will see in the following that the eddy currents tend to concentrate along the sides of the rectangular strand to which the field component is parallel. Thus, the eddy current patterns associated with the two field components do not overlap significantly in any case.

Consider the losses associated with the y component of an external magnetic field as shown in Figure 15.8b, where the coordinate system and geometric parameters are indicated.

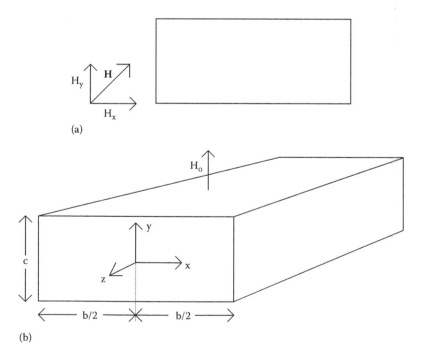

FIGURE 15.8
Geometry for calculating losses in a conducting strand due to an external magnetic field. (a) Magnetic field at the location of a conducting strand. (b) Coordinate system for loss calculation.

We assume an idealized geometry where the strand is infinitely long in the z direction. This implies that none of the electromagnetic fields have a z dependence. We further assume that the magnetic field, both external and internal, has only a y component. Applying Maxwell's equations in this coordinate system and with these assumptions, we obtain in the SI system

$$\nabla \times \mathbf{E} = -\frac{\partial \mathbf{B}}{\partial t} \quad \Rightarrow \quad \frac{\partial E_z}{\partial x} = \mu \frac{\partial H_y}{\partial t}$$

$$\nabla \times \mathbf{H} = \mathbf{J} \quad \Rightarrow \quad \frac{\partial H_y}{\partial x} = J_z \tag{15.11}$$

$$\nabla \cdot \mathbf{B} = 0 \quad \Rightarrow \quad \frac{\partial H_y}{\partial y} = 0$$

where
 μ is the permeability
 $\mathbf{E}$ is the electric field
 $\mathbf{H}$ is the magnetic field

In the case of metallic conductors with a relative permeability of 1, $\mu = \mu_o = 4\pi \times 10^{-7}$ H/m in the SI system. We have ignored the displacement current term, which is only important at extremely high frequencies. In the metallic conductor, we have Ohm's law in the form

$$\mathbf{J} = \sigma \mathbf{E} \quad \Rightarrow \quad J_z = \sigma E_z \tag{15.12}$$

where σ is the electrical conductivity. There is only a z component to $\mathbf{J}$ and $\mathbf{E}$ as (15.11) shows. The return paths for the eddy currents are infinitely far along the conductor's length in both the + and −z directions. Combining (15.11) and (15.12), we obtain

$$\frac{\partial^2 H_y}{\partial x^2} = \mu\sigma \frac{\partial H_y}{\partial t} \tag{15.13}$$

where H_y is a function of x and t.
  Let H_y have a sinusoidal time dependence of the form

$$H_y(x,t) = H_y(x)e^{j\omega t} \tag{15.14}$$

Then (15.13) becomes

$$\frac{\partial^2 H_y}{\partial x^2} = j\omega\mu\sigma H_y = k^2 H_y, \quad k^2 = j\omega\mu\sigma \tag{15.15}$$

H_y is only a function of x in (15.15). Solving (15.15) with the boundary condition that $H_y = H_o$ at $x = \pm b/2$, where b is the strand width normal to the field direction and H_o is the peak amplitude of the external field, we get

$$H_y(x) = H_o \frac{\cosh(kx)}{\cosh(kb/2)} \tag{15.16}$$

and from (15.11)

$$J_z = -kH_o \frac{\sinh(kx)}{\cosh(kb/2)} \tag{15.17}$$

The eddy current loss per unit length in the z direction is

$$\frac{\text{E.C. loss}}{\text{Unit length}} = \frac{c}{2\sigma} \int_{-b/2}^{b/2} |J_z|^2 \, dx = \frac{cH_o^2 |k|^2}{\sigma |\cosh(kb/2)|^2} \int_0^{b/2} |\sinh(kx)|^2 \, dx \tag{15.18}$$

where c is the strand dimension along the y direction. The factor of 1/2 comes from taking a time average and using peak values of the field. The integration through only half the thickness is possible because of the symmetry of the integrand. The factor of 2 involved cancels the factor of 2 in the denominator. Note that, from (15.15),

$$k = (1+j)\sqrt{\frac{\omega\mu\sigma}{2}} = (1+j)q, \quad q = \sqrt{\frac{\omega\mu\sigma}{2}} \tag{15.19}$$

Using this expression for k, we have the identities

$$\begin{aligned}
|\sinh(kx)|^2 &= \frac{1}{2}\left[\cosh(2qx) - \cos(2qx)\right] \\
|\cosh(kx)|^2 &= \frac{1}{2}\left[\cosh(2qx) + \cos(2qx)\right]
\end{aligned} \tag{15.20}$$

Substituting into (15.18), integrating, and dividing by the cross-sectional area = cb, we get the specific eddy current loss (loss/unit volume) as

$$P_{ec} = \frac{H_o^2 q}{\sigma b}\left[\frac{\sinh(qb) - \sin(qb)}{\cosh(qb) + \cos(qb)}\right] \tag{15.21}$$

This is in Watts/m^3 in the SI system. Divide by the density in this system to get the loss per unit mass or weight.

This last equation applies over a broad frequency range, up to where radiation effects start becoming important. At the low-frequency end, which applies to transformers at power frequencies (small qb), this reduces to

$$P_{ec} \xrightarrow[\text{small qb}]{} \frac{H_o^2 q^4 b^2}{6\sigma} = \left(\frac{\pi^2}{6}\right) f^2 \mu^2 b^2 \sigma H_o^2 = \left(\frac{\pi^2}{6}\right)\frac{f^2 b^2 B_o^2}{\rho} \tag{15.22}$$

where we have used $\sigma = 1/\rho$, where ρ is the resistivity, $\omega = 2\pi f$, where f is the frequency, and $B_o = \mu H_o$.

As a numerical example, let $B_o = 0.05$ T, which is a typical leakage induction value in the coil region, $\rho = 2 \times 10^{-8}$ Ω m for copper at its operating temperature, $\sigma = 5 \times 10^7$ (Ω m)$^{-1}$,

$\omega = 2\pi f = 2\pi(60)$ rad/s, b = 6.35×10^{-3} m, and $\mu = 4\pi \times 10^{-7}$ H/m. Then q = 108.8 m^{-1} and qb = 0.691. This is small enough that the small qb limit should apply. Thus, we get from (15.22) $P_{ec} = 2.985 \times 10^4$ W/m^3. The exact formula (15.21) yields $P_{ec} = 2.955 \times 10^4$ W/m^3. Using the density of copper $d_{Cu} = 8933$ kg/m^3, we obtain $P_{ec} = 3.35$ W/kg on a per unit mass basis. The I^2R loss on a per volume basis associated with an rms current of 3×10^6 A/m^2 (a typical value) in a copper winding of the resistivity given earlier is 1.8×10^5 W/m^3 so that the eddy current loss amounts to about 17% of the I^2R losses in this case.

The losses given by (15.21) or (15.22) must be combined with the losses given by a similar formula with H_o or B_o referring to the peak value of the x component of the field and with b and c interchanged. This will give the total eddy current loss density at the location of the strand. A method of calculating the magnetic field or induction at various locations in the coils is needed. From the axisymmetric field calculation (flux map) given in Figures 15.6 and 15.7, we obtain values of the radial and axial components of the field. These replace the x and y components in the loss formulas given earlier. These loss densities will differ in different parts of the winding. To obtain the total eddy current loss, an average loss density can be obtained for the winding, and this is multiplied by the total weight or volume of the winding. However, in determining local winding temperatures and especially the hot spot temperature, we need to know how these losses are distributed along the winding.

From (15.17), (15.19), and (15.20), we can obtain the eddy current loss density as a function of position in the strand:

$$\frac{|J_z|^2}{2\sigma} = \frac{H_o^2 q}{\sigma}\left[\frac{\cosh(2qx) - \cos(2qx)}{\cosh(qb) + \cos(qb)}\right] \qquad (15.23)$$

This vanishes at the center of the strand (x = 0) and is a maximum at the surface (x = ±b/2). The parameter q measures how fast this drops off from the surface. The more rapid is the falloff, the larger the value of q. The reciprocal of q is called the skin depth δ and is given by

$$\delta = \frac{1}{q} = \sqrt{\frac{2}{\omega\mu\sigma}} \qquad (15.24)$$

For copper at ~60°C, $\sigma = 5 \times 10^7$ (Ω m)$^{-1}$, and f = 60 Hz, we get δ = 9.2 mm. For aluminum at ~60°C, $\sigma = 3 \times 10^7$ (Ω m)$^{-1}$, and f = 60 Hz, we get δ = 11.9 mm. Thus, the skin depth is smaller for copper than for aluminum, which means that the eddy currents concentrate more toward the surface of copper than that of aluminum.

The high-frequency limit of (15.21) (large qb) is

$$P_{ec} \xrightarrow[\text{large qb}]{} \frac{H_o^2 q}{\sigma b} \qquad (15.25)$$

This increases as the square root of the frequency.

15.3.2.2 Tie Plate Losses

The tie plate (also called flitch plate) is located just outside the core in the space between the core and innermost winding. It is a structural plate that connects the upper and lower clamps. Tension in this plate provides the clamping force necessary to hold the transformer together if a short circuit occurs. It is usually made of magnetic steel or stainless steel and

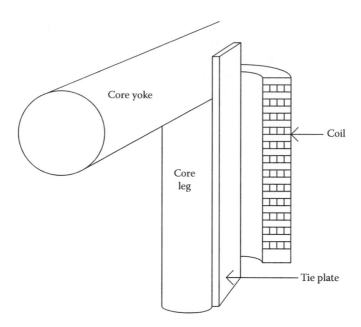

FIGURE 15.9
Tie plate location in a transformer.

could be subdivided into several side-by-side vertical plates to help reduce the eddy current losses. Figure 15.9 shows a schematic diagram of one of the tie plates associated with one leg. There is another on the opposite side of the core leg. These generally have a rectangular cross section.

Since the flux plots in Figures 15.6 and 15.7 are for a 2D axisymmetric geometry, it was not possible to include the tie plate. (This would have made it a solid cylinder around the core.) However, the flux pattern shown in the figures should not be greatly altered by their presence since they occupy a fairly small fraction of the core's circumference. As the flux pattern in the figures shows, the flux is primarily radial at the location of the tie plate. However, with an actual tie plate present, there will be some axial flux carried by the tie plate. This will depend on the permeability of the tie plate relative to that of the core. We can estimate the axial tie plate flux by reference to Figure 15.10 where we show two side-by-side solids of permeabilities μ_1, μ_2 and cross-sectional areas A_1, A_2 carrying flux. We assume that the coils producing this flux create a common magnetic field **H** at the location of the solids. (The field inside an ideal solenoid is a constant axial field.)

The induction inside each solid is given by

$$B_1 = \mu_1 H, \quad B_2 = \mu_2 H \tag{15.26}$$

so that

$$\frac{B_1}{B_2} = \frac{\mu_1}{\mu_2} \tag{15.27}$$

Using this last expression, we can estimate the induction in the tie plate due to the vertical field. We can use the ratio of relative permeabilities in this formula. These are a.c.

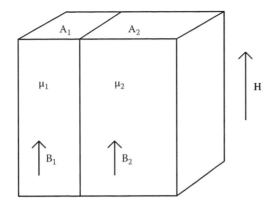

FIGURE 15.10
Dissimilar magnetic materials in a common magnetic field.

permeabilities, which we take to be ~5000 for the core and ~200 for a magnetic steel tie plate. With a core induction of 1.7 T, which is typical, we find that $B_{\text{mag t.p.}} = 0.068$ T. For a stainless steel tie plate of relative permeability = 1, we obtain $B_{\text{s.s. t.p.}} = 0.00034$ T. Thus, the axial induction is not insignificant for a magnetic steel tie plate but ignorable for a stainless steel one.

Let us first look at the losses due to the radial induction since these are common to both magnetic and stainless steel tie plates. We have studied these losses using a 2D finite element analysis. Figure 15.11 shows a flux plot for a magnetic steel tie plate, assuming a uniform 60 Hz sinusoidal flux density of 1 Tesla peak value far from the plate. The plate is assumed to be infinitely long in the dimension perpendicular to the page. Only 1/2 the geometry is modeled, taking advantage of symmetry about the left-hand axis. There is a space between the tie plate and the high-permeability core surface, which is along the bottom boundary of the figure. This space allows for insulation and cooling duct. The loss density contours are shown in Figure 15.12. This shows that the eddy currents are

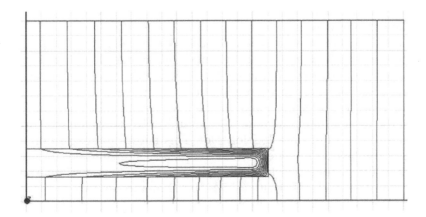

FIGURE 15.11
Flux lines for a magnetic steel tie plate in a uniform 60 Hz magnetic field of 1 Tesla peak value normal to its surface. The tie plate of 9.5 mm thickness is separated from the high-permeability core along the bottom by a space for insulation and cooling. Only the right half of the geometry is modeled.

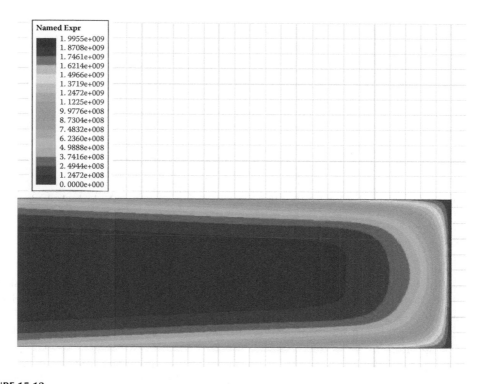

FIGURE 15.12
Loss density contours for the magnetic steel tie plate shown in Figure 15.11. Only the right half of the tie plate is shown. The key is in W/m³.

concentrated near the surface due to the skin effect. Figure 15.13 shows a similar flux plot for a stainless steel tie plate. Note that there is little eddy current screening so the flux penetrates the plate without much distortion. Figure 15.14 shows the loss density contours for the stainless steel tie plate. There is much less surface concentration of the eddy currents. Although we only calculate eddy current losses with a finite element program, the hysteresis losses in a magnetic steel tie plate make up only a small fraction of the total losses for typical tie plate dimensions.

The finite element study was repeated for different tie plate widths (perpendicular to the flux direction) while keeping the thickness (along the flux direction) constant at 9.5 mm. The results are shown in Figure 15.15, where the losses per unit length in the infinitely long direction and per T² are plotted. To get the actual loss, multiply the ordinate by the tie plate length or portion of the tie plate length affected by the losses and by the square of the peak radial induction in T². Figure 15.16 shows the same information as plotted in Figure 15.15 but on a log–log plot. This allows the extraction of the power dependence of the loss on the width of the tie plate. The material parameters assumed were as follows: for magnetic steel, resistivity = 25×10^{-8} Ω m and relative permeability = 200, and for stainless steel, resistivity = 75×10^{-8} Ω m and relative permeability = 1. We obtain for the losses

$$\text{Loss}_{\text{mag steel}} \left(\text{W/mm} \right) = 1.953 \times 10^{-3} \, w^{2.3} B_p^2, \quad w \text{ in mm, } B_p \text{ in T}$$

$$\text{Loss}_{\text{stainless}} \left(\text{W/mm} \right) = 7.407 \times 10^{-5} \, w^3 B_p^2, \quad w \text{ in mm, } B_p \text{ in T}$$

$$(15.28)$$

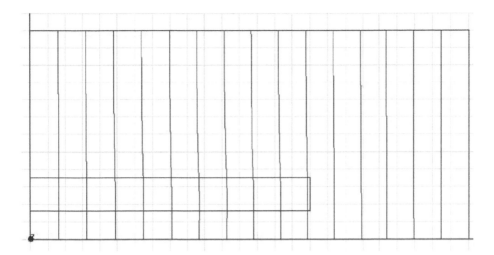

FIGURE 15.13
Flux lines for a stainless steel tie plate in a uniform 60 Hz field of 1 Tesla peak value normal to its surface. The situation is otherwise the same as described in Figure 15.11.

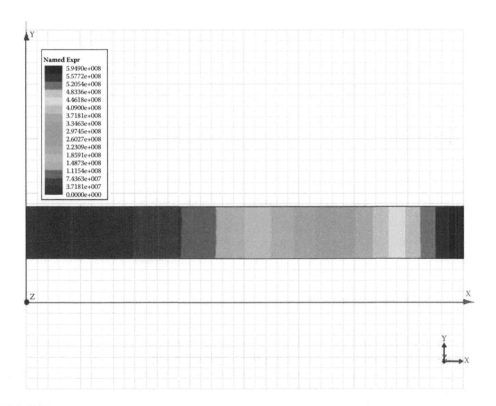

FIGURE 15.14
Loss density contours for the stainless steel tie plate shown in Figure 15.13. The key is in W/m^3.

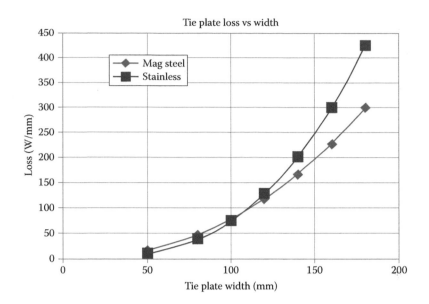

FIGURE 15.15
Losses for 9.5 mm thick tie plates made of magnetic and stainless steel versus the tie plate width. The tie plates are in a uniform 60 Hz magnetic field of 1 Tesla peak value directed normal to the tie plate surface.

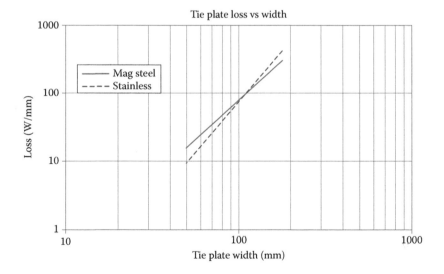

FIGURE 15.16
Same as Figure 15.15 but on a log–log scale.

with the loss in Watts/mm, the tie plate width w in mm, and the peak induction B_p in Tesla. These are losses/unit length in mm along the tie plate length.

Although this study was performed for a specific tie plate thickness of 9.5 mm, it is an example of the kind of study that can be done with a finite element code and that can be parameterized for future design use without the necessity of repeating it for each new case.

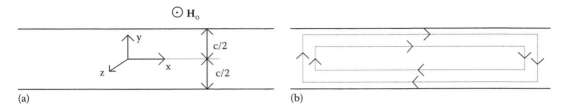

FIGURE 15.17
Geometry of an idealized plate or sheet driven by a uniform sinusoidal magnetic field parallel to its surface (perpendicular to page): (a) sheet geometry and (b) eddy current paths.

We can estimate the losses in the tie plate due to axial flux by resorting to an idealized geometry. We assume the tie plate is infinitely long and the flux is driven by a uniform axial magnetic field parallel to the tie plate's surface. We are also going to assume that its width is much greater than its thickness. In fact, we assume an infinite width. Thus, as shown in Figure 15.17, the only relevant dimension is the y-dimension through the sheet's thickness. The eddy currents will flow primarily in the x direction. We ignore their return paths in the y direction, which are small compared to the x-directed paths.

Applying Maxwell's equations and Ohm's law to this geometry, we obtain in SI units the same equation as (15.13) except that the fields are z-directed:

$$\frac{\partial^2 H_z}{\partial y^2} = \mu\sigma \frac{\partial H_z}{\partial t} \tag{15.29}$$

Thus, using the previous results but altering the notation to fit the present geometry, we get

$$P_{tp,axial} = \frac{H_o^2 q}{\sigma c}\left[\frac{\sinh(qc)-\sin(qc)}{\cosh(qc)+\cos(qc)}\right], \quad k=(1+j)q, \quad q=\sqrt{\frac{\omega\mu\sigma}{2}} \tag{15.30}$$

This is a loss per unit volume (W/m^3) in the tie plate due to axial flux. Since the eddy currents generating this loss are at right angles to the eddy currents associated with the radial flux normal to the surface, there is no interference between them and the two losses can be added.

For magnetic steel at 60 Hz with $\mu_r = 200$, $\mu = \mu_r 4\pi \times 10^{-7}$ H/m, $\sigma = 4 \times 10^6$ (Ω m)$^{-1}$, and $c = 9.5 \times 10^{-3}$ m, we have $qc = 4.14$. This is large enough that (15.30) must be used without taking its small qc limit. Substituting the parameters just given for magnetic steel and using $B_o = 0.07$ T ($H_o = B_o/\mu$), we obtain $P_{tp,axial} = 925$ W/m$^3 = 0.925 \times 10^{-6}$ W/mm^3.

To compare with the radial flux loss formula for magnetic steel given previously, Equation 15.28, we must specify the tie plate width and use the same thickness of $c = 9.5$ mm. Using $w = 127$ mm and $B_p = 0.14$ T and the same value of c yields $P_{tp,radial} = 2.64$ W/mm. This must be divided by the tie plate area in mm = $9.5 \times 127 = 1206$ mm^2. This yields $P_{tp,radial} = 0.0022$ W/mm^3. Thus, it appears that for a magnetic steel tie plate, the loss associated with the axial flux is much less than the loss associated with the radial flux. In any event, these two losses must be added to get the total loss.

Since the tie plate loss is mainly due to radial flux and this depends on the width to a power ~2–3, the losses can be reduced by reducing the width of an individual tie plate while keeping their total cross-sectional area constant, since this area is necessary to provide the required mechanical strength. Thus, subdividing the tie plates either with axial slots or into separate plates can reduce the loss while maintaining the necessary axial tensile strength.

15.3.2.3 Tie Plate and Core Losses due to Unbalanced Currents

There is a more specialized type of loss, which can occur in transformers that have a large unbalanced net current flow. The net current is the algebraic sum of the currents flowing in all the windings. This, in contrast with the net ampere-turns, which are always nearly exactly balanced, can be unbalanced. To visualize the magnetic effect of a winding's current, consider a cylindrical (solenoidal) winding carrying a current that flows in at the bottom and out at the top. Outside the winding, the net upward current appears equally distributed around the cylinder. The magnetic field outside the cylinder associated with this current is the same as that produced by the current flowing along the centerline of the cylinder, ignoring end effects. In a core-form transformer, some of the windings carry current up and some down. The algebraic sum of all these currents can be considered as being carried by one cylinder of radius equal to a weighted average of the contributing windings, weighted by their current magnitudes. Outside this radius, the field is that of a straight wire along the centerline carrying the algebraic sum of the currents.

The field around a long straight wire carrying a current $\mathbf{I}$ is directed along concentric circles about the wire and has the magnitude in SI units

$$H_\varphi = \frac{\mathbf{I}}{2\pi r} \tag{15.31}$$

where
  r is the radial distance from the centerline
  φ, the azimuthal angle, indicates that the field is azimuthally directed

We are using boldface type here to indicate that the current is a phasor quantity, since we are considering a 3-phase transformer on a single 3-phase core. This field cuts through the transformer core windows as shown in Figure 15.18. The alternating flux passing through the core window induces a voltage around the core structure that surrounds the window. The tie plates, including their connection to the upper and lower clamps, make a similar circuit around the core windows so that voltage is induced in them as well.

We can calculate the flux due to the coil in Figure 15.18a through the left-hand window, labeled 1 in the figure, Φ_{a1}, using (15.31)

$$\Phi_{a1} = \int B_\varphi dA = \frac{\mu_o h \mathbf{I}_a}{2\pi} \int_{r_w}^{d} \frac{dr}{r} = \frac{\mu_o h \mathbf{I}_a}{2\pi} \ln\left(\frac{d}{r_w}\right) \tag{15.32}$$

where
  μ_o is the permeability of oil or air $= 4\pi \times 10^{-7}$ H/m
  h is the effective winding height
  d is the leg center-to-center distance
  r_w is the effective winding radius

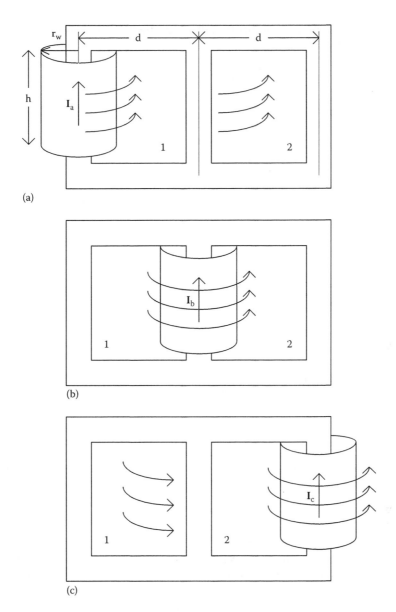

FIGURE 15.18
Field around a cylindrical winding located on a transformer leg and carrying a net upward current: (a) phase a coil, (b) phase b coil, and (c) phase c coil.

h can be obtained as a weighted average of the contributing windings, as was done for r_w. We have integrated all the way to the centerline of the center leg. This is an approximation since the flux lines will no doubt deviate from the ideal radial dependence given in (15.31) near the center leg. The flux through window 2 due to the phase a current, Φ_{a2}, is similarly

$$\Phi_{a2} = \int B_\varphi dA = \frac{\mu_o h I_a}{2\pi} \int_d^{2d} \frac{dr}{r} = \frac{\mu_o h I_a}{2\pi} \ln 2 \tag{15.33}$$

Using the same procedure, we find the flux through the two windows due to phases b and c shown in Figure 15.18b and c:

$$\Phi_{b1} = -\frac{\mu_o h I_b}{2\pi} \ln\left(\frac{d}{r_w}\right), \quad \Phi_{b2} = \frac{\mu_o h I_b}{2\pi} \ln\left(\frac{d}{r_w}\right)$$

$$\Phi_{c1} = -\frac{\mu_o h I_c}{2\pi} \ln 2, \quad \Phi_{c2} = -\frac{\mu_o h I_c}{2\pi} \ln\left(\frac{d}{r_w}\right)$$

$$(15.34)$$

Thus, the net fluxes through windows 1 and 2 are

$$\Phi_1 = \Phi_{a1} + \Phi_{b1} + \Phi_{c1} = \frac{\mu_o h}{2\pi} \ln\left(\frac{d}{r_w}\right)\left[I_a - I_b - \frac{\ln 2}{\ln(d/r_w)} I_c\right]$$

$$\Phi_2 = \Phi_{a2} + \Phi_{b2} + \Phi_{c2} = \frac{\mu_o h}{2\pi} \ln\left(\frac{d}{r_w}\right)\left[\frac{\ln 2}{\ln(d/r_w)} I_a + I_b - I_c\right]$$

$$(15.35)$$

Considering I_a, I_b, I_c to be a positive sequence set of currents and performing the phasor sums given earlier, we get

$$\Phi_1 = \frac{\mu_o h I_a}{2\pi} \ln\left(\frac{d}{r_w}\right)\sqrt{3+g^2}\, e^{j\theta_1}, \quad \theta_1 = \tan^{-1}\left[\frac{\sqrt{3}(1-g)}{3+g}\right]$$

$$\Phi_2 = \frac{\mu_o h I_a}{2\pi} \ln\left(\frac{d}{r_w}\right)\sqrt{3+g^2}\, e^{j\theta_2}, \quad \theta_2 = -\tan^{-1}\left[\frac{\sqrt{3}}{g}\right]$$

$$(15.36)$$

$$\text{where } g = \frac{\ln 2}{\ln(d/r_w)}$$

From Faraday's law, the voltages induced by the two fluxes are

$$V_1 = -\frac{\mu_o h \omega I_a}{2\pi} \ln\left(\frac{d}{r_w}\right)\sqrt{3+g^2}\, e^{j\theta_1}, \quad V_2 = V_1 e^{j\theta_2}$$

$$(15.37)$$

The direction of these voltages or emf's is given by Lenz's law, that is, they try to oppose the driving flux. We can assume that V_1 is the reference phasor and thus drop the minus sign and phase factor from its expression. Then V_2 is given by (15.37) with a-phase of $\theta_2 - \theta_1$ relative to the reference phasor.

The induced voltages will attempt to drive currents through the tie plates and core in loops surrounding the two windows. However, the current flow will be opposed by the tie plates' and core's resistances and by the self- and mutual inductances of the metallic window frames, whether formed of core sections or of tie plate and clamp sections. We can treat these as lumped parameters, organized into the circuits of Figure 15.19.

There are three circuits involved since there are two tie plate circuits on either side of the core plus the core circuit. These are essentially isolated from each other except for coupling through the mutual inductances. Although the tie plate circuits on either side of the core share the top and bottom clamps in common, they are sufficiently symmetric that they can be regarded as separate circuits. We assume that magnetic coupling exists only between window 1 loops or window 2 loops but not between a window 1 loop and a window 2 loop. To make the circuit equations more symmetric, we have positioned V_2 in the circuit in such a way that it is necessary to take the negative of the expression in (15.37).

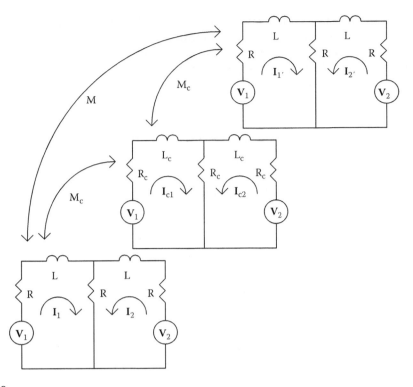

FIGURE 15.19
Equivalent circuits for tie plates and core driven by voltages induced through unbalanced currents. Here 1 and 2 refer to the tie plate circuits on one side of the core, 1′ and 2′ to the tie plate circuits on the other side of the core, and c_1 and c_2 to the core circuits.

Thus, we have for the voltage sources in the circuits of Figure 15.19, assuming $\mathbf{V}_1$ is the reference phasor,

$$\mathbf{V}_1 = \mu_0 h f I_a \ln\left(\frac{d}{r_w}\right)\sqrt{3+g^2}, \quad \mathbf{V}_2 = -\mathbf{V}_1 e^{j\theta}$$

$$\theta = -\tan^{-1}\left(\frac{\sqrt{3}}{g}\right) - \tan^{-1}\left[\frac{\sqrt{3}(1-g)}{3+g}\right], \quad g = \frac{\ln 2}{\ln(d/r_w)} \tag{15.38}$$

where f is the frequency in Hz.

Using the notation of Figure 15.19 and assuming sinusoidal conditions, we can write the circuit equations as

$$\begin{pmatrix} \mathbf{V}_1 \\ \mathbf{V}_2 \\ \mathbf{V}_1 \\ \mathbf{V}_2 \\ \mathbf{V}_1 \\ \mathbf{V}_2 \end{pmatrix} = \begin{pmatrix} Z & R & j\omega M & 0 & j\omega M_c & 0 \\ R & Z & 0 & j\omega M & 0 & j\omega M_c \\ j\omega M & 0 & Z & R & j\omega M_c & 0 \\ 0 & j\omega M & R & Z & 0 & j\omega M_c \\ j\omega M_c & 0 & j\omega M_c & 0 & Z_c & R_c \\ 0 & j\omega M_c & 0 & j\omega M_c & R_c & Z_c \end{pmatrix} \begin{pmatrix} \mathbf{I}_1 \\ \mathbf{I}_2 \\ \mathbf{I}_{1'} \\ \mathbf{I}_{2'} \\ \mathbf{I}_{c1} \\ \mathbf{I}_{c2} \end{pmatrix} \tag{15.39}$$

where $Z = 2R + j\omega L$, $\quad Z_c = 2R_c + j\omega L_c$

Note that R is the resistance of the tie plates associated with one leg side and R_c the core resistance of one leg. The mutual and self-inductances refer to one window of the core. These can be solved for the currents using complex arithmetic. However, it is possible to solve them using real arithmetic by separating the vectors and matrix into real and imaginary parts. Thus, given a matrix equation,

$$\mathbf{V} = M\mathbf{I} \tag{15.40}$$

where $\mathbf{V}$ and $\mathbf{I}$ are complex column vectors and M a complex matrix, write

$$\mathbf{V} = \mathbf{V}_{Re} + j\mathbf{V}_{Im}, \quad \mathbf{I} = \mathbf{I}_{Re} + j\mathbf{I}_{Im}, \quad M = M_{Re} + jM_{Im} \tag{15.41}$$

Here, boldface type indicates vector quantities. Substituting this separation into real and imaginary parts into (15.40), we get

$$\begin{aligned}
\left(\mathbf{V}_{Re} + j\mathbf{V}_{Im}\right) &= \left(M_{Re} + jM_{Im}\right)\left(\mathbf{I}_{Re} + j\mathbf{I}_{Im}\right) \\
&= \left(M_{Re}\mathbf{I}_{Re} - M_{Im}\mathbf{I}_{Im}\right) + j\left(M_{Re}\mathbf{I}_{Im} + M_{Im}\mathbf{I}_{Re}\right)
\end{aligned} \tag{15.42}$$

This reduces to two separate equations:

$$\mathbf{V}_{Re} = \left(M_{Re}\mathbf{I}_{Re} - M_{Im}\mathbf{I}_{Im}\right), \quad \mathbf{V}_{Im} = \left(M_{Re}\mathbf{I}_{Im} + M_{Im}\mathbf{I}_{Re}\right) \tag{15.43}$$

which can be organized into a larger matrix equation

$$\begin{pmatrix} \mathbf{V}_{Re} \\ \mathbf{V}_{Im} \end{pmatrix} = \begin{pmatrix} M_{Re} & -M_{Im} \\ M_{Im} & M_{Re} \end{pmatrix} \begin{pmatrix} \mathbf{I}_{Re} \\ \mathbf{I}_{Im} \end{pmatrix} \tag{15.44}$$

Here, the separate entries are real vectors and matrices. Thus, we have doubled the dimension of the original matrix Equation 15.39 from 6 to 12, but this is still small, considering the power of modern computers.

The remaining issues concern how to evaluate the resistances and the self- and mutual inductances in (15.39). The resistance of a tie plate, R, can simply be taken as its d.c. resistance since it has a relatively small thickness. We can ignore the resistance of the clamps since these should have a much larger cross-sectional area than the tie plates. The core, with its fairly large radius and high permeability, will have an enhanced a.c. resistance relative to its d.c. value. It can be estimated from Figure 15.5, using an effective a.c. permeability and conductivity.

Since we do not expect extreme accuracy in this calculation, in view of the approximations already made, we can use approximate formulas for the self- and mutual inductances. For example, [Gro73] gives a formula for the inductance of a rectangle of sides a and b made of wire with a circular cross section of radius r and relative permeability μ_r, which is, in SI units,

$$\begin{aligned}
L = 4 \times 10^{-7} \Bigg[& a\ln\left(\frac{2a}{r}\right) + b\ln\left(\frac{2b}{r}\right) + 2\sqrt{a^2 + b^2} - a\sinh^{-1}\left(\frac{a}{b}\right) \\
& - b\sinh^{-1}\left(\frac{b}{a}\right) - 2(a+b) + \frac{\mu_r}{4}(a+b) \Bigg]
\end{aligned} \tag{15.45}$$

L is in Henrys and lengths are in meters in this formula. This formula can be applied directly to calculate the inductance of the core window. By defining an effective radius, it can be applied to the tie plate loop as well. In calculating the core inductance, remember that the core current generates magnetic field lines in the shape of concentric circles about the core centerline so that the appropriate relative permeability is roughly the effective permeability perpendicular to the laminations. For a stacking factor of 0.96 and infinitely permeable laminations, the effective perpendicular permeability is $\mu_r = 25$. Similarly, the effective tie plate relative permeability is close to 1.0 for both magnetic and stainless steel tie plates.

The mutual inductance terms are not quite so important so that an even cruder approximation may be used. Reference [Gro73] gives an expression for the mutual inductance between two equal coaxial squares of thin wire that are close together. Letting s be the length of the side of the squares and d their separation, the mutual inductance, in SI units, is

$$M = 8 \times 10^{-7} s \left[\ln\left(\frac{s}{d}\right) - 0.7740 + \left(\frac{d}{s}\right) - 0.0429\left(\frac{d}{s}\right)^2 - 0.109\left(\frac{d}{s}\right)^4 \right] \quad (15.46)$$

In this equation, M is in Henrys and lengths in meters. With a little imagination, this can be applied to the present problem. Once (15.39) or its equivalent (15.44) is solved for the currents, the losses, which are I^2R-type losses, can be calculated. As a numerical example, we found the core and tie plate losses due to an unbalanced current of 20,000 amps rms at 60 Hz in a transformer with the following geometric parameters:

Winding height (h)	0.813 m
Winding radius (r_w)	0.508 m
Leg center-to-center distance (d)	1.727 m
Core radius	0.483 m
Tie plate and core height	5.08 m
Tie plate width	0.229 m
Tie plate thickness	9.525×10^{-3} m

The tie plates were actually subdivided into three plates in the width direction, but this does not affect the calculation. For magnetic steel tie plates, the calculated core loss was 11 W and the total tie plate loss was 1182 W. For stainless steel tie plates, the calculated core loss was 28 W and the total tie plate loss was 2293 W. For the magnetic steel case, the current in the core legs was about 200 amps and in the tie plates about 600 amps. For the stainless steel case, the current in the core legs was about 300 amps and in the tie plates about 500 amps. The stainless steel losses are higher mainly because of the higher resistivity of the material coupled with the fact that the impedances, which limit the currents, are mainly inductive and hence nearly the same for the two cases.

15.3.2.4 Tank and Clamp Losses

Tank and clamp losses are very difficult to calculate accurately. Here, we are referring to the tank and clamp losses produced by the leakage flux from the coils, examples of which are shown in Figures 15.6 and 15.7. The eddy current losses can be obtained from finite element calculations, especially 3D ones that can model the rectangular tank as well as leakage flux from all three phases simultaneously. These tend to be fairly time consuming and not very routine.

However, much can be done with a 2D approach. We refer to a study in [Pav93] where several projection planes in the real 3D geometry were chosen for analysis with a 2D finite element program. The losses calculated with this approach agreed very favorably with test results. As that study and our own show, 2D models allow one to quickly assess the impact on losses of design changes such as the addition of tank and/or clamp shunts made of laminated electrical steel or the effect on losses of aluminum or copper shields at various locations. In fact, the losses in Figure 15.7 with tank and clamp shunts present were dramatically reduced compared with those in Figure 15.6 where no shunts are present. While the real losses may not show as quantitative a reduction, the qualitative effect is real. Another study, which was done very quickly, using the parametric capability of the 2D modeling, was to assess the impact of extending the clamp shunts beyond the top surface of the clamp. We found that some extension was useful in reducing the losses caused by stray flux hitting the vertical side of the clamp.

An example of where a 3D approach is crucial in understanding the effect of design options on losses concerns the laminated steel shunts on the clamps. The side clamps extend along all three phases of a transformer as shown in the top view of Figure 15.20. Should the clamp shunts be made of laminations stacked flat on top of the clamps or should the laminations be on edge, that is, stacked perpendicularly to the top surface of the clamp? In addition, should the shunts extend uninterrupted along the full length of the clamps or can they be subdivided into sections that cover a region opposite each phase but with gaps in between? 2D models cannot really answer these questions. In fact, in a 2D model the flux in the clamp shunts is forced to return to the core eventually, whereas in a 3D model the clamp shunt flux from the three phases can cancel within the shunts for shunts that are continuous along the sides. Because of the laminated nature of the shunt material, the magnetic permeability and electrical conductivity are both anisotropic. This will affect both the flux and eddy current patterns in the shunts in a way that only a 3D model that allows for these anisotropies can capture. We will present some results obtained with a 3D finite element program later in this section, which will examine this clamp shunt issue as well as obtain tank and clamp losses that can be compared with test data. But first it is worthwhile to consider analytic methods of obtaining these stray losses.

It is useful to have analytic formulas for tank or clamp losses. Once obtained, these enable a quick calculation of such losses at the design stage. By comparing these calculated

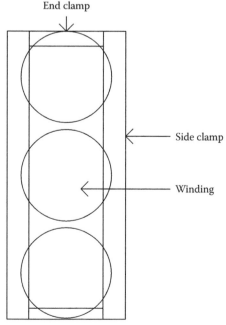

FIGURE 15.20
Geometry of clamp arrangement from a top view.

losses with test data, one can assess their reliability. Often, simple correction factors can be applied to obtain better agreement with test data. Here, we present an analytic method for calculating tank losses based on solving Maxwell's equations in a simplified geometry [Lei99]. To appreciate the approximations, which are made, consider the flux pattern near the tank wall produced by a generic 2-winding transformer as shown in Figure 15.21a. The windings have no taps and uniform current density. Clamps are also absent.

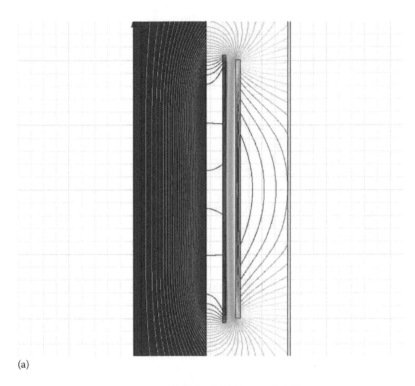

(a)

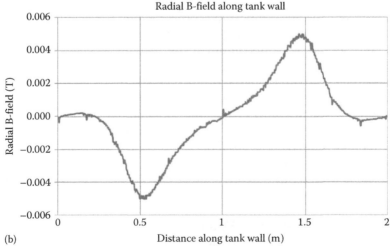

(b)

FIGURE 15.21
Flux pattern near the tank wall for a simple 2-winding transformer with uniform current densities in the windings (a). Radial B-field along a vertical line next to the tank wall (b).

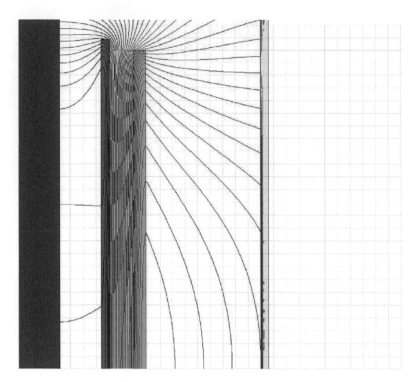

FIGURE 15.22
Flux pattern in the tank wall from accumulation of stray flux from the top of the windings.

Essentially, the flux enters the tank wall from the top of the windings and exits from the bottom of the windings. These directions could be reversed without changing the analysis. Figure 15.22 shows some details of the flux in the tank wall near the top of the windings. The flux enters the wall and is directed downward within the tank wall, increasing toward the winding center from the gradual accumulation of stray flux entering the tank wall. After passing through the vertical position corresponding to the windings' center, the flux begins to leave the tank wall and reenter the windings near their bottom. Thus, the vertical flux in the tank wall reaches a maximum at the windings' center. This is also the place where the losses in the tank wall are highest.

We will make the approximation that the flux (B-field) enters the tank wall perpendicularly and that this perpendicular flux has a sinusoidal distribution along the tank wall with the vertical center at a zero of the sine function. As Figure 15.21b shows, this is approximately true for the simple geometry shown. Further refinement could be achieved through Fourier analyzing the flux component perpendicular to the tank wall and applying the present analysis to each Fourier component. We will assume that the only other B-field component is vertically directed along and within the tank wall. The B-field is assumed to be independent of the azimuthal direction (into the plane of the paper in Figure 15.21a).

The geometry analyzed is shown in Figure 15.23. This shows the tank wall in a horizontal position. The flux from the windings is entering and leaving the tank wall in the z direction. The flux within the wall is directed mainly in the y direction with some in the z direction. The wall is assumed to be infinite in the x direction, that is, the flux distribution is independent of this dimension. The losses calculated will be per unit length in this direction so that

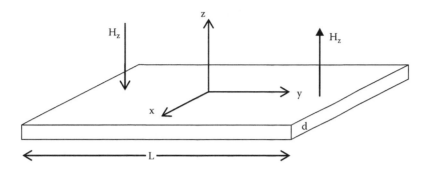

FIGURE 15.23
Geometry of tank wall with flux entering normal to its surface facing the windings. The wall is oriented horizontally in this depiction. L is the effective height of the tank wall and will be taken as the wavelength of the sinusoidal B or H field normal to the tank wall.

the actual losses can be obtained by multiplying the calculated losses by an effective length in the x direction.

We assume that the tank is a homogeneous solid with permeability μ and conductivity σ. We assume the fields are sinusoidal with angular frequency $\omega = 2\pi f$, with f being the actual frequency in Hz. The relevant Maxwell's and Ohm's law equations to solve are

$$\nabla \times \mathbf{E} = -\frac{\partial \mathbf{B}}{\partial t}, \quad \mathbf{B} = \mathbf{B}e^{j\omega t} \quad \Rightarrow \quad \nabla \times \mathbf{E} = -j\omega\mathbf{B}$$

$$\mathbf{E} = \frac{\mathbf{J}}{\sigma} \quad \Rightarrow \quad \nabla \times \mathbf{J} = -j\omega\sigma\mathbf{B}$$

$$\nabla \times \mathbf{H} = \mathbf{J} \quad \Rightarrow \quad \nabla \times (\nabla \times \mathbf{H}) = \nabla \times \mathbf{J} = -j\omega\sigma\mathbf{B} \qquad (15.47)$$

$$\nabla \times (\nabla \times \mathbf{H}) = \nabla(\nabla \cdot \mathbf{H}) - \nabla^2 \mathbf{H}$$

$$\mathbf{B} = \mu\mathbf{H}, \nabla \cdot \mathbf{B} = 0 \quad \Rightarrow \quad \nabla \cdot \mathbf{H} = 0 \quad \Rightarrow \quad \nabla^2 \mathbf{H} = j\omega\mu\sigma\mathbf{H}$$

We are assuming that

$$\mathbf{H} = H_y(y,z)\mathbf{j} + H_z(y,z)\mathbf{k} \qquad (15.48)$$

where **j** and **k** are unit vectors in the y and z directions, respectively. Using this, the fifth line equation in (15.47) becomes

$$\left(\frac{\partial^2 H_y}{\partial y^2} + \frac{\partial^2 H_y}{\partial z^2}\right)\mathbf{j} + \left(\frac{\partial^2 H_z}{\partial y^2} + \frac{\partial^2 H_z}{\partial z^2}\right)\mathbf{k} = j\omega\mu\sigma\left(H_y\mathbf{j} + H_z\mathbf{k}\right) \qquad (15.49)$$

Let $k^2 = j\omega\mu\sigma$. The **j** and **k** component equations in (15.49) can be solved separately. They both have the same form and differ only in the boundary conditions. Beginning with the **k** equation, we have

$$\frac{\partial^2 H_z}{\partial y^2} + \frac{\partial^2 H_z}{\partial z^2} = k^2 H_z \qquad (15.50)$$

With L the wavelength of the sinusoidal flux impinging normal to the tank wall, we assume a solution of (15.50) of the form

$$H_z(y,z) = H_o e^{\beta z} \sin\left(\frac{2\pi y}{L}\right) \tag{15.51}$$

Substituting into (15.50), we obtain for β

$$\beta^2 = k^2 + \left(\frac{2\pi}{L}\right)^2 = j\omega\mu\sigma + \left(\frac{2\pi}{L}\right)^2 = \sqrt{\left(\frac{2\pi}{L}\right)^4 + (\omega\mu\sigma)^2} \, e^{j\tan^{-1}\left(\frac{\omega\mu\sigma L^2}{4\pi^2}\right)}$$

$$\beta = \left[\left(\frac{2\pi}{L}\right)^4 + (\omega\mu\sigma)^2\right]^{\frac{1}{4}} e^{\frac{j}{2}\tan^{-1}\left(\frac{\omega\mu\sigma L^2}{4\pi^2}\right)} \tag{15.52}$$

Noting that the skin depth, δ, in the tank wall is given by (15.24), we can express β in terms of this skin depth:

$$\beta = \frac{\sqrt{2}}{\delta}\left[1 + 4\left(\frac{\pi\delta}{L}\right)^4\right]^{\frac{1}{4}} e^{\frac{j}{2}\tan^{-1}\left(\frac{L}{\sqrt{2}\pi\delta}\right)^2} \tag{15.53}$$

For magnetic steel tank walls, the skin depth is very small especially when compared with L, the length of the surface flux wave, so that $\pi\delta/L \approx 0$. Using this approximation,

$$\beta = \frac{\sqrt{2}}{\delta} e^{j\frac{\pi}{4}} = \frac{(1+j)}{\delta} \tag{15.54}$$

Thus, the exponential term in (15.51) becomes

$$e^{\beta z} = e^{\frac{z}{\delta}} e^{j\frac{z}{\delta}} = e^{\frac{z}{\delta}}\left[\cos\left(\frac{z}{\delta}\right) + j\sin\left(\frac{z}{\delta}\right)\right] \tag{15.55}$$

Noting that z is negative into the tank wall, according to Figure 15.23 where z is zero at the tank surface facing the windings and becomes negative into the tank wall, we see that the exponential term falls off rapidly with negative z.

We can obtain the solution for H_y by noting that, according to (15.47),

$$\nabla \cdot \mathbf{H} = 0 \quad \Rightarrow \quad \frac{\partial H_y}{\partial y} + \frac{\partial H_z}{\partial z} = 0 \quad \Rightarrow \quad \frac{\partial H_y}{\partial y} = -\beta H_o \sin\left(\frac{2\pi y}{L}\right) e^{\beta z} \tag{15.56}$$

Integrating (15.56), we get

$$H_y(y,z) = H_o\left(\frac{\beta L}{2\pi}\right) e^{\beta z} \cos\left(\frac{2\pi y}{L}\right) \tag{15.57}$$

Thus, we get for **H**

$$\mathbf{H} = H_o e^{\beta z}\left[\left(\frac{\beta L}{2\pi}\right)\cos\left(\frac{2\pi y}{L}\right)\mathbf{j} + \sin\left(\frac{2\pi y}{L}\right)\mathbf{k}\right] \tag{15.58}$$

We see from this that the y component of flux along the tank wall peaks at y = 0, whereas the z component of flux is zero at this point. This is what we expect from the finite element figures, Figures 15.21 and 15.22.

We can obtain the current density $\mathbf{J}$ by referring to (15.47):

$$\mathbf{J} = \nabla \times \mathbf{H} = \left(\frac{\partial H_z}{\partial y} - \frac{\partial H_y}{\partial z} \right) \mathbf{i}$$

$$= H_o e^{\beta z} \left[\left(\frac{2\pi}{L} \right) \cos \left(\frac{2\pi y}{L} \right) - \beta^2 \left(\frac{L}{2\pi} \right) \cos \left(\frac{2\pi y}{L} \right) \right] \mathbf{i}$$

$$= H_o e^{\beta z} \left(\frac{2\pi}{L} \right) \cos \left(\frac{2\pi y}{L} \right) \left[1 - \left(\frac{\beta L}{2\pi} \right)^2 \right] \mathbf{i} \tag{15.59}$$

Using the expression for β given in (15.52) and the definition of skin depth, we obtain for $\mathbf{J}$

$$\mathbf{J} = -j H_o e^{\beta z} \left(\frac{L}{\pi \delta^2} \right) \cos \left(\frac{2\pi y}{L} \right) \mathbf{i} \tag{15.60}$$

The current density is directed along the unit vector $\mathbf{i}$, that is, in the x direction.

From this, we can get the loss density/unit volume, P_{vol}:

$$P_{vol} = \frac{\mathbf{J} \cdot \mathbf{J}^*}{2\sigma} = \frac{1}{2\sigma} H_o^2 e^{(\beta + \beta^*)z} \left(\frac{L}{\pi \delta^2} \right)^2 \cos^2 \left(\frac{2\pi y}{L} \right)$$

$$= \frac{1}{2\sigma} H_o^2 e^{2z \operatorname{Re}\beta} \left(\frac{L}{\pi \delta^2} \right)^2 \cos^2 \left(\frac{2\pi y}{L} \right) \tag{15.61}$$

Note that the z exponent is now real, indicating that the loss density drops off very rapidly with z with no oscillations. It is useful to integrate the z dependence out so that the loss density would represent an effective surface loss density, P_{surf}. Because of the small skin depth and rapid exponential falloff, the z integral can be taken from $z = -\infty$ to $z = 0$. Thus, we get for P_{surf}

$$P_{surf} = \frac{1}{\sigma \operatorname{Re}\beta} H_o^2 \left(\frac{L}{2\pi \delta^2} \right)^2 \cos^2 \left(\frac{2\pi y}{L} \right) \tag{15.62}$$

We see that this loss density peaks at y = 0, that is, at the winding center.

To get the total losses per unit length in the x direction, we need to integrate (15.62) over the cosine wave of flux from y = −L/2 to L/2. Doing this, the $\cos^2$ integral equals L/2 so we obtain

$$\frac{\text{Loss}}{\text{Unit} \times \text{Width}} = H_o^2 \left(\frac{L}{2\sigma \operatorname{Re}\beta} \right) \left(\frac{L}{2\pi \delta^2} \right)^2 \tag{15.63}$$

Using the approximation for β given in (15.54), which should be very accurate in this context, this last formula can be simplified to

$$\frac{\text{Loss}}{\text{Unit} \times \text{Width}} = H_o^2 \left(\frac{\pi}{\sigma} \right) \left(\frac{L}{2\pi \delta} \right)^3 = B_o^2 \left(\frac{\pi}{\mu^2 \sigma} \right) \left(\frac{L}{2\pi \delta} \right)^3 \tag{15.64}$$

where the relationship between H and B has been used. Noting the definition of skin depth, we see that these losses increase with frequency to the 3/2 power, increase with conductivity to the ½ power (inversely with the resistivity to the ½ power), and inversely with the permeability to the ½ power. It has the usual dependence on the square of the amplitude of the B-field but a surprising dependence on the cube of the surface B-field wavelength, which is a little larger than the winding height.

As a numerical example, consider a magnetic steel tank wall with $\mu_r = 200$, $\mu = \mu_r 4\pi \times 10^{-7}$ H/m, $\sigma = 4 \times 10^6$ $(\Omega\,\text{m})^{-1}$, $\omega = 2\pi f = 2\pi(60\,\text{Hz})$, L = 2 m, and $B_o = 0.05$ T. For these parameters, we get $\delta = 2.3 \times 10^{-3}$ m. We find the loss/meter in the x direction = 82.4 kW/m. This should be multiplied by the effective length these losses are present in the x direction, which should be ≈ the diameter of the outer winding in meters, to get the total loss/leg at the tank wall. For a 3-phase unit, this loss should be multiplied by 3 to account for the losses on one tank side, although this factor could be changed by 3-phase effects. For the other tank side and the tank ends, the B-field amplitude and wavelength need to be calculated for them and the same formula applied. Some additional effort must be applied to estimate the wavelength and amplitude of the approximate B-field wave at the tank surface, based on the Rabins' method of finding the B-field or on a finite element calculation. To reduce these losses, tank shunts are often used.

15.4 Tank and Shield Losses due to Nearby Busbars

When busbars carrying relatively high current pass close to the tank wall, their magnetic field induces eddy currents in the wall, creating losses. The busbars are usually parallel to the tank wall over a certain length. Because of the magnetic field direction, laminated magnetic shunts positioned near the busbar are not as effective at reducing these losses as are metallic shields made of aluminum or copper. Since busbars are usually present from all three phases, the question arises as to how the grouping of 2 or 3 busbars from different phases would affect the losses. Intuitively, we expect a reduction in the losses due to some cancellation in the magnetic field from the different phases.

15.4.1 Losses Obtained with 2D Finite Element Study

These loss issues can be studied by means of a 2D finite element program if we assume the busbars are infinitely long in the direction along their length. This geometry allows one to calculate losses per unit length that can be multiplied by the total busbar length to get a reasonable approximation to the total loss. This can be made into a parametric study by varying the distance of the busbars from the tank wall. Other parameters such as busbar dimensions could be varied, but this would greatly complicate the study. We chose rather standard sized busbars (76.2 mm × 12.7 mm) with the long side perpendicular to the tank wall and separation distances of 82.6 mm when a grouping of busbars from different phases was studied. Of course, a particular geometry can always be studied if desired. Because eddy current losses are proportional to the square of the current when linear magnetic materials are involved, it is only necessary to calculate the losses at one current. These losses were studied with and without the presence of an aluminum shield, which was 12.7 mm thick and 230–300 mm wide placed flat on the tank wall. The losses in the busbars themselves were not considered in the study. We quote some of the qualitative results of this study here as an example of the useful insight that can be gained by such finite element analyses.

For the case of a single busbar, the losses in the tank wall without a shield drop off with the distance from the busbar to the tank wall, d, while the losses with shield are relatively constant with this distance. The losses with the shield include any losses in the tank as well. We found that shielding reduces the losses by about a factor of 5 at close distances (<150 mm) and a factor of 2–3 at further distances compared with the unshielded losses.

For the two-busbar case, we found that the losses without shield can be cut nearly in half by pairing two phases compared with leaving them separate. We also found that shielding is very effective in reducing these losses.

For a group of 3 busbars, the losses are considerably reduced relative to the single- or double-busbar case. In this case, shielding does not provide much improvement. This is because the 3-phase currents sum to zero at any instant of time, producing little net magnetic field at distances large relative to the conductor spacings.

It thus appears that loss reduction from busbars near bare tank walls can be achieved by pairing two or three phases together, the latter being preferable. Shielding is very effective in reducing the losses associated with 1 or 2 busbars from different phases but not for 3 busbars where the losses are small anyway.

15.4.2 Losses Obtained Analytically*

To further facilitate the calculation of the tank or shield losses due to the busbars, it is useful to have an analytic formula. Such a formula can be obtained if the tank or shield is allowed to be infinitely wide. Naturally, the losses would concentrate near the vicinity of the busbars so that the infinite approximation should have little influence on the real physical situation but makes the calculation feasible.

Previous work in this area focused on plate losses due to a current sheet [Jai70] and on plate losses due to a delta function current [Kul99]. These results can be extended by considering an arbitrary collection of busbars approximated by a collection of delta function currents. The busbars can carry currents of differing magnitudes and phases, and these properties are inherited by their associated delta function current approximations. Different busbar cross sections can be approximated by strategically positioning the delta function current filaments. For example, a rectangular shape can be approximated by a central current filament and one at each corner of the rectangle for a total of five filaments, each one carrying 1/5 of the total current. Other shapes, such as circular, can be approximated by a central filament and others along the circumference. Additional filaments can be positioned for greater accuracy. It should be noted that we consider the currents in the busbars to be uniformly distributed. Thus, we do not take into account the current redistribution in solid busbars, which would result from the a.c. currents within the busbar or any of its neighbors.

15.4.2.1 Current Sheet

[Jai70] solved the problem of a current sheet positioned a height h above a conducting plate of thickness d and carrying a current/unit width of

$$\mathbf{J} = J_0 \cos(ax) e^{j\omega t} \mathbf{k} \tag{15.65}$$

as shown in Figure 15.24. The current is directed along the positive z-axis as indicated by the unit vector $\mathbf{k}$ in (15.65). ω is the angular frequency, $\omega = 2\pi f$, where f is the frequency in Hz,

---

* This section is based on [Del03], *IEEE Trans. Magn.*, 39(1), January 2003, 549–552. © 2003 IEEE.

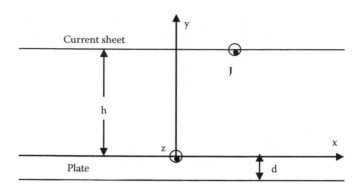

FIGURE 15.24
Current sheet geometry. J indicates the direction of the current in the sheet, which is spread out over the entire sheet. (From Del Vecchio, R.M., *IEEE Trans. Magn.*, 39(1), January 2003, 549–552. © 2003 IEEE.)

and j is the imaginary unit. a is a parameter that plays the role of an inverse wavelength. J_0 has units of current/unit sheet width or amps/m in SI units that are used here.

The geometry in Figure 15.24 extends to infinity in all directions so that the plate and sheet are assumed to be infinitely long and wide. The plate and sheet are surrounded by air or any medium of relative permeability 1 and 0 conductivity. Because of uniformity in the z direction, the solution only depends on the x and y coordinates. The solution will be given in SI units. It must be found throughout all space in order to satisfy all the boundary conditions.

Here, we are only interested in the solution in the plate, in particular the electric field E. Since it is z-directed like the current, we will drop the vector notation. The E-field in the plate due to the current sheet is given by

$$E_{sheet}\left(-d \le y \le 0\right) = -j\omega\mu_0 J_0 e^{j\omega t} e^{-ah} \cos\left(ax\right)$$

$$\cdot \left[\frac{\left(a+\dfrac{\gamma}{\mu_r}\right)e^{\gamma(y+d)} - \left(a-\dfrac{\gamma}{\mu_r}\right)e^{-\gamma(y+d)}}{\left(a+\dfrac{\gamma}{\mu_r}\right)^2 e^{\gamma d} - \left(a-\dfrac{\gamma}{\mu_r}\right)^2 e^{-\gamma d}}\right] \qquad (15.66)$$

$$\text{where } \gamma = \sqrt{a^2 + \frac{2j}{\delta^2}} \quad \text{and} \quad \delta = \sqrt{\frac{2}{\omega\mu_0\mu_r\sigma}}$$

δ is the skin depth and j is the imaginary unit. In SI units, $\mu_0 = 4\pi \times 10^{-7}$ is the vacuum permeability, μ_r the relative permeability, and σ the conductivity of the plate. The plate thickness is d.

15.4.2.2 Delta Function Current

The dependence of E on x is that of a cosine function. This suggests that it could be part of a Fourier expansion of a general function. In particular, the delta function of x, centered at the origin, has the expansion

$$\delta\left(0\right) = \frac{1}{\pi}\int_0^\infty \cos\left(ax\right)da \qquad (15.67)$$

Thus, a current filament in the z direction carrying a current I_0 can be written as

$$\mathbf{I} = \frac{I_0}{\pi} \int_0^\infty \cos(ax)da \cdot e^{j\omega t} \mathbf{k} \tag{15.68}$$

Basically, to get I, we are summing over an infinite number of current sheets, each one contributing a current/unit width of

$$\frac{I_0}{\pi} \cos(ax)da \cdot e^{j\omega t} \tag{15.69}$$

Since Maxwell's equations, including the constitutive relations as assumed here, are linear and since the E-fields are unidirectional, the solution in the plate for the current in (15.68) is given in terms of (15.66) by

$$E_{\delta\text{-function}}\left(-d \le y \le 0\right) = \frac{I_0}{\pi J_0} \int_0^\infty E_{\text{sheet}}\left(-d \le y \le 0\right)da \tag{15.70}$$

Now consider a δ function offset from the origin by an amount b. This is given by the expansion

$$\delta(x - b) = \frac{1}{\pi} \int_0^\infty \cos\left[a(x - b)\right]da \tag{15.71}$$

Had we assumed the sheet current in (15.65) depended on cos[a(x − b)], the solution would have gone through unchanged except that cos[a(x − b)] would replace cos(ax) in (15.66). Therefore, the corresponding δ function solution would be given by (15.70), again with the cosine function replacement. More generally, if the current phase were different from 0 so that the time dependence was given by exp[j(ωt + ϕ)], the same solution would have resulted with exp[j(ωt + ϕ)] replacing exp(jωt). The two types of delta function current filament are shown in Figure 15.25.

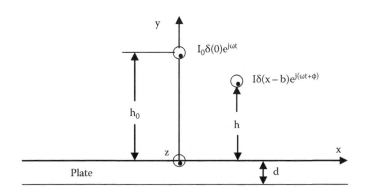

FIGURE 15.25
Delta function currents (filaments). (From Del Vecchio, R.M., *IEEE Trans. Magn.*, 39(1), January 2003, 549–552. © 2003 IEEE.)

The electric field in the plate from a general filament carrying current I, offset in space and time phase, is therefore given by

$$E_\delta = -j\frac{\omega\mu_0 I}{\pi}e^{j(\omega t+\phi)}\int_0^\infty dae^{-ah}\cos\left[a(x-b)\right]$$

$$\cdot\left[\frac{\left(a+\dfrac{\gamma}{\mu_r}\right)e^{\gamma(y+d)}-\left(a-\dfrac{\gamma}{\mu_r}\right)e^{-\gamma(y+d)}}{\left(a+\dfrac{\gamma}{\mu_r}\right)^2 e^{\gamma d}-\left(a-\dfrac{\gamma}{\mu_r}\right)^2 e^{-\gamma d}}\right]\qquad(15.72)$$

15.4.2.3 Collection of Delta Function Currents

Since we can approximate any shaped busbar or busbar collection with a number of current filaments, we now consider the electric field in the plate due to such a collection. Again, because Maxwell's equations and the constitutive relations are linear, the net E-field in the plate, E_T, is simply the vector sum of all the individual E-fields. Since these vectors are collinear, we can take the scalar sum, $E_T = \sum_i E_i$, where E_i are the individual fields due to delta function currents and have the form of (15.72), with the individual current filaments and their parameters labeled with i. Thus, we have h_i = the height of the filament above the plate, b_i its offset from the origin, I_i its current, and ϕ_i its phase displacement in time.

The eddy current loss/unit length in the z direction in the plate for such a collection of current filaments is given by

$$\text{Loss/unit length} = \sigma\int_{-d}^0 dy\int_{-\infty}^\infty dx E_T E_T^*\qquad(15.73)$$

where * denotes complex conjugation. The integral is over the cross-sectional area of the plate. We are assuming that the current and E-field magnitudes are rms quantities so that a factor of ½ does not appear in (15.73). We note that

$$E_T E_T^* = \left(\sum_i E_i\right)\left(\sum_k E_k^*\right) = \sum_i\sum_k E_i E_k^*\qquad(15.74)$$

A typical term in the integrand is $E_i E_k^*$. The time exponential term in this product becomes $\exp[j(\phi_i-\phi_k)]$. However, the $E_k E_i^*$ term, which also occurs in the sum, will have as a time exponential term $\exp[j(\phi_k-\phi_i)]$. Both of these terms are multiplied by the same factor and can therefore be added. Thus, in their sum, only the cosine part of this complex exponential survives since the cos function is an even function and the sine function an odd function of its argument. Thus,

$$e^{j(\phi_i-\phi_k)}+e^{j(\phi_k-\phi_i)} = \cos\left(\phi_i-\phi_k\right)+j\sin\left(\phi_i-\phi_k\right)$$
$$+\cos\left(\phi_k-\phi_i\right)+j\sin\left(\phi_k-\phi_i\right) = 2\cos\left(\phi_i-\phi_k\right)\qquad(15.75)$$

Thus, a $\cos(\phi_i-\phi_k)$ term should multiply each term in the $E_i E_k^*$ product so that when the double sum is performed, a factor of 2 will occur as required by (15.75).

The remaining factor involves integrals over x, y, a, and a'. Both a and a' integrals are needed to express the two different E's. An example of an E, expressed in terms of an a integral, is given in (15.72). An E expressed in terms of an a' integral would have a replaced by a'. The x integral in (15.73) involves only the cos functions:

$$\int_{-\infty}^{\infty} \cos\big[a(x-b_i)\big]\cos\big[a'(x-b_k)\big]dx \tag{15.76}$$

Letting $\xi = x - b_i$, this can be rewritten as

$$\int_{-\infty}^{\infty} \cos(a\xi)\cos\big[a'(\xi+b_i-b_k)\big]d\xi \tag{15.77}$$

Expanding the second cosine term, we get for the product in (15.77)

$$\cos(a\xi)\cos\big[a'(\xi+b_i-b_k)\big] = \cos(a\xi)\times\big\{\cos(a'\xi)\cos\big[a'(b_i-b_k)\big]-\sin(a'\xi)\sin\big[a'(b_i-b_k)\big]\big\}$$

$$= \cos(a\xi)\cos(a'\xi)\cos\big[a'(b_i-b_k)\big]-\cos(a\xi)\sin(a'\xi)\sin\big[a'(b_i-b_k)\big] \tag{15.78}$$

Making use of

$$\int_{-\infty}^{\infty} \cos(a\xi)\cos(a'\xi)d\xi = \pi\delta(a-a')$$

$$\int_{-\infty}^{\infty} \cos(a\xi)\sin(a'\xi)d\xi = 0 \tag{15.79}$$

we obtain for the integral in (15.76)

$$\int_{-\infty}^{\infty} \cos\big[a(x-b_i)\big]\cos\big[a'(x-b_k)\big]dx = \pi\delta(a-a')\cos\big[a'(b_i-b_k)\big] \tag{15.80}$$

Thus, the a' integral in (15.73), when (15.72) is used for the individual E_δ terms with their proper indices, can be done immediately, removing the delta function, and we are left with

$$\text{Loss/unit length} = \frac{\sigma(\omega\mu_0)^2}{\pi}\sum_i\sum_k\Big\{I_iI_k\cos(\phi_i-\phi_k)\int_0^a dae^{-a(h_i+h_k)}\cos\big[a(b_i-b_k)\big]$$

$$\times\int_{-d}^0 dy\left[\frac{\left(a+\dfrac{\gamma}{\mu_r}\right)e^{\gamma(y+d)}-\left(a-\dfrac{\gamma}{\mu_r}\right)e^{-\gamma(y+d)}}{\left(a+\dfrac{\gamma}{\mu_r}\right)^2 e^{\gamma d}-\left(a-\dfrac{\gamma}{\mu_r}\right)^2 e^{-\gamma d}}\right]$$

$$\times\left[\frac{\left(a+\dfrac{\gamma^*}{\mu_r}\right)e^{\gamma^*(y+d)}-\left(a-\dfrac{\gamma^*}{\mu_r}\right)e^{-\gamma^*(y+d)}}{\left(a+\dfrac{\gamma^*}{\mu_r}\right)^2 e^{\gamma^*d}-\left(a-\dfrac{\gamma^*}{\mu_r}\right)^2 e^{-\gamma^*d}}\right]\Big\} \tag{15.81}$$

Making use of

$$\gamma\gamma^* = \eta^2$$
$$\gamma + \gamma^* = 2\eta\cos\theta \tag{15.82}$$
$$\gamma - \gamma^* = j2\eta\sin\theta$$

where

$$\eta = \left[a^4 + \frac{4}{\delta^4}\right]^{1/4}$$
$$\theta = \frac{1}{2}\tan^{-1}\left(\frac{2}{a^2\delta^2}\right) \tag{15.83}$$

we get for (15.81), after performing the y integral,

$$\text{Loss/unit length} = \frac{\sigma(\omega\mu_0)^2}{\pi}\sum_i\sum_k\left\{I_iI_k\cos(\phi_i - \phi_k)\right.$$
$$\left.\times\int_0^a da e^{-a(h_i+h_k)}\cos\left[a(b_i - b_k)\right]\frac{\text{Numer}}{\text{Denom}}\right\} \tag{15.84}$$

where

$$\text{Numer} = \left(a^2 + \frac{2a\eta\cos\theta}{\mu_r} + \frac{\eta^2}{\mu_r^2}\right)\left(\frac{e^{2\eta d\cos\theta} - 1}{2\eta\cos\theta}\right)$$
$$+ \left(a^2 - \frac{2a\eta\cos\theta}{\mu_r} + \frac{\eta^2}{\mu_r^2}\right)\left(\frac{1 - e^{-2\eta d\cos\theta}}{2\eta\cos\theta}\right)$$
$$- \left(a^2 - \frac{\eta^2}{\mu_r^2}\right)\frac{\sin(2\eta d\sin\theta)}{\eta\sin\theta} - \frac{2a}{\mu_r}\left[\cos(2\eta d\sin\theta) - 1\right] \tag{15.85}$$

and

$$\text{Denom} = \left(a^2 + \frac{2a\eta\cos\theta}{\mu_r} + \frac{\eta^2}{\mu_r^2}\right)^2 e^{2\eta d\cos\theta}$$
$$+ \left(a^2 - \frac{2a\eta\cos\theta}{\mu_r} + \frac{\eta^2}{\mu_r^2}\right)^2 e^{-2\eta d\cos\theta}$$
$$- 2\left[\left(a^2 - \frac{\eta^2}{\mu_r^2}\right)^2 - \left(\frac{2a\eta\sin\theta}{\mu_r}\right)^2\right]\cos(2\eta d\sin\theta)$$
$$+ \frac{8a\eta\sin\theta}{\mu_r}\left(a^2 - \frac{\eta^2}{\mu_r^2}\right)\sin(2\eta d\sin\theta) \tag{15.86}$$

The a integral is left to be performed numerically. The integral is well behaved at a = 0 and → 0 as a → ∞. Therefore, it need only be evaluated to large enough a so that the integrand becomes negligible. Also, since the i, k term in the double sum equals the k, i term for i ≠ k, we need to evaluate only one of these terms and multiply the result by 2.

15.4.2.4 Model Studies

The methods outlined here were used to obtain the plate losses due to several configurations of rectangular busbars. Each of the busbars was modeled with a current filament in the center and one at each corner. Thus, the filament current associated with a busbar is 1/5 the busbar current. In the cases studied here, all the busbars carried the same current but with different phases. We used different plate materials with characteristics given in Table 15.1.

The losses obtained from (15.84) through (15.86) were compared with plate losses obtained with a 2D finite element program [Ansoft]. We simulated the infinite geometry by choosing a plate width large compared with the busbar–plate separation and by using balloon boundary conditions.

The first configuration studied was that of a single busbar positioned as shown in Figure 15.26. The busbar dimensions here and in all the subsequent configurations studied were t = 12.7 mm and w = 76.2 mm. The plate thickness d = 12.7 mm and was kept the same in all the studies. The current details, separation distance H, and the loss comparisons are shown in Table 15.2. The agreement is very good.

The 3-phase busbar configuration studied is shown in Figure 15.27. An additional parameter, the separation distance between busbars, s, is shown in the figure. All the busbars are a distance H from the plate and currents from a balanced 3-phase system. For the parameters chosen, the analytic losses are compared with the finite element results in Table 15.3. Again, the agreement is very good.

TABLE 15.1

Plate Characteristics

Plate Material	Relative Permeability	Conductivity $(\Omega\,m)^{-1}$
Mag steel	200	4×10^6
Stainless	1	1.333×10^6
Aluminum	1	3.6×10^7
Copper	1	5×10^7

Source: Del Vecchio, R.M., *IEEE Trans. Magn.*, 39(1), January 2003, 549–552. © 2003 IEEE.

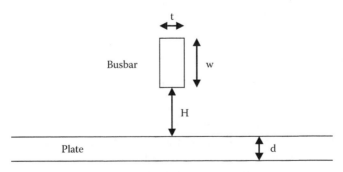

FIGURE 15.26
Single-busbar geometry. (From Del Vecchio, R.M., *IEEE Trans. Magn.*, 39(1), January 2003, 549–552. © 2003 IEEE.)

TABLE 15.2

Loss Comparisons for a Single Busbar

Plate Material	H (mm)	Analytic (W/m)	Finite Element (W/m)
Mag steel	127	1183	1197
Stainless	127	987	985
Aluminum	127	61.1	60.8
Copper	127	48.9	48.7
Mag steel	254	859	878
Stainless	254	664	670
Aluminum	254	34.3	34.6
Copper	254	27.4	27.7

Source: Del Vecchio, R.M., *IEEE Trans. Magn.*, 39(1), January 2003, 549–552. © 2003 IEEE.
The busbar current was 5000 amps rms at 60 Hz. d = 12.7 mm.

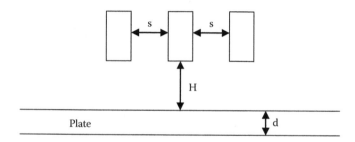

FIGURE 15.27
3-phase busbar geometry. (From Del Vecchio, R.M., *IEEE Trans. Magn.*, 39(1), January 2003, 549–552. © 2003 IEEE.)

TABLE 15.3

Loss Comparisons for a 3-Phase Busbar System

Plate Material	H (mm)	s (mm)	Analytic (W/m)	Finite Element (W/m)
Mag steel	127	101.6	214	213
Stainless	127	101.6	236	230
Aluminum	127	101.6	34.1	32.4
Copper	127	101.6	27.2	26.1

Source: Del Vecchio, R.M., *IEEE Trans. Magn.*, 39(1), January 2003, 549–552. © 2003 IEEE.
The currents in the busbars were 5000 amps rms at 60 Hz with phases of 0°, 120°, and 240°. d = 12.7 mm.

15.5 Tank Losses Associated with the Bushings

Current enters and leaves a transformer tank via the bushings. The bushings are designed to handle the voltage stresses associated with the voltage on the leads without breakdown as well as to dissipate heat due to the losses in the conductor that passes through the bushings. The conductor or lead, which must pass through the tank wall, creates a magnetic field, which can generate eddy currents and accompanying losses in the tank wall near the lead. These losses must be calculated and appropriate steps taken to reduce them if necessary.

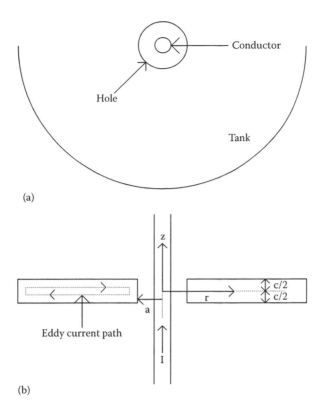

FIGURE 15.28
Idealized geometry and parameters used to calculate losses due to a lead passing through the tank: (a) top view and (b) side view.

We can obtain a reasonable estimate of the tank losses due to a lead penetrating the tank wall by resorting to an idealized geometry as shown in Figure 15.28. Thus, we assume an infinitely long circular cross section lead passing perpendicularly through the center of a circular hole in the tank wall. We can assume that the tank wall itself is a circle of large radius centered on the hole. Since the losses are expected to concentrate near the hole, the actual radial extent of the tank would not matter much. Also since most of the magnetic field, which generates eddy currents in the tank, comes from the portion of the lead near the tank, the infinite extent of the lead does not greatly affect the calculation.

The geometry in Figure 15.28 is axisymmetric. Thus, we need to work with Maxwell's equations in a cylindrical coordinate system. We assume the tank wall has permeability μ and conductivity σ. Combining Maxwell's equations with Ohm's law inside the tank wall and assuming that $\mathbf{H}$ varies harmonically in time as

$$\mathbf{H}(\mathbf{r},t) = \mathbf{H}(\mathbf{r})e^{j\omega t} \tag{15.87}$$

we get

$$\nabla^2 \mathbf{H} = j\omega\mu\sigma\mathbf{H} \tag{15.88}$$

where $\mathbf{H}$ in (15.88) is only a function of position since the time dependence has been factored out.

Because the problem is axisymmetric, **H** does not depend on φ, the azimuthal angle. We also assume that **H** has only a φ component. This is all that is needed to produce the expected eddy current pattern indicated in Figure 15.28b, where the eddy currents approach and leave the hole radially and the paths are completed along short sections in the z direction. The approximate solution we develop here will neglect these z-directed eddy currents, which should not contribute much to the total loss if the plate is thin enough. Thus, we have

$$\mathbf{H}(r) = H_\varphi(r,z)\mathbf{a}_\varphi \tag{15.89}$$

where $\mathbf{a}_\varphi$ is the unit vector in the azimuthal direction. Expressing (15.88) in cylindrical coordinates and using (15.89), we get

$$\frac{1}{r}\frac{\partial}{\partial r}\left(r\frac{\partial H_\varphi}{\partial r}\right) - \frac{H_\varphi}{r^2} + \frac{\partial^2 H_\varphi}{\partial z^2} = j\omega\mu\sigma H_\varphi \tag{15.90}$$

We need to solve this equation subject to the boundary conditions

$$\begin{aligned} H_\varphi &= \frac{I}{2\pi a} \quad \text{at } r = a \\ H_\varphi &= 0 \quad \text{at } r = \infty \\ H_\varphi &= \frac{I}{2\pi r} \quad \text{at } z = \pm\frac{c}{2} \end{aligned} \tag{15.91}$$

where
   a is the radius of the hole in the tank wall
   c is the tank wall thickness
   I is the current in the lead

Once a solution is found, the current density **J** is given by

$$\mathbf{J} = \nabla \times \mathbf{H} = -\frac{\partial H_\varphi}{\partial z}\mathbf{r} + \frac{1}{r}\frac{\partial}{\partial r}(rH_\varphi)\mathbf{k} \tag{15.92}$$

where **r** and **k** are unit vectors in the r and z directions.
To solve (15.90), we use a separation of variables technique and write

$$H_\varphi(r,z) = R(r)Z(z) \tag{15.93}$$

Substituting into (15.90) and dividing by RZ, we obtain

$$\frac{1}{R}\left[\frac{1}{r}\frac{\partial}{\partial r}\left(r\frac{\partial R}{\partial r}\right) - \frac{R}{r^2}\right] + \frac{1}{Z}\frac{\partial^2 Z}{\partial z^2} = j\omega\mu\sigma \tag{15.94}$$

Thus, we have two terms, which are separately a function of r and z and whose sum is a constant. Hence, each term can be separately equated to a constant so long as their sum is $j\omega\mu\sigma$. We choose

$$\frac{1}{Z}\frac{\partial^2 Z}{\partial z^2} = j\omega\mu\sigma, \quad \frac{1}{R}\left[\frac{1}{r}\frac{\partial}{\partial r}\left(r\frac{\partial R}{\partial r}\right) - \frac{R}{r^2}\right] = 0 \tag{15.95}$$

Letting $k^2 = j\omega\mu\sigma$ so that

$$k = (1+j)q, \quad q = \sqrt{\frac{\omega\mu\sigma}{2}} \tag{15.96}$$

we can solve the first equation in (15.95) up to an overall multiplicative constant, Z_o, by

$$Z(z) = Z_o \cosh(kz) = Z_o\left[\cosh(qz)\cos(qz) + j\sinh(qz)\sin(qz)\right] \tag{15.97}$$

This equation takes into account the symmetry about the $z = 0$ plane. The second equation in (15.95) is solved up to an overall multiplicative constant, R_o, by

$$R(r) = R_o \frac{1}{r} \tag{15.98}$$

Thus, the complete solution to (15.94), using the boundary conditions (15.91), is

$$H_\varphi = \frac{I}{2\pi r}\left[\frac{\cosh(qz)\cos(qz) + j\sinh(qz)\sin(qz)}{\cosh(qc/2)\cos(qc/2) + j\sinh(qc/2)\sin(qc/2)}\right] \tag{15.99}$$

This does not exactly satisfy the first boundary condition in (15.91) except in the limit of small qc/2.

Solving for the eddy currents, Equation 15.92, we see that the z-directed currents are zero when we substitute the solution (15.99). We expect that the losses contributed by these short paths will be small. Solving for the r-directed eddy currents, we obtain

$$J_r = \frac{Iq}{2\pi r} \cdot \left\{\frac{\sinh(qz)\cos(qz) - \cosh(qz)\sin(qz) + j\left[\cosh(qz)\sin(qz) + \sinh(qz)\cos(qz)\right]}{\cosh(qc/2)\cos(qc/2) + j\sinh(qc/2)\sin(qc/2)}\right\}$$

$$\tag{15.100}$$

and the loss density as a function of position, assuming I is an rms current, is

$$\frac{|J_r|^2}{\sigma} = \frac{2}{\sigma}\left(\frac{Iq}{2\pi r}\right)^2\left[\frac{\cosh(2qz) - \cos(2qz)}{\cosh(qc) + \cos(qc)}\right] \tag{15.101}$$

This drops off with radius as $1/r^2$ and so is highest near the opening. The total loss is given by

$$\text{Loss}_{\text{bush}} = 2\pi \int_a^b \int_{-c/2}^{c/2} \frac{|J_r|^2}{\sigma} \, r\,dr\,dz \tag{15.102}$$

The upper limit of the r integration, b, is chosen to be a large enough radius that the tank area of interest is covered. The result will not be too sensitive to the exact value chosen. Performing the integrations, we obtain

$$\text{Loss}_{\text{bush}} = \frac{I^2 q}{\pi\sigma} \ln\left(\frac{b}{a}\right)\left[\frac{\sinh(qc) - \sin(qc)}{\cosh(qc) + \cos(qc)}\right] \tag{15.103}$$

For a magnetic steel tank wall, using $\mu_r = 200$, $\sigma = 4 \times 10^6$ $(\Omega\,m)^{-1}$, $f = 60$ Hz, and $c = 9.52 \times 10^{-3}$ m, we get qc = 4.15. Thus, the small qc approximation to (15.103) cannot be used. However, for a stainless steel tank wall of the same thickness and frequency, using $\mu_r = 1$, $\mu = 4\pi \times 10^{-7}$ H/m, and $\sigma = 1.33 \times 10^6$ $(\Omega\,m)^{-1}$, we have qc = 0.17 so the low qc limit may be used. This low-frequency limit is

$$\text{Loss}_{\text{bush}} \xrightarrow[\text{small qc}]{} \left(\frac{\pi}{6}\right)\frac{I^2 f^2 \mu^2 c^3}{\rho}\ln\left(\frac{b}{a}\right) \tag{15.104}$$

where $\omega = 2\pi f$ and $\sigma = 1/\rho$ have been substituted. Applying these last two equations to the case where I = 1000 amps rms (for normalization purposes), the hole radius a = 165 mm, the outer radius b = 910 mm, and using the parameters given earlier for the two types of steel, we get $\text{Loss}_{\text{bush}}$(mag steel) = 61.9 W/$(kA_{\text{rms}})^2$ and $\text{Loss}_{\text{bush}}$(stainless) = 5.87×10^{-3} W/$(kA_{\text{rms}})^2$. These losses are associated with each bushing. Applying this to a situation where the lead is carrying 10 kA_{rms}, the loss in a magnetic steel tank wall would be 6190 W per bushing and in a stainless steel tank wall 0.587 W per bushing. These need to be multiplied by the number of bushings carrying the given current to get the total tank loss associated with the bushings. Thus, when high currents flow in the leads, it might be worthwhile to insert a stainless steel section of tank around the bushings, especially since these losses are concentrated in the part of the tank wall near the opening.

Another method of reducing these losses is to use a stainless steel insert only around part of the opening. This will reduce the effective permeability as seen by the magnetic field, which travels in concentric circles about the center of the opening. We can estimate the effective permeability by using a magnetic circuit approach. This could then be substituted into the previous formulas to calculate the loss. A finite element 3D calculation, however, shows that this underestimates the loss. Apparently, the flux is not confined to the tank wall as it essentially is when the material is all magnetic steel, but can travel around the stainless insert in the surrounding air or oil.

15.5.1 Comparison with a 3D Finite Element Calculation

In order to check the analytic result given earlier, a 3D finite element analysis was performed, using a large number of elements within the tank wall so that the eddy current distribution could be captured and used to obtain the losses. The losses were also calculated using the impedance boundary method, which requires far fewer elements. The geometry, shown in Figure 15.29, consists of a circular disk with a hole in it. The outer diameter of the disk is 305 mm and the diameter of the hole is 132 mm. The disk is 9.5 mm thick and made of low-carbon magnetic steel with a relative permeability of 200 and a conductivity of 4×10^6 $(\Omega\,m)^{-1}$. The lead is 1270 mm long and has a diameter of 25.4 mm. The rms current in the lead is 5800 amps at 60 Hz.

Figure 15.30 shows the surface loss densities on the plate obtained with the surface impedance boundary method. The figure shows that the highest losses are near the center hole as expected.

The total loss, using the impedance boundary method, was 1095 W. When the disk was internally subdivided into a large number of 3D finite elements and the losses calculated by integrating the eddy current losses in the standard manner, a loss of 998 W was obtained. The two methods are in near agreement on this problem. Using the analytic formula developed earlier for the losses, the losses obtained were 1019 W. This agrees

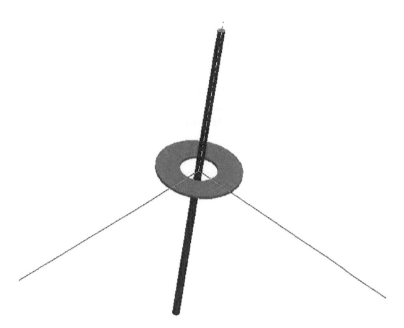

FIGURE 15.29
Geometry of lead passing through a plate with a hole in it to simulate a lead passing through a tank top.

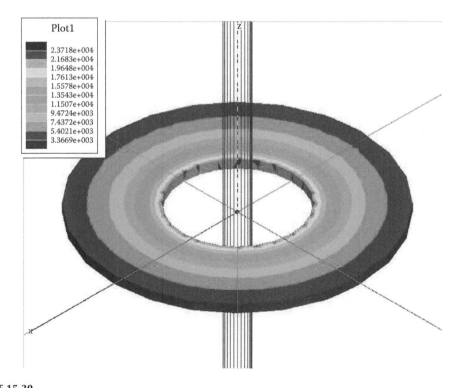

FIGURE 15.30
Surface losses on the magnetic steel disk. The scale is in W/m². Peak currents were used in generating this plot, so the loss densities include the factor of 2, which is necessary, if rms currents are used.

with both types of finite element analyses. We should note, however, that all the methods used assume that the material is linear. This could require some correction for saturation effects as discussed in the next chapter.

It is worth noting here that an exact analytic solution of (15.88) for the lead passing through a hole in the tank top has been found, without assuming that the currents are all azimuthally directed [Max15]. In addition, the temperature distribution using the loss distribution in (15.101) or the exact loss distribution has also been found [Max16].

16

Stray Losses from 3D Finite Element Analysis

16.1 Introduction

Modern methods of analysis, such as finite element or boundary element methods, have facilitated the calculation of stray flux losses in complex geometries. These methods are not yet routine in design because they require a fair amount of geometric input for each new geometry. However, some of these programs support the use of macros that can quickly generate a standard geometry based on key parameters, which can be changed for different designs. These methods can provide useful insights in cases where analytic methods are not available or are very crude.

Typical power transformers have 3-phase cores so that the stray flux produced arises from the currents in all 3 phases. Analytic formulas are generally based on single-phase assumptions. Although these formulas are based on physical principles, they generally require modifications obtained by comparison with test data. The necessity for such modifications can arise from the idealized geometries assumed in the calculation or from the ignored effects of 3-phase flux. 3D finite element methods are now available for modeling all 3 phases, including the effects of 3-phase flux. Even here, some approximations must be made to make the calculations tractable. We will explore this approach in this chapter and compare it with other methods and test results.

16.2 Stray Losses on Tank Walls and Clamps*

Stray loss studies were performed for a 3-phase, 3-legged core. Because all 3 legs can be modeled and 3-phase currents used, one would expect greater accuracy for these losses as compared with analytical or 2D finite element analyses. Because anisotropic conductivity and permeability can be modeled in 3D, this presents the opportunity to study the relative effectiveness of shunts made of laminated core steel, which can be stacked flat or on edge. We also compare the 3D losses to losses obtained by analytic methods and test results for transformers of various ratings and types.

Transformer structural elements such as clamps and tank, where much of the stray loss occurs, are generally made of magnetic steel, which has a small skin depth relative to the thickness of the material. In order to capture the eddy currents, a large number of finite elements would be required. In 2D this is not a problem, but in 3D when modeling a large

---

* This section is based largely on [Ahu06].

item such as a clamp or tank, the number of elements and solution times would be prohibitively large. Therefore, a method called the impedance boundary method has been developed to address this problem [Hol92], [Lam66]. This method only requires that the surface of the material be covered with finite elements. These are special elements, called surface impedance elements. The properties of the material such as its relative permeability and conductivity are used in specifying these elements. Then after running the 3D problem, the solution is processed to obtain the loss in Watts, which is given by the formula

$$\text{Loss} = \sqrt{\frac{\omega \mu_r \mu_o}{8\sigma}} \int\limits_{\text{Surface}} \mathbf{H}_t \cdot \mathbf{H}_t^* \, dS \tag{16.1}$$

where
    ω is the angular frequency = $2\pi f$, where f is the frequency in Hz
    σ is the conductivity in $(\Omega\,m)^{-1}$
    μ_r is the relative permeability
    μ_o is the permeability of free space, which is $4\pi \times 10^{-7}$ H/m in SI units
    $\mathbf{H}_t$ is the tangential component of the magnetic field vector at the surface in amps/m and
        * denotes complex conjugate

Note that if the currents in the windings are rms currents, the loss in the last equation must be multiplied by 2. We used Ansoft's software, Maxwell 3D, to do the finite element calculations and obtain the losses [Ansoft].

Of course, if materials having a relatively large skin depth such as aluminum, stainless steel, or copper are present, they are modeled with interior elements and the losses are calculated via the eddy currents generated inside them.

In order to get good accuracy and reproducibility, a reasonably large number of elements must be used, especially concentrated near the surfaces of interest. We found that typically ~100,000 3D elements were needed for most problems.

16.2.1 Shunts on the Clamps

In order to lower the stray losses in the clamps with a flat top, it has been the practice to place shunts on the top surface. These shunts extend continuously along the entire clamp, straddling all three phases. The idea is that the flux from the 3 balanced phases will cancel within the shunt. Thus, much of the leakage flux will be trapped within the shunt and therefore be unavailable to induce losses in the conducting structural material.

An autotransformer with an FA or top rating of 105 MVA was modeled. The clamps were L-shaped and therefore had a large horizontal surface area, which was fairly close to the windings and their leakage flux. End clamps with a similar shape were also modeled. For the end clamps, the shunts cover only one phase so that the leakage flux must return to the core. A loss analysis with and without clamp shunts was performed. The 3-phase model, excluding the tank, is shown in Figure 16.1.

By giving the shunts the material properties of air or vacuum, the losses can be obtained in the clamps without the shunts. These are shown on the high-voltage (HV) side bottom clamp in Figure 16.2.

The shunts were 38 mm thick and made of high-permeability core laminations. The shunt material is assumed to be linear and therefore does not saturate. If an isolated spot in the material saturates, then it can be assumed that the flux density will redistribute itself

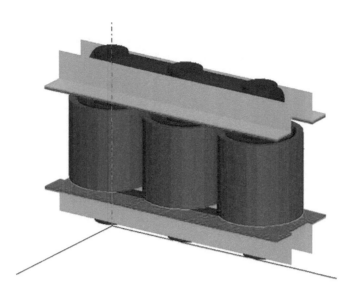

FIGURE 16.1

Transformer geometry showing the 3 phases and the clamps. End clamps were modeled only on the bottom clamp structure for simplicity. There are 2 windings per leg, shown in dark green and red. The clamp shunts are shown in coral and the clamps in light blue green. The core, partially shown, is in blue.

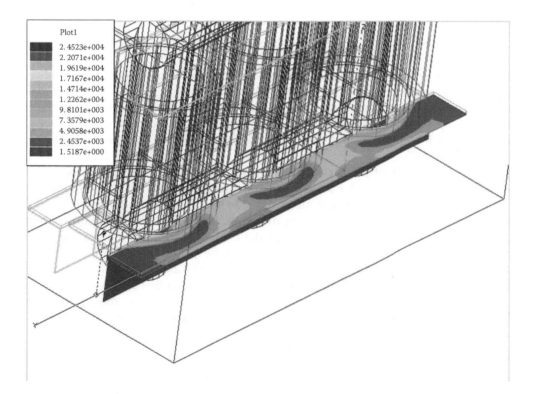

FIGURE 16.2

Surface loss densities on the HV side bottom clamp without shunts. Note that the highest values are at the two end phases. There is more flux cancellation at the center phase than at the outer phases. The units are W/m².

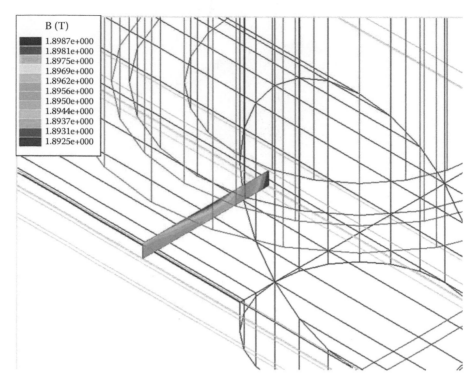

FIGURE 16.3
Flux densities in the bottom LV side clamp shunt at the center of the center leg. The units are Teslas.

within the material. However, if an entire cross section of the material saturates, then this is an indication that the flux will spill out of the material and create losses in the material that the shunts are supposed to protect. The flux densities in the shunts, having their core steel material properties, were obtained by looking at the shunt cross sections at various positions along the shunt. A contour plot of the flux density at the center leg position is shown in Figure 16.3.

The flux density is about 1.9 T at the leg centers, which is below the 2.1 T saturation limit. However, it averages about 2.1 T between legs, which is just at saturation. What is interesting is that the flux is fairly uniform along the side shunt. A vector diagram would show that it is directed along the shunts' long direction, indicating that it remains within the shunt and does not leak out to any great extent. The surface loss densities on the side clamps are about 100 times lower than the loss densities without the shunts.

The flux density in the end clamp shunts is only about 0.2 T mainly because this flux is forced to return to the core through an air gap. They are therefore less effective in reducing the end clamp losses. This is another indication of how the flux from the 3 phases cancels within the side clamp shunts since it is not forced to return to the core.

16.2.2 Shunts on the Tank Wall

Shunts on the tank walls are frequently used to reduce losses and/or to eliminate hot spots. We look at the effectiveness of these shunts when the laminations are placed flat on the tank wall or stacked on edge. Figure 16.4 shows a 160 MVA, 2-winding power transformer with shunts on the low-voltage (LV) side tank wall. The tank itself is not shown.

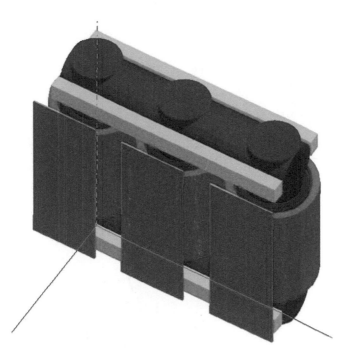

FIGURE 16.4
Power transformer model with LV side tank shunts.

The 25 mm thick shunts are centered on the 3 phases. In practice, if the shunts were stacked flat to the tank wall, they would be subdivided into several vertical sections to reduce the losses. Here, for simplicity, this has not been done. The shunts are made of core steel with the anisotropic properties shown in Table 16.1.

The surface loss densities in the LV side tank wall without the shunts are shown in Figure 16.5.

Notice that the losses are somewhat higher on the end phases than in the middle, due to flux cancellation in the middle phase. Notice also that, on the end wall, the losses are higher at the winding center than at the winding ends. This was a feature of the calculation performed in the last chapter for the analytic tank losses from a single phase.

The tank wall surface loss densities with the shunts present are shown in Figure 16.6. This figure would look the same whether the shunts are placed flat or on edge.

TABLE 16.1

Characteristics of Stacked Core Laminations Used in the Finite Element Studies

Anisotropic Properties of Stacked Core Laminations		
	Relative AC Permeability	Resistivity ($\mu\Omega$ cm)
Rolling direction	5000	50
Cross direction	2500	50
Stacking direction	25	50×10^4

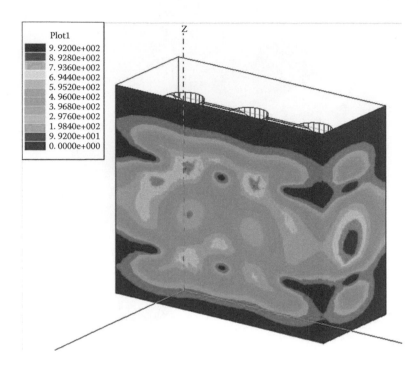

FIGURE 16.5
Surface loss densities in the LV side tank wall and end without shunts. The units are W/m².

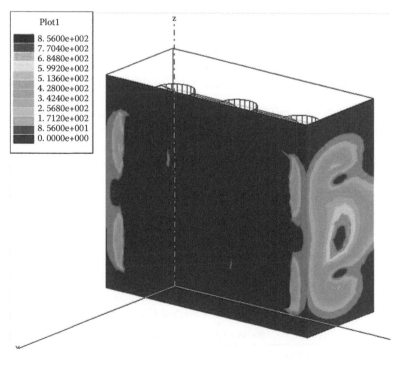

FIGURE 16.6
Surface loss densities on the LV side and tank walls when the shunts are present. There are no shunts on the end walls. The units are W/m².

TABLE 16.2

Losses in the Tank Walls When the LV Side Shunts Are
Stacked Flat or on Edge to the Tank Wall

	Losses with and without 25 mm Shunts (kW)		
	No Shunts	Flat Shunts	On-Edge Shunts
Tank—LV side	13.57	1.14	1.14
Tank—HV side	5.97	5.87	5.87
Tank—left end	4.30	4.33	4.32
Tank—right end	4.85	4.30	4.29

Tank wall and shunt losses are shown in Table 16.2 for flat and on-edge shunts and compared with losses without the shunts. Since the shunts are on the LV side, they primarily lower the tank losses on this wall. The reduction in the LV side losses is about a factor of 10 for both shunt stackings. The flux densities in both types of shunt are well below saturation, and the losses in the shunts are negligible. It would appear from this study that there is virtually no advantage to using shunts stacked on edge as compared with flat shunts. Other considerations, such as convenience, economics, or possibly noise reduction, should dictate the choice.

16.2.3 Effects of 3-Phase Currents on Losses

The effects of having 3-phase currents in the windings were examined. Because all three phases can be included, one might expect the losses to be influenced by how the fluxes add (as phasors) in regions of flux overlap from the different phases. Losses were calculated with 3-phase currents in the windings and with single-phase currents in each phase separately. The sum of the losses with the separate phases carrying current is then compared with the 3-phase current losses. Transformers with several MVA ratings were examined, and this comparison is given in Table 16.3. The single end phase loss is multiplied by 2 and added to the center phase loss to get the separate phase loss. This is because the end phase geometry was identical for the two ends. As can be seen, there is a loss reduction of about 10% when 3-phase currents are used to compute the losses.

16.2.4 Stray Losses from 3D Analysis versus Analytical and Test Losses

An important reason for considering a 3D analysis is to improve the calculation of the losses. The rectangular shapes of tanks and shunts can be modeled in 3D, whereas they

TABLE 16.3

Separate Phases versus 3-Phase Losses

	Tank + Clamp Losses (kW)			
	3 Phase	1 End Phase	Center Phase	3 Phase/Separate
67 MVA	12.76	5.04	3.53	0.94
64 MVA	9.97	4.16	2.70	0.90
96 MVA	13.76	5.89	3.65	0.89

present a problem for 2D analyses where a single phase, treated axisymmetrically, is generally modeled. The increased model accuracy along with the inclusion of 3-phase effects should hopefully improve the accuracy of the loss calculation. Since most of the stray losses, aside from eddy current losses in the windings, occur in materials that have a small skin depth, it is essential to rely on the impedance boundary method to compute these losses.

One of the problems with evaluating the accuracy of the 3D loss calculation is that test stray losses are determined very indirectly. Only the total load losses are measured directly. From these, the calculated I^2R winding losses are subtracted. Then the eddy current losses in the windings are estimated from a 2D flux plot together with analytic formulas that make use of the radial and axial field components throughout the windings. These are subtracted and what remains are the stray losses. Sometimes, the eddy current losses in the windings are considered part of the stray losses but not in this context. Since the I^2R + winding eddy losses comprise the major part of the total load losses, there can be considerable error in the result of subtracting two large numbers. Nevertheless, the losses were determined for several units using 3D finite element analysis and compared with losses extracted from the test data. Losses based on analytic formulas were also determined for a similar comparison.

The losses were calculated separately for the clamps and tank. The miscellaneous losses include the tie plates and pressfeet (small structural supporting projections). Table 16.4 shows the comparisons.

As can be seen from the table, the analytic formulas do a better job of predicting the test results. This is not unexpected, since these formulas were specifically adjusted or tweaked to get good agreement with the measured losses for many units over a long period of time. The 3D losses are about 50% below the measured losses. The likely explanation can be found in [Kul04] where it is pointed out that the impedance boundary method is based on linear magnetic characteristics for the tank or clamp steel. In reality, because of the small skin depth, the material saturates magnetically within this depth. This requires a correction factor, which is estimated to be in the range of 1.3–1.5 times the 3D losses. This would produce much better agreement with the test losses in the table given earlier. A method for including such a nonlinear correction to the impedance boundary losses is given in the next section.

TABLE 16.4

Comparison of Stray Losses Determined by 3D FE and Analytic Methods with Losses Extracted from Test Data

MVA/Type	Analysis Type	Clamps	Tank	Misc	Total Calc	Test
60/Auto	Analytic	8.22	5.26	1.0	14.48	21.7
	3D	4.96	3.98	0.33	9.27	
64/Auto	Analytic	10.25	4.29	1.54	16.08	22.6
	3D	6.44	3.52	1.75	11.71	
96/Auto	Analytic	13.81	6.29	1.71	21.81	21.6
	3D	8.86	4.90	2.04	15.80	
60/Power	Analytic	5.17	2.72	1.17	9.06	9.30
	3D	3.62	0.63	0.57	4.82	
225/Auto	Analytic	2.05	14.67	1.03	17.75	19.6
	3D	1.32	3.29	1.83	6.44	

16.3 Nonlinear Impedance Boundary Correction for the Stray Losses*

The surface impedance boundary condition as implemented in finite element programs is based on a loss formula for linear magnetic materials. These materials generally have high enough permeability that the losses are confined to a thin region near the surface. The losses can then be regarded as only a function of the surface tangential magnetic field (H field). Thus, the eddy current losses do not have to be calculated from the eddy current distribution, which would require a large number of finite elements concentrated near the surface of the material. This becomes impractical for large high-permeability structures such as transformer tank walls or clamps due to the extremely small skin depths of these materials at 50/60/Hz. Using this linear formula, systematic discrepancies have been found between the calculation and measured losses as discussed earlier.

Several authors have proposed corrections to take into account the nonlinear magnetics present in low-carbon structural steels. An approach similar to that presented here was given in [Lab89]. However, an approximate B–H curve was assumed and Maxwell's equations were solved by means of a phasor approximation, assuming fictitious material properties. This led to a fairly complicated procedure to extract the desired losses. Effective permeability approaches have also been proposed, leading to approximate results [Kos02]. Reference [Gue96] uses a step function to approximate the B–H curve and an effective permeability to estimate the nonlinear losses.

In contrast to the method proposed here, many of the previous approaches involve implementing the nonlinear impedance boundary terms directly into the formulation of the finite element or boundary element method, usually by means of an effective permeability. A method based on the nonlinear solution of Maxwell's equations, using Agarwal's approximate B–H curve [Aga59], and extracting an equivalent surface impedance as a function of the surface magnetic field, H, was coupled to boundary element or finite element codes [Kra97]. This requires an iterative procedure and access to the relevant software.

Extensions have also been made to geometries other than plane surfaces and to materials with properties varying not only in the direction normal to the surface but in the tangential direction as well [Yuf99], [Yuf10].

The approach developed here uses a commercial finite element program, which implements the linear surface impedance boundary condition, namely, ANSYS's Maxwell 3D [ANS], although it should work for similar software. A nonlinear correction to the linear formula is developed, which depends on the magnetic properties of the metal used in the structures of interest. This correction depends on the B–H curve and conductivity of the material and is a function of the surface tangential H field as is the formula that computes the linear losses. Thus, it is applied in the same manner as the linear loss calculation.

Two low-carbon steels having different properties are studied, and the losses with and without the nonlinear correction are compared with test values of the stray losses for a variety of transformers.

16.3.1 Linear Loss Calculation for an Infinite Slab

The linear loss formula, given in (16.1), was developed for an infinitely thick slab of material, which is also infinite in the two directions parallel to the surface. It is derived here to compare with the nonlinear loss calculation that will be derived next. Thus, the fields are

---

* This section is based on [Del13], *IEEE Trans. Magn.*, 49(12), December 2013, 5687–5691. © 2013 IEEE.

assumed to vary only in the direction into the slab. These assumptions also apply to a slab of finite thickness, provided that the skin depth is sufficiently smaller than the slab's thickness. In the coordinate system shown in Figure 16.7, the fields vary only in the z direction, which is into the slab. The surface H field vector is in the x–y plane, that is, parallel to the surface of the slab.

In this geometry, Maxwell's equations in SI units simplify to

$$\frac{\partial^2 H(z)}{\partial z^2} = \sigma \frac{\partial B(z)}{\partial t} \tag{16.2}$$

where σ is the electrical conductivity. For a linear magnetic material $B = \mu H$, with $\mu = \mu_r \mu_0$, where μ_r is the relative permeability and $\mu_0 = 4\pi \times 10^{-7}$ in SI units. With this linear assumption, Equation 16.2 can be solved by means of phasors. Thus, all the fields are multiplied by $e^{-j\omega t}$, where j is the imaginary unit and $\omega = 2\pi f$ and f is the frequency in Hz. For an infinite slab with an H field tangential to the surface given by a phasor of magnitude H_0 and an H field at $z = \infty$, which is infinitely far into the slab, equal to 0, the solution to (16.2) is

$$H = H_0 e^{-kz} \tag{16.3}$$

where $k^2 = j\omega\mu_r\mu_0\sigma$.

The current density, J, is given by

$$\mathbf{J} = \nabla \times \mathbf{H} \tag{16.4}$$

which, in this coordinate system, becomes

$$J = -\frac{\partial H}{\partial z} = kH_0 e^{-kz} \tag{16.5}$$

The current density is in the x–y plane but perpendicular to the H field. Using this in the loss formula

$$\text{Loss/unit area} = \frac{1}{2\sigma} \int_0^\infty |J|^2 \, dz \tag{16.6}$$

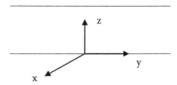

FIGURE 16.7
Coordinate system for solving Maxwell's equations. (From Del Vecchio, R.M. and Ahuja, R., *IEEE Trans. Magn.*, 49(12), December 2013, 5687–5691. © 2013 IEEE.)

we get

$$\text{Loss}/\text{m}^2 = H_0^2 \sqrt{\frac{\omega \mu_r \mu_0}{8\sigma}} \tag{16.7}$$

16.3.2 Nonlinear Loss Calculation for a Finite Slab

For nonlinear B–H curve materials, Equation 16.2 can be solved numerically, using the relation between B and H at each time step. The first step is to obtain the B–H curve for the structural steel used in the analysis. Such a curve was obtained from ANSYS's material properties library with the addition of a few intermediate points to facilitate the analysis. This curve is shown in Figure 16.8. For comparison, a linear B–H curve for a material with a relative permeability of 200 is also shown. This corresponds to the material used for the linear impedance boundary analysis. Both materials have a conductivity of $4 \times 10^6 \ (\Omega \ m)^{-1}$.

This curve was fit with cubic splines, using a mathematical library, which also allowed the derivative of the curve to be obtained. For definiteness, let

$$B = f(H) \tag{16.8}$$

where f is the cubic spline fit or any, possibly analytic, fit to the B–H curve.

For this numerical nonlinear analysis, we need to replace the infinitely thick slab by one of finite thickness of about 5–10 times the skin depth. This finite sheet is then supplied with an equally spaced grid of points along a line in the z direction. About 50–100 points for the grid should suffice. Denote their values on these points by B(i), H(i), where i = 1, …, n for an n point grid. We also need a $B_{new}(i)$ and $H_{new}(i)$ for the field values at the next time step. We do not need to save the field values on all the time steps.

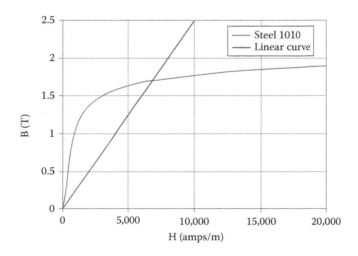

FIGURE 16.8
B–H curve for low-carbon steel compared with a linear curve. (From Del Vecchio, R.M. and Ahuja, R., *IEEE Trans. Magn.*, 49(12), December 2013, 5687–5691. © 2013 IEEE.)

Equation 16.2 can be cast in discrete form on this grid by

$$\frac{H(i+1)-2H(i)+H(i-1)}{(\Delta z)^2}=\sigma\frac{B_{new}(i)-B(i)}{\Delta t} \tag{16.9}$$

where
Δz is the distance between space points
Δt the time step interval

Rewriting (16.9),

$$B_{new}(i)=B(i)+\frac{\Delta t}{\sigma(\Delta z)^2}\Big[H(i+1)-2H(i)+H(i-1)\Big] \tag{16.10}$$

It turns out that in order to achieve a stable converged solution, it is necessary for the inequality to hold:

$$\frac{\Delta t}{\sigma(\Delta z)^2}\le\frac{\mu_{r,diff}\mu_0}{2} \tag{16.11}$$

where $\mu_{r,diff}$ is the relative differential permeability [Gil65]. Given that a fixed grid size is chosen, this fixes Δz. Thus, the time step must be chosen to satisfy inequality (16.11). This time step may vary over the duration of the signal, depending on the differential permeability at that time step.

The surface boundary conditions at $i=1$ and $i=n$ are

$$H(1)=H_0\sin(\omega t),\quad H(n)=0 \tag{16.12}$$

Initially, all fields are set to zero at $t=0$. At $t=\Delta t$, $H(1)$ and $H(n)$ are set according to (16.12). All the other $H(i)$ are still zero at this point. Then the $B_{new}(i)$ are calculated according to (16.10). Given the $B_{new}(i)$ at all points i, the $H_{new}(i)$ can be found at all $i=2,\ldots,n-1$. This is done by inverting (16.8).

For the B–H curve f(H) given in (16.8), this inversion is best done by Newton–Raphson iteration. This is where it is useful to have the derivative of the B–H curve. As discussed in Section 2.7, this iteration procedure involves finding the zero of a function, F(H). Thus, given a B_{new}, a new function is defined by $F(H)=f(H)-B_{new}=B(H)-B_{new}$, so that the zero of F(H) yields the H_{new} corresponding to B_{new}. To begin this process, a starting value of H is chosen by a reasonable guess. Then an iteration is set up so that a better approximation is given by $H+\Delta H$, where

$$\Delta H=-\frac{F(H)}{dF(H)/dH} \tag{16.13}$$

Since B_{new} is considered a constant in this process, the derivative of F(H) is just the derivative of f(H) in (16.8). The iteration terminates when ΔH is small enough.

The solution process continues for more cycles of the sine wave. Thus, at a given time step $t+\Delta t$, $H(1)$ and $H(n)$ are set according to (16.12). The other H values would now, in general, be nonzero. Then the $B_{new}(i)$ are calculated according to (16.10) and the $H_{new}(i)$ for the nonboundary H's are determined by inverting (16.8).

At the end of a cycle, the eddy current losses are determined. The process terminates when the eddy current losses are the same within a given tolerance between two successive cycles. The eddy current loss density is given by (16.4), which, in this geometry, becomes

$$J = -\frac{\partial H}{\partial z} \tag{16.14}$$

It is necessary to discretize (16.14). For points away from the boundaries, a central difference approximation may be used:

$$\frac{\partial H(i)}{\partial z} = \frac{H(i+1) - H(i-1)}{2(\Delta z)} \tag{16.15}$$

For points on the z = 0 surface where i = 1,

$$\frac{\partial H(1)}{\partial z} = \frac{-3H(1) + 4H(2) - H(3)}{2(\Delta z)} \tag{16.16}$$

At the other surface where i = n,

$$\frac{\partial H(n)}{\partial z} = \frac{3H(n) - 4H(n-1) + H(n-2)}{2(\Delta z)} \tag{16.17}$$

These second-order difference approximations are necessary for accuracy.

The losses over a cycle can then be calculated from

$$\text{Loss/Unit area} = \frac{1}{T\sigma} \sum_{\text{cycle}} \Delta t \sum_{i=1}^{n} J^2(i, t) \Delta z \tag{16.18}$$

where T is the cycle time and J has been labeled with the space index i and the time t during the cycle. The inner summation in (16.18) is calculated at each time step, and when multiplied by Δt, a cumulative summation can be obtained over a cycle.

Normally, over a cycle, the smallest differential permeability occurs at the peak surface H field, which, from (16.12), is H_0. Thus, the time step can be set once for the whole cycle, based on (16.11). For some margin, it should be set lower by about a factor of 2, but the relative differential permeability should never go below 1.

As a check, the linear B–H curve was also fit with cubic splines, and the nonlinear analysis described earlier was applied to this linear case. The losses obtained agreed with the analytic formula given in (16.7) within a few tenths of a percent.

16.3.3 Application to Finite Element Loss Calculations

Losses for the low-carbon steel 1010 with the B–H curve given in Figure 16.8 were calculated by the method outlined earlier for a number of peak surface sinusoidal H fields at 60 Hz. Convergence usually occurred after four cycles although the convergence time increased with increasing H_0 field due to the drop in differential permeability caused by the onset of saturation. The nonlinear losses were divided by the losses given by the linear loss analytic formula (16.7) with permeability corresponding to the straight line B–H curve in Figure 16.8. The ratio of the nonlinear to linear losses versus the peak surface H field is plotted in Figure 16.9.

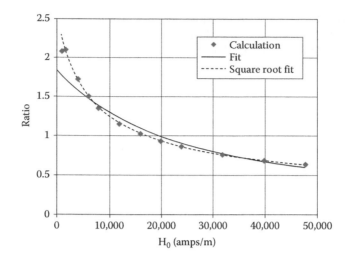

FIGURE 16.9
Ratio of nonlinear to linear losses for low-carbon steel 1010. The calculated points are fit by two methods corresponding to the solid and dotted lines. (From Del Vecchio, R.M. and Ahuja, R., *IEEE Trans. Magn.*, 49(12), December 2013, 5687–5691. © 2013 IEEE.)

The dotted line in Figure 16.9, which gave the best fit, corresponds to the formula

$$\text{Ratio} = \frac{1}{\sqrt{0.149334 + 4.98508 \times 10^{-5} H_0}} \tag{16.19}$$

where H_0 is the peak surface H field. This formula was difficult to implement in the finite element calculator because of the H field value under a square root. An alternative but reasonable fit is given by the solid line in Figure 16.9, which could be implemented by the finite element program. This solid line ratio is expressed by

$$\text{Ratio} = \frac{1}{0.54197 + 2.34934 \times 10^{-5} H_0} \tag{16.20}$$

This ratio then multiplies the linear loss formula given in (16.7) so that the nonlinear loss formula is a product of two terms that are both a function of H_0. Thus, for comparison purposes, both linear and nonlinear losses can be obtained by the finite element program, depending on whether the correction factor is included or not.

Figure 16.9 shows that the nonlinear loss correction factor is >1 for smaller surface H fields, which means that the nonlinear losses are above the linear losses for smaller H fields. This probably because, based on Figure 16.8, the slope of the nonlinear curve (differential permeability) is higher than the slope of the linear curve at smaller H values, which makes the nonlinear skin depth smaller, leading to higher losses. The correction then falls off and drops below 1 as the H field increases. In this case, the nonlinear losses are below the linear losses for these larger H field values. This is probably happening because the nonlinear curve slope is smaller than the slope of the linear curve at higher H values, resulting in a larger skin depth and hence lower losses than the linear skin depth for these larger H fields.

16.3.3.1 Comparison with Test Losses

The measurement of stray losses in a transformer is an indirect process. Generally, only the total load loss is measured. These include the winding d.c. losses, eddy current losses in the windings, and stray losses. The core losses are not included. Therefore, the winding + winding eddy losses must be subtracted from the measured losses to get the stray losses. Since the stray losses are generally a small percentage of the total losses, they are arrived at by subtracting two large numbers. This can result in significant errors. The winding losses consist of d.c. losses (I^2R losses) + eddy losses. The d.c. losses are calculated from the winding resistance, which is measured, and the winding current also measured. The winding eddy current losses are obtained from an analytic formula, which utilizes the values of the leakage field of the energized transformer throughout the windings, together with the conductor dimensions. These calculated eddy losses are also subject to some uncertainty. Nevertheless, by comparing measured stray losses with calculated stray losses for a number of transformers, one can get a statistical sense of the level of agreement between the measured and calculated values.

Thus, 10 transformers with tested losses were used in this study. Their MVA ratings varied from 60 to 420 and included power and autotransformers. The stray losses included losses on all tank walls, on all clamps, and on all tie or flitch plates. In short, the losses on all significant structural components were included. Most had clamp shunts and/or tank shunts on some tank walls. A typical transformer model is shown in Figure 16.10.

The total losses on the various structural components were obtained in the calculator. However, the loss distribution can be displayed graphically on the individual surfaces. An example of this impedance boundary surface loss distribution is shown on tank walls in Figure 16.11.

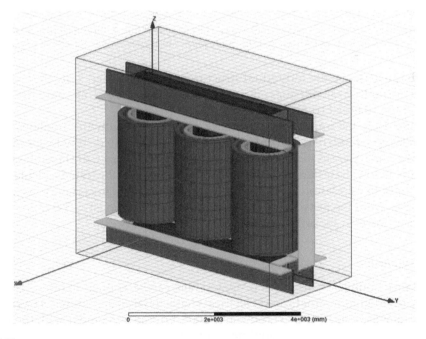

FIGURE 16.10
3-phase transformer model showing the clamps, shunts, and windings. The tank walls are semitransparent. (From Del Vecchio, R.M. and Ahuja, R., *IEEE Trans. Magn.*, 49(12), December 2013, 5687–5691. © 2013 IEEE.)

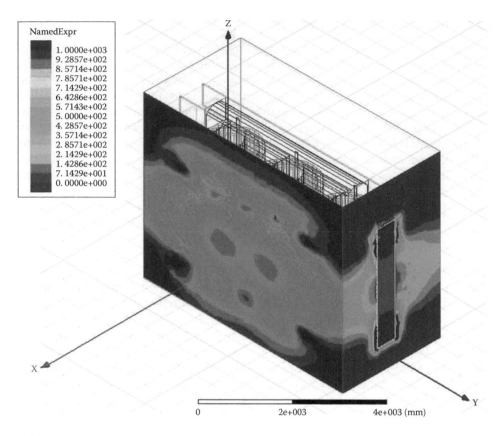

FIGURE 16.11
Impedance boundary surface losses on two tank walls. The key is in W/m². In this unit, only end tank shunts are used. (From Del Vecchio, R.M. and Ahuja, R., *IEEE Trans. Magn.*, 49(12), December 2013, 5687–5691. © 2013 IEEE.)

Although the losses on the individual structures can be obtained, only the total stray loss from test is normally available for comparison with calculation. This comparison is shown in Table 16.5 for 10 tested transformers.

From Table 16.5, the impedance boundary stray losses with the nonlinear correction described here give much better overall agreement with the tested stray losses than with the impedance boundary losses, which used a linear B–H curve. In fact, the linear loss calculation is consistently below the tested loss. The standard deviation in the tested–calculated losses is fairly high for both calculation types. This is possibly due to the uncertainty in extracting the winding losses from the tested total losses as discussed previously. The ratio of nonlinear losses to linear losses is about 1.6 on average.

16.3.3.2 Conclusion

The impedance boundary correction proposed here can be applied to other types of structural steel provided its magnetic and electrical characteristics are known. For example, the same analysis was done for low-carbon steel 1008, but the agreement with the tested data was not as good. On average, the nonlinear corrected losses are about a factor of 1.6 times the linear impedance boundary losses. This factor is about the same as the loss discrepancy reported by various sources in the literature. Moreover, the correction depends on the

TABLE 16.5

Comparison of Tested Stray Losses with Impedance Boundary Losses with and without the Nonlinear Correction

3 Phase (MVA)	Test Losses	Calculated Nonlinear	Calculated Linear	Nonlinear Test	Linear Test
60	4.25	3.92	2.24	−0.33	−2.01
102	12.7	15.4	9.22	2.7	−3.48
120	5.41	5.21	2.96	−0.02	−2.45
142	31.0	24.9	13.1	−6.1	−17.9
180	22.5	26.6	16.3	4.1	−6.2
240	12.9	15.7	9.1	2.8	−3.8
290	83.4	57.9	36.2	−25.5	−47.2
300	52.0	56.7	34.9	4.7	−17.1
336	21.6	23.5	13.8	1.9	−7.8
420	74.0	93.5	61.7	19.5	−12.3
			Ave	0.357	−12.02
			Std dev	10.60	12.98

Source: Del Vecchio, R.M. and Ahuja, R., *IEEE Trans. Magn.*, 49(12), December 2013, 5687–5691. © 2013 IEEE.

Losses are in kW.

surface H field distribution, as does the linear loss formula. It is not simply a single number that multiplies the total surface loss. It multiplies the linear loss formula point by point over the surface where the losses are being evaluated.

The correction described here is not only dependent on the B–H curve of the steel used but also on the linear permeability of the material used to implement the linear impedance boundary condition in the finite element program. It also assumes that they have the same electrical conductivity.

17

Thermal Design

17.1 Introduction

A thermal model of an oil-cooled power transformer is presented here, along with details of the computer implementation and experimental verification. Any such model, particularly of such a complex system, is necessarily approximate. Thus, the model assumes that the oil flows in definite paths and ignores local circulation or eddy patterns that may arise. The most detailed model assumes that the oil flow through the disk coils is guided by means of oil flow washers. However, we also discuss oil flow through vertical ducts in less detail. Recent studies have shown that irregular eddy flow patterns may exist in nondirected oil flow cooling in vertical ducts [Pie92]. Such patterns may also occur in the bulk tank oil. We assume these are small compared with the major or average convective cooling flow in such coils or in the tank. We further assume that the convective flow in the tank results in a linear temperature profile from the bottom of the radiators to the top of the coils in the tank oil external to the coils. The model likewise ignores localized heating that may occur, for example, due to high current-carrying leads near the tank wall. It accounts for these types of stray losses in only an average way. However, a localized distribution of eddy current losses in the coils is allowed for, along with the normal I^2R losses.

The radiators we model consist of a collection of radiator plates spaced equally along inlet and outlet pipes. The plates contain several vertical ducts in parallel. The ducts have oblong-shaped, nearly rectangular cross-sections. Fans, vertically mounted, may or may not be present (or turned on). In addition to radiator cooling, cooling also occurs from the tank walls by natural convection to the surrounding air and by radiation. Although the oil may be pumped through the radiators, most of our designs are without pumps so that the oil flow in the radiators and coils is assumed to be laminar. Even with pumps, the flow is often laminar except through the pipes entering and exiting the tank to the radiators. Laminar flow determines the expressions used for the heat transfer and friction coefficients along the oil flow paths.

Our model of a disk winding is similar to that of Oliver [Oli80]. However, since we consider the whole transformer, we need to reconcile the heat generated by the individual coils and stray losses with the radiator and tank heat dissipation or cooling in an overall iteration scheme in order to arrive at a steady-state condition. When we consider transient heating, steady state is only reached gradually.

Previous thermal models of whole transformers have focused on developing analytic formulas with adjustable parameters to predict overall temperature rises of the oil and coils [IEE81], [Blu51], [Eas65], [Tay58], [Aub92], and [Pie92a]. While these produce acceptable results on average or for a standard design, they are less reliable when confronting a new or untried design. The approach taken here is to develop a model to describe the basic physical processes occurring in the unit so that reliance on parameter fitting is minimized. Such an approach can accommodate future improvements in terms of a more detailed description of the basic processes or the addition of new features as a result of a design change or different types of oil.

17.2 Thermal Model of a Disk Coil with Directed Oil Flow*

The disk coil is assumed to be subdivided into directed oil flow cooling paths as shown in Figure 17.1. A smaller section is shown in Figure 17.2, which shows the disks, nodes, and paths numbering scheme. The geometry is really cylindrical with the inner radius R_{in}. Only one section (region between two oil flow washers) is shown, but there can be as many sections as desired. Each section can contain different numbers of disks, and the number of turns per disk, insulation thickness, etc., can vary from section to section. The duct sizes can vary within a section as well as from section to section.

The unknowns, which must be solved for include the nodal oil temperatures T, the nodal pressures P, the path oil velocities v, the path oil temperature rises ΔT, and the disk temperatures T_C. These are labeled with their corresponding node, path, or disk number. Note that the surface heat transfer coefficients from the coil surfaces are associated with the oil path with which the surfaces are in contact. We are not allowing for a temperature profile along a single disk but are assuming that each disk is at a uniform temperature. This is an approximation, which could be refined if more detail is required.

17.2.1 Governing Equations and Solution Process

Along a given path, the oil velocities are uniform since the cross-sectional area is assumed to remain constant. Ignoring gravitational effects, which do not influence the oil flow, the pressure drop along a given path is only required to overcome friction. Considering a generic path and labeling the pressures at the beginning and end of the paths P_1 and P_2, respectively, we can write, using standard notation [Dai73],

$$P_1 - P_2 = \frac{1}{2}\rho f \frac{L}{D} v^2 \qquad (17.1)$$

where
   ρ is the fluid density
   f is the friction coefficient
   L is the path length
   D is the hydraulic diameter
   v is the fluid velocity

We note that the hydraulic diameter is given by

$$D = 4 \times \text{cross-sectional area/wetted perimeter}$$

---

* This section draws on our published papers [Del99], *1999 IEEE Transmission and Distribution Conference*, Now Orleans, LA, April 11–16, 1999, pp. 914–919. © 1999 IEEE and [Del14], *IEEE Trans. Power Deliv.*, 29(5), October 2014, 2279–2286. © 2014 IEEE.

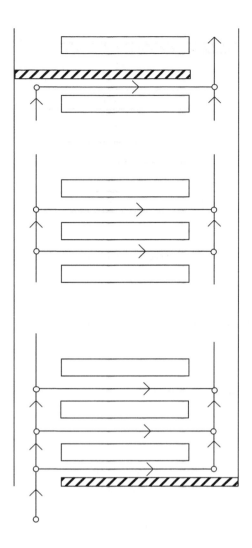

FIGURE 17.1
Oil flow paths for a disk coil with directed oil flow. (From Del Vecchio, R.M. and Feghali, P., *1999 IEEE Transmission and Distribution Conference*, Now Orleans, LA, April 11–16, 1999, pp. 914–919. © 1999 IEEE.)

SI units are used throughout this chapter. For laminar flow in circular ducts, $f = 64/\mathrm{Re}_D$, where Re_D is the Reynolds number, given by

$$\mathrm{Re}_D = \frac{\rho v D}{\mu} \tag{17.2}$$

where μ is the fluid viscosity. For laminar flow in noncircular ducts, the number 64 in the expression for f changes. In particular, for rectangular ducts with sides a and b with a < b, we can write

$$f = \frac{K(a/b)}{\mathrm{Re}_D} \tag{17.3}$$

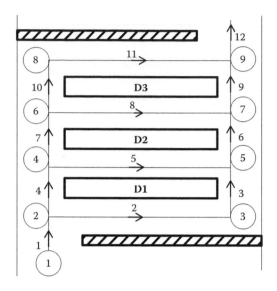

FIGURE 17.2
This shows the numbering scheme used for the disks, the flow paths, and the nodes. Only three disks are shown but sections can include many more disks. (From Del Vecchio, R.M., *IEEE Trans. Power Deliv.*, 29(5), October 2014, 2279–2286. © 2014 IEEE.)

where K(a/b) is given approximately by

$$K(a/b) = 56.91 + 40.31\left(e^{-3.5a/b} - 0.0302\right) \tag{17.4}$$

This expression is based on a fit to a table given in Reference [Ols80].
  Substituting (17.2) and (17.3) into (17.1), we obtain

$$P_1 - P_2 = \frac{1}{2}\frac{\mu KL}{D^2}v \tag{17.5}$$

where $K = K(a/b)$ is implied. This equation is linear in the pressures and velocities. However, the oil viscosity is temperature dependent and this will necessitate an iterative solution. Based on a table of transformer oil viscosities versus temperature given in [Kre80], we achieved a good fit to the table with the expression

$$\mu = \frac{6900}{(T+50)^3} \tag{17.6}$$

with T in °C and μ in N s/m².

For non-laminar flow ($Re_D > 2000$), the expression for f is more complicated and (17.5) would no longer be linear in v. We are ignoring the extra friction arising from flow branching and direction changing at the nodes. There is considerable uncertainty in the literature as to what these additional frictional effects are for laminar flow. Instead, we have chosen to allow the friction coefficient in (17.3) to be multiplied by a correction factor, if warranted by test results. This correction factor has been found to be close to 1.

The additional equations needed to solve for the pressures and velocity unknowns come from conservation of mass at the nodes. A mass of fluid of velocity v flowing through a duct of cross-sectional area A per unit time, dM/dt, is given by

$$\frac{dM}{dt} = \rho A v \tag{17.7}$$

Since the fluid is nearly incompressible, we can consider conservation of volume instead, where $Q = Av$ is the volume flow per unit time.

The overall pressure drop through the coil is produced by the difference in buoyancy between the hot oil inside the coil's cooling ducts and the cooler tank oil outside the coil. Thus,

$$\Delta P_{coil} = \left(\rho_{ave,out} - \rho_{ave,in} \right) gH \tag{17.8}$$

where
$\rho_{ave,in}$ is the average oil density inside the coil
$\rho_{ave,out}$ is the average density outside the coil
g is the acceleration of gravity
H is the coil height

Letting β be the volume coefficient of thermal expansion, we have

$$\frac{1}{\rho} \frac{d\rho}{dT} = -\beta \tag{17.9}$$

and thus, since $\beta = 6.8 \times 10^{-4}/K$ for transformer oil, we have, to a good approximation over a reasonably large temperature range, $\Delta \rho = \beta \rho \Delta T$ so that (17.8) can be rewritten as

$$\Delta P_{coil} = \beta \rho g H \left(T_{ave,in} - T_{ave,out} \right) \tag{17.10}$$

Because we are only considering pressures that produce oil flows, the pressure at the top of the coil when steady state is achieved should be zero. This means that ΔP_{coil} = the pressure at the input node = P_0.

The oil velocity at the entrance of the input path of the coil, v_0, is determined by the overall energy balance: The energy per unit time acquired by the oil must equal the energy per unit time lost by the coil in steady state. The latter loss is just the total resistive loss of the coil. Thus,

$$\rho c v_0 A_{in} \Delta T_{coil} = \sum_{i=1}^{\# \, disks} I^2 R_i \tag{17.11}$$

where
c is the specific heat of the oil
ΔT_{coil} is the increase in the temperature of the oil after passing through the coil
R_i is the temperature-dependent resistance of disk i, including eddy current effects
A_{in} is the input path area

For the single-disk resistance, we use

$$R_i = \gamma_o \left(1 + \alpha \Delta T_i\right)\left(1 + ecf_i\right) \frac{\ell_i}{A_{turn}} \tag{17.12}$$

where
  γ_o is the resistivity at some standard temperature T_o
  ΔT_i is the temperature rise of the disk above the standard temperature
  α is the temperature coefficient of resistivity
  ecf_i is the fraction of the normal losses due to eddy currents for disk i
  ℓ_i is the length of the cable or wire in disk i
  A_{turn} is the cross-sectional current-carrying area of the cable or wire

ecfi can vary from disk to disk to account for the effects of nonuniform stray flux along the coil.

From (17.12), we see that the coil resistances are temperature dependent. Other quantities such as the oil viscosity and density are also temperature dependent. These temperatures are set initially to some starting temperature and will be redetermined by solving the various conservation equations. This will necessitate an iterative solution process where the recalculated temperatures are used to update the resistances, viscosities, etc., before resolving the conservation equations. This process will continue until some convergence criterion is met.

Pressures and temperatures are defined on the nodes in Figure 17.2. Oil velocities and temperature changes (usually rises) are defined on the paths. The pressure and temperature at node 1 of the section at the bottom of the winding are taken as given initial conditions. The velocity along path 1, v_0, is also assumed given initially. Node 1 at the bottom of the next section occurs at the end of path 12 in Figure 17.2. The temperature and pressure at this node and the velocity along path 12 will be determined after solving the equations for the bottom section. It is only necessary to reverse the inner and outer vertical ducts in the next section to apply the same network methodology to the next section.

The temperature at the bottom of the winding, T_0, is the oil temperature at the bottom of the tank in which the windings are situated. This is determined by the overall heat balance in the transformer between the windings and radiators or cooling system and will be discussed later. The initial oil velocity along path 1 at the bottom of the section is determined by an overall heat balance in the winding, since the oil flow and oil temperature rise through the winding must be sufficient to remove the heat generated by I^2R and eddy losses.

At each iteration, (17.11) is used to determine a new v_0 after updating the winding losses, ρ and c, for the new temperature distributions in the oil and winding. Also, (17.10) is used to determine a new P_0, assuming the top oil pressure is 0. Since this is usually not the case, we have adopted an iteration strategy as follows: v_0 and P_0 from the previous iteration are saved as $v_{0,save}$ and $p_{0,save}$. We then take $v_{0,new}$ for the next iteration to be

$$v_{0,new} = Fv_0 + \left(1 - F\right)\left(\frac{P_0}{P_{0,save} - P_{top}}\right)v_{0,save} \tag{17.13}$$

where F is a fraction between 0 and 1 that can vary throughout the iterations. F is a relaxation factor used to make the successive iterations less abrupt and thus facilitate convergence. $F = 1$ would remove the second term and not allow the top pressure to approach 0, while $F = 0$ would not take into account the new velocity calculation. $P_{0,new}$ is taken as P_0 as determined by (17.12). Iterations proceed until v_0 and $v_{0,save}$ are sufficiently close as well as P_0 and $P_{0,save}$ and until P_{top} is close to 0. Convergence is also required for the nodal and path variables.

17.2.2 Oil Pressures and Velocities

We must have mass conservation at the nodes, that is, the mass flow of oil into a node must equal the mass flow out of the node. From the mass flow equation (17.7), assuming ρ is nearly constant, we can assume that volume flow Av is conserved at the nodes. Thus, at node 6 in Figure 17.2, we have

$$A_7 v_7 = A_8 v_8 + A_{10} v_{10} \tag{17.14}$$

Other nodes such as 3 or 8 would have only two terms in the volume conservation equation.

This is where graph theory can be of use in keeping track of all the indices for the nodes and paths involved in solving a set of equations like (17.14) or (17.5) for the entire winding section [Deo74]. An important matrix for a directed graph is the directed incidence matrix. This is a matrix whose rows are the graph nodes and whose columns are the graph paths or edges. For a given node or row, an edge directed into the node gets a matrix entry of -1, while an edge directed out of the node gets a matrix entry of 1. Thus, the directed incidence matrix for the graph in Figure 17.2, A_{DI} (matrix quantities will be italicized), is

$A_{DI} = \text{Node} \backslash \text{Edge}$

$$
\begin{array}{c|cccccccccccc}
 & \mathbf{1} & \mathbf{2} & \mathbf{3} & \mathbf{4} & \mathbf{5} & \mathbf{6} & \mathbf{7} & \mathbf{8} & \mathbf{9} & \mathbf{10} & \mathbf{11} & \mathbf{12} \\
\mathbf{1} & 1 & 0 & 0 & 0 & 0 & 0 & 0 & 0 & 0 & 0 & 0 & 0 \\
\mathbf{2} & -1 & 1 & 0 & 1 & 0 & 0 & 0 & 0 & 0 & 0 & 0 & 0 \\
\mathbf{3} & 0 & -1 & 1 & 0 & 0 & 0 & 0 & 0 & 0 & 0 & 0 & 0 \\
\mathbf{4} & 0 & 0 & 0 & -1 & 1 & 0 & 1 & 0 & 0 & 0 & 0 & 0 \\
\mathbf{5} & 0 & 0 & -1 & 0 & -1 & 1 & 0 & 0 & 0 & 0 & 0 & 0 \\
\mathbf{6} & 0 & 0 & 0 & 0 & 0 & 0 & -1 & 1 & 0 & 1 & 0 & 0 \\
\mathbf{7} & 0 & 0 & 0 & 0 & 0 & -1 & 0 & -1 & 1 & 0 & 0 & 0 \\
\mathbf{8} & 0 & 0 & 0 & 0 & 0 & 0 & 0 & 0 & 0 & -1 & 1 & 0 \\
\mathbf{9} & 0 & 0 & 0 & 0 & 0 & 0 & 0 & 0 & -1 & 0 & -1 & 1 \\
\end{array}
\tag{17.15}
$$

The matrix column and row numbers are given as the leftmost column and the top row of the matrix in boldfaced type for convenience, but these are not part of the matrix itself. Note that the standard incidence matrix, A_I, has a 1 for every edge incident on a given node so that the −1's in (17.15) are replaced by 1's for A_I. For further reference, the incidence matrix is given by

$$A_I = \text{Node} \backslash \text{Edge}$$

	1	2	3	4	5	6	7	8	9	10	11	12
1	1	0	0	0	0	0	0	0	0	0	0	0
2	1	1	0	1	0	0	0	0	0	0	0	0
3	0	1	1	0	0	0	0	0	0	0	0	0
4	0	0	0	1	1	0	1	0	0	0	0	0
5	0	0	1	0	1	1	0	0	0	0	0	0
6	0	0	0	0	0	0	1	1	0	1	0	0
7	0	0	0	0	0	1	0	1	1	0	0	0
8	0	0	0	0	0	0	0	0	0	1	1	0
9	0	0	0	0	0	0	0	0	1	0	1	1

The incidence or directed incidence matrix for this graph has 9 rows or nodes and 12 columns or edges. In general, if the section has n disks, there will be 2n + 3 nodes and 3n + 3 edges to the graph. For sections having more disks, the middle section of the matrix in (17.15) will be expanded, keeping the same repetitive pattern.

In order to implement an equation like (17.14) on all the nodes, define an oil velocity vector along the oil paths (graph edges)

$$\mathbf{V} = \begin{pmatrix} \mathbf{v}_1 \\ \mathbf{v}_2 \\ \vdots \\ \mathbf{v}_{12} \end{pmatrix} \tag{17.16}$$

Vectors will be denoted by boldface type. Also define a duct cross-sectional area diagonal matrix, A_{duct}, which contains the areas of the ducts along the different paths. Many of these will be the same, for instance the duct areas along the vertical paths. The horizontal duct areas could be different depending on the key spacer thickness, but even here, most of them will be the same. A_{duct} will be an edge × edge diagonal matrix

$$A_{duct} = \begin{pmatrix} A_1 & 0 & \cdots & 0 \\ 0 & A_2 & 0 & 0 \\ \vdots & 0 & \ddots & 0 \\ 0 & 0 & 0 & A_{12} \end{pmatrix} \tag{17.17}$$

Using (17.15) through (17.17), the mass conservation equations can be implemented on the entire graph according to

$$A_{DI} A_{duct} \mathbf{V} = \begin{pmatrix} A_1 v_0 \\ 0 \\ \vdots \\ 0 \end{pmatrix} \tag{17.18}$$

Thus, (17.18) expresses equations like (17.14). When all terms are placed on the left side of the equality sign, the right-hand side equals 0 except for the entrance node at the bottom of the section. At this bottom node, the input mass (volume) flow occurs as an initial condition on the right-hand side of (17.18). At the other nodes, mass or volume balance is built into the incidence matrix, so no entries are needed on the right-hand side of (17.18).

To implement (17.5) on all the paths, we need the transpose of the directed incidence matrix. For the 3-disk graph, this is

$$A_{DI}{}^T = \text{Edge}\backslash\text{Node}$$

$$
\begin{array}{c|ccccccccc}
 & 1 & 2 & 3 & 4 & 5 & 6 & 7 & 8 & 9 \\
\hline
1 & 1 & -1 & 0 & 0 & 0 & 0 & 0 & 0 & 0 \\
2 & 0 & 1 & -1 & 0 & 0 & 0 & 0 & 0 & 0 \\
3 & 0 & 0 & 1 & 0 & -1 & 0 & 0 & 0 & 0 \\
4 & 0 & 1 & 0 & -1 & 0 & 0 & 0 & 0 & 0 \\
5 & 0 & 0 & 0 & 1 & -1 & 0 & 0 & 0 & 0 \\
6 & 0 & 0 & 0 & 0 & 1 & 0 & -1 & 0 & 0 \\
7 & 0 & 0 & 0 & 1 & 0 & -1 & 0 & 0 & 0 \\
8 & 0 & 0 & 0 & 0 & 0 & 1 & -1 & 0 & 0 \\
9 & 0 & 0 & 0 & 0 & 0 & 0 & 1 & 0 & -1 \\
10 & 0 & 0 & 0 & 0 & 0 & 1 & 0 & -1 & 0 \\
11 & 0 & 0 & 0 & 0 & 0 & 0 & 0 & 1 & -1 \\
12 & 0 & 0 & 0 & 0 & 0 & 0 & 0 & 0 & 1 \\
\end{array}
\tag{17.19}
$$

Let F = a diagonal edge × edge matrix with elements

$$
F = \frac{1}{2}
\begin{pmatrix}
\dfrac{\mu_1 K_1 L_1}{D_1^2} & 0 & \cdots & 0 \\
0 & \dfrac{\mu_2 K_2 L_2}{D_2^2} & 0 & 0 \\
\vdots & 0 & \ddots & \vdots \\
0 & 0 & \cdots & \dfrac{\mu_{12} K_{12} L_{12}}{D_{12}^2}
\end{pmatrix}
\tag{17.20}
$$

Define a nodal pressure vector

$$
\mathbf{P} =
\begin{pmatrix}
P_1 \\
P_2 \\
\vdots \\
P_9
\end{pmatrix}
\tag{17.21}
$$

Using (17.19) through (17.21), (17.5) can be expressed for the entire graph by

$$\Delta \mathbf{P} = A_{DI}^T \mathbf{P} = F\mathbf{V} \tag{17.22}$$

Since F is a diagonal matrix, it is inverted by taking the reciprocal of all the diagonal elements. Solving (17.22) for **V**,

$$\mathbf{V} = F^{-1}A_{\mathrm{DI}}^{\mathrm{T}}\mathbf{P} \tag{17.23}$$

Applying (17.18) to (17.23), we obtain

$$A_{\mathrm{DI}}A_{\mathrm{duct}}\mathbf{V} = A_{\mathrm{DI}}A_{\mathrm{duct}}F^{-1}A_{\mathrm{DI}}^{\mathrm{T}}\mathbf{P} = \begin{pmatrix} A_1 v_0 \\ 0 \\ \vdots \\ 0 \end{pmatrix} \tag{17.24}$$

The matrix multiplying **P** is a square node × node matrix and is invertible so that (17.24) can be solved for **P**. **V** can then be obtained from (17.23). We could not solve (17.18) directly for **V**, since the matrix in that equation is not invertible.

In spite of (17.22) not applying strictly to the last edge, v_{12} will still be determined by the last mass conservation equation. Solving (17.24) for **P** does not take into account the bottom pressure P_0. This can be done by using (17.22) to determine $\Delta\mathbf{P}$, $\Delta\mathbf{P} = A_{\mathrm{DI}}^{\mathrm{T}}\mathbf{P}$. The pressures P at all except the bottom input nodes are determined indirectly. Set $P_1 = P_0$, the input pressure, then we can find $P_2 = P_1 - \Delta P_1$. Then $P_3 = P_2 - \Delta P_2$, $P_4 = P_2 - \Delta P_4$, $P_5 = P_4 - \Delta P_5$, etc., until $P_9 = P_8 - \Delta P_{11}$. Thus, the top nodal pressure is determined by the input pressure P_0 and the pressure drops along select paths leading to the top node. These paths essentially constitute a spanning tree for the graph. The pressure drop on the top path 12 is not used in this process.

17.2.3 Disk Temperatures

We assume that each disk has a uniform temperature. Label it T_C. The heat generated by the I^2R loss in each disk is removed by heat transfer to the surrounding oil in equilibrium. The I^2R loss is assumed to contain the eddy current loss as well. Let h be the overall heat transfer coefficient to the surrounding oil. This includes conductive heat transfer through any paper covering as well as surface heat transfer to the surrounding oil.

The energy lost per unit time through a surface of a conductor is given by

$$hA_C\left(T_C - T_b\right) \tag{17.25}$$

where
  h is the surface heat transfer coefficient
  A_C is the surface area
  T_C is the conductor temperature
  T_b is the average (bulk) oil temperature in the adjacent duct

h can be expressed as follows:

$$h = \frac{h_{\mathrm{conv}}}{1 + \dfrac{h_{\mathrm{conv}}\tau_{\mathrm{insul}}}{k_{\mathrm{insul}}}} \tag{17.26}$$

where
  h_{conv} is the convection heat transfer coefficient
  τ_{insul} is the insulation or paper thickness
  k_{insul} is the thermal conductivity of the insulation

While k_{insul} is nearly constant over the temperature range of interest, h_{conv} varies with temperature and oil velocity. For laminar flow in ducts, we use [Kre80]

$$h_{conv} = 1.86 \frac{k}{D} \left(Re_D \, Pr \frac{D}{L} \right)^{0.33} \left(\frac{\mu}{\mu_s} \right)^{0.14} \tag{17.27}$$

where
  k is the thermal conductivity of the oil
  D is the hydraulic diameter
  L is the duct length
  Re_D is the Reynolds number (17.2)
  Pr is the Prandtl number ($Pr = \mu c/k$)
  μ is the viscosity of the bulk oil
  μ_s is the oil viscosity at the conductor surface

Equation 17.27 applies when $Re_D PrD/L > 10$. A correction must be applied for smaller values. The major temperature variation comes from the viscosity (17.6) and the velocity dependence comes from the Reynold's number. The other parameters are nearly constant over the temperatures of interest. For transformer oil, we use [Kre80] $\rho = 867$ kg/m^3, c = 1880 J/kg ρ °C, and k = 0.11 W/m °C.

Let S be the area of the disk surface through which the heat is being removed. Each disk has four such surfaces labeled 1, 2, 3, and 4. Let T_B be the bulk (average) oil temperature in the oil duct adjacent to the surface S. Then the thermal equation for each disk is given by

$$I^2 R = h_1 S_1 \left(T_C - T_{B1} \right) + h_2 S_2 \left(T_C - T_{B2} \right) + h_3 S_3 \left(T_C - T_{B3} \right) + h_4 S_4 \left(T_C - T_{B4} \right)$$
$$= T_C \left(h_1 S_1 + h_2 S_2 + h_3 S_3 + h_4 S_4 \right) - h_1 S_1 T_{B1} - h_2 S_2 T_{B2} - h_3 S_3 T_{B3} - h_4 S_4 T_{B4} \tag{17.28}$$

To implement (17.28), a circuit matrix can be used. Each disk defines a circuit so that Figure 17.2 has three circuits. The edges included in a circuit are the paths surrounding the disk. Thus, the circuit for disk D2 in Figure 17.2 has the paths 5, 6, 7, and 8. The circuit matrix has as many rows as circuits and as many columns as paths in the graph. All paths are not necessarily part of a circuit and some paths can belong to more than one circuit. For each circuit (row), a 1 is placed in a column if the edge is included in the circuit, a 0 otherwise. For the graph in Figure 17.2, the circuit matrix is

$$C = Circ \backslash Edge$$

	1	2	3	4	5	6	7	8	9	10	11	12
1	0	1	1	1	1	0	0	0	0	0	0	0
2	0	0	0	0	1	1	1	1	0	0	0	0
3	0	0	0	0	0	0	0	1	1	1	1	0

$$\tag{17.29}$$

This shows that paths 5 and 8 are included in two circuits. This is important because in the ducts corresponding to those paths, heat transfer occurs from two adjacent disks. In paths 2 and 11, which are next to the oil flow washers as well as from all the side duct paths, heat transfer occurs from only one disk surface to the duct oil because the path from only one circuit occurs in those ducts.

We need to define an edge × edge diagonal matrix, H, which contains the products hS corresponding to the different paths. Thus,

$$
\text{Edge / Edge}
$$

$$
H = \begin{pmatrix} h_1S_1 & 0 & \cdots & 0 \\ 0 & h_2S_2 & 0 & 0 \\ \vdots & 0 & \ddots & \vdots \\ 0 & 0 & \cdots & h_{12}S_{12} \end{pmatrix} \tag{17.30}
$$

Heat conduction will not occur along paths 1 and 12, so these matrix elements will be set to 0. Along the other paths the individual heat transfer coefficients must be calculated. These will depend on the oil temperature and velocity in the duct and are calculated iteratively. These velocities have been calculated in Section 17.2.2 and should be available for the disk temperature calculation and later use. The surfaces S will depend on the disk dimensions and will be the same for identical disks.

Let E be an edge × circ matrix with all 1's as entries. Using E, along with C and H, we can obtain a new circ × circ matrix, which for the 3-disk section is

$$
G = CHE = \begin{pmatrix} g_1 & 0 & 0 \\ 0 & g_2 & 0 \\ 0 & 0 & g_3 \end{pmatrix} \tag{17.31}
$$

with entries

$$
g_1 = h_2S_2 + h_3S_3 + h_4S_4 + h_5S_5
$$
$$
g_2 = h_5S_5 + h_6S_6 + h_7S_7 + h_8S_8
$$
$$
g_3 = h_8S_8 + h_9S_9 + h_{10}S_{10} + h_{11}S_{11}
$$

Let $\mathbf{T}_C$ be a vector of disk temperatures. The dimension of this vector = the number of circuits. Thus, an element of this vector, T_{Ci}, is defined on circuit i since they are in 1–1 correspondence with the number of circuits. Letting G operate on $\mathbf{T}_C$, we obtain

$$
G\mathbf{T}_C = \begin{pmatrix} g_1 & 0 & 0 \\ 0 & g_2 & 0 \\ 0 & 0 & g_3 \end{pmatrix} \begin{pmatrix} T_{C1} \\ T_{C2} \\ T_{C3} \end{pmatrix} = \begin{pmatrix} g_1T_{C1} \\ g_2T_{C2} \\ g_3T_{C3} \end{pmatrix} \tag{17.32}
$$

This reproduces the first term after the second equality sign in (17.28) for all the disks.

For the second term, we need to obtain the bulk or average oil temperatures in the ducts. We note that the temperature of the oil entering a duct is one of the nodal temperatures, T_{node}, and the temperature rise of the oil in the duct is ΔT_{duct}. Thus, the average duct oil temperature is given by $T_{duct,ave} = T_{node} + \frac{1}{2}\Delta T_{duct}$.

We see from the transpose of the directed incidence matrix in (17.19) that of the two nodes associated with an edge, the first one corresponds to the entrance node for the oil flow into the edge (duct). Its matrix entry is 1. The exit node has a matrix entry of −1. To eliminate the −1's from this matrix, we can add the transpose of the directed incidence matrix to the transpose of the incidence matrix (where all the −1's are replaced by 1's so that these entries will end up being 0 when the sum is taken) and take ½ the sum. Label this new matrix A_+^T. Thus,

$$A_+^T = \frac{1}{2}\left(A_{DI}^T + A_I^T\right) \tag{17.33}$$

Let $\mathbf{T}_N$ be a vector of nodal temperatures. Then we see that $A_+^T\mathbf{T}_N$ is an edge vector whose entries are the nodal temperatures at the nodes corresponding to the oil flow entry into the edge (duct). It is only necessary to add ½ of the vector corresponding to the temperature rise along the duct, call it $\Delta\mathbf{T}$, to this so that the average duct temperature vector, $\mathbf{T}_B$, is given by

$$\mathbf{T}_B = A_+^T\mathbf{T}_N + \frac{1}{2}\Delta\mathbf{T} \tag{17.34}$$

The nodal vector, $\mathbf{T}_N$, and the edge vector, $\Delta\mathbf{T}$, will be determined in the next section. The matrix CH applied to $\mathbf{T}_B$ gives us the second set of terms after the second equality sign in (17.28). Thus, letting $\mathbf{I}^2\mathbf{R}$ be a vector of I^2R losses for the disks, (17.28) can be expressed in matrix form as

$$\mathbf{I}^2\mathbf{R} = G\mathbf{T}_C - CHA_+^T\mathbf{T}_N - \frac{1}{2}CH\Delta\mathbf{T} \tag{17.35}$$

The I^2Rs for the disks are input quantities. Although the current I may be common for all the disks, the resistance R is temperature dependent and includes any eddy losses due to the particular magnetic field distribution in the transformer and can vary from disk to disk.

17.2.4 Nodal Temperatures and Duct Temperature Rises

The rise in oil temperature in a given duct is due to heat transfer from the disk or disks contiguous with the duct. The heat transferred to a mass of oil flowing through a duct per unit time is given by

$$c\Delta T\frac{dM}{dt} = c\rho Av\Delta T \tag{17.36}$$

where (17.7) has been used. The heat transferred per unit time from a disk surface, S, in contact with the flowing oil is $hS(T_C - T_B)$. The symbols for the disk temperature, T_C, and average duct oil temperature, T_B, have been defined previously. Implementing this heat balance equation for horizontal duct 5 in Figure 17.2, for example, we get

$$c\rho A_5 v_5 \Delta T_5 = h_5 S_5 \left[(T_{C1} - T_{B5}) + (T_{C2} - T_{B5}) \right] \tag{17.37}$$

On the vertical ducts and the top and bottom ducts next to the oil flow washers, only one term would appear on the right-hand side of (17.37).

To implement the left-hand side of (17.37) in matrix form, we need to create a diagonal matrix containing the products of duct areas × duct oil velocities such as $A_5 v_5$. Call this edge × edge matrix U so that

$$U = \begin{pmatrix} A_1 v_1 & 0 & \cdots & 0 \\ 0 & A_2 v_2 & 0 & 0 \\ \vdots & 0 & \ddots & \vdots \\ 0 & 0 & \cdots & A_{12} v_{12} \end{pmatrix} \tag{17.38}$$

On the right-hand side of (17.37), we need to associate one or two disks with a given duct. This can be accomplished by means of the transpose of the circuit matrix

Edge\Circ

$$C^T = \begin{pmatrix} & \mathbf{1} & \mathbf{2} & \mathbf{3} \\ \mathbf{1} & 0 & 0 & 0 \\ \mathbf{2} & 1 & 0 & 0 \\ \mathbf{3} & 1 & 0 & 0 \\ \mathbf{4} & 1 & 0 & 0 \\ \mathbf{5} & 1 & 1 & 0 \\ \mathbf{6} & 0 & 1 & 0 \\ \mathbf{7} & 0 & 1 & 0 \\ \mathbf{8} & 0 & 1 & 1 \\ \mathbf{9} & 0 & 0 & 1 \\ \mathbf{10} & 0 & 0 & 1 \\ \mathbf{11} & 0 & 0 & 1 \\ \mathbf{12} & 0 & 0 & 0 \end{pmatrix} \tag{17.39}$$

When this operates on the vector of disk temperatures, which are associated with the circuits, we see that ducts (edges) 2, 3, and 4 will pick out the disk temperature T_{C1}, whereas duct 5 will pick out both T_{C1} and T_{C2}. So this matrix, when operating on $\mathbf{T}_C$, will automatically contain the appropriate disk or disks required by (17.37) for the given duct. We only need to multiply it by the matrix H, as given in (17.30), to get the expressions containing T_C on the right-hand side of (17.37). This same matrix H can multiply $\mathbf{T}_B$ to get one of the bulk oil terms in (17.37). To get two T_B terms for the horizontal ducts, excepting the top and bottom ones, we need to introduce another matrix, which contains all 1's on the diagonal except for the appropriate horizontal ducts where the diagonal elements equal 2. Call this matrix Q. For the 3-disk example in Figure 17.2, this matrix is

Edge\Edge

$$Q = \begin{pmatrix}
 & 1 & 2 & 3 & 4 & 5 & 6 & 7 & 8 & 9 & 10 & 11 & 12 \\
1 & 1 & 0 & 0 & 0 & 0 & 0 & 0 & 0 & 0 & 0 & 0 & 0 \\
2 & 0 & 1 & 0 & 0 & 0 & 0 & 0 & 0 & 0 & 0 & 0 & 0 \\
3 & 0 & 0 & 1 & 0 & 0 & 0 & 0 & 0 & 0 & 0 & 0 & 0 \\
4 & 0 & 0 & 0 & 1 & 0 & 0 & 0 & 0 & 0 & 0 & 0 & 0 \\
5 & 0 & 0 & 0 & 0 & 2 & 0 & 0 & 0 & 0 & 0 & 0 & 0 \\
6 & 0 & 0 & 0 & 0 & 0 & 1 & 0 & 0 & 0 & 0 & 0 & 0 \\
7 & 0 & 0 & 0 & 0 & 0 & 0 & 1 & 0 & 0 & 0 & 0 & 0 \\
8 & 0 & 0 & 0 & 0 & 0 & 0 & 0 & 2 & 0 & 0 & 0 & 0 \\
9 & 0 & 0 & 0 & 0 & 0 & 0 & 0 & 0 & 1 & 0 & 0 & 0 \\
10 & 0 & 0 & 0 & 0 & 0 & 0 & 0 & 0 & 0 & 1 & 0 & 0 \\
11 & 0 & 0 & 0 & 0 & 0 & 0 & 0 & 0 & 0 & 0 & 1 & 0 \\
12 & 0 & 0 & 0 & 0 & 0 & 0 & 0 & 0 & 0 & 0 & 0 & 1
\end{pmatrix} \qquad (17.40)$$

Using these matrices, we can cast (17.36) and (17.37) in a form that applies to all the ducts in the section

$$\rho c \mathcal{U} \Delta \mathbf{T} = HC^{\mathsf{T}}\mathbf{T_C} - HQ\mathbf{T_B} = HC^{\mathsf{T}}\mathbf{T_C} - HQA_+^{\mathsf{T}}\mathbf{T_N} - \frac{1}{2}HQ\Delta\mathbf{T} \qquad (17.41)$$

As shown in (17.41), the nodal temperatures and duct temperature rises are interrelated. We need more information to determine them uniquely. This requirement is an energy balance at the nodes. The energy per unit time of a mass of fluid flowing through a duct of cross-sectional A is given by $\rho c A v(T - T_{ref})$. This is taken relative to some reference temperature T_{ref} and assumes that the specific heat, c, remains relatively constant from the reference temperature up to some temperature T. T_{ref} will cancel out in the energy flow balance equations at the nodes because of mass balance at the nodes and so will be dropped. The ρc term will also cancel out in the equations, since it is assumed to be constant.

Thus, the energy flow balances at the two types of nodes (left and right), say nodes 6 and 7 in Figure 17.2, are

$$A_7 v_7 \left(T_4 + \Delta T_7\right) = A_8 v_8 T_6 + A_{10} v_{10} T_6$$
$$A_6 v_6 \left(T_5 + \Delta T_6\right) + A_8 v_8 \left(T_6 + \Delta T_8\right) = A_9 v_9 T_7 \qquad (17.42)$$

To implement this in matrix form, we note that the Av's for all the ducts are contained in the matrix U given in (17.38). We also note that the matrix A_+^{T} defined in (17.33) picks out the entrance node for a given duct or edge. Thus, $A_+^{\mathsf{T}}\mathbf{T_N}$ is an edge vector containing the nodal temperatures of the nodes at the entrance to the duct or edge. Therefore, (17.42), excluding the ΔT terms, can be expressed in matrix form as $A_{DI}U A_+^{\mathsf{T}}\mathbf{T_N}$. To pick out the ΔT terms, we need a node × edge matrix operating on an edge vector, such as $\Delta\mathbf{T}$ and picks out only the edge or edges entering the node. Such a matrix can be obtained from the directed and standard incidence matrix according to

$$A_- = \frac{1}{2}(A_{DI} - A_I) \tag{17.43}$$

This is the counterpart to A_+ whose transpose is defined in (17.33). Thus, the ΔT terms in (17.42) can be picked out by $A\_U\Delta\mathbf{T}$.

Although operating on an edge vector, the result is a nodal vector, which can be added to the vector term containing the nodal temperatures. Because (17.42) is an energy balance equation, we must allow for an entrance energy term at node 1 so that the resulting matrix equation is

$$A_{DI}UA_+^T\mathbf{T}_N + A\_U\Delta\mathbf{T} = \begin{pmatrix} A_1 v_0 T_0 \\ 0 \\ \vdots \\ 0 \end{pmatrix} \tag{17.44}$$

The nodal temperature, duct temperature rise, and disk temperature equations can all be combined into a single matrix equation so that they can all be solved simultaneously. By defining an overall vector containing $\Delta\mathbf{T}$, $\mathbf{T}_N$, and $\mathbf{T}_C$, we can write

$$\begin{pmatrix} \rho c U + \frac{1}{2}HQ & HQA_+^T & -HC^T \\ A\_U & A_{DI}UA_+^T & 0 \\ -\frac{1}{2}CH & -CHA_+^T & G \end{pmatrix} \begin{pmatrix} \Delta\mathbf{T} \\ \mathbf{T}_N \\ \mathbf{T}_C \end{pmatrix} = \begin{pmatrix} 0 \\ \vdots \\ A_1 v_0 T_0 \\ \vdots \\ \mathbf{I}^2\mathbf{R} \end{pmatrix} \tag{17.45}$$

Most of the matrix elements are temperature and velocity dependent, so (17.45), along with (17.23) and (17.24), which determine **P** and **V**, must be solved iteratively.

17.2.5 Comparison with Test Data

Many comparisons were made between the graph theory approach and the previous one, as discussed in [Del99] and [Del10]. The previous approach used separate indexing schemes to label the disks, paths, and nodes in a section. Then appropriate physical hydraulic or thermal equations are written for particular disks, paths, and nodes. In this scheme, the disk equations depend on whether the disk is at the bottom, top, or middle of the section, since the equations are different for these cases. Likewise the path equations depend on whether the path is horizontal or vertical and the nodal equations depend on whether the node is on the left or right. This complicates the computer coding compared to the graph theory approach. However, both approaches gave closely similar results, since they used the same physical input.

Only a sampling of these comparisons will be given here. For simplicity, we chose cases with only HV and LV directed oil flow disk windings and no tap windings, since these are not usually directed oil flow windings. The new disk winding calculation was embedded in a program, which considered the full transformer, including the radiators. Table 17.1 shows the comparison between the previous method and the graph theory method for the

TABLE 17.1

Comparison of the Graph Theory Calculation of Key Transformer Temperatures with the Previous Method and with Test Data

	Previous	Graph	Test
25 MVA ONAF			
Top oil	56.2	56.1	54.9
Bot oil	29.5	29.4	28.7
LV Ave	57.4	57.1	58.8
HV Ave	54.1	53.7	54.4
Hot spot	77.6	79.5	74.4
100 MVA ONAF			
Top oil	49.6	49.5	47.4
Bot oil	12.1	12.2	11.6
LV Ave	46.0	45.3	47.9
HV Ave	51.6	50.7	52.8
Hot spot	73.2	71.7	71.9
300 MVA Auto ONAF			
Top oil	45.6	45.4	42.4
Bot oil	15.6	15.8	13.8
LV Ave	43.1	42.5	43.0
HV Ave	40.9	40.6	40.1
Hot spot	61.4	61.7	60.5

Source: Del Vecchio, R.M., *IEEE Trans. Power Deliv.*, 29(5), October 2014, 2279–2286. © 2014 IEEE.

disk winding cooling and with heat run test data. The heat run test data acquisition method complies with the IEEE standards, C57.12.00 and C57.12.90, and uses state-of-the-art equipment for loss and temperature measurements [IEE10], [IEEE10a]. Based on repetitive tests, the winding temperatures are accurate to ±2°C and the oil temperatures to ±1°C. The calculated temperatures generally agree with the test temperatures to within ±5°C.

Figure 17.3 shows the oil velocity in the horizontal ducts along an LV winding calculated with the new graph theory program. Brief spacing separates the different winding sections between oil flow washers. The key spacer thickness is uniform at 3 mm along the winding except for two short sections where it is 12 mm. These short sections are where the velocity dips in the graph. This velocity profile is nearly identical to that produced by the previous calculation method.

In the graph theory approach, the details of the network are handled by standard graph theory matrices or combinations of these matrices. Naturally, the specific details of the winding such as duct sizes, paper thicknesses, the number of key spacers, and the number of disks per section must be inputted. Also, physical parameters such as the friction factors and heat transfer coefficients must be inputted or calculated for the specific disk geometry. These winding-specific or physical parameters appear in diagonal matrices. These matrices are reasonably easy to organize at the beginning of the calculation. This should be compared with the older method where these parameters are inserted in individual equations.

The velocities, pressures, and temperatures obtained by the present procedure and with those obtained by previous methods are in good agreement, but this approach offers organizational advantages and greater simplicity in coding.

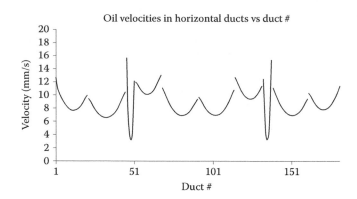

FIGURE 17.3

Oil velocities in the horizontal ducts along an LV winding. (From Del Vecchio, R.M., *IEEE Trans. Power Deliv.*, 29(5), October 2014, 2279–2286. © 2014 IEEE.)

17.3 Thermal Model for Coils without Directed Oil Flow

Our treatment of nondirected oil flow coils is fairly simplistic. As shown in Figure 17.4, there are inner and outer vertical oil flow channels with cross-sectional areas A_1 and A_2 and hydraulic diameters D_1 and D_2. The oil velocities in the two channels can differ. They are labeled v_1 and v_2 in the figure. The oil temperature is assumed to vary linearly from T_o at the coil bottom to $T_o + \Delta T_1$ and $T_o + \Delta T_2$ at the top of the inner and outer channel, respectively. The conductor temperatures are also assumed to vary linearly with an average value of T_c.

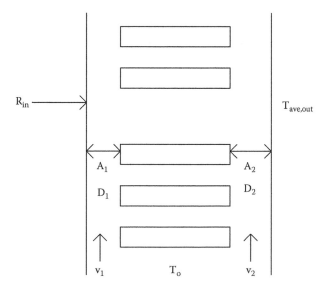

FIGURE 17.4

Thermal model of a nondirected oil flow coil.

The thermal pressure drops in the two channels are given by (17.8), which becomes, in terms of the variables defined earlier,

$$\Delta P_1 = \beta \rho g H \left(T_o + \frac{\Delta T_1}{2} - T_{ave,out} \right)$$

$$\Delta P_2 = \beta \rho g H \left(T_o + \frac{\Delta T_2}{2} - T_{ave,out} \right)$$

(17.46)

These pressure drops need only overcome the fluid friction in the two channels, which, for laminar flow, is given by (17.5). In terms of the present parameters, this becomes

$$\Delta P_1 = \frac{1}{2} \frac{\mu_1 K_1 H}{D_1^2} v_1, \quad \Delta P_2 = \frac{1}{2} \frac{\mu_2 K_2 H}{D_2^2} v_2$$

(17.47)

Equating (17.46) and (17.47), we obtain

$$\frac{\mu_1 K_1}{\beta \rho g D_1^2} v_1 - \Delta T_1 = 2 \left(T_o - T_{ave,out} \right)$$

$$\frac{\mu_2 K_2}{\beta \rho g D_2^2} v_2 - \Delta T_2 = 2 \left(T_o - T_{ave,out} \right)$$

(17.48)

The overall energy balance equation, analogous to (17.11), is

$$\rho c v_1 A_1 \Delta T_1 + \rho c v_2 A_2 \Delta T_2 = I^2 R$$

(17.49)

where R is the total resistance of the coil and is a function of ΔT by a formula analogous to (17.12). Further thermal equations come from the energy balance between the surface heat loss by the coil to the separate oil flow paths. Thus, we have approximately

$$h_1 A_{c,1} \left[T_c - \left(T_o + \frac{\Delta T_1}{2} \right) \right] = \rho c A_1 v_1 \Delta T_1$$

$$h_2 A_{c,2} \left[T_c - \left(T_o + \frac{\Delta T_2}{2} \right) \right] = \rho c A_2 v_2 \Delta T_2$$

(17.50)

where
  h_1 and h_2 are the surface heat transfer coefficients as given by (17.26)
  $A_{c,1}$ and $A_{c,2}$ are the conductor surface areas across which heat flows into the two oil channels

We take these areas to be half the total surface area of the conductor. This assumes the oil meanders into the horizontal spaces between the conductors on its way up the coil. We use (17.27) for h_{conv} with a smaller effective value for L than the coil height.

Equations 17.48 through 17.50 are five equations in the five unknowns v_1, v_2, ΔT_1, ΔT_2, and T_c. The quantities T_o and $T_{ave,out}$ will be determined from an overall energy balance for the transformer and are considered as known here. These equations are nonlinear because the μ's and h's depend on temperature and velocity and also because products of unknowns such as $v\Delta T$ occur. We use a Newton–Raphson iteration scheme to solve these equations. The pressure drops can be obtained from (17.47).

17.4 Radiator Thermal Model

The radiators modeled here are fairly typical in that they consist of a series of vertical plates containing narrow oil channels. The plates are uniformly spaced and attached to inlet and outlet pipes at the top and bottom. These pipes are attached to the transformer tank and must be below the top oil level. Fans may be present. They blow air horizontally through one or more radiators stacked side by side. The oil flow paths for a radiator are similar to those of a disk winding section turned on its side as shown in Figure 17.5. For this analysis, a node and path numbering scheme is used. In some of the paths, however, the positive flow direction is reversed. The analysis is very similar to that given previously for disk coils with directed oil flow, which did not use graph theory, and will only be sketched here.

The pressure differences along a path are balanced by the frictional resistance in the steady state so that, for laminar flow, Equation 17.5 holds along each path, where the velocity and pressure unknowns will be labeled by the appropriate path and node numbers. For n plates along a radiator, there are 2n nodes and 3n − 1 oil paths as shown in Figure 17.5. Thus, we need 2n more equations. These are given as before by mass continuity at the nodes, similar to Equation 17.7. Some differences will occur because the positive flow direction is changed for some of the paths. Similar to what was done for the coils, we determine the overall pressure difference across the radiator from buoyancy considerations so that an equation such as (17.10) holds. However, $T_{ave,out}$ will differ from that used for the coils, since the tank oil adjacent to the radiators will be hotter than that adjacent to the coils because the average radiator vertical position is usually above that of the coils. An overall energy balance is needed to obtain the input oil velocity so that an equation similar to (17.11) holds with the right-hand side replaced by the total heat lost by the radiators.

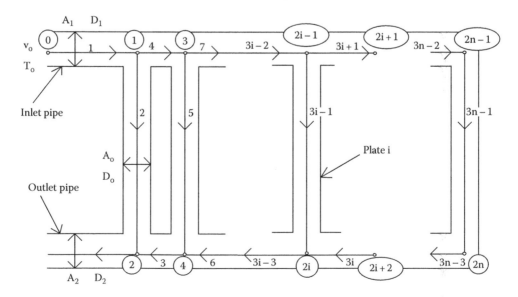

FIGURE 17.5
Node and path numbering scheme for a radiator containing n plates.

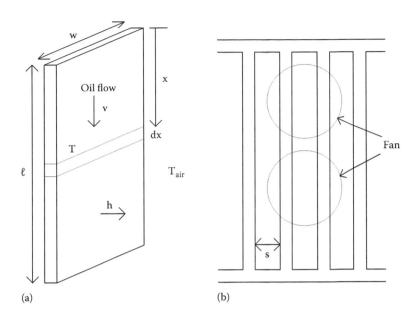

FIGURE 17.6
Parameters used in radiator cooling calculation: (a) one plate and (b) stacked plates separated by distance s.

In the radiator cooling process, the oil temperature drops as it passes downward through a radiator plate, giving up its heat to the air through the radiator surface. Figure 17.6a shows a simplified drawing of a plate with some of the parameters labeled.

We consider a thin horizontal strip of radiator surface of area $2w\,dx = 2w\ell(dx/\ell) = 2A_s(dx/\ell)$, where A_s is the area of one side of the plate and the factor of 2 accounts for both sides. ℓ is the plate height and w is its width. The heat lost through this surface is $2hA_s(T - T_{air})(dx/\ell)$, where h is the heat transfer coefficient, T is the oil temperature at position x, and T_{air} is the ambient air temperature. The heat lost by the oil in flowing past the distance dx is $-\rho cvA_R\,dT$, where v is the oil velocity and A_R the cross-sectional area through which the oil flows through the plate. Equating these expressions and rearranging, we get

$$\frac{2hA_s}{\rho cvA_R\ell}dx = -\frac{dT}{T - T_{air}} \qquad (17.51)$$

Assuming that h is constant here (it will be evaluated at the average temperature), we can integrate to obtain

$$\Delta T = \left(T_{top} - T_{air}\right)\left[1 - \exp\left(-\frac{2hA_s}{\rho cvA_R}\right)\right] \qquad (17.52)$$

where
  ΔT is the temperature drop across the plate
  T_{top} is the top oil temperature

We get an equation like this for each plate so that for plate i, v should be labeled v_{3i-1}, T_{top} as T_{2i-1}, h as h_i, and ΔT as ΔT_i. This gives n equations. The new unknowns are the $n\Delta T$'s and 2n nodal temperatures. However, the top nodal temperatures are all equal to the oil input temperature T_o, since we are neglecting any cooling along the input and output pipes. At the bottom nodes, we obtain for node 2i

$$A_R v_{3i-1}\left(T_o - \Delta T_i\right) + A_2 v_{3i} T_{2i+2} = A_2 v_{3i-3} T_{2i} \tag{17.53}$$

A slight modification of this is needed for node 2n. Thus, we obtain sufficient equations to solve for all the unknowns. Iteration is required between the pressure–velocity equations and $T - \Delta T$ equations, since they are interdependent. Iteration is also required because of the nonlinearities.

The expression for the surface heat transfer coefficient, h, depends on whether or not fans are used. For natural convection (no fans), we use an expression, which applies to a row of vertical plates separated by a distance s [Roh85]:

$$h = \frac{k_{air}}{s}\left[\left(\frac{24}{Ra}\right)^{1.9} + \left(\frac{1}{0.62 Ra^{1/4}}\right)^{1.9}\right]^{-\frac{1}{1.9}} \tag{17.54}$$

Ra is the Rayleigh number given by the product of the Grashof and Prandtl numbers:

$$Ra = Gr\,Pr = \frac{g\beta_{air}c_{air}\rho_{air}^2 \Delta T}{k_{air}\mu_{air}}\left(\frac{s^4}{\ell}\right) \tag{17.55}$$

where the compressibility β, specific heat c, density ρ, thermal conductivity k, and viscosity μ all apply to air at the temperature $(T_s + T_{air})/2$ where T_s is the average surface temperature of the plate. Also, $\Delta T = T_s - T_{air}$. The quantities c, ρ, k, and μ for air vary with temperature [Kre80]. This must be taken into account in the said formulas since T_s can differ from plate to plate. However, we find in practice that a similar expression given later for tank cooling, (17.58) with ℓ replacing L, works better for our radiators, so it is used.

When fans are blowing, we use the heat transfer coefficient for the turbulent flow of a fluid through a long narrow channel of width s:

$$h = 0.023\frac{k_{air}}{D}\,Re_D^{0.8}\,Pr^{0.33} \tag{17.56}$$

where the hydraulic diameter D = 2s. The Reynolds number is given by (17.2) with ρ and μ for air, D = 2s, and v is an average velocity determined by the characteristics of the fans and the number of radiators stacked together. The Prandtl number for air is nearly constant throughout the temperature range of interest and is Pr = 0.71.

A simple way of parameterizing the radiator air velocity with fans present, which works well in practice, is

$$v_n = \left(\frac{1}{n}\right)^p v_1 \tag{17.57}$$

where
    n is the number of stacked radiators cooled by a given fan bank
    v_n is the air velocity flowing past the radiator surfaces
    p is an exponent to be determined by test results
    v_1 is the fan velocity produced when only one radiator is present

We find in our designs that p = 0.58 works well. In addition, v_1 was taken to be the nominal fan velocity as specified by the manufacturer. (This can be determined from the flow capacity and the fan's area.) We assumed that the radiator surface cooled by fans was proportional to the fraction of the fan's area covering the radiator side (Figure 17.6b). The remaining surface was assumed cooled by convective cooling.

17.5 Tank Cooling

Cooling from the tank occurs by means of natural convection and radiation. Of course, the radiators also cool to some extent by radiation; however, this is small compared with convective cooling. This is because the full radiator surface does not participate in radiative cooling. Because of the nearness of the plates to each other, much of a plate's radiant energy is reabsorbed by a neighboring plate. The effective cooling area for radiation is really an outer surface envelope and as such is best lumped with the tank cooling. Thus, the effective tank area for radiative cooling is an outer envelope, including the radiators. (A string pulled tautly around the tank plus radiators would lie on the cooling surface.) However, for convective cooling, the normal tank surface area is involved.

For natural convection in air from the tank walls of height L, we use the heat transfer coefficient [Roh85]:

$$h_{conv,tank} = \frac{k_{air}}{L} \left\{ \left[\frac{2.8}{\ln\left(1 + \frac{5.44}{Ra_L^{1/4}}\right)} \right]^6 + 1.18 \times 10^{-6} Ra_L^2 \right\}^{1/6} \quad (17.58)$$

where Ra is the Rayleigh number, which, in this context, is given by

$$Ra_L = \frac{g \beta_{air} c_{air} \rho_{air}^2 L^3 \Delta T}{k_{air} \mu_{air}} \quad (17.59)$$

where the temperature dependencies are evaluated at $(T_s + T_{air})/2$ and $\Delta T = T_s - T_{air}$ as before. The Rayleigh number in this formula should be restricted to the range $1 < Ra_L < 10^{12}$. This is satisfied for the temperatures and tank dimensions of interest. T_s is the average tank wall temperature and T_{air} the ambient air temperature. Thus, the heat lost from the tank walls per unit time due to convection, $W_{conv,tank}$, is given by

$$W_{conv,tank} = h_{conv,tank} A_{conv,tank} \left(T_s - T_{air} \right) \quad (17.60)$$

where $A_{conv,tank}$ is the area of the tank's lateral walls.

The top surface contributes to the heat loss according to a formula similar to (17.60) but with a heat transfer coefficient given by [Kre80]

$$h_{conv,top} = 0.15 \frac{k_{air}}{B} Ra_B^{0.333} \quad (17.61)$$

and with the tank top area in place of the side area. In (17.61), B is the tank width, which must also be used in the Rayleigh number (17.59) instead of L. In this formula, the Rayleigh number should be restricted to the range $8 \times 10^6 < \text{Ra}_B < 10^{11}$, which is also satisfied for typical tank widths in power transformers. On the bottom, we assume the ground acts as an insulator.

Radiant heat loss per unit time from the tank, $W_{rad,tank}$, is given by the Stephan–Boltzmann law

$$W_{rad,tank} = \sigma E A_{rad,tank} \left(T_{K,s}^4 - T_{K,air}^4 \right) \tag{17.62}$$

where
$\sigma = 5.67 \times 10^{-8} \ W/m^2 \ K^4$ is the Stephan–Boltzmann constant
E is the surface emissivity ($E \cong 0.95$ for gray paint)
$A_{rad,tank}$ is the effective tank area for radiation
$T_{K,s}$ is the average tank surface temperature in K
$T_{K,air}$ is the ambient air temperature in K

Equation 17.62 can be written to resemble (17.60) with a convection coefficient given by

$$h_{rad,tank} = \sigma E \left(T_{K,s} + T_{K,air} \right) \left(T_{K,s}^2 + T_{K,air}^2 \right) \tag{17.63}$$

This is temperature dependent, but so is $h_{conv,tank}$.

The total power loss by the tank is given by the sum of (17.60), together with the corresponding expression for the top heat loss, and (17.62). This is then added to the radiator loss to get the total power loss from the transformer. The radiator loss is the sum from all the radiators. Although we modeled only one radiator, assuming they are all identical, the total radiator loss is just the number of radiators times the loss from one. Otherwise, we must take into account differences among the radiators. In equilibrium, the total power loss must match the total power dissipation of the coils + core + the stray losses in the tank walls, brackets, leads, etc.

17.6 Oil Mixing in the Tank

Perhaps the most complex part of the oil flow in transformers occurs in the tank. The lack of constraining channels or baffles means that the oil is free to take irregular paths such as along localized circulations or eddies. Nevertheless, there is undoubtedly some overall order in the temperature distribution and flow pattern. As occurs with any attempt to model a complicated system, we make some idealized assumptions here in an effort to describe the average behavior of the tank oil.

As shown in Figure 17.7, we assume that the cold oil at the bottom of the tank has a uniform temperature, T_{bot}, between the bottom radiator discharge pipe and the tank bottom and that the top oil is at a uniform temperature, T_{top}, from the top of the coils to the top oil level. We further assume that the temperature variation between T_{bot} and T_{top} is linear. This allows us to calculate the average oil temperature along a column of oil adjacent to and of equal height to the coils and likewise for the radiators, thus determining $T_{ave,out,coils}$ and $T_{ave,out,rads}$. These two temperatures will differ because the average radiator vertical position

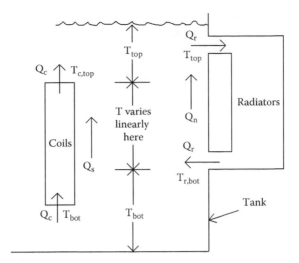

FIGURE 17.7
Assumed oil temperature distribution inside a tank. The oil flows, Q, as well as the flow-weighted temperatures are also indicated.

is above that of the coils. These average temperatures are used in determining the thermal pressure drop across the coils and radiators as discussed previously.

In Figure 17.7, we have labeled the volumetric oil flows Q with subscripts c for coils, r for radiators, s for stray loss oil flows, and n for net. Thus,

$$Q_n = Q_r - Q_c - Q_s \tag{17.64}$$

Q_s accounts for the oil flow necessary to cool the core, brackets, tank walls, etc. Since there is no constraint, mechanical or otherwise, to force the radiator and coil + stray loss flows to be identical, Q_n can differ from zero. The flow Q_c refers to the sum of the flows from all the coils and Q_r the sum from all the radiators.

We have also indicated temperatures associated with some of the flows. The oil flowing into the bottom of the coils is at temperature T_{bot} and the oil flowing into the top of the radiators is at temperature T_{top}. The temperature of the oil flowing out of the top of the coils, $T_{c,top}$, is a flow-weighted temperature of the oil from all of the coils. Thus,

$$Q_c T_{c,top} = \sum_{i=1}^{\text{\# coils}} Q_{c,i} T_{c,top,i} \tag{17.65}$$

where i labels the individual coil flows and top temperatures of the oil emerging from the coils. Since we are assuming that all radiators are identical, $T_{r,bot}$ is the bottom temperature of the oil exiting a radiator. If the radiators were not all alike, we would use a flow-weighted average for this temperature also.

We assume, for simplicity, that the volumetric flow Q_s, associated with the stray power loss W_s, results in a temperature change of $\Delta T = T_{top} - T_{bot}$ for the oil participating in this flow. Thus, Q_s is given by

$$Q_s = \frac{W_s}{\rho c \left(T_{top} - T_{bot} \right)} \tag{17.66}$$

This is inherently an upward flow and, like the coil flow, is fed by the radiators. We do not attribute any downward oil flow to the tank cooling but assume that this merely affects the average oil temperature in the tank.

We assume that any net flow, Q_n, if positive, results in the transport of cold radiator oil at temperature $T_{r,bot}$ to the top of the tank and, if negative, results in the transport of hot coil oil at temperature $T_{c,top}$ to the bottom of the tank. These assumptions imply that for $Q_n > 0$,

$$T_{top} = \frac{T_{c,top}Q_c + T_{r,bot}Q_n}{Q_r - Q_s}$$
$$T_{bot} = T_{r,bot} \tag{17.67}$$

while for $Q_n < 0$,

$$T_{top} = T_{c,top}$$
$$T_{bot} = \frac{T_{r,bot}Q_r - T_{c,top}Q_n}{Q_c + Q_s} \tag{17.68}$$

Another way of handling the net flow Q_n is to assume that it raises the level of the bottom oil layer at temperature T_{bot} if positive and that it allows the top oil layer at temperature T_{top} to drop downward if negative. This will continue until $Q_n = 0$ at equilibrium.

After each calculation of the coil and radiator flows and temperatures, the quantities T_{top}, T_{bot}, $T_{ave,out,coils}$, $T_{ave,out,rads}$, and Q_s can be determined. Also, the losses in the coils and from the radiators can be obtained. From these, updated pressure drops across the coils and radiators as well as updated values of the oil velocities flowing into the coils and radiators can be calculated. The iterations continue until the temperatures reach their steady-state values and the losses generated equal the losses dissipated to the atmosphere to within some acceptable tolerance. This requires several levels of iteration. The coil and radiator iterations assume that the tank oil temperature distribution is known and these in turn influence the tank oil temperatures. A relaxation technique is required to keep the iteration process from becoming unstable. This simply means that the starting parameter values for the new iteration are some weighted average of the previous and newly calculated values.

17.7 Time Dependence

The basic assumption we make in dealing with time-dependent conditions is that at each instant of time the oil flow is in equilibrium with the heat (power) transferred by the conductors to the oil and with the power loss from the radiators at that instant. The velocities and oil temperatures will change with time but in such a way that equilibrium is maintained at each time step. This is referred to as a quasi-static approximation. The conductor heating, or cooling, on the other hand, is transient. We assume, however, that the temperature of a disk, which is part of a directed oil flow coil, is uniform throughout the disk so that only the time dependence of the average disk temperature is treated. A similar assumption is made for nondirected oil flow coils.

The heat generated in a conductor per unit time is I^2R, where R is its resistance (including eddy current effects) at the instantaneous temperature T_c and I its current. Here, "conductor" refers to a single disk for a directed oil flow coil and to the entire coil for a nondirected oil flow coil. The transient thermal equation for this conductor is

$$\rho_{cond} c_{cond} V \frac{dT_C}{dt} = I^2R - \sum_i h_i A_{C,i} \left(T_C - T_{b,i} \right) \tag{17.69}$$

The left-hand side is the heat stored in the conductor of volume V per unit time. This equals the heat generated inside the conductor per unit time minus the heat lost through its surface per unit time. The sum is over all surfaces of the conductor. The other symbols have their usual meaning with the subscript cond indicating that they refer to the conductor properties. The usual temperature and velocity dependencies occur in some of the parameters in (17.69). We solve these equations via a Runge–Kutta technique. These equations replace the steady-state disk coil equations for the T_C.

The previous equation accounts for heat (energy) storage in the current-carrying conductors as time progresses. Thermal energy is also stored in the rest of the transformer. We assume, for simplicity, that the remainder of the transformer is at a temperature given by the average oil temperature. This includes the core, the tank, the radiators, the brackets or braces, coils not carrying current, the insulation, the main tank oil, etc. Thus, the conservation of energy (power) requires that

$$\left(\sum_i c_i m_i \right) \frac{dT_{ave,oil}}{dt} = \rho c Q_c \left(T_{c,top} - T_{bot} \right) + W_s - W_{rads+tank} \tag{17.70}$$

where c_i is the specific heat and m_i the mass of the parts of the transformer, apart from the current-carrying coils, which store heat. Equation 17.70 states that the heat absorbed by the various parts of a transformer per unit time, except the current-carrying coils, at a particular instant equals the power flowing out of the coils in the form of heated oil plus the stray power losses minus the power dissipated by the radiators and tank to the atmosphere. Here, the stray losses include core losses, tank losses, and any losses occurring outside the coils. Equation 17.70 is solved using a trapezoidal time-stepping method.

Thus, transient cooling can be handled by making relatively minor modifications to the steady-state treatment, using the assumptions given previously. So far, the computer program we have developed treats the case where the mega volt amperes in SI units (MVA) of the transformer is suddenly changed from one level to another. A steady-state calculation is performed at the first MVA level. Then the transient calculation begins with currents and stray losses appropriate to the second MVA level. Also the fans may be switched on or off for the transient calculation. As expected, the solution approaches the steady-state values appropriate to the second MVA level after a sufficiently long time. With a little extra programming, one could input any desired transformer transient loading schedule and calculate the transient behavior.

17.8 Pumped Flow

The type of pumped flow considered here is one in which the radiator oil is pumped and some type of baffle arrangement is used to channel some or all of the pumped oil through the coils. Thus, the oil velocity into the entrance radiator pipes due to the pumps, v_o, is given by

$$v_o = \frac{Q_{pump}}{A_{rad}N_{rad}} \tag{17.71}$$

where

Q_{pump} is the volume of oil per unit time, which the pump can handle under the given conditions
A_{rad} is the area of the radiator entrance pipe
N_{rad} is the number of radiators

In addition to the pressure drop across the radiators due to the pump, the thermal pressure drop is also present and this adds a contribution, although small, to the velocity in (17.71).

We assume that a fraction f of the pumped oil through the radiators bypasses the coils and does not participate in cooling. This could be due to inefficient baffling or it could be by design. We also assume that part of the oil flow from the radiators is used to cool the stray losses. This flow, Q_s, is given by

$$Q_s = \frac{W_s}{W_t}(1-f)Q_{pump} \tag{17.72}$$

where

W_s is the stray power loss
W_t is the total loss
f is the fraction of the flow bypassing the coils

In addition to the thermal pressure drop across the coils, a pressure drop due to the pump is present. At equilibrium, this is determined by the requirement that

$$Q_c = (1-f)Q_{pump} - Q_s \tag{17.73}$$

Thus after each iteration, the flow from all the coils is determined and the pressure drop across the coils is adjusted until (17.73) is satisfied at steady state.

For pumped flow, it is necessary to check for non-laminar flow conditions and to adjust the heat transfer and friction coefficients accordingly.

17.9 Comparison with Test Results

Computer codes, based on the analysis given in this chapter, were written to perform steady-state, transient, and pumped oil flow calculations. Although we did not measure detailed temperature profiles along a coil, the codes calculate temperatures of all coil disks and of the oil in all the ducts as well as duct oil velocities. Figures 17.8 through 17.10 show these profiles

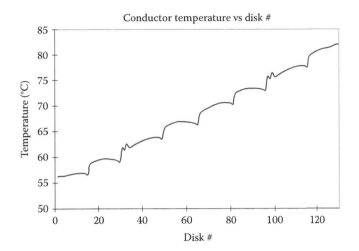

FIGURE 17.8
Calculated temperatures of the disks along a disk coil with directed oil flow washers. (From Del Vecchio, R.M. and Feghali, P., *1999 IEEE Transmission and Distribution Conference*, Now Orleans, LA, April 11–16, 1999, pp. 914–919. © 1999 IEEE.)

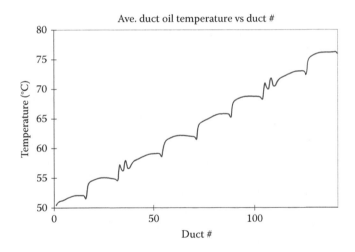

FIGURE 17.9
Calculated average oil temperatures in the horizontal ducts along a disk coil with directed oil flow washers. The temperature is assumed to vary linearly in the ducts. (From Del Vecchio, R.M. and Feghali, P., *1999 IEEE Transmission and Distribution Conference*, Now Orleans, LA, April 11–16, 1999, pp. 914–919. © 1999 IEEE.)

for one such coil having directed oil flow washers. The oil ducts referred to in the figures are the horizontal ducts and the conductor temperature is the average disk temperature. There is thinning in this coil, that is, increased duct size at two locations near disk numbers 35 and 105, and this can be seen in the profiles. The kinks in the profiles are an indication of the location of the oil flow washers.

As mentioned previously, some leeway was allowed for in the friction factors for the coils and radiators by including an overall multiplying factor. This accounts, in an average way, for nonideal conditions in real devices and must be determined experimentally.

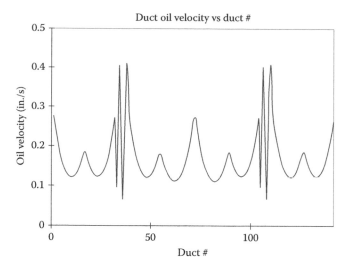

FIGURE 17.10

Calculated velocities of the oil in the horizontal ducts along a disk coil with directed oil flow washers. (From Del Vecchio, R.M. and Feghali, P., *1999 IEEE Transmission and Distribution Conference*, Now Orleans, LA, April 11–16, 1999, pp. 914–919. © 1999 IEEE.)

By comparing test data with calculations, we determined that the best agreement is achieved with a multiplying factor of 1.0 for the coils and 2.0 for the radiators. Thus, the coil friction is close to the theoretical value, whereas that for the radiators is twice as high.

Temperature data normally recorded in our standard heat runs are (1) mean oil temperature rise, (2) top oil temperature rise, (3) temperature drop across the radiators, and (4) average temperature rise of the windings. These are usually measured under both OA (OA is a common transformer designation for natural air cooling without fans) and FA conditions. The rises are with respect to the ambient air temperature. A statistical analysis of such data, taken on a number of transformers with MVA's ranging from 12 to 320, was performed to determine how well the calculations and test results agreed. This is shown in Table 17.2 where the mean and standard deviations refer to the differences between the calculated and measured quantity. Thus, a mean of 0 and a standard deviation of 0 would indicate a perfect fit. Nonzero results for these statistical measures reflect both the limitations of the model and some uncertainty in the measured quantities. Table 17.2 shows generally good agreement between the calculated and measured results. Essentially, only one parameter was adjusted in order to improve the agreement, namely, the overall friction factor multiplier for the radiators.

TABLE 17.2

Comparison of Calculated and Measured Temperatures

	OA		FA	
	Mean	**Std. Dev.**	**Mean**	**Std. Dev.**
Mean oil rise	0.86	2.58	0.19	5.11
Top oil rise	1.56	2.36	2.11	4.82
Drop across radiators	−2.23	4.51	−0.83	3.44
Average winding rise	−0.56	3.19	−0.14	2.38

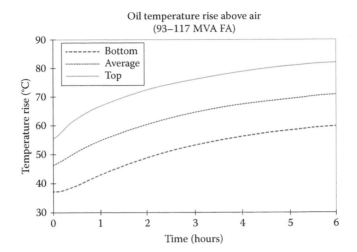

FIGURE 17.11
Time evolution of the bottom, average, and top tank oil temperatures starting from a steady-state loading of 93 MVA at time 0 when a sudden application of a 117 MVA load is applied.

Figures 17.11 and 17.12 show representative output from the transient program in graphical form. In this example, the transformer is operating at steady state at time 0 at the first MVA value (93 MVA) and suddenly, the loading corresponding to the second MVA (117 MVA) is applied. The time evolution of the oil temperature is shown in Figure 17.11, while that of one of the coils is shown in Figure 17.12. The program calculates time constants from this information. Figure 17.13 shows a direct comparison of the calculated and measured hot spot temperature of a coil in which a fiber optics temperature probe was imbedded. Although absolute temperatures are off by a few degrees, the shapes of the curves are very similar, indicating that the time constants are nearly the same.

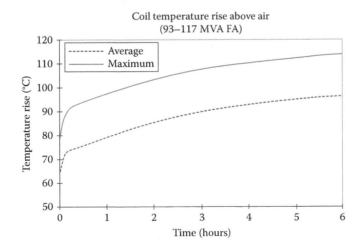

FIGURE 17.12
Time evolution of the average and maximum coil temperature starting from a steady-state loading of 93 MVA at time 0, when a sudden application of a 117 MVA load is applied.

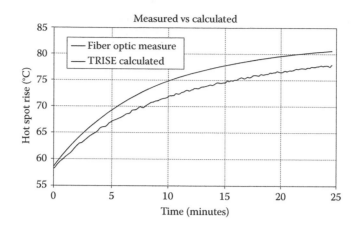

FIGURE 17.13

Comparison of calculated and measured temperature versus time of the coil's hot spot. The coil is operating at steady state at time 0 when a sudden additional loading is applied. The hot spot is measured by means of a fiber optics probe. The top curve is calculated and the bottom is the fiber optic curve.

17.10 Determining m and n Exponents

In order to estimate the temperature rises, which occur under overload conditions, approximate empirical methods have been developed, which make use of the m and n exponents. Although a detailed thermal model such as the one presented previously should make these approximations unnecessary, these exponents are widely used to obtain quick estimates of the overload temperatures for transformers in service. Although they are frequently measured in heat run tests, they can also be obtained from either the steady-state or transient calculations of the type presented earlier.

The n exponent is used to estimate the top oil temperature rise above the ambient air temperature, ΔT_{top}, under overload conditions, based on the value of this quantity under rated conditions. For this estimation, the transformer total losses under rated conditions must be known, as well as the losses at the overload condition. The latter losses can be estimated from knowing how the losses change with increased loading and will be discussed later. The formula used to obtain the overload top oil temperature rise above the ambient temperature is

$$\frac{\Delta T_{top,2}}{\Delta T_{top,1}} = \left(\frac{\text{Total Loss}_2}{\text{Total Loss}_1} \right)^n \tag{17.74}$$

Thus if 2 refers to the overload condition and 1 to the rated condition, the top oil temperature rise above ambient air temperature under overload conditions, $\Delta T_{top,2}$, can be obtained from a knowledge of the other quantities in (17.74).

In order to estimate the losses under overload conditions, we must separate the losses into I^2R resistive losses in the coils and stray losses, which include the core, tank, clamp, and additional eddy current losses that occur in the coils due to the stray flux. The I^2R loss can be calculated fairly accurately. By subtracting it from the total measured loss, a good estimate can be obtained for the stray loss. Alternatively, core loss, eddy current loss, tank loss, and clamp loss can be obtained from design formulas or other calculation methods. All of these losses depend on the square of the load current in either winding. (The stray losses depend on the square of the stray flux, which is proportional to the square of the current, assuming the materials are operating in their linear range.) Since the load current is proportional to the MVA of the unit, these losses are proportional to the MVA squared. The I^2R losses are proportional to the resistivity of the winding conductor, which is temperature dependent, while the stray losses are inversely proportional to the resistivity of the material in which they occur. Thus, the total loss at the higher loading can be obtained from the rated loss by

$$\text{Total loss}_2 = \left(\frac{MVA_2}{MVA_1}\right)^2 \left[I^2R \text{ loss}_1 \left(\frac{\rho_2}{\rho_1}\right)_{coil} + \text{Stray loss}_1 \left(\frac{\rho_1}{\rho_2}\right)_{stray}\right] \tag{17.75}$$

where ρ is the resistivity in this context. The resistivity ratios are temperature dependent so that some iteration may be required with (17.74) to arrive at a self-consistent solution. The stray resistivity ratio is meant to be a weighted average over the materials involved in the stray loss. However, since the temperature dependence of the coil, tank, and clamp materials is nearly the same, this ratio will be nearly the same as that of the coils.

We should note that the n coefficient is generally measured or calculated under conditions where the cooling is the same for the rated and overload conditions. Thus if fans are turned on for the rated loading, they are also assumed to be on for the overload condition. Similarly, if pumps are turned on at the 1 rating, they must also be on for the 2 rating for the forenamed formulas to work. Although n can be determined from measurements or calculations performed at two ratings, it is often desirable to determine it from three different loadings and do a best fit to (17.74) based on these. The three ratings generally chosen are 70%, 100%, and 125%. The 70%, 100% or 100%, 125% combinations are chosen when only two ratings are used. Measurements and calculations show that n based on the 70%, 100% combination can be fairly different from n based on the 100%, 125% combination. This suggests that n is not strictly a constant for a given transformer, and it should be determined for loadings in the range where it is expected to be used. A typical value of n for large power transformers is 0.9.

The m exponent is defined for each winding and relates the winding gradient to the winding current. The winding gradient at a particular steady-state loading is the temperature difference between the mean winding temperature and the mean oil temperature in the tank, that is,

$$\text{Gradient} = \text{Mean winding temperature} - \text{Mean tank oil temperature} \tag{17.76}$$

Letting 1 and 2 designate rated and overload conditions, the m coefficient for a particular winding relates the gradients to the winding currents by

$$\frac{\text{Gradient}_2}{\text{Gradient}_1} = \left(\frac{\text{Current}_2}{\text{Current}_1}\right)^{2m} \tag{17.77}$$

Since the currents are proportional to the MVAs, the MVA ratio could be substituted in formula (17.77) for the current ratio. The m exponents can be determined from two or three MVA loadings and, like the n exponent, depend to some extent on the MVA range covered. They also can vary considerably for different windings. A typical value for large power transformers is about 0.8. The factor of 2 in the exponent in (17.77) reflects the fact that the losses in the winding are proportional to the current squared and, ideally, so would the temperature rises, so that m takes into account deviations from this expectation.

The winding gradient is often used to estimate the winding's maximum temperature. A common procedure is to add some multiple of the gradient to the top oil temperature to arrive at the maximum temperature for that winding, for example,

$$\text{Max winding temp} = \text{Top oil temp} + 1.1 \times \text{Winding gradient} \qquad (17.78)$$

The factor of 1.1 can differ among manufacturers or transformer types and should be determined experimentally. Formulas such as (17.78) become unnecessary when detailed temperature calculations are available or maximum temperatures are measured directly by fiber optics or possibly other types of temperature probes inserted into the winding at its most probable location. Detailed temperature calculations such as (17.78) can provide some guidance as to where these probes should be inserted.

17.11 Loss of Life Calculation

Although transformers can fail for a number of reasons, such as from the application of excessive electrical or mechanical stress, even a highly protected unit will eventually fail due to the aging of its insulation. While the presence of moisture and oxygen affects the rate of insulation aging, these are usually limited to acceptable levels by appropriate maintenance practices so that the main factor affecting insulation aging is temperature. The insulation's maximum temperature, called the hot spot temperature, is thus the critical temperature governing aging. This is usually the highest of the maximum temperatures of the different windings, although it could also be located on the leads, which connect the windings to the bushings or to each other.

Numerous experimental studies have shown that the rate of insulation aging as measured by various parameters such as tensile strength or degree of polymerization (DP) follows an Arrhenius relationship:

$$K = Ae^{-B/T} \qquad (17.79)$$

Here
 K is a reaction rate constant (fractional change in quantity per unit time)
 A and B are parameters
 T is the absolute temperature

For standard cellulose-based paper insulation, $B = 15{,}000$ K. This value for B is an average from different studies using different properties to determine aging [McN91]. If X is the property used to measure aging, for example, tensile strength or degree of polymerization, DP, then we have

$$\frac{dX}{X} = -Kdt \Rightarrow X(t) = X(t=0)e^{-Kt} \tag{17.80}$$

The time required for the property X to drop to some fraction f of its initial value at time 0 if held at a constant temperature T, t_T, is thus determined by

$$\frac{X(t=t_T)}{X(t=0)} = f = e^{-Kt_T} \tag{17.81}$$

Taking logarithms, we obtain

$$t_T = -\frac{\ln f}{K} = -\frac{\ln f}{A}e^{B/T} \tag{17.82}$$

where (17.79) has been used.

There are several standard or normal insulation lifetimes, which have found some acceptance. For these, the insulation hot spot is assumed to age at a constant temperature of 110°C. The insulation is also assumed to be well dried and oxygen-free. Under these conditions, the time required for the insulation to retain 50% of its initial tensile strength is 7.42 years (65,000 hours), to retain 25% of its initial tensile strength is 15.41 years (135,000 hours), and to retain a DP level of 200 is 17.12 years (150,000 hours). Any of these criteria could be taken as a measure of a normal lifetime, depending on the experience or the degree of conservatism of the user. However, if the transformer is operated at a lower temperature continuously, the actual lifetime can be considerably longer than this. For example, if the hot spot is kept at 95°C continuously, the transformer insulation will take 36.6 years to be left with 50% of its initial tensile strength. Conversely, if operated at higher temperatures, the actual lifetime will be shortened.

Depending on the loading, the hot spot temperature will vary over the course of a single day so that this must be taken into account in determining the lifetime. A revealing way of doing this is to compare the time required to produce a fractional loss of some material property such as tensile strength at a given temperature with the time required to produce the same loss of the property at the reference temperature of 110°C. Thus, the time required to produce the fractional loss f in the material property at temperature T is given by (17.82) and the time to produce the same fractional loss at the reference temperature, designated T_o, is given by the same formula with T_o replacing T. This latter time will be designated t_{T_o}. Hence, we have for the given fractional loss,

$$\frac{t_{T_o}}{t_T} = e^{(B/T_o - B/T)} = AAF(T) \tag{17.83}$$

We have defined the aging acceleration factor, AAF, in (17.83). Thus, it requires AAF(T) times as much time at the reference temperature than at temperature T to produce the same fractional loss of life as determined by property X. Since the T's are in K, we have, converting to °C,

$$AAF(T) = \exp\left\{\frac{15,000}{T_o(°C) + 273} - \frac{15,000}{T(°C) + 273}\right\} \tag{17.84}$$

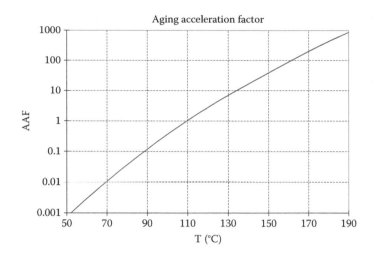

FIGURE 17.14
Aging acceleration factor (AAF) relative to 110°C vs. temperature.

AAF is > or <1 depending on whether T(°C) is > or <T_o(°C). This is plotted in Figure 17.14 for T_o = 110°C. Thus, from the figure, a unit operating at 95°C would have an AAF $\cong$ 0.2 so that the same unit operating at 110°C would require 20% of the time to age as much.

A method of using the daily variations of hot spot temperature to compute aging is to subdivide the day into (not necessarily equal) time intervals over which the hot spot temperature is reasonably constant. Let AAF_i be the aging acceleration factor for time interval Δt_i measured in hours. Then for a full day,

$$\mathrm{AAF}_{\text{day ave}} = \frac{1}{24} \sum_{i=1}^{N} \mathrm{AAF}_i \Delta t_i \tag{17.85}$$

where N is the total number of time intervals for that day.

The actual loss of life will depend on the definition of lifetime. If the 50% retained tensile strength criterion is used, then in 1 day at 110°C insulation hot spot temperature, the transformer loses the fraction $24/65{,}000 = 3.69 \times 10^{-4}$ of its life. For the 25% retained tensile strength criterion, the fractional loss of life for 1 day = 1.78×10^{-4}.

A cautionary note must be sounded before too literal a use is made of this procedure. Gas bubbles can start to form in the oil next to the insulation when the insulation temperature reaches about 140°C. These bubbles can lead to dielectric breakdown that could end the transformer's life, rendering the calculations inapplicable.

There is bound to be some uncertainty involved in determining the hot spot temperature unless fiber optics or other probes are used and properly positioned. Short of this, it should be estimated as best as possible, based on top oil temperature, winding gradients, etc. Also, a winding temperature indicator, if properly calibrated, can be used.

Since the AAF is independent of the choice made for the normal transformer lifetime, the fractional loss of life can be easily recalculated if a different choice of normal lifetime is made based on additional experience or knowledge.

17.12 Cable and Lead Temperature Calculation

Although the conductors in the coils are usually cooled sufficiently to meet the required hot spot limits, it is also necessary to insure that the lead and cable temperatures remain below these limits as well. Even though these are generally not in critical regions of electric stress to cause a breakdown, if gassing occurs, the gassing itself can trigger alarms, which could put the transformer out of service. It is thus necessary for the design engineer to insure that the leads and cables are sized to meet the temperature requirements.

In general, no more insulation (paper) should be used on the leads and cables than is necessary for voltage standoff. Otherwise, excess paper wrapping could lead to elevated cable temperatures. Given this minimum paper thickness, the current-carrying area of the lead or cable should then be chosen so that the temperature limits are met when rated current is flowing. A method for calculating the lead or cable temperature rise is given here. It allows for the possibility that the lead is brazed to a cable and that some heat conduction can occur to the attached cable, acting as a heat sink. It also considers the case of a lead inside of a duct or tube in addition to a lead in the bulk transformer oil.

We treat a cylindrical geometry. In the event the lead is not cylindrical, an effective diameter should be calculated so that a cylindrical approximation may be used. The cylinder is assumed to be long so that end effects may be neglected. The geometric parameters are shown in Figure 17.15.

The convective surface heat transfer coefficient of the oil is denoted by h. The thermal conductivities of the conductor and paper are denoted k_c and k_p, respectively, and the power generated per unit volume inside the conductor is denoted by q_v. This is given by the Joule loss density, ρJ^2, where ρ is the resistivity and J the rms current density. In addition, to account for extra losses due to stray flux, we should multiply this by $(1 + f)$, where f is the stray flux loss contribution expressed as a fraction of the Joule losses.

There are two cases to consider for the convective heat transfer coefficient of the oil. They are the case of a horizontal cylinder in free tank oil and the case of a horizontal cylinder inside of a channel. (The heat transfer coefficient for a vertical cylinder is higher, so the resulting temperature rises will be lower than for a horizontal cylinder.) These are given by [Kre80].

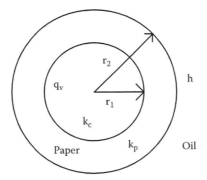

FIGURE 17.15
Cylindrical geometry of a paper-wrapped conductor surrounded by oil.

Natural convection from a horizontal cylinder of diameter D in bulk fluid is given by

$$
\mathrm{Nu_D} = \left[0.60 + 0.387 \left\{ \frac{\mathrm{Gr_D}\,\mathrm{Pr}}{\left[1 + \left(0.56\,/\,\mathrm{Pr} \right)^{9/16} \right]^{16/9}} \right\}^{1/6} \right]^2 \tag{17.86}
$$

Mixed natural convection and laminar flow in horizontal ducts are given by

$$
\mathrm{Nu_D} = 1.75 \left[\mathrm{Gz} + 0.012 \left(\mathrm{Gz}\,\mathrm{Gr_D^{1/3}} \right)^{4/3} \right]^{1/3} \left(\frac{\mu_b}{\mu_s} \right)^{0.14} \tag{17.87}
$$

where the symbols have the following meaning:

Nusselt number	$\mathrm{Nu_D} = \dfrac{hD}{k}$
Reynold's number	$\mathrm{Re_D} = \dfrac{\rho_m vD}{\mu}$
Prandtl number	$\mathrm{Pr} = \dfrac{\mu c}{k}$
Grashof number	$\mathrm{Gr} = \dfrac{g\beta\rho_m^2 D^3 \left(T_s - T_b \right)}{\mu^2}$
Graetz number	$\mathrm{Gz} = \mathrm{Re_D}\,\mathrm{Pr}\left(\dfrac{D}{L} \right)$

v is the fluid velocity

D is the hydraulic diameter for flow inside ducts = 4 × flow area/wetted perimeter = outer cylinder diameter for external flow

L is the duct or cylinder length

g is the acceleration of gravity = 9.8 m/s²

h is the surface convective heat transfer coefficient

k is the thermal conductivity of transformer oil = 0.11 W/m °C

c is the specific heat of transformer oil = 1880 J/kg °C

μ is the oil viscosity = $6900.0/(T + 50)^3$ N s/m² , T = temp in °C

ρ_m is the oil mass density = $867 \exp[-0.00068 (T - 40)]$ kg/m³ , T = temp in °C

β is the volume expansion coefficient of oil = 0.00068/°C

Subscripts s and b label the oil at the outside surface of the paper and the bulk oil

In steady state, the thermal equations governing this situation are as follows. Inside conductor,

$$
\frac{1}{r} \frac{d}{dr} \left(r \frac{dT_c}{dr} \right) + \frac{q_v}{k_c} = 0 \tag{17.88}
$$

Inside the paper,

$$\frac{1}{r}\frac{d}{dr}\left(r\frac{dT_p}{dr}\right) = 0 \tag{17.89}$$

The boundary conditions are

At r = 0,	$\dfrac{dT_c}{dr} = 0$
At r = r_1,	$T_c = T_p$ and $k_c\dfrac{dT_c}{dr} = k_p\dfrac{dT_p}{dr}$
At r = r_2,	$-k_p\dfrac{dT_p}{dr} = h(T_s - T_b)$

where c labels the temperature inside the conductor and p inside the paper. The solution is given by

$$T_c(r) = \frac{q_v}{4k_c}\left(r_1^2 - r^2\right) + \frac{q_v r_1^2}{2k_p}\left[\ln\left(\frac{r_2}{r_1}\right) + \frac{k_p}{hr_2}\right] + T_b \tag{17.90}$$

$$T_p(r) = \frac{q_v r_1^2}{2k_p}\left[\ln\left(\frac{r_2}{r}\right) + \frac{k_p}{hr_2}\right] + T_b \tag{17.91}$$

The highest paper temperature occurs at r = r_1 and is given, in terms of temperature rise above the bulk oil, by

$$T_c(r_1) - T_b = \frac{q_v r_1}{2}\left[\frac{r_1}{k_p}\ln\left(\frac{r_2}{r_1}\right) + \frac{r_1}{hr_2}\right] \tag{17.92}$$

The quantity $q_v r_1/2$ is the surface heat flux, q_s, since

$$q_s = \frac{q_v \pi r_1^2 L}{2\pi r_1 L} = \frac{q_v r_1}{2} \tag{17.93}$$

where L is the cylinder length that cancels. Thus, (17.92) can be expressed as

$$\dot{q}_s = h_{eff}\left[T_c(r_1) - T_b\right] \tag{17.94}$$

where

$$h_{eff} = \frac{1}{\dfrac{r_1}{hr_2} + \dfrac{r_1}{k_p}\ln\left(\dfrac{r_2}{r_1}\right)} = \frac{h}{\dfrac{r_1}{r_2} + \dfrac{r_1 h}{k_p}\ln\left(\dfrac{r_2}{r_1}\right)} \tag{17.95}$$

is an effective heat transfer coefficient, which takes into account conduction through the paper. In the case of a thin paper layer of thickness τ where $\tau \ll r_1$ and $r_2 = r_1 + \tau$, (17.95) reduces to

$$h_{eff} = \frac{h}{1 + \dfrac{\tau h}{k_p}} \tag{17.96}$$

This last expression also applies to a planar geometry. Note that the surface heat flux in (17.93) is based on the surface area of a cylinder of radius r_1, which is the radius of the metallic part of the cable.

We also need the temperature rise of the surface oil in order to compute the Grashof number. This is given by

$$T_p\left(r_2\right)-T_b = \frac{q_v r_1}{2}\left(\frac{r_1}{hr_2}\right) \tag{17.97}$$

so that, taking the ratio of (17.97) to (17.92), we get

$$\frac{T_p\left(r_2\right)-T_b}{T_c\left(r_1\right)-T_b} = \frac{1}{1+\dfrac{hr_2}{k_p}\ln\left(\dfrac{r_2}{r_1}\right)} \tag{17.98}$$

so that the temperature rise of the surface paper can be found once the maximum paper temperature rise is calculated.

In case another cable is brazed to the conductor, which may act as a heat sink, we must consider heat conduction along the conductor to the other cable. The situation is depicted in Figure 17.16.

Letting W = the total power dissipated inside the conductor of length L, we have

$$\begin{aligned} W &= h_{eff}A_{side}\left(T_c - T_b\right)+h_{cond}A_{cond}\left(T_c - T_{cable}\right) \\ &= h_{eff}A_{side}\left(T_c - T_b\right)+h_{cond}A_{cond}\left(T_c - T_b\right)-h_{cond}A_{cond}\left(T_{cable} - T_b\right) \end{aligned} \tag{17.99}$$

where

T_c is the conductor surface temperature
T_b is the bulk fluid temperature
T_{cable} is the brazed-on cable temperature
A_{side} = pL is the conductor's side area through which heat flows to the cooling fluid
    p is the perimeter of the conductor's cooling surface
A_{cond} is the conductor's current-carrying area
h_{cond} = k_{cu}/L is the heat transfer coefficient for heat to flow from the conductor to the brazed-on cable. For copper, k_{cu} = 400 W/m °C.

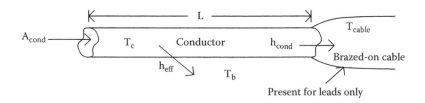

FIGURE 17.16
Geometry and thermal parameters for a cable or lead with another brazed to it.

In order to use the said expression to obtain the lead temperature rise, the temperature rise of the cable must be calculated first. Since the cable is normally very long, its temperature rise can be obtained by neglecting its brazed connection.

Although we have indicated the numerical value of most of the material parameters in the preceding section, the missing ones are given as follows:

Thermal conductivity of paper, $k_p = 0.16$ W/m °C

Resistivity of copper, $\rho = 1.72 \times 10^{-8} \{1 + 0.004 [T (°C) - 20]\}$

For cooling in a horizontal duct, (17.87) is used and this involves the velocity via the Reynold's number. We use v = oil velocity in duct = 2.54 mm/s. (This is a conservative value. If a better value can be obtained, e.g., from an oil flow calculated in a cooling program, then it should be used.)

In addition, it should be noted that many of these material parameters are temperature dependent so that it is necessary to specify a bulk oil temperature and to iterate the calculations. The appropriate temperature to use for evaluating the oil parameters is $(T_s + T_b)/2$, that is, the average of the surface and bulk oil temperatures. For the conductor, its temperature is nearly uniform, so its maximum temperature can be used. The Grashof number depends on the temperature difference $T_s - T_b$, so it must be recalculated during the iteration process along with the other material parameters.

These formulas have been applied to some standard cable sizes to obtain the maximum continuous currents permissible for a hot spot temperature rise of about 25°C. We assumed an oil temperature rise of 55°C above an ambient air temperature of 30°C, resulting in a bulk oil temperature of 85°C. Thus, a rise of 25°C would result in a hot spot temperature of 110°C. The results are shown in Table 17.3. Note that the conductor area may not fill 100% of the geometrical area of the cable if the conductor is stranded.

TABLE 17.3

Maximum Continuous rms Currents for Standard Cable Sizes Producing a Temperature Rise above the Surrounding Oil of 25°C

Cable Size (AWG/MCM)	Cond Area (mm²)	Cond Diameter (mm)	Max Continuous Current (rms amps)		
			(2.4 mm paper)	(6.35 mm paper)	(12.7 mm paper)
6	13.30	4.72	125	100	85
4	21.15	5.97	170	135	114
2	33.63	7.52	230	180	152
1/0	53.48	9.50	320	245	205
2/0	67.42	10.67	375	285	235
3/0	85.03	13.54	465	345	285
4/0	107.23	15.22	545	405	330
300	152.00	18.14	700	515	415
350	177.35	19.63	780	570	460
400	202.71	20.96	860	625	505
500	253.35	23.44	1010	730	585
600	304.06	25.96	1155	830	660

17.13 Tank Wall Temperature Calculation

Heating in the tank wall is due to eddy currents induced by the stray leakage flux from the main coils and flux from any nearby leads or busses. Generally, a chief cause for concern is the presence of high current-carrying leads near the tank wall. We have indicated how these losses may be calculated in Chapter 15. Here, we are going to assume that these losses are known and that they are uniformly distributed throughout the tank wall thickness. We will also assume that the tank wall surface dimensions involved are large compared with the tank wall thickness so that only one spatial dimension through the wall thickness is important. Figure 17.17 shows the geometric and other relevant parameters.

Here, L is the wall thickness, q_v the losses generated in the tank wall per unit volume, k the thermal conductivity of the tank wall material, h_1 the heat transfer coefficient from the inner tank wall to the oil, h_2 the heat transfer coefficient from the outer tank wall to the air, and T(x) the temperature distribution in the tank wall.

The steady-state thermal equation for this situation is given by

$$\frac{d^2T}{dx^2} + \frac{q_v}{k} = 0 \tag{17.100}$$

with the boundary conditions

$$k\frac{dT}{dx}\bigg|_{x=0} = h_1\big[T(x=0) - T_{oil}\big], \quad k\frac{dT}{dx}\bigg|_{x=L} = -h_2\big[T(x=L) - T_{air}\big] \tag{17.101}$$

The different signs in the two boundary conditions are necessary to properly account for the direction of heat flow. The solution can be expressed as

$$T(x) = T_{oil} - \frac{q_vL^2}{2k}\left(\frac{x}{L}\right)^2 + \left[\frac{\dfrac{q_vL^2}{2k}\left(1 + \dfrac{2k}{h_2L}\right) - \left(T_{oil} - T_{air}\right)}{1 + \dfrac{k}{h_1L}\left(1 + \dfrac{h_1}{h_2}\right)}\right]\left(\frac{x}{L} + \frac{k}{h_1L}\right) \tag{17.102}$$

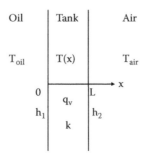

FIGURE 17.17
Tank wall geometry and thermal parameters.

This expression can be simplified by noticing that thermal transfer within the tank wall is much larger than that at the surface. This means that k is much larger than h_1L or h_2L so that, ignoring terms ≤ 1 relative to k/h_1L or k/h_2L and the $(x/L)^2$ term, which is also small, (17.102) becomes

$$T(x) = T_{tank} = \frac{q_v L}{(h_1 + h_2)} + \frac{h_1 T_{oil} + h_2 T_{air}}{(h_1 + h_2)} \tag{17.103}$$

This is independent of x so that the tank wall temperature is essentially uniform throughout, and we have written T_{tank} for it. The last term on the right is the weighted average of the oil and air temperatures, weighted by their respective heat transfer coefficients. The first term on the right accounts for the heat generated in the tank wall. We can use the heat transfer coefficients given in Section 17.5 on tank cooling. Another expression, which could be used for heat transfer due to natural convection from one side of a vertical plate of height H immersed in a fluid (air or oil), is given by [Kre80]

$$h = 0.021 \frac{k}{H} \left(Gr_H Pr \right)^{2/5} \tag{17.104}$$

where the Grashof and Prandtl numbers have been defined in Section 17.12. This expression holds for $10^9 < Gr_H Pr < 10^{13}$. To this, one would have to add the radiation term to the air heat transfer coefficient.

As a numeric example, let

$h_1 = h_{oil} = 70$ W/(m² °C),	$h_2 = h_{air} + h_{rad} = 13$ W/(m² °C)
$k = k_{steel} = 40$ W/m °C,	$L = 9.525 \times 10^{-3}$ m
$q_v = 2 \times 10^5$ W/m³,	$q_v L^2/2k = 0.23$

From these, we obtain $k/h_1L = 60$ and $k/h_2L = 323$. Both of these are much larger than 1. From the values given earlier, we see that the $(x/L)^2$ term is small and the approximate formula (17.103) may be used. For $T_{oil} = 85°C$, $T_{air} = 30°C$, we obtain $T_{tank} = 99°C$.

17.14 Tie plate Temperature Calculation

Tie plate losses were calculated in Chapter 15. These losses must be obtained by some method, such as that described there, in order to calculate the tie plate's surface temperature. In order to make this temperature calculation tractable analytically, we make some idealized assumptions that appear to be reasonable. We assume the plate is infinite in two dimensions and that the losses fall off exponentially from the surface facing the coils into the plate. The exponential drop-off is determined by the skin depth and has the form $e^{-x/\tau}$ where x is the distance from the free surface and τ is the skin depth. The surface facing the coils is assumed convectively cooled by the oil with a heat transfer coefficient h. The boundary condition on the other surface facing the core will depend on the construction details, which may differ among manufacturers. We will assume that there is a cooling gap here but that the heat transfer coefficient may be different from that at the other surface.

To allow for an insulated core-facing surface, this heat transfer coefficient may simply be set to zero. This geometry and the cooling assumptions are identical to that assumed for the tank wall temperature calculation. However, here we are allowing for an exponential drop-off of the loss density, which is probably more realistic in the case of magnetic steel tie plates. Thus, we are using the geometric and thermal parameters of Figure 17.17, except that air now refers to oil cooling on the surface at a distance x = L, where L is the tie plate thickness. Tie plates have a finite width and we are ignoring the enhanced loss density at the corners, which is evident in the finite element study for mag steel of Chapter 15. However, there is also extra cooling at the corner from the side surface, which could mitigate the effect of this extra loss density. One could address this problem by means of a thermal finite element program. However, the analytic solution presented in (17.105) should apply to temperatures away from the corner, and these may possibly be higher or comparable to the corner temperature.

This is a 1-dimensional steady-state heat transfer problem. The differential equation to solve is the same as (17.100) in the previous section. However, here we have $q_v = q_{vo}e^{-x/\tau}$, where q_{vo} is the loss density at the surface (x = 0) and τ is the skin depth, rather than the constant assumption for q_v used for tank cooling. Thus, the side facing the coils is at x = 0. The boundary conditions are the same as (17.101) with the air subscript replaced by oil. The solution is

$$T(x) - T_{oil} = \frac{q_{vo}\tau}{h_1}\left\{1 + \frac{h_1\tau}{k}\left(1 - e^{-x/\tau}\right) - \left(1 + \frac{h_1 x}{k}\right)\left[\frac{e^{-L/\tau}\left(1 - \frac{h_2\tau}{k}\right) + \frac{h_2\tau}{k} + \frac{h_2}{h_1}}{1 + \frac{h_2 L}{k} + \frac{h_2}{h_1}}\right]\right\} \quad (17.105)$$

We have expressed the solution in this form so that the case of an insulated back side, ($h_2 = 0$), could be easily obtained. From (17.105), the surface temperature rise at the x = 0 surface is

$$T(x = 0) - T_{oil} = \frac{q_{vo}\tau}{h_1}\left\{1 - \left[\frac{e^{-L/\tau}\left(1 - \frac{h_2\tau}{k}\right) + \frac{h_2\tau}{k} + \frac{h_2}{h_1}}{1 + \frac{h_2 L}{k} + \frac{h_2}{h_1}}\right]\right\} \quad (17.106)$$

As a numerical example, let us consider the surface temperature rise above that of the surrounding oil for mag steel and stainless steel tie plates using the following parameters:

$h_1 = 70$ W/(m² °C)

$h_2 = 20$ W/(m² °C)

$L = 9.525 \times 10^{-3}$ m

$k = 40$ W/m °C (mag steel)

$k = 15$ W/m °C (stainless steel)

$\tau = 2.297 \times 10^{-3}$ m (mag steel)

$\tau = 5.627 \times 10^{-2}$ m (stainless steel)

We would like to compare these two materials when the total losses in them are the same. This means that the losses per unit surface area should be the same since we are assuming the geometry is unlimited in the plane of the surface. Letting q_A be the loss per unit area obtained by integrating q_v over the thickness of the tie plate, L, we can obtain q_{vo} in terms of this. Thus,

$$q_{vo} = \frac{q_A}{\tau\left(1 - e^{-L/\tau}\right)} \qquad (17.107)$$

Using (17.106) and (17.107) and the numerical data given previously, we obtain the surface temperature rise above the ambient oil for the two types of tie plates in terms of q_A:

$T(x = 0) - T_{oil} = 0.0111q_A$ (mag steel)

$T(x = 0) - T_{oil} = 0.0111q_A$ (stainless steel)

with q_A in W/m^2 and temperatures in °C. Thus, we see that for the same total loss, which is simply q_A times the lateral dimensions of the tie plate, magnetic iron tie plates have the same surface temperature rise as stainless steel tie plates. We also find that the surface temperature rise at the back surface ($x = L$) is nearly the same as the front surface temperature rise for the two materials. Even though the losses are distributed quite differently in the two materials with the mag steel having a higher concentration of losses near the surface, the higher thermal conductivity of the mag steel rapidly equalizes the temperature within the material so that extreme temperature differences cannot develop.

If the back surface were insulated ($h_2 = 0$), all the losses would have to leave the front surface in the steady state. In this case, also the surface temperatures of the mag steel and stainless steel tie plates would be nearly identical for the same total loss. As h_2 increases up to h_1, the front surface temperatures gradually differ for the two cases but not significantly.

17.15 Core Steel Temperature Calculation

Because core steel is made up of thin, stacked insulated laminations, it has an anisotropic thermal conductivity. The thermal conductivity in the plane of the laminations is much higher than the thermal conductivity perpendicular to this plane or in the stacking direction. The steady-state heat conduction equation for this situation is, in rectangular coordinates,

$$k_x \frac{\partial^2 T}{\partial x^2} + k_y \frac{\partial^2 T}{\partial y^2} + q_v = 0 \qquad (17.108)$$

where the thermal conductivities k, which can differ in the x and y directions, are labeled accordingly. The loss per unit volume, q_v, is assumed to be a constant here. Again, there are finite element programs that can solve this equation for complex geometries, such as a stepped core. We will derive a simple but approximate analytical solution here for a rectangular geometry as shown in Figure 17.18.

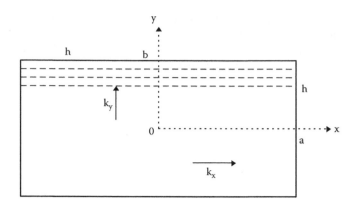

FIGURE 17.18
Rectangular geometry for anisotropic thermal calculation.

This could apply to one step of a core, say the central step, with cooling ducts on the large flat sides or to all or part of a core between two cooling ducts by suitably calculating effective dimensions by approximating a rectangle from the steps involved.

We look for a solution of (17.108) of the form

$$T(x,y) = A + Bx + Cx^2 + Dy + Ey^2 \qquad (17.109)$$

satisfying the boundary conditions at the center:

$$\frac{\partial T}{\partial x} = \frac{\partial T}{\partial y} = 0 \quad \text{at } x = y = 0 \qquad (17.110)$$

This is required by symmetry. At the outer surfaces, we only approximately satisfy the boundary conditions for convective cooling:

$$
\begin{aligned}
-k_x \frac{\partial T}{\partial x} &= h(T - T_{\text{oil}}) \quad \text{at } x = a,\ y = 0 \\
-k_y \frac{\partial T}{\partial y} &= h(T - T_{\text{oil}}) \quad \text{at } x = 0,\ y = b
\end{aligned}
\qquad (17.111)
$$

This means that we are satisfying the convective boundary condition exactly only at the surface points on the two axes. However, this is where we will evaluate the surface temperature. With these assumptions, the solution is

$$T(x,y) - T_{\text{oil}} = \frac{q_v}{h} \left[\frac{1 - \left(\dfrac{\frac{ha}{2k_x}}{1 + \frac{ha}{2k_x}}\right)\left(\dfrac{x}{a}\right)^2 - \left(\dfrac{\frac{hb}{2k_y}}{1 + \frac{hb}{2k_y}}\right)\left(\dfrac{y}{b}\right)^2}{\dfrac{1}{a\left(1 + \frac{ha}{2k_x}\right)} + \dfrac{1}{b\left(1 + \frac{hb}{2k_y}\right)}} \right] \qquad (17.112)$$

Note that a and b are 1/2 the rectangle dimensions. We can estimate the thermal conductivity in the direction perpendicular to the lamination plane (the y-direction here) from knowing the coating thickness or stacking factor and the thermal properties of the coating. For stacked silicon steel, we use

$k_x = 30$ W/m °C and $k_y = 4$ W/m °C

$h = 70$ W/m² °C for convective oil cooling

Assuming an effective stack or rectangle size of 0.762 m × 0.254 m and using $q_v = 10^4$ W/m³ (1.3 W/kg), we find for the maximum internal temperature rise above the surrounding oil

$$T(x = 0, y = 0) - T_{oil} = 25.8°C.$$

The surface temperature rise in the x or long dimension is

$$T(x = a, y = 0) - T_{oil} = 17.8°C$$

The surface temperature rise at the surface at y = b in the short direction perpendicular to the stack is

$$T(x = 0, y = b) - T_{oil} = 12.2°C$$

Note that the surface in the direction of lower thermal conductivity is actually cooler than the surface toward which heat is more easily conducted. This is because, the central temperature being the same, there is a smaller thermal gradient in the high-conductivity direction relative to the lower one.

The temperatures calculated with these formulas were compared with temperatures obtained with a 2D finite element thermal code, keeping the uniform loss density and thermal parameters the same for the two methods. The actual core step pattern and core ducts were used for the 2D model. The maximum internal core temperature and maximum surface temperature were compared for each step. They were the same within a few degrees for the two methods.

18

Load Tap Changers

18.1 Introduction

The flicker of the house lights while having dinner is a sign of a load tap changer (LTC) tap change. When most people get home from work, they start using electricity for cooking and lighting. This increases the electrical load on the distribution network, which causes the voltage to sag below its nominal value. The latter signals the LTC to change taps to adjust the voltage back to its nominal value with some tolerance.

There are many applications of transformers in modern electric power systems: from generator step-up to system interconnection to distribution to arc furnace to HVDC converters to mention just a few. The role of a transformer is to convert the electrical energy from one voltage level to another. As power systems become larger and more complex, power transformers play a major role in how efficient and stable the system is. For each transformer installed in a network, there is an ideal (optimal) voltage ratio for an optimal operation of the system. Unfortunately, this optimal voltage ratio varies depending on the operating conditions of the total network. Early in the history of electric power systems, it became evident that for power systems to operate satisfactorily, transformer voltage ratios needed to be adjustable without interrupting the flow of energy. This is the role of an LTC.

18.2 General Description of LTC

An LTC is a device that connects different taps of tapped windings of transformers without interrupting the load. It must be capable of switching from one tap position to another without at any time interrupting the flow of the current to the load and without at any time creating a short circuit between any two taps of the transformer winding. Tap changing transformers are used to control the voltage or the phase angle or both in a regulated circuit.

An LTC is made of four elements:

1. A selector switch, which allows the selection of the active tap
2. A changeover switch, referred to as a reversing switch when it reverses the polarity of the tapped winding, used to double the number of positions available
3. A transition mechanism, including an arcing or diverter switch, which effects the transition from one tap to the other
4. A driving mechanism, which includes a motor and gear box and controls to drive the system

There are two generic types of tap changers:

1. *In-tank*: The cover-mounted, in-tank tap changer, known as the Jansen type, sits in the main transformer oil together with the core-and-coil assembly. Its selector and changeover switches are at the bottom of the tap changer in the main oil. The arcing (diverter) switch is located in a separate compartment at the top, usually within a sealed cylinder made of fiberglass or other similar material. All arcing is confined to this compartment. They exist in single-phase or three-phase neutral end Wye (Y)-connected versions. For three-phase fully insulated applications, three single-phase tap changers must be used. This type of tap changer is used for higher voltages or current levels.

2. *Separate compartment*: The side-mounted, separate compartment types have their own box and are assembled separately from the transformer. They are bolted to the side of the tank and connected to the transformer tapped windings through a connecting board. Their selector, changeover, and arcing switches are located in an oil compartment completely isolated from the main transformer oil. Some of them have two compartments, one for the arcing switch and one for the selector and changeover switches. Others have everything in one compartment only. They are available in three-phase assembly, either Wye connected for application at the neutral end of a three-phase transformer or fully insulated for applications at the line end.

18.3 Types of Regulation

The main types of regulation are illustrated in Figure 18.1 and are discussed in the following.

Linear: In linear switching, tapped turns are added in series with the main winding and their voltage adds to the voltage of the main winding. No changeover switch is needed for this type. The tapped winding is totally bypassed in the minimum voltage position. The rated position can be any one of the tap positions.

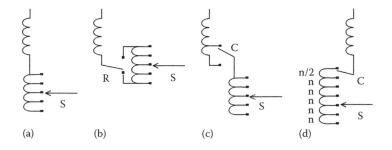

(a)   (b)   (c)   (d)

FIGURE 18.1
LTC main regulation types. S, selector switch; C, changeover switch; R, reversing switch. (a) Linear, (b) reversing, (c) coarse–fine, and (d) bias winding.

Plus–minus (reversing): In a reversing type of regulation, the whole tapped winding can be connected in additive or reversed polarity with respect to the main winding. The tapped turns can add or subtract their voltage with respect to the main winding. The tapped winding is totally bypassed in the neutral (midrange) voltage position. The rated position is normally the mid one. The total number of positions available is twice the number of sections in the tapped winding plus one.

Coarse–fine: The coarse–fine regulation can be defined as a two-stage linear regulation where the first or coarse stage contains a large number of turns, which can be totally bypassed, by the changeover selector. These turns are shown as a single loop in the top coil of the figure. Fine regulation is achieved with the selector switch. Normally, the coarse section contains as many turns as the tapped winding plus one section. In that way, the total number of positions available is twice the number of sections in the tapped winding plus one.

Bias winding: The bias winding type of regulation is similar to the coarse–fine except that the number of turns in the bias winding is half the turns of one section of the tapped winding. It is used to provide half steps between the main tap steps. Thus, the total number of positions available is twice the number of sections in the tapped winding plus one. The bias winding technique can be combined with the reversing scheme to provide twice the number of positions (four times the number of sections in the tapped winding plus one) at the expense of adding one more switch and increasing the complexity of the switching and driving mechanism.

Although not indicated in the figure, the tap selector switch is often a circular type of switch to allow a smooth transition between the tap voltages when the changeover or reversing switch operates.

18.4 Principle of Operation

An LTC must be capable of switching from one tap position to another without interrupting the flow of current to the load at any time. It must therefore follow a "make-before-break" switching sequence. On the other hand, it cannot at any time create a short circuit between any two taps of the transformer winding. This means that during this make-before-break interval, there must be something to prevent the shorting of the turns. Two ways are primarily used to accomplish this.

18.4.1 Resistive Switching

Figure 18.2 shows an example of a six-step switching sequence. This particular type of tap changer has two selector switches and an arcing switch with four contacts.

1. Step 1 of this figure shows the steady state of the TC just before the switching operation. The load current is flowing through the contact #1 and the selector #1. Let us call that position tap #2.
2. At step 2, the nonconducting selector has moved 2 steps down, from position 1 to position 3. The current is still flowing in contact #1 and the selector #1.

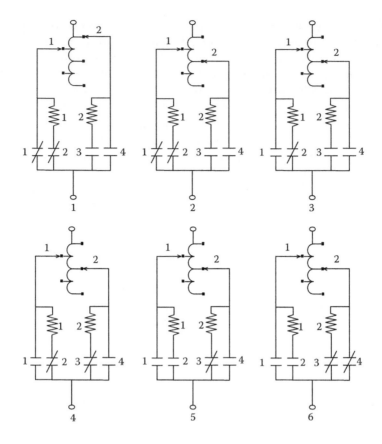

FIGURE 18.2
Sequence of operations involved in tap changers using resistive switching. The selector switches are the topmost switches. The arcing switches are closed when a diagonal (shorting) line is present and open otherwise.

3. At step 3, the contact #1 opens and the current flows through resistor #1 and selector #1.

4. At step 4, the contact #3 closes and the current splits between the 2 resistors and the 2 selectors. There is also a circulating current flowing into the loop limited by the 2 resistors and the reactance of the loop.

5. At step 5, the contact #2 opens, breaking the current in the resistor and selector #1. It is at that moment that arcing occurs. The arc remains on until the current crosses the zero line. At worst, it can last one half cycle, which, at 60 Hz, is 8 ms. The average arcing time is around 5–6 ms.

6. At step 6, the contact #4 closes and the load current bypasses the resistor #2 and goes directly to the selector #2. The TC has reached the steady state for tap position #3.

This type of transition requires that the resistors be capable of withstanding the full load current plus the circulating current during the transition (from step 3 to step 5). In order to reduce the energy absorption requirement for the resistors, the time of the complete transition has to be minimized. These tap changers have normally very fast transitions.

18.4.2 Reactive Switching with Preventative Autotransformer

Another widely used way of handling the transition from one tap to another is to use reactors instead of resistors. These reactors do not have to dissipate as much energy as the resistors. They mainly use reactive energy, which does not produce any heat. Therefore, they can be designed to withstand the full load plus the circulating current for long periods of time, even continuously. Two reactors per phase are needed. They are normally wound on a common gapped core, making them mutually coupled. When they are connected in series, they act as an autotransformer. When they are connected in parallel, they act as a single reactor. Transformer designers take advantage of this feature. They use the reactors not only to prevent the load current from being interrupted or to prevent sections of a tapped winding from being shorted but also to act as a transformer and provide intermediate voltage steps in between two consecutive tap sections of the main transformer. The total number of positions available with this scheme is twice the number of sections in the tapped winding plus one. Here is how it works.

The method of operation is diagrammed in Figure 18.3. The tap changer has two selectors, two reactors (actually one reactor with two windings), two bypass switches (#1 and #2), and an arcing switch (#3).

1. In step 1 of this figure, the tap changer is in the steady-state mode on tap position #2. The load current flows in the 2 selectors, the 2 reactors, and the bypass switches. Although the arcing switch is closed, no current flows through it.

2. In step 2, the bypass switch opens. The current flows through the switch #1 and then splits up between the 2 reactors and the 2 selectors.

3. In step 3, the arcing switch opens. All current flows through the reactor 1 and the selector 1. Arcing occurs at that step because the current in the reactor is interrupted.

4. In step 4, the right selector moves one step to the right, while it carries no current.

5. In step 5, the arcing switch closes, causing current to flow again in the reactor #2 and in selector #2. Note that the two selectors are on different taps. This causes a circulating current to flow into the loop.

6. In step 6, the bypass switch #2 closes and a steady-state condition is reached, which is in this case tap position #1. Because the 2 ends of the reactors are on 2 different taps, they are in series between the 2 taps. The voltage at their midpoint is halfway between the 2 taps. This state is called the "bridging position," where the reactors bridge between the 2 taps.

When combined with a reversing switch, this scheme provides four times as many voltage positions as tapped sections in the winding plus one. The price to pay is that the reactor has to be designed to withstand continuously the full load current plus the circulating current; the tapped winding has also to withstand the full load plus the circulating current; and the reactor introduces losses and draws more magnetizing current from the source, especially on bridging positions. It might also add audible noise to the transformer.

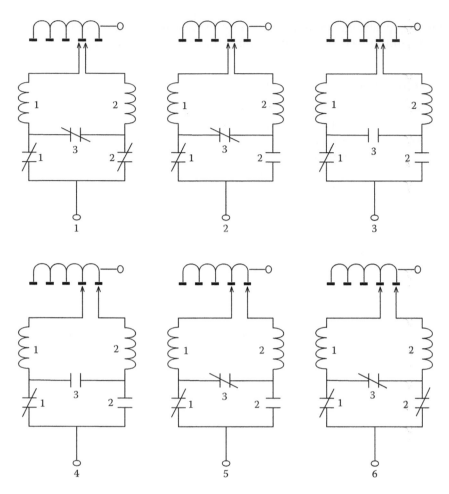

FIGURE 18.3
Sequence of operations involved in tap changers using reactive switching.

18.5 Connection Schemes

18.5.1 Power Transformers

Normally, the primary winding of transformers is fed at a constant voltage, and the role of the tap changer is to add or subtract turns in order to vary the voltage ratio of the transformer to maintain its output at a constant voltage despite fluctuations in the load current. In principle, this can be accomplished by changing the number of active turns either in the primary or secondary winding. There are subtle differences in the two ways of connecting the tap changer, which transformer and system designers have to be aware of.

18.5.1.1 Fixed Volts/Turn

The most natural way to add a tapped winding to a transformer is to connect it in series with the regulated side. As shown in Figure 18.4, the primary winding of the transformer

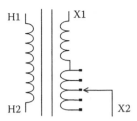

FIGURE 18.4
Fixed volts/turn tap changing scheme.

(H1–H2) is fed at a constant voltage and it has a fixed number of turns. So, the volts/turn of the transformer is constant. The voltage across X1–X2 varies with the number of turns. If all the taps have equal number of turns, then the voltage increase is equal for each step. If the X winding is Wye connected, then the tapped winding and the tap changer can be placed at the low potential neutral end and do not require a high insulation level. The price to pay is that the tap changer has to be capable of carrying the current of the X winding. For high-current windings, the cost of such a tap changer could become prohibitive. Another disadvantage is that low-voltage (LV) windings often have few turns. The design of the tapped winding might become impractical if not impossible considering that fractional turns cannot be used.

18.5.1.2 Variable Volts/Turn

One way to solve the problem of high-current LV windings is to put the tapped winding in the high-voltage (HV) side. The LV side can still be regulated in this way. This is particularly applicable if the HV is Wye connected since the tap changer could be placed at the neutral end and would not require a high insulation level while not carrying a high current. This scheme is illustrated in Figure 18.5.

Although the solution might look very attractive, it has its disadvantages:

1. If we consider that the voltage across H1–H2 is constant, varying the number of turns in that winding implies that the volts/turn and thus the flux in the core vary since the core flux is proportional to the volts/turn. It means that the core has to be designed for the minimum turn position since this position requires the core to carry its maximum flux. The core would be bigger than its fixed volt/turn counterpart and would really be used efficiently only at that minimum turn position. At any other position, it operates at lower flux densities.

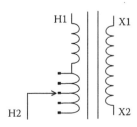

FIGURE 18.5
Variable volts/turn tap changing scheme.

2. If the flux varies in the core, the no-load losses, the exciting current, the imped-ance, and the sound level of the transformer will also vary. If a transformer is designed to meet certain guaranteed losses and sound level at rated position, which is generally the mid position, it is likely that at the minimum turn position, it will exceed these guaranteed values significantly.

3. In a variable flux transformer, the voltage variation per step is not constant even if the number of turns per step is constant. This is because the volts/turn changes with step position.

4. If there is a third winding in the transformer used to feed a different circuit, its voltage will vary as the tap changer moves, which might not be desirable.

18.5.2 Autotransformers

Autotransformers with tap changers present special challenges for the transformer designer. There are three ways of connecting a tap changer in an autotransformer without using an auxiliary transformer. These are illustrated in the following figures.

The connection shown in Figure 18.6 is used when the HV of the transformer is to be varied while keeping the LV constant. In this case, we are assuming that the LV side is sup-plied by a fixed voltage source. The voltage at the HV terminal varies linearly with the number of turns added or subtracted to the series winding. The flux in the core is constant. In this connection, the tap changer and the tapped winding must be rated for the voltage level of the LV line terminal plus the voltage across the tapped section. The tapped wind-ing is directly exposed to any voltage surge coming through the LV line, so extra precau-tions have to be taken at the design stage.

Figure 18.7 shows the connection used when the HV side of the transformer is to be kept constant while the LV varies. In this case, the HV side is fed from a constant voltage source. The voltage at the LV terminal varies linearly with the number of turns added or sub-tracted in series with the LV line. As in the previous case, the flux in the core is constant; the tap changer and the tapped winding must be designed for the voltage level of the LV line terminal plus the voltage across the tapped section. The taps are directly exposed to any voltage surge coming through the LV line, so extra precautions have to be taken at the design stage. Finally, the tap changer must be designed to carry the full load current of the LV terminal.

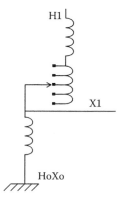

FIGURE 18.6
Taps in the series winding of an autotransformer.

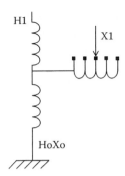

FIGURE 18.7
Taps in the LV line from the autoconnection.

In the previous two cases, if a three-phase transformer is designed, the tap changer must have full insulation between the phases. If the voltage on the LV is above 138 kV, then three single-phase, Jansen-type tap changers must be used.

The connection shown in Figure 18.8 can be used when the LV is to be varied while the HV is to be kept constant. The advantages of this connection are as follows:

1. The level of insulation for the tapped winding and the tap changer is low.

2. A Wye-connected three-phase tap changer can be used for a three-phase transformer.

3. The current in the common winding is lower than the current in the LV line terminal, so a smaller tap changer can be selected.

As usual, there is a price to pay to obtain these benefits. If we assume that the voltage between H1 and HoXo is constant, varying the number of turns implies varying the volts/turn and therefore the flux in the core. It is a variable flux design. As explained earlier, in a variable flux transformer, the losses, the exciting current, the impedance, and the sound level vary with the tap position, and a bigger core must be selected for proper operation at the minimum turn position. If there is a tertiary winding, its voltage will vary as the tap changer changes position.

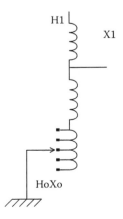

FIGURE 18.8
Taps in the common winding of an autotransformer at the neutral end.

In an autotransformer connection, the turns in the common branch are part of both the HV and LV circuits. This means that if we add turns in the common branch, we add turns to both the HV and LV circuits. Because of this, the number of turns required to achieve a specified regulating range is higher than for the other two cases.

Let us derive the tap winding requirements for regulating the LV or common winding voltage by the positive and negative fractional voltages of f_p and f_m. Let N_S, N_C, and N_T be the turns in the series, common, and tap winding, respectively. These turns are for the individual windings. Let V_H and V_X be the high and low terminal to ground voltages, assuming a Y connection. For a delta connection, these would refer to the phase voltages. Although V_H is assumed to be fixed, V_X will change with tap position and will be labeled V_{Xp}, V_{Xo}, and V_{Xm} for the positive, neutral, and negative tap positions, respectively.

Using VPT_p, VPT_o, and VPT_m to indicate the volts per turn for the positive, neutral, and negative tap settings, we have

$$VPT_p = \frac{V_H}{N_S + N_C + N_T}$$
$$VPT_o = \frac{V_H}{N_S + N_C} \tag{18.1}$$
$$VPT_m = \frac{V_H}{N_S + N_C - N_T}$$

The LV voltages for the different tap positions are

$$V_{Xp} = (N_C + N_T)VPT_p$$
$$V_{Xo} = (N_C)VPT_o \tag{18.2}$$
$$V_{Xm} = (N_C - N_T)VPT_m$$

The fractional voltage changes corresponding to the positive and negative tap settings are

$$f_p = \frac{V_{Xp} - V_{Xo}}{V_{Xo}} = \frac{V_{Xp}}{V_{Xo}} - 1 = \left(\frac{N_C + N_T}{N_C}\right)\frac{VPT_p}{VPT_o} - 1$$

$$= \left(\frac{N_C + N_T}{N_C}\right)\left(\frac{N_S + N_C}{N_S + N_C + N_T}\right) - 1 = \left[\frac{1 + \dfrac{N_S}{N_C}}{1 + \dfrac{N_S}{N_C + N_T}}\right] - 1$$

$$f_m = \frac{V_{Xo} - V_{Xm}}{V_{Xo}} = 1 - \frac{V_{Xm}}{V_{Xo}} = 1 - \left(\frac{N_C - N_T}{N_C}\right)\frac{VPT_m}{VPT_o} \tag{18.3}$$

$$= 1 - \left(\frac{N_C - N_T}{N_C}\right)\left(\frac{N_S + N_C}{N_S + N_C - N_T}\right) = 1 - \left[\frac{1 + \dfrac{N_S}{N_C}}{1 + \dfrac{N_S}{N_C - N_T}}\right]$$

Solving (18.3) for N_T, we find that there are two different values corresponding to f_p and f_m, which we will label N_{Tp} and N_{Tm}. They are

$$N_{Tp} = \frac{N_C + N_S}{\dfrac{1}{f_p}\left(\dfrac{N_S}{N_C}\right) - 1}$$

$$N_{Tm} = \frac{N_C + N_S}{\dfrac{1}{f_m}\left(\dfrac{N_S}{N_C}\right) + 1} \tag{18.4}$$

Here, we might want to take the largest N_T value to get the full regulation in one direction at the expense of providing more than the required regulation in the other direction. If we take an average of the 2, calling it N_{Tave}, we get

$$N_{Tave} = \frac{1}{2}(N_C + N_S)\left(\frac{N_S}{N_C}\right)\left\{\frac{\dfrac{1}{f_p} + \dfrac{1}{f_m}}{\left[\dfrac{1}{f_p}\left(\dfrac{N_S}{N_C}\right) - 1\right]\left[\dfrac{1}{f_m}\left(\dfrac{N_S}{N_C}\right) + 1\right]}\right\} \tag{18.5}$$

In case we have $f_p = f_m = f$, this simplifies to

$$N_{Tave} = (N_C + N_S)\left\{\frac{\dfrac{1}{f}\left(\dfrac{N_S}{N_C}\right)}{\left[\dfrac{1}{f}\left(\dfrac{N_S}{N_C}\right)\right]^2 - 1}\right\} \tag{18.6}$$

As an example, let us suppose that we want a ±10% voltage regulation on the LV winding. Suppose that we have 100 turns in the series winding and 100 turns in the common winding. If the taps were X-line taps, it would require $N_T = 10$ turns. For neutral taps, using (18.4), we find that $N_{Tp} = 22.2$ turns and $N_{Tm} = 18.2$ turns. We might want to use 22 turns in the tap winding to get the full positive regulation and providing more than the required negative regulation. Taking an average and using (18.6) since $f_p = f_m$, we get $N_{Tave} = 20.2$ turns. This means we need about twice as many turns for the neutral taps compared with the X-line taps to achieve the ±10% voltage variation on the X terminal.

Another consideration is the core flux or, for a given size core, the flux density. This is proportional to the volts per turn. Since the negative tap setting has the highest volts per turn, the maximum core flux should not exceed this value. Usually, the maximum core flux is specified for rated conditions. This value would then have to be reduced to allow for the higher flux density at the negative tap setting. The reduction factor, F_R, is given by

$$F_R = \frac{VPT_o}{VPT_m} = \frac{N_C + N_S - N_T}{N_C + N_S} = 1 - \frac{N_T}{N_C + N_S} \tag{18.7}$$

If we use the average value for N_T and we assume that $f_p = f_m = f$, then we get for F_R

$$F_R = 1 - \left\{ \frac{\dfrac{1}{f}\left(\dfrac{N_S}{N_C}\right)}{\left[\dfrac{1}{f}\left(\dfrac{N_S}{N_C}\right)\right]^2 - 1} \right\} \tag{18.8}$$

For the numerical example given earlier, we get $F_R = 0.9$. Thus, the transformer would need to be designed for a flux density 10% below the rated value.

Another consideration is the MVA of the tap winding. For the ±10% X-line taps, since the full LV load current flows in these, the MVA rating is 10% of the terminal MVA. The neutral taps have 20% of the common winding turns. The current in them is the terminal current times the coratio. The neutral coratio is 0.5, and the positive and negative tap coratios are 0.455 and 0.556, respectively. Using the negative tap coratio, the neutral tap rating is $0.2 \times 0.556 = 0.111$ or 11% of the terminal MVA, so it is a bit higher than for the X-line taps.

18.5.3 Use of Auxiliary Transformer

When a transformer with a relatively high current LV winding is specified and regulation is desired on the LV side, it may be economical to use an auxiliary transformer to achieve regulation. This gives the designer a greater flexibility in the design of the tapped winding and the savings on the main transformer often outweigh the cost of the auxiliary transformer. There are two popular ways of using auxiliary transformers.

The series voltage regulator: In this connection, the tapped winding is located on the auxiliary transformer that is inserted in series with the LV line terminal, adding or subtracting its voltage from the LV winding of the main transformer. The main transformer has only two windings and can be optimized independently from the series voltage regulator. This is shown in Figure 18.9.

The series (booster) transformer: In this connection, shown in Figure 18.10, the tapped winding is a separate winding on the main transformer. The advantages of this connection are that there is much more flexibility in the design of the tapped winding. The number of turns per step can be selected so that the tap changer carries a smaller current. The main saving in this case is on the tap changer itself and in the greater flexibility in the optimization of the main transformer.

Auxiliary transformers can also be used with autotransformers pretty much in the same way. There are also plenty of possible connections for using auxiliary transformers. The choice is only limited by the imagination of the designer.

18.5.4 Phase Shifting Transformers

A phase shifting transformer is ideally a transformer with a 1:1 voltage ratio which has the ability to change the phase angle of its output voltage relative to its input. This feature is used by electric power companies to help transfer power more efficiently within the electrical grid. In these transformers, the voltage from the tapped windings is added in quadrature with the input line terminals, producing an effective phase shift in the output.

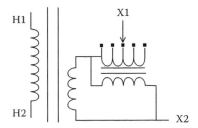

FIGURE 18.9
Tap changing scheme using a series voltage regulator.

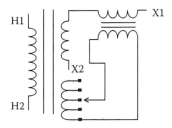

FIGURE 18.10
Tap changing scheme using a series (booster) transformer.

There are quite a few possible connections to achieve this, and some of these are discussed in Chapter 7. Let us just mention here that in a-phase shifting transformer, the tap changer is a key element and the size and rating of the tapped windings are much higher than for conventional transformers.

18.5.5 Reduced versus Full-Rated Taps

Because the current in the tap winding and the tap changer is larger than the rated current at the negative tap setting, it is sometimes desirable to limit it to the rated current at the negative tap setting. Effectively, this amounts to reducing the MVA of the transformer at the negative tap setting. In this case, the taps are referred to as having a reduced rating. Otherwise, the taps are referred to as having a full rating.

18.6 General Maintenance

The reason for the maintenance on a tap changer is to insure its reliability. Any device with moving parts, some of which are used to interrupt currents and voltages, requires maintenance. It would be useless to repeat here what can be found in the manufacturers' instruction books. We will only mention general principles. The factors affecting the reliability of tap changers include the following:

- Oil quality
- Contact pressure

- Contact resistance and temperature
- Timing of movements
- Load currents
- Number of operations

The maintenance of the controls and the drive mechanism are normally not a problem. The gears and other moving parts might wear and require replacement, but in general, they are designed to last very long and do not need frequent inspection. The main areas of concern are the switches and the connecting board if present. The contacts have to be inspected and verified periodically. The frequency of these inspections depends mainly on the number of operations, the current flowing through the contacts, and the cleanliness of the oil.

The inspection and replacement of contacts require that the transformer be taken out of service. Since it is always costly to take a transformer out of service, it is desirable to extend the period between inspections as much as possible. Besides recommendations that can be found in most instruction books for tap changers, here are some of the tactics used by utilities to reduce their maintenance costs and the failure rate of their tap changers.

The process of changing tap position includes arcing, which generates gases and carbon particles. The movement of contacts also promotes erosion of the metal and generates small metal particles. In the long run, these particles can affect the dielectric properties of the oil. Moreover, the carbon particles tend to aggregate on contacts and increase the contact resistance. They also accumulate on connecting boards and may lead to tracking on the surface of the board. For these reasons, the reliability of a tap changer and its overall performance can be improved by a simple filtering device, which would keep the oil clean. This is especially important if the arcing switch is located in the same compartment as the selector and changeover switches. Some utilities have introduced oil filters for LTCs, and they have found that with these filters, they can extend their time between inspections. However, in some tap changers, the arcing occurs in a separate vacuum chamber so that oil contamination is reduced or eliminated.

In a tap changer, contacts that operate frequently remain clean. The ones that do not move for long periods of time tend to oxidize or collect carbon or both. With time, their contact resistance increases, which produces more heat and the situation deteriorates. This is typical of changeover selectors. Some of those operate only twice a year. Many cases of damage to these switches have been reported. Several utilities now force their tap changers to ride over the mid position at least once a month, and they claim a much better performance of their tap changers since they have introduced the procedure.

Instead of taking all transformers out of service for regular and frequent inspections of tap changers, some utilities have introduced monitoring procedures for tap changers without de-energizing their transformers. One of those procedures is dissolved gas analysis. They regularly take oil samples from tap changer compartments and monitor the gases. After a while, they know the pattern of all their tap changers. Whenever one changes pattern, they go and inspect it. They report that this procedure has helped them reduce the frequency of their internal inspections while alerting them to dangerous problems before they lead to a catastrophic failure.

Another trick is to have inspection windows in the tap changer box. When used in conjunction with an oil filter, they can look at their contacts without opening the box and take the transformer out of service if required.

Another possible procedure is to monitor the temperature on the tap changer compartment regularly with infrared measuring devices. It is very inexpensive and easy to do. They claim that if anything goes wrong in a tap changer, it typically increases the temperature of the contacts to a point that you can detect it by checking the temperature on the outside of the wall of the box. Their rule is that if the temperature on the surface of the tap changer compartment is higher than the tank of the transformer by more than $7°$–$10°$, something is going wrong inside the transformer. They would then inspect the tap changer. Quite a few problems have been detected this way.

19

Constrained Nonlinear Optimization with Application to Transformer Design

19.1 Introduction

A transformer must perform certain functions such as transforming power from one voltage level to another without overheating or without damaging itself when certain abnormal events occur, such as lightning strikes or short circuits. Moreover, it must have a reasonable lifetime (>20 years) if operated under rated conditions. Satisfying these basic requirements still leaves a wide latitude in possible designs. A transformer manufacturer will therefore find it in its best economic interest to choose, within the limitations imposed by the constraints, that combination of design parameters, which results in the lowest-cost unit. To the extent that the costs and constraints can be expressed analytically in terms of the design variables, the mathematical theory of optimization with constraints can be applied to this problem.

Optimization is a fairly large branch of mathematics with major specialized subdivisions such as linear programming, unconstrained optimization, and linear or nonlinear equality- or inequality-constrained optimization. Transformer design optimization falls into the most general category of such methods, namely, nonlinear equality- and inequality-constrained optimization. In this area, there are no algorithms or iteration schemes, which guarantee that a global optimum (in our case a minimum) will be found. Most of the algorithms proposed in the literature will converge to a local optimum with varying degrees of efficiency, although even this is not guaranteed if one starts the iterations too far from a local optimum. Some insight is therefore usually required to find a suitable starting point for the iterations. Often, past experience can serve as a guide.

There is, however, a branch of optimization theory called geometric programming, which does guarantee convergence to a global minimum, provided the function to be minimized, called the objective or cost function, and the constraints are expressible in a certain way [Duf67]. These restricted functional forms, called posynomials, will be discussed later. This method is very powerful if the problem functions conform to this type. Approximating techniques have been developed to cast other types of functions into posynomial form. Later versions of this technique removed many of its restrictions but at the expense of no longer guaranteeing that it will converge to a global minimum [Wil67]. One of the earliest applications of geometric programming was to transformer design.

Whatever choice of optimization method is made, there is also the question of how much detail to include in the problem description. Although the goal is to find the lowest cost, one might wish that the solution would provide sufficient information so that an actual design could be produced with little additional work. In general, the more detail one can generate from the optimization process, the less work that will be required later on. However, it would be unrealistic to expect that the optimum cost design for a transformer, for example, would automatically satisfy all the mechanical, thermal, and electrical constraints that require sophisticated design codes to evaluate. Rather, these constraints can be included in an approximate (conservative) manner in the optimization process. When the design codes are subsequently employed, hopefully only minor adjustments would need to be made in the design parameters to produce a workable unit.

We will briefly describe geometric programming, presenting enough of the formalism to appreciate some of its strengths and weaknesses. Then, we will present a more general approach, which is developed in greater detail, since it is applied subsequently to the transformer cost minimization problem considered in this chapter.

19.2 Geometric Programming

Geometric programming requires that the function be minimized, the objective or cost function and all the constraints be expressed as posynomials. Using the notation of Reference [Duf67], these are functions of the form

$$g = u_1 + u_2 + \cdots + u_n \tag{19.1}$$

with

$$u_i = c_i t_1^{a_{i1}} t_2^{a_{i2}} \ldots t_m^{a_{im}} \tag{19.2}$$

where the c_i are positive constants and the variables $t_1, t_2, \ldots, t_m$ are positive. The exponents a_{ij} can be of either sign or zero. They can also be integer or non-integer.

The method makes use of the geometric inequality:

$$\delta_1 U_1 + \delta_2 U_2 + \cdots + \delta_n U_n \geq U_1^{\delta_1} U_2^{\delta_2} \ldots U_n^{\delta_n} \tag{19.3}$$

where the U_i are nonnegative and the δ_i are positive weights that satisfy

$$\delta_1 + \delta_2 + \cdots + \delta_n = 1 \tag{19.4}$$

called the normality condition. Equality is obtained (19.3) if and only if all U_i are equal. Letting $u_i = \delta_i U_i$, Equation 19.3 can be rewritten as

$$g = u_1 + u_2 + \cdots + u_n \geq \left(\frac{u_1}{\delta_1}\right)^{\delta_1} \left(\frac{u_2}{\delta_2}\right)^{\delta_2} \ldots \left(\frac{u_n}{\delta_n}\right)^{\delta_n} \tag{19.5}$$

Substituting (19.2) for the u_i in (19.5), we obtain

$$g = u_1 + u_2 + \cdots + u_n \geq \left(\frac{c_1}{\delta_1}\right)^{\delta_1} \left(\frac{c_2}{\delta_2}\right)^{\delta_2} \ldots \left(\frac{c_n}{\delta_n}\right)^{\delta_n} t_1^{D_1} t_2^{D_2} \ldots t_m^{D_m} \tag{19.6}$$

$$\text{where} \quad D_j = \sum_{i=1}^{n} \delta_i a_{ij}, \ j = 1, \ldots, m$$

If we can choose the weights δ_i in such a way that

$$D_j = \sum_{i=1}^{n} \delta_i a_{ij} = 0, \quad j = 1, \ldots, m \tag{19.7}$$

then the right-hand side of (19.6) will be independent of the variables t_j. Equation 19.7 is called the orthogonality condition. With the D_j's so chosen, the right-hand side of (19.6) is referred to as the dual function, $v(\delta)$, and is given by

$$v(\delta) = \left(\frac{c_1}{\delta_1}\right)^{\delta_1} \left(\frac{c_2}{\delta_2}\right)^{\delta_2} \cdots \left(\frac{c_n}{\delta_n}\right)^{\delta_n} \tag{19.8}$$

where the vector $\delta = (\delta_1, \delta_2, \ldots, \delta_n)$. Using vector notation for the variables t_j, $t = (t_1, t_2, \ldots, t_n)$, Equation 19.5 can be rewritten as

$$g(t) \geq v(\delta) \tag{19.9}$$

subject to (19.7).

As (19.9) shows, the objective function $g(t)$ is bounded below by the dual function $v(\delta)$. Thus, the global minimum of $g(t)$ cannot be less than $v(\delta)$. Similarly, the global maximum of $v(\delta)$ cannot be greater than $g(t)$. We now show that the minimum of $g(t)$ and the maximum of $v(\delta)$ are equal. Let $t = t'$ at the minimum. At this point, the derivative of $g(t)$ with respect to each variable must vanish:

$$\frac{\partial g(t')}{\partial t'_j} = 0, \quad j = 1, \ldots, m \tag{19.10}$$

Since $t'_j > 0$ by assumption, Equation 19.10 can be multiplied by t'_j,

$$t'_j \frac{\partial g(t')}{\partial t'_j} = t'_j \frac{\partial}{\partial t'_j} \sum_{i=1}^{n} u_i(t') = \sum_{i=1}^{n} t'_j \frac{\partial u_i(t')}{\partial t'_j}$$

$$= \sum_{i=1}^{n} t'_j \frac{\partial \left(c_i t_1'^{a_{i1}} t_2'^{a_{i2}} \cdots t_j'^{a_{ij}} \cdots t_m'^{a_{im}}\right)}{\partial t'_j} = \sum_{i=1}^{n} u_i(t') a_{ij} = 0 \tag{19.11}$$

using (19.1) and (19.2). Dividing by $g(t')$ and letting

$$\delta'_i = \frac{u_i(t')}{g(t')} \tag{19.12}$$

we see that (19.11) becomes

$$\sum_{i=1}^{n} \delta'_i a_{ij} = 0, \quad j = 1, \ldots, m \tag{19.13}$$

Thus, the δ_i' satisfies the orthogonality condition (19.7). Moreover, the δ_i' sum to 1 and, thus, satisfy the normality condition (19.4). Therefore, we can write

$$g(t') = \left(g(t')\right)^{\delta_1 + \delta_2 + \cdots + \delta_n} = \left(g(t')\right)^{\delta_1} \left(g(t')\right)^{\delta_2} \cdots \left(g(t')\right)^{\delta_n} \tag{19.14}$$

But using (19.12), (19.14) becomes

$$g(t') = \left(\frac{u_1(t')}{\delta_1'}\right)^{\delta_1'} \left(\frac{u_2(t')}{\delta_2'}\right)^{\delta_2'} \cdots \left(\frac{u_n(t')}{\delta_n'}\right)^{\delta_n'} \tag{19.15}$$

The right-hand side of (19.15) is $v(\boldsymbol{\delta}')$. Thus at the minimum,

$$g(t') = v(\delta') \tag{19.16}$$

This equation, along with (19.9), shows that the minimum of g(t) and the maximum of $v(\boldsymbol{\delta})$ are equal and are in fact a global minimum and a global maximum.

Note that at the minimum, according to (19.12),

$$u_i(t') = g(t')\delta_i' = v(\delta')\delta_i' \tag{19.17}$$

This is also a requirement for the geometric inequality to become an equality. In the geometric programming approach, one looks for a maximum of the dual function $v(\boldsymbol{\delta})$ subject to the normality and orthogonality conditions. Having found this $v(\boldsymbol{\delta}')$ and the corresponding δ_i', Equation 19.17 can be used to find the design variables t_j' at the minimum of g. Equation 19.17 also shows that, at the minimum, the weights δ_i' give the relative importance (cost) of each term in the objective (cost) function.

In geometric programming with constraints, all the constraints must be expressed in the form

$$g_k(t) \le 1, \quad k = 1, \ldots, p \tag{19.18}$$

where the g_k are posynomials. Write

$$g_k = u_{k1} + u_{k2} + \cdots + u_{kn} \tag{19.19}$$

as before with the u_{ki} having the same form as (19.2). Note that kn is not necessarily equal to n and can be a different number for each constraint. In dealing with inequality constraints, the weights are allowed to be unnormalized. Letting λ_k denote their sum, we have

$$\lambda_k = \delta_{k1} + \delta_{k2} + \cdots + \delta_{kn} \tag{19.20}$$

However, in applying the geometric inequality (19.5), we must use normalized weights given by δ_{ki}/λ_k, so that we get

$$g_k = u_{k1} + u_{k2} + \cdots + u_{kn}$$

$$\geq \left(\frac{u_{k1}}{\delta_{k1}/\lambda_k}\right)^{\frac{\delta_{k1}}{\lambda_k}} \left(\frac{u_{k2}}{\delta_{k2}/\lambda_k}\right)^{\frac{\delta_{k2}}{\lambda_k}} \cdots \left(\frac{u_{kn}}{\delta_{kn}/\lambda_k}\right)^{\frac{\delta_{kn}}{\lambda_k}}$$

$$= \left(\frac{u_{k1}}{\delta_{k1}}\right)^{\frac{\delta_{k1}}{\lambda_k}} \left(\frac{u_{k2}}{\delta_{k2}}\right)^{\frac{\delta_{k2}}{\lambda_k}} \cdots \left(\frac{u_{kn}}{\delta_{kn}}\right)^{\frac{\delta_{kn}}{\lambda_k}} \left(\lambda_k^{\frac{\delta_{k1}}{\lambda_k} + \frac{\delta_{k2}}{\lambda_k} + \cdots + \frac{\delta_{kn}}{\lambda_k}}\right)$$

$$= \left(\frac{u_{k1}}{\delta_{k1}}\right)^{\frac{\delta_{k1}}{\lambda_k}} \left(\frac{u_{k2}}{\delta_{k2}}\right)^{\frac{\delta_{k2}}{\lambda_k}} \cdots \left(\frac{u_{kn}}{\delta_{kn}}\right)^{\frac{\delta_{kn}}{\lambda_k}} \lambda_k \tag{19.21}$$

The last equation follows because the exponent of λ_k sums to 1.

Since the g_k are ≤ 1, raising both sides of this inequality to a power does not change it, since 1 to any power is still 1 and any positive number ≤ 1 raised to a power is still ≤ 1. Thus, raising g_k to the power λ_k in (19.21), we obtain

$$1 \geq g_k^{\lambda_k} \geq \left(\frac{u_{k1}}{\delta_{k1}}\right)^{\delta_{k1}} \left(\frac{u_{k2}}{\delta_{k2}}\right)^{\delta_{k2}} \cdots \left(\frac{u_{kn}}{\delta_{kn}}\right)^{\delta_{kn}} \lambda_k^{\lambda_k} \tag{19.22}$$

There is an inequality like (19.22) for each constraint. We can include the objective function, g, along with the constraints by giving it the subscript 0, so that we get

$$g = g_0 = u_{01} + u_{02} + \cdots + u_{0n} \tag{19.23}$$

For the objective function $0n = n$. Therefore, we can treat it along with the constraints by letting $k = 0, 1, \ldots, p$ and where $\lambda_0 = 1$. We assume that both the objective function and the constraints have m positive variables t_i, $i = 1, \ldots, m$. If a variable t_j is missing in any term, its exponent can be set to 0 in that term. Hence,

$$g_k(t) = \sum_{i=1}^{n} c_{ki} t_1^{a_{ki1}} t_2^{a_{ki2}} \ldots t_m^{a_{kim}}, \quad k = 0, \ldots, p \tag{19.24}$$

where $g_0(t) = g(t)$ is the objective function. We now multiply all of the inequalities together, that is, the extreme left and right sides of (19.22) for each constraint, to get

$$1 \geq \prod_{k=1}^{p} \left(\prod_{i=1}^{n} \left(\frac{u_{ki}}{\delta_{ki}}\right)^{\delta_{ki}}\right) \lambda_k^{\lambda_k} \tag{19.25}$$

Multiplying both sides of (19.25) by $g(t) = g_0(t)$ and using inequalities (19.5) and (19.23), we can subsume g_0 in with the inequalities to get

$$g(t) \geq \prod_{k=0}^{p} \left(\prod_{i=1}^{n} \left(\frac{u_{ki}}{\delta_{ki}}\right)^{\delta_{ki}}\right) \lambda_k^{\lambda_k}$$

$$= \prod_{k=0}^{p} \left(\prod_{i=1}^{n} \left(\frac{c_{ki}}{\delta_{ki}}\right)^{\delta_{ki}} t_1^{\delta_{ki} a_{ki1}} t_2^{\delta_{ki} a_{ki2}} \ldots t_m^{\delta_{ki} a_{kim}}\right) \lambda_k^{\lambda_k} \tag{19.26}$$

with the auxiliary conditions,

$$D_j = \sum_{k=0}^{p}\sum_{i=1}^{n} \delta_{ki} a_{kij} = 0, \quad j = 1, \ldots, m \tag{19.27}$$

which eliminates the t_j dependencies so that the right-hand side of (19.27) becomes a dual function, $v(\boldsymbol{\delta})$, expanded to include the constraints:

$$v(\boldsymbol{\delta}) = \prod_{k=0}^{p}\left(\prod_{i=1}^{n}\left(\frac{c_{ki}}{\delta_{ki}}\right)^{\delta_{ki}}\right)\lambda_k^{\lambda_k}$$

In addition, we have the normality condition for $k = 0$,

$$\sum_{i=1}^{n} \delta_{0i} = 1 \tag{19.28}$$

All the weights δ_{ki} are required to be nonnegative,

$$\delta_{ki} \geq 0, \quad i = 1, \ldots, n, \; k = 0, \ldots, p \tag{19.29}$$

The problem of minimizing the objective function subject to its constraints becomes one of maximizing the dual function $v(\boldsymbol{\delta})$ subject to the dual constraints (19.27) through (19.29). These dual constraints, aside from the nonnegativity condition, are linear constraints in contrast to the, in general, nonlinear constraints of the original formulation. There are as many unknowns δ_{ki} as terms in the objective plus constraint functions, namely, $T = n + 1n + 2n + \cdots + pn$, where T is the total number of terms in the objective + constraint functions. There are also $p\lambda_k$ unknowns for a total of $T + p$ unknowns. However, (19.27) and (19.28) must be satisfied and there are $m + 1$ of these equations, where m is the number of design variables. Also there are p equations like (19.20) to determine the λ_k. The degree of difficulty is defined as $T + p - m - 1 - p = T - m - 1$. When this equals 0, this means that there are as many unknowns as equations and the solution is found by solving the set of linear equations (19.27) and (19.28) simultaneously. When this is greater than 0, techniques for maximizing a, in general, nonlinear function subject to linear constraints must be employed. Under very general conditions, it can be shown that the minimum of the objective function subject to its constraints equals the maximum of the dual function subject to the dual constraints [Duf67]. Furthermore, at the maximizing point $\boldsymbol{\delta}'$, the following equations hold:

$$u_i = \begin{cases} \delta_{0i} v(\boldsymbol{\delta}) & i = 1, \ldots, n \\ \dfrac{\delta_{ki}'}{\lambda_k} & k > 1, i = 1, \ldots, n \end{cases} \tag{19.30}$$

which permit the determination of the design variables t_j' at the minimizing point of the objective function via (19.2).

As a simple example, consider minimizing

$$g(\mathbf{t}) = 10 t_1 t_2^{-1} + 20 t_2 t_3^{-2}$$

subject to

$$g_1(t) = 2t_1 + 0.5t_1^{-2}t_3 \leq 1$$

There are four terms in total so that there are four δ_i and three variables, so m = 3. Therefore, $T - m - 1 = 0$. The dual function is

$$v(\delta) = \left(\frac{10}{\delta_1}\right)^{\delta_1} \left(\frac{20}{\delta_2}\right)^{\delta_2} \left(\frac{2}{\delta_3}\right)^{\delta_3} \left(\frac{0.5}{\delta_4}\right)^{\delta_4} (\delta_3 + \delta_4)^{\delta_3 + \delta_4}$$

subject to

$$
\begin{array}{ll}
\text{Normality} & \delta_1 + \delta_2 = 1 \\
t_1 \text{ exponent} = 0 & \delta_1 + \delta_3 - 2\delta_4 = 0 \\
t_2 \text{ exponent} = 0 & -\delta_1 + \delta_2 = 0 \\
t_3 \text{ exponent} = 0 & -2\delta_2 + \delta_4 = 0
\end{array}
$$

This problem has a degree of difficulty = 0, so the solution is determined by solving the enumerated linear equations to obtain

$$\delta'_1 = \frac{1}{2}, \quad \delta'_2 = \frac{1}{2}, \quad \delta'_3 = \frac{3}{2}, \quad \delta'_4 = 1$$

The maximum of the dual function is therefore

$$v(\delta) = \left(\frac{10}{1/2}\right)^{1/2} \left(\frac{20}{1/2}\right)^{1/2} \left(\frac{2}{3/2}\right)^{3/2} \left(\frac{0.5}{1}\right)^{1} \left(\frac{5}{2}\right)^{5/2} = 215.2$$

The design variables can be found from

$$10t_1 t_2^{-1} = \frac{1}{2}(215.2)$$

$$20t_2 t_3^{-2} = \frac{1}{2}(215.2)$$

$$2t_1 = \frac{3}{5}$$

$$\frac{1}{2} t_1^{-2} t_3 = \frac{2}{5}$$

These can be solved by taking logarithms or in this case by substitution. The solution is

$$t'_1 = 0.3, \quad t'_2 = 0.028, \quad t'_3 = 0.072$$

This solution can be checked by substituting these values into the objective function g(t'), which should produce the same value as given by v(**δ**') earlier.

When the degree of difficulty is not zero, the method becomes one of maximizing a non-linear function subject to linear constraints for which various solution strategies are available. For problems formulated in the language of geometric programming, these strategies lead to the global maximum of the dual function (global minimum of the objective function). This desirable outcome is not, however, without a price. In the case of transformer design optimization, the number of terms in a realistic cost function, not to mention the constraints, is quite large, whereas the number of design variables can be kept reasonably small. This means that the degree of difficulty is large. In addition, it can be quite awkward to express some of the constraints in posynomial form. A further difficulty is that the constraints must be expressed as inequalities, whereas in some cases an equality constraint may be desired.

19.3 Nonlinear Constrained Optimization

The general nonlinear optimization problem with constraints can be formulated in the following way:

$$\text{minimize} \quad f(\mathbf{x}), \quad \mathbf{x} \text{ is an n-vector}$$
$$\text{subject to the equality constraints}$$
$$c_i(\mathbf{x}) = 0, \quad i = 1, \ldots, t \tag{19.31}$$
$$\text{and the inequality constraints}$$
$$c_i(\mathbf{x}) \geq 0, \quad i = t+1, \ldots, m$$

where f and the c_i are nonlinear functions in general. We will assume throughout that they are at least twice differentiable. Although we have expressed this optimization in terms of a minimum, maximizing a function f is equivalent to minimizing $-f$. Also, any equality or inequality constraint involving analytic functions can be expressed in the said manner. For example, a constraint of the type $c_i(x) \leq 0$ can be rewritten as $-c_i(x) \geq 0$. Thus, this formulation is quite general.

19.3.1 Characterization of the Minimum

We will be concerned with finding a relative minimum of $f(x)$ subject to the constraints. This may or may not be a global minimum for the particular problem considered. We first consider how to characterize such a minimum, that is, how do we know we have reached a relative minimum while satisfying all the constraints. Let x^* denote the design-variable vector at a minimum. In this section, minimum always means relative minimum. If an inequality constraint is not active at the minimum, that is, $c_i(x^*) > 0$, then it will remain positive in some neighborhood about the minimum. (An inequality constraint is active if $c_i(x) = 0$, and it is said to be violated if $c_i(x) < 0$.) Thus, for an inactive constraint, one could move some small distance away from the minimum without violating this constraint. In effect, near the minimum, a non-active constraint places no restrictions on the design variables. It can therefore be ignored. Only those inequality constraints that are active play a role in characterizing the minimum. We therefore include them in the list of equality constraints in our formulation (19.31). Strategies to add or drop inequalities from the list of equality constraints will be discussed later.

Much of the methodology presented in this section is based on Reference [Gil81]. We adopt their notation, which is briefly reviewed here. First, a distinction is made between column and row vectors. A column vector is denoted by

$$\mathbf{v} = \begin{pmatrix} v_1 \\ v_2 \\ \vdots \\ v_n \end{pmatrix}$$

where the v_i are its components. If there are n components, it can also be referred to as an n-vector. Its transpose is a row vector, that is,

$$\mathbf{v}^T = \begin{pmatrix} v_1 & v_2 & \cdots & v_n \end{pmatrix}$$

where T denotes transpose. The dot or scalar product between two vectors $\mathbf{v}$ and $\mathbf{w}$ is given by

$$\mathbf{v}^T \mathbf{w} = v_1 w_1 + v_2 w_2 + \cdots + v_n w_n$$

It is essentially the matrix product between a row and column vector in that order. If the order were reversed, we would have

$$\mathbf{v}\mathbf{w}^T = \begin{pmatrix} v_1 w_1 & v_1 w_2 & \cdots & v_1 w_n \\ v_2 w_1 & v_2 w_2 & & \\ \vdots & & \ddots & \\ v_n w_1 & & & v_n w_n \end{pmatrix}$$

that is, the matrix product between a column and row vector in that order. A matrix M is called an m × n matrix if it has m rows and n columns. The i, jth matrix element is denoted M_{ij} and these are organized into an array:

$$M = \begin{pmatrix} M_{11} & M_{12} & \cdots & M_{1n} \\ M_{21} & M_{22} & & \\ \vdots & & \ddots & \\ M_{m1} & & & M_{mn} \end{pmatrix}$$

A matrix vector product between an m × n matrix and an n-vector, denoted $M\mathbf{v}$, is an m-vector whose components are

$$(M\mathbf{v})_i = \sum_{j=1}^{n} M_{ij} v_j$$

If we multiply the said m-vector on the left by a row m-vector $\mathbf{w}^T$, we get the scalar product between $\mathbf{w}$ and the said vector

$$\mathbf{w}^T M \mathbf{v} = \sum_{i=1}^{m} \sum_{j=1}^{n} w_i M_{ij} v_j$$

Further notation will be introduced as it is encountered.

Taylor's theorem for the expansion of a continuous function of a single variable, $f(x)$, about a point $x + \Delta x$, where Δx is assumed to be small, to second order in Δx is

$$f(x + \Delta x) = f(x) + \frac{\partial f}{\partial x} \Delta x + \frac{1}{2} \frac{\partial^2 f}{\partial x^2} (\Delta x)^2 + \cdots$$

For a function of n variables, $f(\mathbf{x})$, x becomes an n-vector $\mathbf{x} = (x_1, x_2, \ldots, x_n)$ and Δx becomes $\varepsilon \mathbf{p}$ where ε is assumed to be a small positive constant and $\mathbf{p}$ is a vector indicating the direction in which x is changing. Thus, Taylor's theorem in n-dimensions becomes

$$f(\mathbf{x} + \varepsilon \mathbf{p}) = f(\mathbf{x}) + \varepsilon \mathbf{g}^T(\mathbf{x}) \mathbf{p} + \frac{1}{2} \varepsilon^2 \mathbf{p}^T G(\mathbf{x}) \mathbf{p} + \cdots \tag{19.32}$$

where $\mathbf{g}$ is the gradient vector and is given by

$$\mathbf{g}^T(\mathbf{x}) = \left(\begin{array}{cccc} \dfrac{\partial f}{\partial x_1} & \dfrac{\partial f}{\partial x_2} & \cdots & \dfrac{\partial f}{\partial x_n} \end{array} \right) \tag{19.33}$$

G is the Hessian matrix, which is given by

$$G(\mathbf{x}) = \begin{pmatrix} \dfrac{\partial f}{\partial x_1 \partial x_1} & \dfrac{\partial f}{\partial x_1 \partial x_2} & \cdots & \dfrac{\partial f}{\partial x_1 \partial x_n} \\[2ex] \dfrac{\partial f}{\partial x_2 \partial x_1} & \dfrac{\partial f}{\partial x_2 \partial x_2} & & \\[2ex] \vdots & & \ddots & \\[2ex] \dfrac{\partial f}{\partial x_n \partial x_1} & & & \dfrac{\partial f}{\partial x_n \partial x_n} \end{pmatrix} \tag{19.34}$$

The Hessian matrix is symmetric because of the equality of mixed partial derivatives. Vector dot products or matrix products are implied in (19.32). Higher-order terms in the expansion (19.32) are assumed to be small for the values of ε considered.

We use Taylor's theorem to expand the function to be minimized, f along with the equality constraints, c_i, about the relative minimum $\mathbf{x}^*$:

$$f(\mathbf{x}^* + \varepsilon \mathbf{p}) = f(\mathbf{x}^*) + \varepsilon \mathbf{g}^T(\mathbf{x}^*) \mathbf{p} + \frac{1}{2} \varepsilon^2 \mathbf{p}^T G(\mathbf{x}^*) \mathbf{p} + \cdots$$

$$c_i(\mathbf{x}^* + \varepsilon \mathbf{p}) = c_i(\mathbf{x}^*) + \varepsilon \mathbf{a}_i^T(\mathbf{x}^*) \mathbf{p} + \frac{1}{2} \varepsilon^2 \mathbf{p}^T G_i(\mathbf{x}^*) \mathbf{p} + \cdots \tag{19.35}$$

where $\mathbf{a}_i$ is the gradient vector for c_i and G_i, its Hessian matrix. Note that, similar to (19.33), we have

$$\mathbf{a}_j^T(\mathbf{x}) = \left(\frac{\partial c_j}{\partial x_1} \quad \frac{\partial c_j}{\partial x_2} \quad \cdots \quad \frac{\partial c_j}{\partial x_n} \right)$$

Here, $c_i(\mathbf{x}^*) = 0$ for the equality constraints. We would like to continue to satisfy the constraints as we move away from the minimum, since they are assumed to apply for any choice of design parameters. This means that the first-order term, that is, the term in ε should vanish, which implies

$$\mathbf{a}_i^T(\mathbf{x}^*)\mathbf{p} = 0 \tag{19.36}$$

There is an equation like this for each equality constraint. Let A denote a matrix whose ith row is $\mathbf{a}_i^T$. Then (19.36) can be expressed compactly for all equality constraints as

$$A\mathbf{p} = 0 \tag{19.37}$$

A has t rows and n columns. It also depends on $\mathbf{x}$ or $\mathbf{x}^*$, but we omit this for clarity. We assume that $t \leq n$ and that A has a full-row rank, that is, that the constraint gradient vectors $\mathbf{a}_i$ are linearly independent. If this is not true, then one or more of the constraints is redundant (to first order) and can be dropped from the list. For easier visualization, assume n = 6 and t = 4 then

$$A = \begin{pmatrix} a_{11} & a_{12} & a_{13} & a_{14} & a_{15} & a_{16} \\ a_{21} & a_{22} & a_{23} & a_{24} & a_{25} & a_{26} \\ a_{31} & a_{32} & a_{33} & a_{34} & a_{35} & a_{36} \\ a_{41} & a_{42} & a_{43} & a_{44} & a_{45} & a_{46} \end{pmatrix}$$

From (19.37), we see that $\mathbf{p}$ must be orthogonal to the row space of A. Thus, $\mathbf{p}$ lies in the orthogonal complement of this row space. (Any vector space is decomposable into disjoint subspaces, which are orthogonal complements of each other.) Let Z denote a matrix whose column vectors form a basis of this orthogonal subspace of dimension n − t. Thus, Z has n rows and n − t columns. This orthogonality can be expressed by

$$AZ = 0 \tag{19.38}$$

Since $\mathbf{p}$ lies in this orthogonal subspace, it can be expressed as a linear combination of its basis vectors, that is, a linear combination of the column vectors of Z. Thus,

$$\mathbf{p} = Z\mathbf{p}_Z \tag{19.39}$$

where $\mathbf{p}_Z$ is a vector of dimension n − t.

Probably the best way to obtain Z is via a triangularization of A using Hausholder transformation matrices. These are orthogonal matrices of the form

$$H = I - \frac{2\mathbf{w}\mathbf{w}^T}{|\mathbf{w}|^2} \tag{19.40}$$

where
  $\mathbf{w}$ is a vector of magnitude $|\mathbf{w}|$
  I is the unit diagonal matrix

By suitably choosing **w**, which depends on the matrix to be triangularized, the columns of any matrix can be successively put in upper triangular form [Gol89]. In our case, we apply this procedure to A^T to obtain

$$H_n \cdots H_2 H_1 A^T = Q A^T = \begin{pmatrix} R \\ 0 \end{pmatrix} \tag{19.41}$$

where
  Q is an n × n orthogonal matrix, since it is the product of orthogonal matrices
  R is an upper triangular matrix of dimension t × t
  0 is a zero matrix of dimension (n − t) × t

For easier visualization, using n = 6 and t = 4 as before, we have

$$\begin{pmatrix} R \\ 0 \end{pmatrix} = \begin{pmatrix} R_{11} & R_{12} & R_{13} & R_{14} \\ 0 & R_{22} & R_{24} & R_{24} \\ 0 & 0 & R_{33} & R_{34} \\ 0 & 0 & 0 & R_{44} \\ 0 & 0 & 0 & 0 \\ 0 & 0 & 0 & 0 \end{pmatrix}$$

Taking the transpose of (19.41), we get

$$A Q^T = \begin{pmatrix} L & 0 \end{pmatrix} \tag{19.42}$$

where
  $L = R^T$ is a lower triangular matrix
  0 is the t × (n − t) zero matrix

Again for easier visualization, we have

$$\begin{pmatrix} L & 0 \end{pmatrix} = \begin{pmatrix} R_{11} & 0 & 0 & 0 & 0 & 0 \\ R_{12} & R_{22} & 0 & 0 & 0 & 0 \\ R_{13} & R_{23} & R_{33} & 0 & 0 & 0 \\ R_{14} & R_{24} & R_{34} & R_{44} & 0 & 0 \end{pmatrix}$$

Thus, we can take Z to be the last n − t columns of Q^T since these are orthogonal to the rows of A. Since Q is an orthogonal matrix, its column vectors form an orthonormal basis for the whole space. Thus, the first t columns form an orthonormal basis for the row space of A, and the last n − t columns form an orthonormal basis for its orthogonal complement. We group the first t columns into a matrix called Y so that

$$Q^T = \begin{pmatrix} Y & Z \end{pmatrix} \tag{19.43}$$

With **p** restricted to the form (19.39), the expression (19.32) becomes first order

$$f\left(\mathbf{x}^* + \varepsilon \mathbf{p}\right) = f\left(\mathbf{x}^*\right) + \varepsilon \mathbf{g}^T\left(\mathbf{x}^*\right) Z \mathbf{p}_Z + \cdots \tag{19.44}$$

For f(**x***) to be a minimum, we must have

$$f\left(\mathbf{x}^* + \varepsilon \mathbf{p}\right) \geq f\left(\mathbf{x}^*\right) \tag{19.45}$$

In order for this to hold, the term in ε in (19.44) must be positive or zero. Since ε is positive, if this term is nonzero for some vector $\mathbf{p}_Z$, it can be made negative by possibly changing $\mathbf{p}_Z$ to $-\mathbf{p}_Z$ and thus violate (19.45). Hence, we must require at the minimum

$$\mathbf{g}^{T}\left(\mathbf{x}^*\right)Z = 0 \tag{19.46}$$

This implies by (19.38) that the gradient vector is in the row space of A. Note that, using our previous numerical example for visualization purposes, with n = 6 and n – t = 2,

$$\mathbf{g}^{T}\left(\mathbf{x}^*\right)Z = \begin{pmatrix} g_1 & g_2 & g_3 & g_4 & g_5 & g_6 \end{pmatrix} \begin{pmatrix} Z_{11} & Z_{12} \\ Z_{21} & Z_{22} \\ Z_{31} & Z_{32} \\ Z_{41} & Z_{42} \\ Z_{51} & Z_{52} \\ Z_{61} & Z_{62} \end{pmatrix} = \begin{pmatrix} 0 & 0 \end{pmatrix}$$

Therefore,

$$\mathbf{g}\left(\mathbf{x}^*\right) = A^{T}\lambda^* \tag{19.47}$$

where λ^* is a vector of dimension t called the vector of Lagrange multipliers. (Note that the column space of A^T is the row space of A.) Let us rewrite (19.47) in matrix-vector form using n = 6 and t = 4 as before:

$$\begin{pmatrix} g_1 \\ g_2 \\ g_3 \\ g_4 \\ g_5 \\ g_6 \end{pmatrix} = \begin{pmatrix} a_{11} & a_{21} & a_{31} & a_{41} \\ a_{12} & a_{22} & a_{32} & a_{42} \\ a_{13} & a_{23} & a_{33} & a_{43} \\ a_{14} & a_{24} & a_{34} & a_{44} \\ a_{15} & a_{25} & a_{35} & a_{45} \\ a_{16} & a_{26} & a_{36} & a_{46} \end{pmatrix} \begin{pmatrix} \lambda_1^* \\ \lambda_2^* \\ \lambda_3^* \\ \lambda_4^* \end{pmatrix}$$

Let us now consider that one of our equality constraints at the minimum was originally an inequality constraint. Let this be the c_jth constraint. Then a displacement vector **p** could be chosen so that

$$c_j\left(\mathbf{x}^* + \varepsilon \mathbf{p}\right) \geq 0 \tag{19.48}$$

without violating the constraint. Noting that $c_j(\mathbf{x}^*) = 0$ so that from (19.35) (with j replacing i), we see that to first order in ε, we get

$$\mathbf{a}_j^{T}\left(\mathbf{x}^*\right)\mathbf{p} \geq 0 \tag{19.49}$$

Assume that $\mathbf{p}$ satisfies the other constraints, that is, Equation 19.36 holds for $i \neq j$. Then rewriting (19.47), we have

$$\mathbf{g}(\mathbf{x}^*) = \overset{*}{\lambda_1}\mathbf{a}_1 + \overset{*}{\lambda_2}\mathbf{a}_2 + \cdots + \overset{*}{\lambda_j}\mathbf{a}_j + \cdots + \overset{*}{\lambda_t}\mathbf{a}_t \tag{19.50}$$

where we have omitted the dependence of the $\mathbf{a}$'s on $\mathbf{x}^*$. For the said choice of $\mathbf{p}$,

$$\mathbf{g}^{\mathrm{T}}(\mathbf{x}^*)\mathbf{p} = \overset{*}{\lambda_j}\mathbf{a}_j^{\mathrm{T}}\mathbf{p} \tag{19.51}$$

Substituting into (19.32), we get to first order in ε

$$f(\mathbf{x}^* + \varepsilon\mathbf{p}) = f(\mathbf{x}^*) + \varepsilon\overset{*}{\lambda_j}\mathbf{a}_j^{\mathrm{T}}\mathbf{p} \tag{19.52}$$

If $f(\mathbf{x}^*)$ is a local minimum, then (19.45) shows that the term $\varepsilon\overset{*}{\lambda_j}\mathbf{a}_j^{\mathrm{T}}\mathbf{p} \geq 0$. But $\varepsilon > 0$, together with (19.49), shows that we must have

$$\overset{*}{\lambda_j} \geq 0 \tag{19.53}$$

This says that the Lagrange multipliers associated with the inequality constraints that are active at the minimum must be nonnegative. If the jth Lagrange multiplier were negative, then the direction $\mathbf{p}$ chosen earlier can be used to lower the value of f without violating the jth constraint, thus contradicting the assumption what we are at a minimum. In searching for a minimum, a negative Lagrange multiplier for an active inequality constraint is an indication that the constraint can be dropped (deactivated). This strategy is useful in deciding when to drop a constraint from the active set.

The discussion so far applies equally well to whether we are looking for a minimum or maximum of the objective function. The only difference is that the sign of the Lagrange multiplier of an active inequality constraint would be negative for a maximum. To distinguish a minimum from a maximum, we need to go to higher order. In the case of nonlinear constraints, we can no longer move along a straight line with direction $\mathbf{p}$ when considering higher-order effects because the constraints would eventually be violated. (With linear constraints, one could move freely along the direction $\mathbf{p}$ given by (19.39) without violating the active constraints.) For nonlinear constraints, we need to move along a curve $\mathbf{x}(\theta)$ parametrized by θ chosen in such a way that the constraints are not violated to second order as we move a small amount $\Delta\theta$ away from the minimum. At the minimum, let $\theta = \theta^*$ and $\mathbf{x}^* = \mathbf{x}(\theta^*)$. Let $\mathbf{p}$ be the tangent vector to this curve at the minimum, that is,

$$\mathbf{p} = \left.\frac{d\mathbf{x}}{d\theta}\right|_{\theta=\theta^*} \tag{19.54}$$

Using Taylor's theorem in one variable to second order for the active constraints, we have

$$c_i\big(\mathbf{x}(\theta^* + \Delta\theta)\big) = c_i(\mathbf{x}^*) + \frac{dc_i}{d\theta}\Delta\theta + \frac{1}{2}\frac{d^2c_i}{d\theta^2}\Delta\theta^2 + \cdots \tag{19.55}$$

All derivatives are evaluated at θ^*. Using the chain rule, we obtain

$$\frac{dc_i}{d\theta} = \sum_{j=1}^{n} \frac{\partial c_i}{\partial x_j} \frac{dx_j}{d\theta} = \mathbf{a}_i^T \mathbf{p} \tag{19.56}$$

and, using (19.56),

$$
\begin{aligned}
\frac{d^2 c_i}{d\theta^2} &= \frac{d\mathbf{a}_i^T}{d\theta} \mathbf{p} + \mathbf{a}_i^T \frac{d\mathbf{p}}{d\theta} \\
&= \sum_{j=1}^{n} \left(\frac{d\mathbf{a}_i^T}{dx_j} \frac{dx_j}{d\theta} \mathbf{p} \right) + \mathbf{a}_i^T \frac{d\mathbf{p}}{d\theta} \\
&= \mathbf{p}^T G_i(\mathbf{x}^*) \mathbf{p} + \mathbf{a}_i^T \frac{d\mathbf{p}}{d\theta} \tag{19.57}
\end{aligned}
$$

where G_i is the Hessian of the ith constraint defined previously. Since $c_i(\mathbf{x}^*) = 0$ at the minimum, in order for (19.55) to vanish to second order, we must have both (19.56) and (19.57) vanish. The vanishing of (19.56) yields the first-order results, such as (19.36) previously derived. The vanishing of (19.57) yields the new second-order results. Just as we previously turned (19.56) into a matrix equation so it applied to all the constraints, Equation 19.37, we apply the same procedure to (19.57). Organizing the $\mathbf{a}_i^T$ into rows of the A matrix and setting (19.57) to zero, we get

$$A \frac{d\mathbf{p}}{d\theta} = - \begin{pmatrix} \mathbf{p}^T G_1 \mathbf{p} \\ \mathbf{p}^T G_2 \mathbf{p} \\ \vdots \\ \mathbf{p}^T G_t \mathbf{p} \end{pmatrix} \tag{19.58}$$

Now consider the behavior of the objective function f along this curve, which satisfies the constraints to second order. Using Taylor's theorem again in one variable,

$$f\big(\mathbf{x}(\theta^* + \Delta\theta)\big) = f(\mathbf{x}^*) + \frac{df}{d\theta} \Delta\theta + \frac{1}{2} \frac{d^2 f}{d\theta^2} \Delta\theta^2 + \cdots \tag{19.59}$$

Since by assumption $f(\mathbf{x}^*)$ is a minimum, and since $\Delta\theta$ can have either sign, we must have to first order

$$\frac{df}{d\theta} = \sum_{j=1}^{n} \frac{\partial f}{\partial x_j} \frac{dx_j}{d\theta} = \mathbf{g}^T \mathbf{p} = 0 \tag{19.60}$$

This is the same first-order condition we obtained previously, which led to formulas (19.46) and (19.47) that are still valid in the present context. As mentioned, they would be true, whether we were at a minimum or maximum. Now since we are at a minimum,

$$f\big(\mathbf{x}(\theta^* + \Delta\theta)\big) \geq f(\mathbf{x}^*) \tag{19.61}$$

Since the first-order term in (19.59) is zero, Equation 19.61 implies that the second-order term is nonnegative. Using the same mathematics as was used for this term for the constraints, Equation 19.57, we find using (19.60)

$$
\begin{aligned}
\frac{d^2 f}{d\theta^2} &= \frac{d\mathbf{g}^T}{d\theta}\mathbf{p} + \mathbf{g}^T \frac{d\mathbf{p}}{d\theta} \\
&= \sum_{j=1}^{n}\left(\frac{d\mathbf{g}^T}{dx_j}\frac{dx_j}{d\theta}\mathbf{p}\right) + \mathbf{g}^T \frac{d\mathbf{p}}{d\theta} \\
&= \mathbf{p}^T G(\mathbf{x}^*)\mathbf{p} + \mathbf{g}^T \frac{d\mathbf{p}}{d\theta} \geq 0
\end{aligned}
\tag{19.62}
$$

Substituting for $\mathbf{g}^T$ from (19.47) into (19.62), we obtain

$$
\mathbf{p}^T G(\mathbf{x}^*)\mathbf{p} + \lambda^{*T} A \frac{d\mathbf{p}}{d\theta} \geq 0
\tag{19.63}
$$

Substituting (19.58) into (19.63), we get

$$
\mathbf{p}^T\left(G(\mathbf{x}^*) - \sum_{i=1}^{t}\lambda_i^* G_i(\mathbf{x}^*)\right)\mathbf{p} \geq 0
\tag{19.64}
$$

Using (19.39) for $\mathbf{p}$, Equation 19.64 becomes

$$
\mathbf{p}_Z^T\left[Z^T\left(G(\mathbf{x}^*) - \sum_{i=1}^{t}\lambda_i^* G_i(\mathbf{x}^*)\right)Z\right]\mathbf{p}_Z \geq 0
\tag{19.65}
$$

Since (19.65) must be true for any choice of $\mathbf{p}_Z$, this requires the matrix within the square brackets to be positive semidefinite. Often, minimization with constraints is formulated in terms of a Lagrangian function:

$$
L(\mathbf{x}, \lambda) = f(\mathbf{x}) - \sum_{i=1}^{t}\lambda_i^T c_i(\mathbf{x})
\tag{19.66}
$$

Then (19.47) follows from requiring the Lagrangian to vanish to first order at the minimum and Equation 19.65 follows from requiring the projected Hessian of the Lagrangian to be positive semidefinite. (The Hessian is projected by means of the Z matrix.) The Lagrangian approach does not appear to be as well motivated as the one we have taken.

Conditions (19.65) and (19.47) are necessary for a minimum subject to the constraints. They are not, however, sufficient. This is because if (19.65) were 0 for some nonzero vector $\mathbf{p}_Z$, the point $\mathbf{x}^*$ could be a saddle point. To eliminate this possibility, the inequality in (19.65) must be replaced by a strict inequality, >, so that the matrix is positive definite. This is sufficient to guarantee a minimum. We will assume this requirement to be met in the minimization problems we deal with.

19.3.2 Solution Search Strategy

Because the functions we are dealing with are, in general, nonlinear, iterations are required in order to arrive at a minimum from some starting set of the design variables. Assuming the iterations converge, we can then check that the conditions derived in the previous section hold at the converged point to guarantee that it is a minimum. This may or may not be a global minimum. However, if suitable starting values are chosen based on experience, then for most practical problems, the minimum reached will be, with high probability, a global minimum.

One of the best convergence strategies is Newton–Raphson iteration. Provided one is near the minimum, this method converges very rapidly. In fact, for a quadratic function, it converges in one step. Applying this method to the constraint functions, with $\mathbf{p}$ replacing $\Delta\mathbf{x}$ in the usual formulation of the Newton–Raphson method, we get

$$c_i(\mathbf{x}+\mathbf{p}) \approx c(\mathbf{x}) + \mathbf{a}_i^T\mathbf{p} = 0 \tag{19.67}$$

Collecting these into a vector of constraint functions, $\mathbf{c}(\mathbf{x}) = (c_1\ c_2\ \dots\ c_t)$, and using the matrix A whose rows are $\mathbf{a}_i^T$ as defined previously, Equation 19.67 can be written to include all the active constraints as

$$A(\mathbf{x})\mathbf{p} = -\mathbf{c}(\mathbf{x}) \tag{19.68}$$

where the dependence of A on the value of $\mathbf{x}$ at the current iterate is indicated. Since $\mathbf{p}$ is an n-vector in the space of design variables, we can express it in terms of the basis we obtained from the QR factorization of A. Thus, using (19.43)

$$\mathbf{p} = Y\mathbf{p}_Y + Z\mathbf{p}_Z \tag{19.69}$$

where
  $\mathbf{p}_Y$ is a t-vector in the row space of A
  $\mathbf{p}_Z$ is an $(n - t)$-vector in its orthogonal complement

Using (19.38), (19.68) becomes

$$AY\mathbf{p}_Y = -\mathbf{c}(\mathbf{x}) \tag{19.70}$$

where the dependence of the matrices on $\mathbf{x}$ has been suppressed for clarity. However, from (19.42) and (19.43), we see that $AY = L$. Thus, Equation 19.70 becomes

$$L\mathbf{p}_Y = -\mathbf{c}(\mathbf{x}) \tag{19.71}$$

Since L is a lower triangular matrix, Equation 19.71 can be solved readily using forward substitution. Thus, the QR factorization of A has an additional benefit in facilitating the solution of a matrix equation.

We now apply the Newton–Raphson technique to (19.47), which must be satisfied at the minimum:

$$\mathbf{g}(\mathbf{x}+\mathbf{p}) - A^T(\mathbf{x}+\mathbf{p})\lambda = \mathbf{g}(\mathbf{x}+\mathbf{p}) - \sum_{i=1}^{t}\mathbf{a}_i(\mathbf{x}+\mathbf{p})\lambda_i$$

$$\approx \mathbf{g}(\mathbf{x}) + G(\mathbf{x})\mathbf{p} - A^T\lambda - \sum_{i=1}^{t}\lambda_i G_i(\mathbf{x})\mathbf{p} = 0 \tag{19.72}$$

Rewriting (19.72),

$$\left(G(\mathbf{x}) - \sum_{i=1}^{t} \lambda_i G_i(\mathbf{x})\right)\mathbf{p} = -\left(\mathbf{g}(\mathbf{x}) - A^T\lambda\right) \tag{19.73}$$

Define

$$W(\mathbf{x}) = G(\mathbf{x}) - \sum_{i=1}^{t} \lambda_i G_i(\mathbf{x}) \tag{19.74}$$

This is the Hessian of the Lagrangian function. Then (19.73) can be rewritten as

$$W(\mathbf{x})\mathbf{p} = -\mathbf{g}(\mathbf{x}) + A^T\lambda \tag{19.75}$$

Multiply this last expression by Z^T to obtain

$$Z^T W(\mathbf{x})\mathbf{p} = -Z^T\mathbf{g}(\mathbf{x}) \tag{19.76}$$

In going from (19.75) to (19.76), we have used the fact that the term $Z^T A^T\lambda$ vanishes by the transpose of (19.38). Using (19.69) for $\mathbf{p}$, Equation 19.76 becomes

$$Z^T W(\mathbf{x})Z\mathbf{p}_Z = -Z^T\left(\mathbf{g}(\mathbf{x}) + W(\mathbf{x})Y\mathbf{p}_Y\right) \tag{19.77}$$

Having obtained $\mathbf{p}_Y$ from (19.71), this last equation can be used to solve for $\mathbf{p}_Z$.

The projected Hessian of the Lagrangian function occurs in (19.77). It was previously shown that this matrix must be positive definite to insure a minimum. However, during the iterations involved in finding a minimum, this matrix may not be positive definite. If this occurs, the Newton–Raphson iteration method may have difficulty converging. One method of circumventing this problem is to modify this matrix so that it is positive definite at each iteration step. The procedure must be such that when the matrix eventually becomes positive definite, no modification is made.

One way of producing a positive definite matrix that is not too different from the matrix to be altered is called a modified Cholesky factorization [Gil81]. In this method, a Cholesky factorization is begun as if the matrix were positive definite. For a positive definite matrix, M, a Cholesky factorization would result in the factorization

$$M = PDP^T \tag{19.78}$$

where
   P is a lower triangular matrix
   D a positive diagonal matrix, that is, a diagonal matrix with all positive elements

If, during the course of the factorization of a not necessarily positive definite matrix, an element of D is calculated to be zero or negative, the corresponding diagonal element of M is increased until the element of D becomes positive. Limitations are also placed on how large the values of P can become during the factorization. These limitations can also be achieved by increasing the diagonal values of the original matrix. The result is a Cholesky factorization of a modified matrix M', which is related to the original matrix by

$$M' = PDP^T = M + E \qquad (19.79)$$

where E is a positive diagonal matrix. Using this approach, Equation 19.77 is solved for $\mathbf{p}_Z$ at each iteration.

With $\mathbf{p}_Y$ and $\mathbf{p}_Z$ determined, the complete step given by (19.69) is known. This step may, however, be too large. Some of the equality constraints may be violated to too great an extent or the objective function may not decrease. The step may also be so large as to violate an inequality constraint that is not active. We would like to keep the direction of $\mathbf{p}$ but restrict its magnitude based on these considerations. To achieve this, multiply $\mathbf{p}$ by ε to get a step $\varepsilon\mathbf{p}$. By letting ε vary from 0 to 1, we can check these other criteria at each ε, stopping when they become too invalidated or when $\varepsilon = 1$. One of these criteria would be that we not violate any of the inactive inequality constraints. If this occurs at some value of $\varepsilon \leq 1$, then this restricts the step size and the constraint, which is just violated, is added to the list of active constraints on the next iteration. Another criteria to use to restrict the step size is to define a merit function, which measures to what extent the constraints are satisfied and to what extent the objective function is decreasing in value along the step. One choice for this function is

$$\text{Merit function} = f(\mathbf{x}) + \rho \sum_{i=1}^{t} c_i(\mathbf{x})^2 \qquad (19.80)$$

where ρ is a positive constant. The values of ε are increased toward 1 so long as this merit function decreases or until one of the inactive constraints is violated. When this function starts increasing or an inactive constraint is violated, that value of ε is used to determine the step size $\varepsilon\mathbf{p}$. In this approach, units must be chosen so that the relative magnitudes of f and the c_i are close. This means that, for values of $\mathbf{x}$, which can be expected to violate the constraints during the iterations, the values of f and the c_i should be of the same order of magnitude. Then ρ can really be used to weigh the importance of satisfying the constraints relative to achieving a minimum of the objective function. In our work, we have chosen $\rho = 20$. Thus, we gradually increase ε until either the merit function starts increasing or an inactive constraint is violated or $\varepsilon = 1$, whichever comes first.

Having chosen ε, we solve for new values of the Lagrange multipliers; using (19.75), we obtain

$$A^T\lambda = \mathbf{g}(\mathbf{x}) + W(\mathbf{x})\varepsilon\mathbf{p} \qquad (19.81)$$

which is an equation for the unknown λ, since all other quantities are known. The solution of this equation is also facilitated by the results of the QR factorization of A. From (19.41), we can write

$$A^T = Q^T \begin{pmatrix} R \\ 0 \end{pmatrix} \qquad (19.82)$$

since $Q^TQ = QQ^T = I$ for orthogonal matrices. Here, I is the unit diagonal matrix. Substituting into (19.81) and multiplying by Q, we obtain

$$\begin{pmatrix} R \\ 0 \end{pmatrix} \lambda = Q\big(\mathbf{g}(\mathbf{x}) + W(\mathbf{x})\varepsilon\mathbf{p}\big) \tag{19.83}$$

But using (19.43), this becomes

$$\begin{pmatrix} R \\ 0 \end{pmatrix} \lambda = \begin{pmatrix} Y^T \\ Z^T \end{pmatrix}\big(\mathbf{g}(\mathbf{x}) + W(\mathbf{x})\varepsilon\mathbf{p}\big) \tag{19.84}$$

Only the top t equations in (19.84) are needed to determine λ and these are

$$R\lambda = Y^T\big(\mathbf{g}(\mathbf{x}) + W(\mathbf{x})\varepsilon\mathbf{p}\big) \tag{19.85}$$

Since $R = L^T$ is upper triangular, Equation 19.85 can be directly solved by back substitution. This value of λ is used in the next iterate. At the start, all the λ_i are set to 1.

At the next iterate, the values of the design variables become

$$\mathbf{x}_{new} = \mathbf{x}_{old} + \varepsilon\mathbf{p} \tag{19.86}$$

where
$_{old}$ labels the present values
$_{new}$ the new values for the design variables

This process is continued until successive iterations produce negligible changes in the design variables and the Lagrange multipliers. At this point, it is necessary to check that the conditions, which should hold at a minimum, are satisfied. Thus, the projected Hessian of the Lagrangian should be positive definite and the gradient of the objective function should satisfy (19.47) to some level of accuracy. The equality constraints should also equal 0 to some level of accuracy. The Lagrange multipliers of the inequality constraints, which are active at the minimum, should be positive. None of the inequality constraints should be violated, that is, <0.

During the course of the iterations, the number of equality constraints may change as inequality constraints are added or dropped from the list of active constraints. They are added when a step size threatens to violate an inactive constraint, and they are dropped when a Lagrange multiplier becomes negative. Strictly speaking, the Lagrange multiplier of an active inequality constraint only needs to be positive at the minimum. This does not have to hold away from the minimum as occurs during the iterations. However, there will be some neighborhood about the minimum where they will remain positive if the constraint is active at the minimum. We also noted previously that the gradients of the active constraints are assumed to be linearly independent. If this is not true, then constraints are dropped until a linearly independent set is obtained. This is determined during the QR factorization of A since the process requires the rows of A to be linearly independent.

For certain types of problems, it may be necessary for all or some of the variables to be positive. This can be treated as a special type of inequality constraint. Thus if variable x_j must be positive, we add to the list of constraints

$$c_k(\mathbf{x}) = x_j \geq 0 \tag{19.87}$$

where k labels the inequality constraint. The gradient vector for this constraint is given by

$$\mathbf{a}_k^T = \begin{pmatrix} 0 & \dots & 0 & 1 & 0 & \dots & 0 \end{pmatrix} \tag{19.88}$$

where the 1 is in the jth position. Its Hessian matrix is identically zero. Thus, this type of restriction can be handled like an ordinary inequality constraint. It has no influence on the minimization, unless x_j becomes 0 or negative during the course of the iterations.

19.3.3 Practical Considerations

It is useful to choose units so that all the design variables are comparable in magnitude. It is also desirable to keep the objective function and the constraints (violated to some degree or away from 0 if inequalities) of comparable magnitude. A good choice for this magnitude is unity. For example, if the objective function is cost, one could express it in $, kilo$, or mega$, whichever produces a value of order of magnitude unity at the minimum. Similarly, an equality or inequality constraint of the form

$$A + B + C + D + \cdots \geq 0$$

could be divided by the absolute value of the maximum term so as to bring it closer to 1. This may not be possible in all cases, so some other strategy may be necessary. This procedure will prevent the merit function from favoring one constraint more than another.

Although the method developed here applies to any nonlinear, twice differentiable objective or constraint functions, in practice it is found that these functions are often of the form

$$F(\mathbf{x}) = \sum_{i=1}^{m} b_i x_1^{a_{i1}} x_2^{a_{i2}} \dots x_n^{a_{in}} = \sum_{i=1}^{m} b_i \prod_{j \neq k}^{n} x_j^{a_{ij}} \tag{19.89}$$

where the coefficients b_i and the exponents a_{ij} are arbitrary constants. This differs from the posynomials considered earlier where the coefficients b_i all had to be positive. The kth component of the gradient vector of this function is given by

$$\frac{\partial F(\mathbf{x})}{\partial x_k} = \sum_{i=1}^{m} b_i a_{ik} x_k^{a_{ik}-1} \prod_{j \neq k}^{n} x_j^{a_{ij}} \tag{19.90}$$

The k, qth entry of its Hessian matrix, where $k \neq q$, is given by

$$\frac{\partial F(\mathbf{x})}{\partial x_k \partial x_q} = \sum_{i=1}^{m} b_i a_{ik} a_{iq} x_k^{a_{ik}-1} x_q^{a_{iq}-1} \prod_{\substack{j \neq k \\ j \neq q}}^{n} x_j^{a_{ij}} \tag{19.91}$$

When $k = q$, the diagonal entry of the Hessian matrix is

$$\frac{\partial^2 F(\mathbf{x})}{\partial x_k^2} = \sum_{i=1}^{m} b_i a_{ik} (a_{ik} - 1) x_k^{a_{ik}-2} \prod_{j \neq k}^{n} x_j^{a_{ij}} \tag{19.92}$$

Some care must be exercised in using the forenamed formulas when a variable or an exponent vanishes. Also when a variable is negative and the corresponding exponent is non-integer, the expression becomes imaginary. In the latter case, which could occur during the iterations even if the variables are constrained to be nonnegative, the exponent is simply rounded off to the nearest integer. This allows the iterations to proceed. Eventually as the minimum is approached, the appropriate variables will become nonnegative so that this rounding becomes unnecessary. If a variable is expected to be negative at the minimum, then it should occur with an integer exponent in all the functions.

19.4 Application to Transformer Design

In designing a transformer, we normally wish to minimize the cost. Our objective function, $f(\mathbf{x})$, is therefore a cost function. It will, in general, have many terms. These will include material costs, labor costs, the cost of losses to the customer, and overhead costs. These component costs, as well as the constraint functions, must be expressed in terms of a basic set of design variables. Although the choice of design variables is somewhat arbitrary, they should be chosen in such a way that the cost and constraint functions can be easily expressed in terms of them. Here, we consider a 2-winding, 3-phase, core-form power transformer. We will simplify the details in order to focus more on the method.

19.4.1 Design Variables

Useful design variables are

1. B: Core flux density
2. J_s: OA current density in the secondary or LV winding
3. R_c: Core radius
4. g: HV–LV gap
5. R_s: Mean radius of the secondary or LV winding
6. R_p: Mean radius of the primary or HV winding
7. h_s: Height of the secondary winding
8. t_s: Thickness (radial build) of the secondary winding
9. t_p: Thickness (radial build) of the primary winding
10. M_c: Weight of the core steel
11. M_t: Weight of the tank

Note that the last two weights can be expressed in terms of the other design variables. However, since some of the material and labor costs and losses are easily expressed in terms of them, we find it convenient to include them in the set of basic design variables. Their dependence on the other variables will be expressed in terms of equality constraints. The units chosen for the enumerated variables are used internally in the computer optimization program. As far as input and output units are concerned, that is, what the designer deals with, the units are a matter of familiarity and can differ from the internal units.

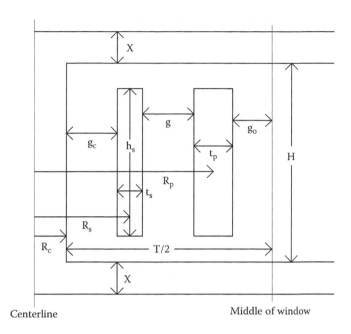

FIGURE 19.1
Geometry of a 2-winding core-form power transformer illustrating some of the design variables.

Figure 19.1 illustrates some of the design variables geometrically. We have not considered the height of the primary winding as a design variable, since in the designs dealt with here, it is taken to be a multiple of the secondary winding height. We express this as $h_p = \alpha h_s$, where h_p is the height of the primary winding and α is typically ≈ 0.95. g_c, g, and g_o are minimum gaps, which are either inputted by the user or based on the BIL of the windings. g_o depends on the phase-to-phase voltages. H is the window height and T the window width. X is the maximum stack width of the yoke $\approx 2R_c$. These are expressible in terms of the other variables.

19.4.2 Cost Function

One of the cost components is the cost of the winding conductor. To obtain this, we need the conductor weight. In terms of the design variables, the weight of the three secondary windings (for three phases) is

$$M_s = 3\rho\eta_s\left(2\pi R_s h_s t_s\right) \tag{19.93}$$

where
  ρ is the density of the metallic conductor
  η_s is the fill factor, that is, the fraction of the coil's cross-sectional area that is metal

If this is in kg, it must then be multiplied by a \$/kg cost to arrive at the cost to include in the objective function. This \$/kg cost must reflect the cost of the raw conductor plus the add-on cost of forming it into an insulated wire or cable plus any overhead or storage costs. The fill factor, although set initially, will be determined during the optimization process when a cable size, which meets constraints on the coil's cross-sectional dimensions and the number of turns, is selected. The weight of the primary winding is similarly

$$M_p = 3\rho\eta_p\alpha\left(2\pi R_p h_s t_p\right) \tag{19.94}$$

where η_p is the primary winding's fill factor. This also gets multiplied by a \$/kg, which will differ from that of the secondary winding because it will generally be made of a different type of wire or cable.

The weight of the core is simply M_c. This is multiplied by a \$/kg for core steel if the core weight is in kg before inclusion in the cost function. This \$/kg also includes overhead and storage costs. The tank cost is similarly M_t times the appropriate \$/kg for tank steel if kg is the chosen weight unit. M_c and M_t will be expressed in terms of the other variables by means of equality constraints.

Other material costs are the cost of the oil; the insulation such as cylinders, key spacers, and lead support structure; and the cost of leads, clamping, bushings, tap gear, conservator if present, radiators, and auxiliary reactors or series transformers. The details are omitted here to simplify the discussion. Some of the said costs, such as the bushing and tap gear costs, are not part of the optimization process but are add-on costs based on their size or other characteristics and are often obtained from tables.

Load and no-load losses are part of the overall cost of operating a transformer and are included in the cost function. A customer buys a transformer not on just its initial manufacturing cost but also on the cost of operating it over many years. These latter costs depend on how efficient the unit is, that is, how high are its losses. The load losses include the Joule heating in the windings (copper losses) and the losses due to the stray flux impinging on tank walls, clamps, etc. (stray losses). The losses in the secondary winding under OA conditions can be expressed as

$$W_s = \gamma\left(1 + ecf_s\right)J_s^2 V_s \tag{19.95}$$

where γ is the resistivity of the winding material, which is evaluated at the appropriate temperature. ecf_s is the eddy current fraction of the I^2R losses, which is due to stray flux and depends on the type of wire or cable making up the winding. V_s is the metallic volume of the winding material, which can be expressed as

$$V_s = 3\eta_s\left(2\pi R_s h_s t_s\right) \tag{19.96}$$

J_s and γ need to be in appropriate units or an overall multiplying factor used to get W_s in kW. This must then be multiplied by the cost of load losses in \$/kW. Similarly, the losses in the primary windings are given by

$$W_p = \gamma\left(1 + ecf_p\right)J_p^2 V_p \tag{19.97}$$

where the conductor volume is

$$V_p = 3\eta_p\alpha\left(2\pi R_p h_s t_p\right) \tag{19.98}$$

and where J_p is the current density in the primary winding under OA conditions. Because the ampere-turns of the primary and secondary windings are equal under balanced conditions, we can obtain J_p in terms of J_s by

$$J_p = \frac{J_s \eta_s t_s}{\alpha \eta_p t_p} \tag{19.99}$$

This is then substituted into (19.97). We then multiply W_p by the \$/kW cost of load losses before adding it to the cost function. The stray losses are also part of the load losses and must be added to the cost function, but we omit discussing them here for simplicity. If the transformer includes a reactor, step voltage regulator, or series transformer, then the cost of losses associated with these components must be added to the cost function.

The no-load losses are essentially the core losses. Generally, the core steel manufacturer provides a curve or polynomial expression for the core losses in W/unit weight as a function of the core flux density, that is,

$$\text{Core loss} \left(\text{W/unit weight}\right) = a_0 + a_1 B + a_2 B^2 + \cdots + a_k B^k$$

Multiplying this by the core weight, M_c, gives a core loss for an ideal core. Actual cores have higher losses due to nonuniform flux in the corners and due to building stresses and other factors. These must be accounted for in the expression for the no-load loss. Having done this, the core loss in kW is then multiplied by the cost of the no-load losses in \$/kW before adding to the cost function.

Labor costs are an important part of the cost of a transformer since power transformers are usually custom-made. The various specialized types of labor must be tied to the appropriate design variables. For instance, coil winding labor costs should be expressed in terms of the coil weight or coil dimensions, while core stacking labor costs should be expressed in terms of the core weight or core dimensions. Labor associated with tank welding or painting should be tied to the tank weight or tank dimensions. Correlations need to be established between the hours required to perform a certain job function and the appropriate design variables. These correlations will differ from manufacturer to manufacturer and are omitted in this discussion. Establishing meaningful correlations requires much effort. Such correlations should be revised periodically as conditions warrant. Once established, they should be multiplied by a labor cost in \$/h, which includes labor overhead, before adding to the cost function.

As can be seen, the cost function can get rather lengthy. However, all the terms mentioned previously are expressions of the form given by formula (19.89). Labor hour correlations can also be expressed this way. Therefore, gradients and Hessians can be obtained by means of the formulas given in Section 19.3.3. Since we wish the overall cost to be close to unity in magnitude, the cost is divided by a typical cost for the transformer of similar power or MVA for purposes of optimization.

19.4.3 Equality Constraints

Equality constraints are those, which must be satisfied, exactly in the final design. Probably the most important such constraint is on the total transferred power or MVA, which is specified by the customer. We need to express it in terms of the chosen design variables.

In the following, we work in terms of phase quantities. The power transferred per phase, P, is given by P = MVA/3. This is also given by

$$P = V_s I_s \qquad (19.100)$$

where
  V_s is the rms secondary phase voltage in kV
  I_s is the rms secondary phase current in kA

Using Faraday's law,

$$V_s = \frac{10^{-3}}{\sqrt{2}} N_s A_{Fe} 2\pi f B \qquad (19.101)$$

where
  B is the peak flux density in Tesla
  N_s is the number of secondary turns
  A_{Fe} is the area of the iron in a cross-section of the core steel in m²
  f is the frequency in Hz

Expressing A_{Fe} in terms of our design variables, we have

$$A_{Fe} = \eta_c \pi R_c^2 \qquad (19.102)$$

where η_c is a fill factor for the core steel, that is, the fraction of actual steel in a circle of radius R_c. The current I_s is given in terms of the current density by

$$I_s = \frac{J_s \eta_s h_s t_s}{N_s} \qquad (19.103)$$

Substituting these into the expression for power, we see that N_s cancels out and we find

$$P = \left[10^{-3} \sqrt{2} \pi^2 f \right] \eta_c \eta_s B R_c^2 J_s h_s t_s \qquad (19.104)$$

Divide by P to get this to be of order 1 and express it in the form of our standard equality constraints, $c_i(x) = 0$,

$$\left[\frac{10^{-3} \sqrt{2} \pi^2 f}{P} \right] \eta_c \eta_s B R_c^2 J_s h_s t_s - 1 = 0 \qquad (19.105)$$

This is an expression of the form (19.89). Note that −1 is of this form with all the $a_{ij} = 0$, $b_1 = -1$, and m = 1. We have kept the numerical constants explicit in (19.105) in order to reduce errors.

  Perhaps the next most important equality constraint is on the per unit reactance between the primary and secondary windings, which is specified by the customer. A simplified expression for this, which works well in practice, is

$$X \text{ (p.u.)} = \frac{(2\pi)^2 \mu_0 f (VI)_b}{(V/N)_b^2} \frac{r^2}{(h+s)} \left[\frac{R_s t_s}{3} + \frac{R_p t_p}{3} + R_g g \right] \qquad (19.106)$$

where r is the co-ratio (=1 for a non-auto transformer). R_s and R_p are mean radii of the windings. R_g is the mean radius of the gap between the primary and secondary windings and is given by

$$R_g = \frac{R_s + R_p}{2} + \frac{t_s - t_p}{4} \qquad (19.107)$$

h is the average height of the two windings:

$$h = \left(\frac{1+\alpha}{2}\right)h_s \qquad (19.108)$$

s is a correction factor for fringing flux:

$$s = 0.32\left(R_p + \frac{t_p}{2} - R_c\right) \qquad (19.109)$$

$(VI)_b$ is the base MVA/phase = P and $(V/N)_b$ is the base kV/turn, which is obtained from (19.101):

$$\frac{V_s}{N_s} = \left[10^{-3}\sqrt{2}\pi^2 f\right]\eta_c R_c^2 B \qquad (19.110)$$

Substituting into (19.106) and converting to standard form, we obtain

$$\left[\frac{Pr^2}{\left[10^{-3}\sqrt{2}\pi^2 f\right]^2 X \,(p.u.)}\right]\frac{1}{\eta_c^2 R_c^4 B^2 h_s}$$
$$\times\left\{\frac{R_s t_s}{3} + \frac{R_p t_p}{3} + \frac{R_s g}{2} + \frac{R_p g}{2} + \frac{t_s g}{4} - \frac{t_p g}{4}\right\} - \left(\frac{1+\alpha}{2}\right)$$
$$-0.032\frac{R_p}{h_s} - 0.16\frac{t_p}{h_s} + 0.32\frac{R_c}{h_s} = 0 \qquad (19.111)$$

This is in the form of Equation 19.89 and has been suitably normalized to order of magnitude 1. The constants are again kept explicit to reduce errors.

We treated R_p as an independent variable thus far, since it appears in many formulas. However, it can really be expressed in terms of other design variables as

$$R_p = R_s + \frac{t_s}{2} + g + \frac{t_p}{2} \qquad (19.112)$$

Converting to standard form and suitably normalizing, we get the equality constraint:

$$\frac{R_s}{R_p} + \frac{t_s}{2R_p} + \frac{g}{R_p} + \frac{t_p}{2R_p} - 1 = 0 \qquad (19.113)$$

An equality constraint is needed for the core weight. In terms of the window height H, window width T, and the maximum sheet width X that make up the core stacks shown in Figure 19.1, this is given by

$$M_c = \rho_{Fe}\eta_c\pi R_c^2\left[3H + 4T + 6X + 0.30235\frac{R_c}{\eta_c}\right] \qquad (19.114)$$

The last term is a correction factor for the joints. ρ_{Fe} is the density of the core iron. In terms of the basic design variables,

$$H = h_s + \text{slack}_s \qquad (19.115)$$

where slack$_s$ is a slack distance in the window, which depends on the voltage or BIL of the winding and is a constant for the unit under consideration:

$$T = 2\left(R_p + \frac{t_p}{2} + g_o - R_c\right) \qquad (19.116)$$

and X will be taken here as $2R_c$, although a more exact formula can be used for greater accuracy. Thus we obtain, in standard normalized form,

$$[\rho_{Fe}\pi]\frac{\eta_c R_c^2}{M_c}\left\{3h_s + 8R_p + 4t_p + 4R_c + 0.30235\frac{R_c}{\eta_c} + 3\text{slack}_s + 8g_o\right\} - 1 = 0 \qquad (19.117)$$

There is an equality expressing the tank weight in terms of the other variables. The tank dimensions depend not only on the size of the 3-phase core and coils but also on clearances based on voltage considerations and on the presence of reactors, tap leads, etc.

19.4.4 Inequality Constraints

We place an inequality constraint on the mean radius of the LV winding, since it must not drop below a minimum value given by

$$R_s \geq R_c + g_c + \frac{t_s}{2} \qquad (19.118)$$

Expressing this in standard form, we have

$$\frac{R_s}{R_c} - \frac{t_s}{2R_c} - \frac{g_c}{R_c} - 1 \geq 0 \qquad (19.119)$$

The HV–LV gap g must not fall below a minimum value given by voltage or BIL considerations. Calling this minimum gap g_{min} leads to the inequality

$$g \geq g_{min} \qquad (19.120)$$

or, in standard form,

$$\frac{g}{g_{min}} - 1 \geq 0 \qquad (19.121)$$

The maximum flux density B is limited by the saturation of iron or by a lower value determined by overvoltage or sound level considerations. Calling the maximum value B_{max} leads to the inequality in standard form

$$\frac{B_{max}}{B} - 1 \geq 0 \qquad (19.122)$$

There is also a reasonable limit on the OA current density, J_s, which we call J_{max}. This is based on cooling considerations, which are most severe under FA conditions. But since the FA current density is related to the OA current density by the ratio of FA and OA MVA's, the current limitation can be placed on the OA current density. Thus we have, in standard form,

$$\frac{J_{max}}{J_s} - 1 \geq 0 \qquad (19.123)$$

It may be necessary to limit the tank height for shipping purposes. Referring to Figure 19.1, the tank height can be expressed as

$$H + 2X + slack_t \leq H_{max} \qquad (19.124)$$

where
    H_{max} is the maximum tank height
    $slack_t$ is the vertical slack of the core in the tank and will depend on the clamping structure, the presence of leads, etc.

$slack_t$ will be a constant for a given unit. H is given by (19.115) and, taking $X \approx 2R_c$, we get

$$h_s + 4R_c + slack_s + slack_t \leq H_{max} \qquad (19.125)$$

or, in standard form,

$$\left(\frac{H_{max} - slack_s - slack_t}{H_{max}} \right) - \frac{h_s}{H_{max}} - \frac{4R_c}{H_{max}} \geq 0 \qquad (19.126)$$

The quantity in parentheses in formula (19.126) is a constant for the unit under consideration.

There are other inequalities involving the radial forces on the windings, which are limited by the tensile strength of the conductor used; the cooling capacity of the windings, which depends on the winding current density and fill factor; and the impulse strength, which depends on the cable type, voltage level, etc. These inequalities are fairly inexact since detailed design codes are ultimately used to determine winding stresses, temperatures, and impulse voltages. Nevertheless, they place some restrictions on the initial design so that later on, in the detailed design phase, difficulties are not encountered. In addition, all the design variables are positive so that a positivity condition on these variables must be included among the inequality constraints.

19.4.5 Optimization Strategy

A transformer cost minimization program has been developed based on the formulation described earlier. It uses the 11 basic design variables discussed earlier for the

2-winding design. Initial values must be selected for these variables to start the iteration process. Because of the equality constraints, of which there are five, we can only select six variables independently. For these, we choose an initial value for B, $B_{init} = 0.95B_{max}$, where B_{max} is the maximum flux density specified in the input; an initial value for J_s, $J_{s,init} = 0.8J_{max}$, where J_{max} is the inputted maximum OA current density; an initial value for g, $g_{init} = g_{min}$, with g_{min} the inputted minimum high–low gap; an initial value for h_s, $h_{s,init}$, that is, a secondary coil height; an initial value for t_s, $t_{s,init}$, that is, a secondary coil width; and an initial value for t_p, $t_{p,init}$, that is, a primary coil width. These initial winding dimensions can be automatically selected based on the MVA of the unit. We also choose, for starting values, $\eta_s = \eta_p = 0.5$, $\eta_c = 0.88$, and $ecf_s = ecf_p = 0.1$. The remainder of the starting values for the design variables are determined by satisfying the equality constraints or equations such as Faraday's law, which was used in their derivation, or by other means based on convenience. For greater flexibility, the designer can override the starting values for the six variables mentioned previously.

The iterations are then started. The cost and constraint functions are evaluated, together with their gradients and Hessians. The Lagrange multipliers are initially set to 1 for the equality constraints. The Hessian of the Lagrangian is then formed. The matrix A is formed and subjected to a QR factorization. During this process, if a row vector of A is found to be linearly dependent on the other row vectors, the corresponding constraint is dropped from the active set. This process produces the Y, Z, L, and $R = L^T$ matrices. The vector $\mathbf{p}_Y$ is then obtained via Equation 19.71. Equation 19.77 is then solved for $\mathbf{p}_Z$ using a modified Cholesky factorization method. The new step direction is then determined, $\mathbf{p} = Y\mathbf{p}_Y + Z\mathbf{p}_Z$. The merit function (19.80) is then evaluated for steps of size $\varepsilon\mathbf{p}$ where ε increases from 0 to 1 in small increments. This process continues until (1) the merit function starts increasing, (2) one of the inequality constraints is violated, or (3) $\varepsilon = 1$. If (2) is true, then the violated constraint is included among the active constraints on the next iteration. For the value of ε determined by the said procedure, new Lagrange multipliers are determined by means of (19.85). If one of these is negative for an active inequality constraint, that constraint is dropped from the active set on the next iteration. A new set of design variables is then determined, using (19.86), and the process repeated until convergence is achieved, that is, until the changes in the design variables and Lagrange multipliers are negligible. At convergence, the projected Hessian of the Lagrangian must be positive definite, the equality constraints must be satisfied to a given level of accuracy, the inequality constraints must not be violated, and (19.47) must be satisfied to a given level of accuracy.

At this stage, details concerning the wire or cable types for the two windings are worked out based on the optimum dimensions h_s, t_s, $h_p = \alpha h_s$, and t_p. From the formula for the given power per phase, P,

$$P = V_s I_s = V_p I_p \tag{19.127}$$

where V_s and V_p are known (inputted) secondary and primary voltages, we can determine the phase currents I_s and I_p for the secondary and primary windings, respectively. From these phase currents and the optimized current densities J_s and J_p, we obtain secondary and primary turn areas $A_s = I_s/J_s$ and $A_p = I_p/J_p$. We use (19.110) to get volts/turn in terms of the optimized variables and, using the known secondary and primary voltages, we obtain the number of secondary and primary turns N_s and N_p. In addition, from the BIL levels of the windings, which are inputted, we obtain, via tables, the required paper and key spacer thicknesses for the two windings. Staying within these constraints, a wire size (magnet wire or cable) is selected by a search procedure from those that are available.

In arriving at a suitable wire type, eddy current loss, thermal gradient, and impulse strength limitations are considered. The possibility of needing to use several parallel wires or cables is also considered in the search process. If a wire (cable) size or parallel package of wires (cables) cannot be found, which would meet all the requirements, the wire area is then allowed to change. This would change the current density on subsequent iterations, but as long as it remains below J_{max}, this is acceptable.

Having arrived at a suitable wire size, from the paper and key spacer thicknesses along with other allowances such as for cross-overs and thinning, the fill factors of the two windings, η_s and η_p, can be determined. Also from the optimized core radius, a better value for the core fill factor η_c can be determined. These fill factors are then used in subsequent iterations.

With optimized design variables as starting values and the newly determined fill factors, iterations are resumed until convergence is achieved. Wire sizes are then determined for these new optimized parameters and new fill factors determined and the process is continued until wire sizes and fill factors no longer change between sets of iterations.

At this point, the core radius may not correspond to one of our standard radii. It is therefore set to the nearest standard value and an equality constraint is imposed on the core radius, $R_c = R_0$, or in standard form,

$$\frac{R_c}{R_0} - 1 = 0 \tag{19.128}$$

where R_0 is a standard radius. The entire iteration process is then restarted with this new equality constraint. Since the starting values for the design variables and the fill factors are the values arrived at by the optimization process up to this point, the iterations with the fixed core radius usually converge very quickly.

On output, the program prints out the total cost of the optimized unit and a detailed breakdown of this cost such as the cost of each coil, the core cost, the tank cost, the oil cost, the cost of radiators, the cost of load and no-load losses, labor costs itemized by each job type, and a summary of the total material, labor, loss, and overhead costs. In addition, the physical values associated with these costs, such as kg of material, kW of loss, or labor hours, are also printed out. The optimum values of the design variables are printed out as well as additional design information such as the tank dimensions, details of the magnet wire or cable used for each winding, the number of turns and the turns/disk for each winding, the paper and key spacer thicknesses for the windings, the number of radiators and their height, and the size of the conservator if needed.

At this point, if the wire selection is not satisfactory for whatever reason, an auxiliary separate program is desirable to allow the designer to redo the wire selection for each winding. This could alter the winding dimensions and the space factor of the windings. These altered values can then be transferred to the optimizer as starting or fixed values for the next round of optimization.

References

[Abr72] M. Abramowitz and I. A. Stegun, Eds., *Handbook of Mathematical Functions*, Dover Publications, Inc., New York, 1972.

[Aco89] J. Acosta and K. J. Cornick, Computer models for complex plant based on terminal measurements, *IEEE Trans. Power Deliv.*, 4(2), April 1989, 1393–1400.

[Aga59] P. D. Agarwal, Eddy current losses in solid and laminated iron, *Trans. AIEE*, 42(Pt. 1), 1959, 169–181.

[Ahu06] R. Ahuja and R. M. Del Vecchio, Transformer stray loss and flux distribution studies using 3D finite element analysis, *Trafotech 2006, Seventh International Conference on Transformers*, Mumbai, India, January 20–21, 2006, pp. II-1–II-8.

[Ahu14] R. Ahuja and R. M. Del Vecchio, Analysis of 2 core transformer designs, CIGRE session paper no. A2_210_2014, *CIGRE 2014 General Session*, Session 45, August 24–30, 2014.

[Ame57] *American Institute of Physics Handbook*, McGraw-Hill Book Co., Inc., New York, 1957.

[ANS] ANSYS, Inc., Canonsburg, PA.

[Ansoft] MAXWELL® is a product of Ansoft Corp., Pittsburgh, PA.

[Aub92] J. Aubin and Y. Langhame, Effect of oil viscosity on transformer loading capability at low ambient temperatures, *IEEE Trans. Power Deliv.*, 7(2), April 1992, 516–524.

[Bel77] W. R. Bell, Influence of specimen size on the dielectric strength of transformer oil, *IEEE Trans. Electr. Insul.*, EI-12(4), August 1977, 281–292.

[Blu51] L. F. Blume, A. Boyajian, G. Camilli, T. C. Lennox, S. Minneci, and V. M. Montsinger, *Transformer Engineering*, 2nd edn., John Wiley & Sons, Inc., New York, 1951.

[Bos72] A. K. Bose, Dynamic response of windings under short circuit, *CIGRE International Conference on Large High Tension Electric Systems, 1972 Session*, August 28–September 6, 1972.

[Bra82] V. Brandwajin, H. W. Dommel, and I. I. Dommel, Matrix representation of three-phase transformers for steady-state and transient studies, *IEEE Trans. Power App. Syst.*, PAS-101(6), June 1982, 1369–1378.

[Chu60] R. V. Churchill, *Complex Variables and Applications*, 2nd edn., McGraw-Hill Book Co., Inc., New York, 1960.

[Cla62] F. M. Clark, *Insulating Materials for Design and Engineering Practice*, John Wiley & Sons, Inc., New York, 1962.

[Cle39] J. E. Clem, Equivalent circuit impedance of regulating transformers, *AIEE Trans.*, 58, 1939, 871–873.

[Cyg87] S. Cygan and J. R. Laghari, Dependence of the electric strength on thickness area and volume of polypropylene, *IEEE Trans. Electr. Insul.*, EI-22(6), December 1987, 835–837.

[Dai73] J. W. Daily and D. R. F. Harleman, *Fluid Dynamics*, Addison-Wesley Publishing Co., Inc., Reading, MA, 1973.

[Dan90] M. G. Danikas, Breakdown of transformer oil, *IEEE Electr. Insul. Mag.*, 6(5), September/October 1990, 27–34.

[deL92] F. de Leon and A. Semlyen, Efficient calculation of elementary parameters of transformers, *IEEE Trans. Power Deliv.*, 7(1), January 1992, 376–383.

[deL92a] F. de Leon and A. Semlyn, Reduced order model for transformer transients, *IEEE Trans. Power Deliv.*, 7(1), January 1992, 361–369.

[Del10] R. M. Del Vecchio, B. Poulin, P. T. Feghali, D. M. Shah, and R. Ahuja, *Transformer Design Principles*, 2nd edn., CRC Press, Boca Raton, FL, 2010.

[Del94] R. M. Del Vecchio, Magnetostatics, in: *Encyclopedia of Applied Physics*, Vol. 9, VCH Publishers, Inc., 1994, pp. 207–227.

[Del98] R. M. Del Vecchio, B. Poulin, and R. Ahuja, Calculation and measurement of winding disk capacitances with wound-in-shields, *IEEE Trans. Power Deliv.*, 13(2), April 1998, 503–509.

[Del99] R. M. Del Vecchio and P. Feghali, Thermal model of a disk coil with directed oil flow, *1999 IEEE Transmission and Distribution Conference*, Now Orleans, LA, April 11–16, 1999, pp. 914–919.

[Del01] R. M. Del Vecchio, B. Poulin, and R. Ahuja, Radial buckling strength calculation and test comparison for core-form transformers, *Proceedings of the 2001 International Conference of Doble Clients—Sec 8-1*, Doble Engineering Company, Watertown, MA, 2001.

[Del02] R. M. Del Vecchio, R. Ahuja, and R. D. Frenette, Determining ideal impulse generator settings from a generator-transformer circuit model, *IEEE Trans. Power Deliv.*, 17(1), January 2002, 142–148.

[Del03] R. M. Del Vecchio, Eddy-current losses in a conducting plate due to a collection of bus bars carrying currents of different magnitudes and phases, *IEEE Trans. Magn.*, 39(1), January 2003, 549–552.

[Del04] R. M. Del Vecchio, Geometric effects in the electrical breakdown of transformer oil, *IEEE Trans. Power Deliv.*, 19(2), April 2004, 652–656.

[Del06] R. M. Del Vecchio, Applications of a multiterminal transformer model using two winding leakage inductances, *IEEE Trans. Power Deliv.*, 21(3), July 2006, 1300–1308.

[Del08] R. M. Del Vecchio, Multiterminal three phase transformer model with balanced or unbalanced loading, *IEEE Trans. Power Deliv.*, 23(3), July 2008, 1439–1447.

[Del13] R. M. Del Vecchio and R. Ahuja, Analytic nonlinear correction to the impedance boundary condition, *IEEE Trans. Magn.*, 49(12), December 2013, 5687–5691.

[Del14] R. M. Del Vecchio, Directed oil flow cooling of disk windings using graph theory, *IEEE Trans. Power Deliv.*, 29(5), October 2014, 2279–2286.

[Deo74] N. Deo, *Graph Theory with Applications to Engineering and Computer Science*, Prentice Hall, Englewood Cliffs, NJ, 1974.

[Des69] C. A. Desoer and E. S. Kuh, *Basic Circuit Theory*, McGraw-Hill Book Co., New York, 1969.

[Duf67] R. J. Duffin, E. L. Peterson, and C. Zener, *Geometric Programming*, Wiley, New York, 1967.

[Dwi61] H. B. Dwight, *Tables of Integrals and Other Mathematical Data*, 4th edn., The MacMillan Co., New York, 1961.

[Eas65] C. Eastgate, Simplified steady-state thermal calculations for naturally cooled transformers, *Proc. IEE*, 112(6), June 1965, 1127–1134.

[End57] H. S. Endicott and K. H. Weber, Electrode area effect for the impulse breakdown of transformer oil, *Trans. IEEE*, August 1957, 393–398.

[For69] J. A. C. Forrest and R. E. James, Improvements in or relating to windings for inductive apparatus, Patent no. 1,158,325, The Patent Office, London, U.K., July 16, 1969.

[Franc] M. A. Franchek and T. A. Prevost, EHV Weidmann, private communication.

[Gall75] T. J. Gallager, *Simple Dielectric Liquids*, Clarendon Press, Oxford, U.K., 1975.

[Gil65] D. H. Gillott and J. F. Calvert, *IEEE Trans. Magn.*, 1, 1965, 126.

[Gil81] P. E. Gill, W. Murray, and M. H. Wright, *Practical Optimization*, Academic Press, NY, 1981.

[Gol89] G. H. Golub and C. F. Van Loan, *Matrix Computations*, 2nd edn., The Johns Hopkins University Press, Baltimore, MD, 1989.

[Grc58] J. A. Greenwood and J. B. P. Williamson, Electrical conduction in solids, II. Theory of temperature-dependent conductors, *Proc. Roy. Soc. Lond. A*, 246, 1958, 13–31.

[Gre66] J. A. Greenwood, Constriction resistance and the real area of contact, *Brit. J. Appl. Phys.*, 17, 1966, 1621–1632.

[Gre71] A. Greenwood, *Electrical Transients in Power Systems*, Wiley-Interscience, a Division of John Wiley & Sons, Inc., New York, 1971.

[Gro73] F. W. Grover, *Inductance Calculations*, Dover Publications, Inc., New York, 1973 (Special edition prepared for the Instrument Society of America).

[Gue96] C. Guerin, G. Meunier, and G. Tanneau, Surface impedance for 3D non-linear eddy current problems—Application to loss computation in transformers, *IEEE Trans. Magn.*, 32(3), May 1996, 808–811.

[Gum58] E. J. Gumbel, *Statistics of Extremes*, Columbia University Press, New York, 1958.

[Hig75] H. Higake, K. Endou, Y. Kamata, and M. Hoshi, Flashover characteristics of transformer oil in large cylinder-plane electrodes up to gap length of 200 mm, and their application to the insulation from outer winding to tank in UHV transformers, *IEEE PES Winter Meeting*, New York, January 26–31, 1975.

[Hir71] K. Hiraishi, Y. Hore, and S. Shida, Mechanical strength of transformer windings under short-circuit conditions, Paper 71 TP 8-PWR, Paper presented at the, *IEEE Winter Power Meeting*, New York, January 31–February 5, 1971.

[Hob39] J. E. Hobson and W. A. Lewis, Regulating transformers in power-system analysis, *AIEE Trans.*, 58, 1939, 874–886.

[Hol92] S. A. Holland, G. P. O'Connell, and L. Haydock, Calculating stray loss in power transformers using surface impedance with finite elements, *IEEE Trans. Magn.*, 28(2), March 1992, 1355–1358.

[Hue72] L. P. Huelsman, *Basic Circuit Theory with Digital Computations*, Prentice-Hall, Inc., Englewood Cliffs, NJ, 1972.

[IEE81] ANSI/IEEE C57.92-1981, Guide for loading mineral oil immersed transformers up to and including 100 MVA with 55°C or 65°C average winding rise, 1981.

[IEE93] IEEE Std. C57.12.00-1993, IEEE standard general requirements for liquid-immersed distribution, power, and regulating transformers, IEEE, August 27, 1993.

[IEE10] IEEE Std. C57.12.00-2010, IEEE standard for general requirements for liquid-immersed distribution, power, and regulating transformers, IEEE, New York, September 10, 2010.

[IEEE10a] IEEE Std. C57.12.90-2010, IEEE standard test code for liquid-immersed distribution, power, and regulating transformers, IEEE, New York, October 15, 2010.

[Jai70] M. P. Jain and L. M. Ray, Field pattern and associated losses in aluminum sheet in presence of strip bus bars, *IEEE Trans. Power App. Syst.*, PAS-89(7), September/October 1970, 1525–1539.

[Jes96] S. Jeszenszky, History of transformers, *IEEE Power Eng. Rev.*, 16(12), December 1996, 9–12.

[Kau68] R. B. Kaufman and J. R. Meador, Dielectric tests for EHV transformers, *IEEE Trans. PAS*, PAS-87(1), January 1968, 135–145.

[Kok61] J. A. Kok, *Electrical Breakdown of Insulating Liquids*, Interscience Publishers, Inc., New York, 1961.

[Kos02] A. Kost, J. P. A. Bastos, K. Miethner, and L. Janicke, Improvement of nonlinear impedance boundary conditions, *IEEE Trans. Magn.*, 38(2), March 2002, 573–576.

[Kra88] R. F. Krause and R. M. Del Vecchio, Low core loss rotating flux transformer, *J. Appl. Phys.*, 64(10), November 15, 1988, 5376–5378.

[Kra97] L. Krahenbuhl, O. Fabregue, S. Wanser, M. De Sousa Dias, and A. Nicolas, Surface impedances, BIEM and FEM coupled with 1D non linear solutions to solve 3D high frequency eddy current problems, *IEEE Trans. Magn.*, 33(2), March 1997, 1167–1172.

[Kra98] A. Kramer and J. Ruff, Transformers for phase angle regulation considering the selection of on-load tap-changers, *IEEE Trans. Power Deliv.*, 13(2), April 1998, 518–525.

[Kre80] F. Kreith and W. Z. Black, *Basic Heat Transfer*, Harper & Row Publishers, New York, 1980.

[Kuf84] E. Kuffel and W. S. Zaengl, *High Voltage Engineering*, Pergamon Press, New York, 1988.

[Kul99] S. V. Kulkarni, Stray loss in power transformer—Evaluation and experimental verification, Thesis, Department of Electrical Engineering, Indian Institute of Technology, Mumbai, India, 1999.

[Kul04] S. V. Kulkarni and S. A. Khaparde, *Transformer Engineering*, Marcel Dekker, Inc., New York, 2004.

[Lab89] D. Labridis and P. Dokopoulos, Calculation of eddy current losses in nonlinear ferromagnetic materials, *IEEE Trans. Magn.*, 35(3), May 1989, 2665–2669.

[Lam66] J. Lammeraner and M. Stafl, *Eddy Currents*, Iliffe Books, Ltd., London, U.K., 1966.

[Lei99] L. Leites, Lecture notes from presentation at North American Transformer Plant, Milpitas, CA, April 1999.

[Les02] O. Lesaint and T. V. Top, Streamer initiation in mineral oil. Part I: Electrode surface effect under impulse voltage, *IEEE Trans. Dielectr. Electr. Insul.*, 9(1), February 2002, 84–91.

[Lyo37] W. V. Lyon, *Applications of the Method of Symmetrical Components*, McGraw-Hill Book Co., Inc., New York, 1937.

[Max15] S. Maximov, J. C. Olivares-Galvan, S. Magdaleno-Adame, R. Escarela-Parez, and E. Campero-Littlewood, New analytical formulae for electromagnetic field and eddy current losses in bushing regions of transformers, *IEEE Trans. Magn.*, 51(4), April 2015, 1–10.

[Max16] S. Maximov, R. Escarela-Perez, J. C. Olivares-Galvan, J. Guzman, and E. Campero-Littlewood, New analytical formula for temperature assessment on transformer tanks, *IEEE Trans. Power Deliv.*, 31(3), June 2016, 1122–1131.

[McN91] W. J. McNutt, Insulation thermal life considerations for transformer loading guides, Paper presented at the *IEEE/PES Summer Meeting*, San Diego, CA, July 28–August 1, 1991.

[Meh97] S. P. Mehta, N. Aversa, and M. S. Walker, Transforming transformers, *IEEE Spectrum*, 34, July 1997, 43–49.

[Mik78] A. Miki, T. Hosoya, and K. Okuyama, A calculation method for impulse voltage distribution and transferred voltage in transformer windings, *IEEE Trans. PAS*, PAS-97(3), May/June 1978, 930–939.

[MIT43] Members of the Staff of the Department of Electrical Engineering of the Massachusetts Institute of Technology, *Magnetic Circuits and Transformers*, The MIT Press, Massachusetts Institute of Technology, Cambridge, MA, 1943.

[Mom02] E. E. Mombello, Impedances for the calculation of electromagnetic transients within transformers, *IEEE Trans. Power Deliv.*, 17(2), April 2002, 479–488.

[Mos79] H. P. Moser, Transformerboard, Special print of *Scientia Electrica*, 1979.

[Mos87] H. P. Moser and V. Dahinden, Transformerboard II, H. Weidmann AG, Ch-8640, Rapperswil, Switzerland, 1987.

[Nel89] J. K. Nelson, An assessment of the physical basis for the application of design criteria for dielectric structures, *IEEE Trans. Electr. Insul.*, 24(5), October 1989, 835–847.

[Nor48] E. T. Norris, The lightening strength of power transformers, *J. Inst. Electr. Eng.*, 95, 1948, 389–406.

[Nuy78] R. van Nuys, Interleaved high-voltage transformer windings, *IEEE Trans. Power App. Syst.*, PAS-97(5), September/October 1978, 1946–1954.

[Oli80] A. J. Oliver, Estimation of transformer winding temperatures and coolant flows using a general network method, *IEE Proc. C*, 127(6), November 1980, 395–405.

[Ols80] R. M. Olson, *Essentials of Engineering Fluid Mechanics*, 4th edn., Harper & Row Publishers, New York, 1980.

[Pal69] S. Palmer and W. A. Sharpley, Electric strength of transformer insulation, *Proc. IEE*, 116, 1969, 1965–1973.

[Pat80] M. R. Patel, Dynamic stability of helical and barrel coils in transformers against axial short-circuit forces, *IEE Proc. C*, 127(5), September 1980, 281–284.

[Pav93] D. Pavlik, D. C. Johnson, and R. S. Girgis, Calculation and reduction of stray and eddy losses in core-form transformers using a highly accurate finite element technique, *IEEE Trans. Power Deliv.*, 8(1), January 1993.

[Pee29] F. W. Peek, Jr., *Dielectric Phenomena in High Voltage Engineering*, 3rd edn., McGraw-Hill Book Co., New York, 1929.

[Pie92] L. W. Pierce, An investigation of the thermal performance of an oil filled transformer winding, *IEEE Trans. Power Deliv.*, 7(3), July 1992, 1347–1358.

[Pie92a] L. W. Pierce, Predicting liquid filled transformer loading capability, *IEEE/IAS Petroleum and Chemical Industry Technical Conference*, San Antonio, TX, September 28–30, 1992.

[Pry58] R. H. Pry and C. P. Bean, Calculation of the energy loss in magnetic sheet materials using a domain model, *J. Appl. Phys.*, 29(3), 1958, 532–533.

[Pug62] E. M. Pugh and E. W. Pugh, *Principles of Electricity and Magnetism*, Addison-Wesley Publishing Co., Inc., Reading, MA, 1962.

[Rab56] L. Rabins, Transformer reactance calculations with digital computers, *AIEE Trans.*, 75(Pt. I), July 1956, 261–267.

[Ree73] J. A. Rees, Ed., *Electrical Breakdown in Gasses*, John Wiley & Sons (A Halsted Press Book), New York, 1973.

[Roh85] W. M. Rohsenow, J. P. Hartnett, and E. N. Ganic, Eds., *Handbook of Heat Transfer Fundamentals*, 2nd edn., McGraw-Hill Book Co., New York, 1985.

[Rud40] R. Rudenberg, Performance of traveling waves in coils and windings, *Trans. AIEE*, 59, 1940, 1031–1040.

[Rud68] R. Rudenberg, *Electrical Shock Waves in Power Systems*, Harvard University Press, Cambridge, MA, 1968.

[Sam15] K. Samarawickrama, N. D. Jacob, and B. Kordi, Impulse generator optimum setup for transient testing of transformers using frequency-response analysis and genetic algorithm, *IEEE Trans. Power Deliv.*, 30(4), August 2015, 1949–1957.

[Sar00] M. P. Saravolac, P. A. Vertigen, C. A. Sumner, and W. H. Siew, Design verification criteria for evaluating the short circuit withstand capability of transformer inner windings, *CIGRE*, 2000 Session, 12-208, 2000.

[Shi63] R. B. Shipley, D. Coleman, and C. F. Watts, Transformer circuits for digital studies, *AIEE Trans. III*, 81, February 1963, 1028–1031.

[Smy68] W. R. Smythe, *Static and Dynamic Electricity*, 3rd edn., McGraw-Hill Book Co., New York, 1968.

[Ste62] E. Stenkvist and L. Torseke, Short-circuit problems in large transformers, *CIGRE*, Report no. 142, Appendix II, 1962.

[Ste62a] W. D. Stevenson, Jr., *Elements of Power System Analysis*, McGraw-Hill Book Co., Inc., New York, 1962.

[Ste64] G. M. Stein, A study of the initial serge distribution in concentric transformer windings, *IEEE Trans. PAS*, PAS-83, September 1964, 877–893.

[Ste72] R. B. Steel, W. M. Johnson, J. J. Narbus, M. R. Patel, and R. A. Nelson, Dynamic measurements in power transformers under short-circuit conditions, *International Conference on Large High Tension Electric Systems, CIGRE, 1972 Session*, August 28–September 6, 1972.

[Tay58] E. D. Taylor, B. Berger, and B. E. Western, An experimental approach to the cooling of transformer coils by natural convection, Paper no. 2505 S, IEE, April 1958.

[Tho79] H. A. Thompson, F. Tillery, and D. U. von Rosenberg, The dynamic response of low voltage, high current, disk type transformer windings to through fault loads, *IEEE Trans. Power App. Syst.*, PAS-98(3), May/June 1979, 1091–1098.

[Tim55] S. Timoshenko, *Strength of Materials, Parts I and II*, 3rd edn., D. Van Nostrand Co., Inc., Princeton, NJ, 1955 (Part II 1956).

[Tim70] S. P. Timoshenko and J. N. Goodier, *Theory of Elasticity*, 3rd edn., McGraw-Hill Book Co., New York, 1970.

[Top02] T. V. Top, G. Messala, and O. Lesaint, Streamer propagation in mineral oil in semi-uniform geometry, *IEEE Trans. Dielectr. Electr. Insul.*, 9(1), February 2002, 76–83.

[Top02] T. V. Top and O. Lesaint, Streamer initiation in mineral oil. Part II: Influence of a metallic protrusion on a flat electrode, *IEEE Trans. Dielectr. Electr. Insul.*, 9(1), February 2002, 92–96.

[Tri82] N. Giao Trinh, C. Vincent, and J. Regis, Statistical dielectric degradation of large-volume oil-insulation, *IEEE Trans. Power App. Syst.*, PAS-101(10), October 1982, 3712–3721.

[Wat66] M. Waters, *The Short-Circuit Strength of Power Transformers*, Macdonald & Co. (Publishers) Ltd., London, U.K., 1966.

[Web56] K. H. Weber and H. S. Endicott, Area effect and its extremal basis for the electric breakdown of transformer oil, *AIEE Trans.*, June 1956, 371–381.

[Wes64] Central Station Engineers of the Westinghouse Electric Corp, *Electrical Transmission and Distribution Reference Book*, Westinghouse Electric Corp., E. Pittsburgh, PA, 1964.

[Wil53] W. R. Wilson, A fundamental factor controlling the unit dielectric strength of oil, *AIEE Trans. PAS*, 72, February 1953, 68–74.

[Wil55] R. W. Wilson, The contact resistance and mechanical properties of surface films on metals, *Proc. Phys. Soc. B*, 68, 1955, 625–641.

[Wil67] D. J. Wilde and C. S. Beightler, *Foundations of Optimization*, Prentice-Hall, Inc., Englewood Cliffs, NJ, 1967.

[Wil87] J. B. P. Williamson, Intensive course on electrical contacts, Course notes, sponsored by the *IEEE Holm Conference on Electrical Contacts*, 1987.

[Wil88] D. J. Wilcox, M. Conlon, and W. G. Hurley, Calculation of self and mutual impedances for coils on ferromagnetic cores, *IEE Proc. A*, 135(7), September 1988, 470–476.

[Wil89] D. J. Wilcox, W. G. Hurley, and M. Conlon, Calculation of self and mutual impedances between sections of transformer windings, *IEE Proc. C*, 136(5), September 1989, 308–314.

[Yuf99] S. Yuferev and N. Ida, Selection of the surface impedance boundary conditions for a given problem, *IEEE Trans. Magn.*, 35(3), May 1999, 1486–1489.

[Yuf10] S. Yuferev and L. Di Rienzo, Surface impedance boundary conditions in terms of various formalisms, *IEEE Trans. Magn.*, 46(9), September 2010, 3617–3628.

Index